A COURSE OF MATHEMATICAL ANALYSIS

For B.A./B.Sc. (Pass and Honours) and M.Sc. Students of all Indian Universities

REVISED EDITION

SHANTI NARAYAN

Formerly
Dean of Colleges
University of Delhi, and
Principal
Hansraj College, Delhi

Revised by

P.K. MITTAL

M.Sc., Ph.D.
Head
Department of Mathematics
Government Post Graduate College, Rishikesh
Uttarakhand

S. CHAND
PUBLISHING

S Chand And Company Limited
(ISO 9001 Certified Company)

S Chand And Company Limited

(ISO 9001 Certified Company)

Head Office: Block B-1, House No. D-1, Ground Floor, Mohan Co-operative Industrial Estate, New Delhi – 110 044 | Phone: 011-66672000

Registered Office: A-27, 2nd Floor, Mohan Co-operative Industrial Estate, New Delhi – 110 044 | Phone: 011-49731800

www.**schandpublishing.com**; e-mail: **info@schandpublishing.com**

Branches

Chennai	:	Ph: 23632120; chennai@schandpublishing.com
Guwahati	:	Ph: 2738811, 2735640; guwahati@schandpublishing.com
Hyderabad	:	Ph: 40186018; hyderabad@schandpublishing.com
Jalandhar	:	Ph: 4645630; jalandhar@schandpublishing.com
Kolkata	:	Ph: 23357458, 23353914; kolkata@schandpublishing.com
Lucknow	:	Ph: 4003633; lucknow@schandpublishing.com
Mumbai	:	Ph: 25000297; mumbai@schandpublishing.com
Patna	:	Ph: 2260011; patna@schandpublishing.com

First Edition 1945
Subsequent Editions and Reprints 1949, 53, 55, 58, 59, 62, 64, 66, 68, 71, 73, 76, 79, 81, 83, 84, 86, 87, 89, 91, 94, 95, 96, 98, 2000, 2002, 2004, 2007, 2008, 2009 (Twice), 2011, 2013, 2014, 2016 (Twice), 2018 (Twice), 2020 (Twice)
Reprint 2021

ISBN : 978-81-219-0472-8 **Product Code :** H6MTA61MATH10ENZX0XR

PRINTED IN INDIA

By Vikas Publishing House Private Limited, Plot 20/4, Site-IV, Industrial Area Sahibabad, Ghaziabad – 201 010 and Published by S Chand And Company Limited, A-27, 2nd Floor, Mohan Co-operative Industrial Estate, New Delhi – 110 044.

PREFACE TO THE REVISED EDITION

This book originally written about 56 years ago, has during the intervening period, been revised and reprinted several times. Since its last revision, tremendous changes in the trend of studies have taken place which necessitated thorough revision of the book. But, it was really a very difficult task to revise this perfect well-written book of late Dr. Shanti Narayan . Since a long time undoubtedly, it was a book loved by teachers and students alike.

I took up the task to revise the book with great devotion to meet the rapidly changing demands of students interested in self-study and appearing in different competitive examinations. Keeping this in view, a large variety of illustrative solved examples have been included in every chapter.

This book, in the present form, is a humble effort to make it more useful to students and teachers.

I owe my special gratitude to Shri Ravindra Kumar Gupta, Managing Director and Shri Himanshu Gupta, Director of S.Chand & Co. Ltd., for giving me opportunity to revise the books of late Dr. Shanti Narayan, an eminent Indian Mathematician.

As the need for improvement is never ending, I will look forward to receive valuable and useful suggestions from our users for further improvement in the book.

P.K. MITTAL

PREFACE TO THE FIRST EDITION

The book is designed for students who, after an introductory course on Calculus based on geometrically and intuitively perceived notion of the Continuum, desire to proceed to a course preparatory to a serious study of the purely arithmetical theory of functions of real variables in its general aspects and is based on the lectures which the author has from year to year, been delivering to the students preparing for the M.A. (Mathematics) examination of the Punjab University.

The treatment of the various topics in essentially rigorous; at the same time, an attempt has been made to present the subject in a clear, lucid and intelligible manner.

This book contains a large number of examples some of which have been solved.

I have been during my study of the subject, influeneed by the works of a large number of authors including Goursat Garslaw. Pierpont, Courant, Landau; Kowalewski, and Perron and I feel deeply greatful to them all.

I take this opportunity of also thanking the publishers and the printers who did all they could to give the book the best form possible under the present very difficult circumstances.

D.A.V. College, Lahore,
July, 1945,

SHANTI NARAYAN

मेरे जेयेष्ठ रविन्द्र की धर्मपत्नी

सुश्री मीना

की पुण्य स्मृति में सस्नेह समर्पित

CONTENTS

1. Real Numbers ... **1–36**

1.1 Introduction. Set-Theoretic notation and terminology ... 1
1.2. Finite and infinite sets ... 3
1.3. Fundamental properties of the set of rational numbers ... 5
1.4. Dedekind's construction of the system of real numbers sections of rational numbers ... 8
1.5 Set of real numbers ... 12
1.6 Order structure of the set of real numbers ... 13
1.7 Algebraic structure of the set R ... 17
1.8 The sets of rational and real rational numbers. Isomorphism ... 24
1.9 Restatement of the basic properties of R. ... 26
1.10 Representation of real numbers by points along a straight line ... 33

2. Bounded Sets, Open and Closed Sets ... **37–55**

2.1. Closed intervals, open intervals, semi-closed or semi-open intervals ... 37
2.2. Sets bounded above and sets bounded below. Bounded sets ... 37
2.3. Limit points of a set ... 42
2.4. Closed sets ... 48
2.5. Compact sets. Heine-Borel theorem ... 49

3. Real Sequences ... **56–100**

3.1. Notion of a function. Domain and range of a function ... 56
3.2. Bounded sequences ... 57
3.3. Convergent sequences ... 62
3.4. Non-convergent sequences ... 67
3.5. Algebraic operations and limits ... 68
3.6. Sub-sequences of a sequence ... 74
3.7. Theorem on nested intervals ... 75
3.8. Infinite series : its convergence and sum ... 76
3.9. The meaning of a^x when $a > 0$, and x is any rational number ... 82

4. Real Valued Functions of a Single Real Variable. Limit and Continuity ... **101–149**

4.1. Real valued functions of a single real variable ... 101
4.2. Limits ... 105
4.3. Continuity, uniform continuity, properties of functions continuous in closed intervals ... 116
4.4. Monotonic functions ... 128
4.5. Functions of bounded variation ... 130

5. Real Valued Functions of a Single Real Variable Derivability ... **150–194**

5.1. Derivability ... 150
5.2. Meaning of the sign of derivative at a point ... 161
5.3. Darboux's theorem ... 162

5.4. Rolle's theorem ... 163
5.5. Higher derivatives ... 173
5.6. Young's form of taylor's theorem ... 177
5.7. Applications of taylor's theorem ... 178
5.8. The indeterminate form 0/0 ... 181
5.9. The indeterminate form ∞/∞ ... 183

6. Real Valued Functions of a Single Real Variable. Riemann Integrability ... 195–252
6.1. Introduction ... 195
6.2. Riemann integrability and integral of a bounded functions over finite domain ... 196
6.3. Darboux's theorem ... 202
6.4. Another equivalent definition of integrability and integral ... 203
6.5. Conditions for integrability ... 208
6.6. Particular classes of bounded integrable functions ... 210
6.7. Properties of integrable functions ... 216
6.8. Functions defined by definite integrals ... 227
6.9. Mean value theorem of integral calculus ... 229
6.10. Change of variable in an integral ... 234
6.11. Integration by parts ... 235

7. Sequences of Functions. Point-wise and Uniform Convergence ... 253–276
7.1. Introduction ... 253
7.2. Necessary and sufficient condition for uniform convergence ... 255
7.3. Properties of uniformly convergent sequences of functions ... 258
7.4. Infinite series of functions ... 264
7.5. Test of the uniform convergence of a series ... 265

8. Elementary Functions ... 277–293
8.1. Introduction ... 277
8.2. Trigonometric functions ... 278
8.3. The number p. The smallest positive root of the equation $\cos x = 0$ ... 280
8.4. The functions $\sin x$, $\cos x$ and their properties ... 281
8.5. The function $\tan x$ ... 283
8.6. Logarithmic function ... 285
8.7. Inverse of logarithmic function. Exponential function, Generalised power function ... 287
8.8. Power series expansions of e^x, $\log(1+x)$, $(1+x)^m$ ... 288

9. Improper Integrals ... 294–335
9.1. Introduction ... 294
9.2. Definitions ... 295
9.3. Test for convergence at 'a'. Positive integrand ... 300
9.4. $f(x)$, not necessarily positive. General test for convergence ... 310
9.5. Convergence at ∞ ... 311

9.6. Convergence at ∞ ; the integrand being not necessarily positive. General test for convergence ... 317
9.7. Absolute convergence ... 318
9.8. Tests for conditional convergence ... 319

10. Fourier Series ... 336–364

10.1. Fourier series ... 336
10.2. Theorem ... 337
10.3. Preliminary theorem ... 339
10.4. Main theorem ... 340
10.5. Dirichlet's condition ... 342
10.6. Fourier series for odd and even functions ... 347
10.7. Half range series ... 347
10.8. Other forms of fourier series ... 349

11. Euclidean Spaces, Open and Closed Sets. Compact Sets ... 365–380

11.1. Introduction ... 365
11.2. The set R^n ... 365
11.3. Neighbourhoods of a point. Spheres in R^n ... 368
11.4. Limit points of a set ... 370
11.5. Bolzano-Weierstrass theorem ... 372
11.6. Theorem ... 374
11.7. Heine-Borel property ... 374
11.8. Heine-Borel theorem ... 377

12. Real Valued Functions of Several Real Variables. Limit. Continuity ... 381–395

12.1. Real valued functions ... 381
12.2. Limit ... 381
12.3. Algebra of limits ... 385
12.4. Continuity ... 386
12.5. Continuity in a domain ... 386
12.6. Functions continuous on compact sets ... 389
12.7. Intermediate value theorem ... 391
12.8. Vector-valued functions ... 392

13. Partial Derivatives ... 396–434

13.1. Partial derivatives ... 396
13.2. Directional derivatives ... 398
13.3. Composite functions ... 400
13.4. Derivability of composites ... 404
13.5. Differentiability at a point ... 405
13.6. Partial derivatives of higher orders ... 413
13.7. Second and higher order derivatives of composites ... 419
13.8. Taylor's theorem in R^n ... 421
13.9. Extreme values of a function ... 422

14. Invertible Functions Implicit Functions ... 435–453

14.1. Invertible functions ... 435

14.2. Locally invertible functions ... 436
14.3. Jacobian of a transformation ... 437
14.4. Linear functions. Linear transformations ... 439
14.5. Inverse function theorem ... 441
14.6. Implicitly defined functions ... 446
14.7. Implicit function determined by a single functional equation ... 449

15. Integrals as Functions of a Parameter ... 454–472

15.1. Definite integral as a function of parameter ... 454
15.2. Uniform convergence of improper integrals ... 460
15.3. The test for uniform convergence ... 461
15.4. Properties of uniformly convergent improper integrals ... 463

16. Integration in R^2 Line Integrals. Double Integrals ... 473–519

16.1. Introduction ... 473
16.2. The concept of a plane curve. Line integral ... 473
16.3. The area of a sub-set of R^2 integrability of a bounded function over a rectangle ... 478
16.4. Conditions for integrability ... 480
16.5. Particular classes of bounded integrable functions ... 481
16.6. The calculation of a double integral ... 482
16.7. Simple closed curves. Jordan curves ... 487
16.8. Integrability and integral of a bounded function *f* over any bounded domain E. Area of a set ... 489
16.9. Green's theorem ... 498
16.10. Double integral as a limit ... 501
16.11. Change of variable in a double integral ... 501

17. Curve Lengths. Surface Areas ... 520–537

17.1. Rectifiability of a curve ... 520
17.2. Properties of rectifiable curves ... 522
17.3. Integral expression for the length of a curve ... 525
17.4. Surfaces ... 528

18. Integration in R^3 Gauss's and Stoke's Theorems ... 538–578

18.1. Line Integral ... 538
18.2. Oriented curves and surfaces ... 542
18.3. Surface integrals ... 544
18.4. First generalisation of green's theorem. Stoke's theorem ... 548
18.5. Volume integrals ... 550
18.6. Gauss's theorem. Second generalisation of green's theorem ... 554
18.7. Change of variables in triple integral ... 556
18.8. Evaluation of volumes by triple integration and by double integration ... 565

Miscellaneous exercises ... 571
Answers ... 581
Appendix ... 589

1

Real Numbers

1.1. INTRODUCTION

Set-Theoretic Notation and Terminology. What is called Real Analysis is a development of the set of real numbers which is reached through a series of successive extensions and generalisations starting from the set of natural numbers. As a matter of fact, starting from the set of natural numbers, we pass on successively to the set of integers, set of rational numbers and that of real numbers. Of course, from the set of real numbers, we can also pass on to the set of complex numbers but since Real Analysis is not concerned with complex numbers, we shall have nothing to do with complex numbers in this course. What is known as *Complex Analysis* is a development of the set of complex numbers.

It is no part of the plan of this book to construct the sets of Natural numbers, Integers and Rational numbers starting from some given system of Postulates. It is only proposed to construct the set of real numbers and obtain its properties on the basis of the set of rational numbers and its properties as known. While several methods have been proposed for this construction, we shall give the method, with some modifications of details, as given by Dedekind.

For the sake of brevity and clarity of exposition, we propose to make use of a few notations and adopt some new terminology which we shall now describe.

If S is a set of objects and x a member of S, we shall write

$$x \in S,$$

Also we shall write

$$S = \{x : P(x)\}$$

which states that S is a set of objects x which have a given property P.

*The reader may refer to 'Number systems' by the author for axiomatic presentation.

In case a set S consists of only a finite number of elements, we may list the elements, enclosing the same by parentheses. For example

$$S = \{1, 2, 3\}$$

denotes the set consisting of 3 elements *viz.*, 1, 2, 3.

We shall use the symbols

$$N, Z, Q, R$$

to denote the sets of Natural numbers, Integers, Rational numbers and Real numbers respectively.

If A and B are two sets such that any element of A is as well an element of B, we say that A is a sub-set of B and denote this by writing

$$A \subset B \text{ or } B \supset A$$

If A is a subset of B, we also say that A is contained in B. It may be seen that each set is a sub-set of itself.

Thus if A is a sub-set of B, there is no element of A which is not as well an element of B.

Sometimes we also say that A is a proper sub-set of B, if A is a sub-set of B without being the same as B. We have

$$N \subset Z \subset Q \subset R$$

Equality of Sets. If S and T are two sets, we write

$$S = T$$

to indicate that each element of S is an element of T and that each element of T is an element of S.

Union and Intersection of two sets. The symbols $A \cup B$, $A \cap B$.

Union. Let A and B be two sets.

The set consisting of elements which belong to A or to B or to both is called the Union of A and B and is denoted by the symbol $A \cup B$.

Thus $A \cup B$ is a set such that

$$A \cup B = \{x : x \in A \text{ and / or } x \in B\}.$$

Intersection. The set consisting of elements which belong to A as well as to B is called the Intersection of A and B and is denoted by

$$A \cap B.$$

Thus we have

$$A \cap B = \{x : x \in A \text{ and } x \in B\}$$

Void Set. It is found convenient to utilize the notion of a set containing no element. This set is denoted by the symbol ϕ and is called the *void set*, *empty set* or *null set*.

The Symbols. $\exists, \forall, \Rightarrow, \Leftrightarrow$

We shall use the symbol

$$\exists$$

for 'There exists' and the symbol

$$\forall$$

for 'For all'.

If P and Q are two propositions such that P implies Q *i.e.*, if P then Q, we shall symbolically write

$$P \Rightarrow Q$$

and read it as 'P implies Q'.

If the propositions P and Q are such that each implies the other *i.e.*, if P then Q and if Q then P, we shall write

$$P \Leftrightarrow Q.$$

Also in this case, we say that P if and only if Q.

Note. The symbols $\exists$ and $\forall$ are known as *Quantifiers* and the symbols $\Rightarrow, \Leftrightarrow$ as *Connectives*.

1.2. FINITE AND INFINITE SETS

A set S is finite if there exists a natural number n such that S contains n members; otherwise S is infinite.

Countable and Uncountable sets. Let S be any infinite set. We say that S is countable or is a countable set, if there exists a correspondence which associates to each member of S a natural number *i.e.*, a member of N such that

(*i*) to two different members of S are associated different members of N.

(*ii*) each member of N is associated with some member of S.

Surely every countable set is infinite.

A set is said to be *uncountable* if it is infinite and not countable.

A countable set is also sometimes referred to as *Enumerable*.

A set is said to be at *most countable* if it is finite or countable.

The following are examples of countable sets :

(*i*) the set N of natural numbers,

(*ii*) the set of all even (odd) natural numbers,

(*iii*) the set I of all integers.

EXAMPLES

1. *Show that the union of two countable sets is countable.*

Sol. Let X and Y be two countable sets. Also suppose that

$$X \cup Y = \phi$$

Given that X and Y are countable sets. We can write the elements of X and Y as :

$$X = \{x_1, x_2, x_3, \ldots x_n\}$$
$$Y = \{y_1, y_2, y_3, \ldots y_n\}$$

Now we arrange the correspondence between the elements of the set X and set of odd integers, *i.e.,* we put

$$f_n = x_n : n = 1, 3, 5, \ldots (2n + 1)$$

Similarly we also have

$$f_n = y_n : n = 2, 4, 6, \ldots (2n + 2)$$

Thus we can arrange the elements of union of two countable sets.

2. *The set of all real numbers in the interval* [0, *1*] *is uncountable.*

Sol. Let S be the set of all real numbers in [0, 1]. If possible, let us assume, that the set S of all real numbers in the open interval [0, 1] is countable which means that all such numbers can be put in the form of a sequence given by $< f_n > \equiv < x_n >$, where $n = 1, 2, 3, \ldots, n, \ldots$.

Every real number is expressible in the decimal form having an infinite terms either in terminating or non-terminating form.

By writing $x_1, x_2, x_3, \ldots, x_n, \ldots$ in the decimal form we have

$$x_1 = 0 \,.\, a_{11} \quad a_{12} \quad a_{13} \;\cdots\; a_{1n} \;\cdots$$
$$x_2 = 0 \,.\, a_{21} \quad a_{22} \quad a_{23} \;\cdots\; a_{2n} \;\cdots$$
$$x_3 = 0 \,.\, a_{31} \quad a_{32} \quad a_{33} \;\cdots\; a_{3n} \;\cdots$$
$$\cdots\cdots\cdots\cdots\cdots\cdots\cdots\cdots$$
$$\cdots\cdots\cdots\cdots\cdots\cdots\cdots\cdots$$
$$x_n = 0 \,.\, a_{n1} \quad a_{n2} \quad a_{n3} \;\cdots\; a_{nn} \;\cdots$$
$$\cdots\cdots\cdots\cdots\cdots\cdots\cdots\cdots$$
$$\cdots\cdots\cdots\cdots\cdots\cdots\cdots\cdots$$

where the a's are any one of the digits 0, 1, 2, ..., 9.

Now, we determine a real number which is not the element of the sequence $< x_n >$. Let x be such number given by

$$x = 0.\alpha_1 \, \alpha_2 \, \alpha_3 \ldots \alpha_n \ldots$$

Here α_n is any one of the digits 0, 1, 2, ..., 8 and also differs from a_{nn}. Here we have excluded the digit 9. It is obvious that the number x is different from any one of the numbers $x_1, x_2, \ldots, x_n, \ldots$. As such the above sequence $< x_n >$ does not include that all the numbers of [0, 1]. Hence the set in [0, 1] is not countable.

3. *Show that the set of all irrational numbers is uncountable.*

Sol. Let S be the set of all irrational numbers. If possible let us assume that S is countable.

We know that the union of the sets of irrational numbers (S) and the set of rational numbers (Q) *i.e.*, $S \cup Q$ is countable.

But $S \cup Q = R$, where R is set of real numbers. We also know that R is uncountable.

This contradicts the assumption that S is countable. Hence the set of all irrational numbers is uncountable.

Ex. 4. *Show that a sub-set of a countable set is at most countable.*

Ex. 5. *Show that the set Q of rational numbers is countable.*

Ex. 6. *Show that the set of all algebraic numbers is countable.*

Ex. 7. *Show that the set of all transedental numbers is uncountable.*

1.3. FUNDAMENTAL PROPERTIES OF THE SET OF RATIONAL NUMBERS

We shall now describe the fundamental laws of the order structure and the Algebraic structure possessed by the set Q of rational numbers.

Order structure.

0.1 If a, b are two different rational numbers, then

$$\text{either } a > b \text{ or } b > a.$$

This property is expressed by saying that the Order relation is total or that it satisfies the law of Trichotomy.

0.2 If a, b, c denote three rational numbers such that

$$a > b \text{ and } b > c,$$

then $\quad a > c,$

which may also be expressed as

$$a > b,\ b > c \Rightarrow a > c.$$

The property is expressed by saying that Order relation is transitive.

Algebraic structure. This refers to the addition and multiplication compositions in Q.

Addition. If a, b be two rational numbers, then $a + b$ is also a rational number. Addition obeys the following laws a, b, c denoting arbitrary rational numbers :

A-1. *Commutative Law* :

$$a + b = b + a.$$

A-2. *Associative Law* :

$$(a + b) + c = a + (b + c).$$

A-3. *Existence of Additive identity called zero* :

$$a + 0 = a, \ \forall \ a \in Q.$$

A-4. *Existence of Additive Inverse called negative.* To each rational number, a, there corresponds another, *viz.*, $-a$, such that

$$a + (-a) = 0$$

Multiplication. If a, b be two rational numbers, then, ab, is also a rational number. Multiplication obeys the following laws; a, b, c denoting arbitrary rational numbers :

M-1. *Commutative Law* :

$$ab = ba.$$

M-2. *Associative Law* :

$$(ab)\, c = a\, (bc).$$

M-3. *Existence of multiplicative identity, called unity* :

$$a_1 = a, \ \forall \ a \in Q.$$

M-4. *Existence of multiplicative inverse :*

To each non-zero rational number a, there corresponds another, *viz.*, a^{-1}, such that

$$a\, a^{-1} = 1.$$

AM *Distributive Law* :

$$a\,(b + c) = ab + bc.$$

This last law relates the two compositions of addition and multiplication.

Compatibility of the two structures. The following laws relate the order and the algebraic structures; a, b, c being rational numbers;

OA Order and addition.

$$a < b \Rightarrow a + c < b + c.$$

OM Order and multiplication.

$$a < b, \ c > 0 \Rightarrow ac < bc.$$

In view of the above two laws, we say that the order structure of the set of rational numbers is compatible with its algebraic structure.

EXAMPLES

1. *Employing strictly the properties of the set of rational numbers listed above, show that if a, b, c ∈ Q.*

$$a(b-c) = ab - ac.$$

We have
$$\begin{aligned} a(b-c) &= a[b+(-c)] \\ &= ab + a(-c) && \text{[Left Distributive Law]} \\ &= ab + [-(ac)] \\ & && [\text{since } a(-c) = -(ac)] \\ &= ab - ac. \end{aligned}$$

2. *There exists one and only one rational number x such that*

$$ax = b;$$

a, b being two given rational numbers and a ≠ 0.

Since $a \neq 0$, hence $a^{-1} \in Q$

Hence
$$\begin{aligned} ax = b &\Rightarrow a^{-1}(ax) = a^{-1}b \\ &\Rightarrow (a^{-1}a)\,x = a^{-1}b && \text{[Associative Law]} \\ &\Rightarrow 1.x = a^{-1}b\ [a^{-1} \text{ is multiplicative inverse of } a] \\ &\Rightarrow x = a^{-1}b \in Q \end{aligned}$$

Hence x is a rational number.

If this solution is not unique, then let its solutions are $x = x_1$ and $x = x_2$.

$$\begin{aligned} \therefore\quad ax_1 &= b \text{ and } ax_2 = b, \\ \Rightarrow\quad ax_1 &= ax_2 \\ \Rightarrow\quad x_1 &= x_2 && \text{[By Left Cancellation Law]} \end{aligned}$$

Therefore, solution is unique.

Ex 3. *Employing strictly the properties of the set of rational numbers listed above, show that if a, b, c ∈ Q.*

(*i*) $a(-b) = -(ab)$. (*ii*) $(-a)(-b) = ab$.

Ex 4. *There exists one and only one rational number x such that a, b being two given rational numbers.*

Def. *If a, b be two different rational numbers such that a < b, then any rational number c such that a < c < b is said to be* **between** *a, b.*

Theorem. *Between two different rational numbers, there lie an infinite number of rational numbers.*

Let a, b be two rational numbers such that $a < b$.

Now $a < b \Rightarrow a + b < b + b = 2b$

$$\Rightarrow \frac{1}{2}(a+b) < b.$$

Again $\quad a < b \Rightarrow a + a < a + b$

$$\Rightarrow a < \frac{1}{2}(a+b).$$

We thus see that $\frac{1}{2}(a+b)$ is a rational number between a and b.

We have thus shown that between two different rational numbers there lies a *third* rational number. From this it follows that there lie an *infinite* number of rational numbers between two different rational numbers.

Def. *The property of the set* **Q** *of rational numbers proved above is expressed by saying that the set* **Q** *is* **order-dense.**

Theorem. *If a, b, be two arbitrary positive rational numbers then there exists a positive integer n such that nb > a.*

Let, if possible,

$$nb \leq a, \ \forall \ n \in N.$$

$$nb \leq a \Rightarrow nba^{-1} \leq 1 \Rightarrow nba^{-1} \leq m \ \forall \ m \in N$$

$$\Rightarrow nm^{-1}b \leq a$$

$$\Rightarrow nm^{-1} \leq ab^{-1}$$

$$\Rightarrow x \leq ab^{-1} \ \forall \ x \in \mathbf{Q}_+$$

where $\mathbf{Q}_+$ denotes the set of positive rational numbers.

Thus we arrive at a contradiction.

Note. The property of the set of rational numbers established above is referred to by saying that the set of rational numbers is *Archimedean Ordered* or simply Archimedean.

Dedekind's Construction of the System of Real Numbers

1.4. SECTIONS OF RATIONAL NUMBERS

Let L and U be any two *non-empty* sub-sets of the set Q of rational numbers such that

(*i*) $L \cup U = Q$.

(*ii*) $x \in L$ and $y \in U \Rightarrow x < y$.

so that each rational number either belongs to L or to U and that each member of L is less than each member of U.

We may see that, as a consequence of the conditions satisfied by L and U, we have,

$$L \neq Q, U \neq Q, L \cap U = \phi$$

We may say that the subset U is the **complement** of the subset L in Q and *vice-versa* in as much as U consists of all those rational numbers which are not members of L. Also L is the complement of U in Q.

Such a pair as L, U of sub-sets of Q is called a *Section of rational numbers* and is denoted as (L, U).

Moreover, L is called the Lower class and U the Upper class of the section.

Three types of sections :

(*i*) Consider any rational number, say, 3.

Let $L = \{x : x < 3\}$, $U = \{x : x \geq 3\}$,

Then (L, U) is a section of rational numbers such that the lower class L has no greatest member and the upper class has a smallest member, *viz.*, 3.

(*ii*) Let $L = \{x : x \leq 3\}$, $U = \{x : x > 3\}$.

Again (L, U) is a section of rational numbers whose lower class has a greatest member, *viz.*, 3 and whose upper class has no smallest member.

(*iii*) Let $L = \{x : x \leq 0\} \cup \{x > 0 \text{ and } x^2 < 2\}$,

$U = \{x : x > 0 \text{ and } x^2 > 2\}$.

To be sure that there exists no rational number which escapes classification, we have to show that there exists no rational number whose square is 2.

If possible, let p/q be a rational number such that

$$(p/q)^2 = 2 \Rightarrow p^2 = 2q^2 \qquad ...(i)$$

We suppose that p, q have no common factor, for, such factors, if any, can be cancelled to begin with.

From (*i*), we see that p^2 is an even number. Therefore p must also be even. Let, then, $p = 2m$, where m is an integer. Therefore

$$4m^2 = 2q^2 \Rightarrow q^2 = 2m^2$$

$$\Rightarrow q^2 \text{ is even} \Rightarrow q \text{ is even.}$$

Hence p, q have a common factor 2 and this conclusion contradicts the supposition that p, q have no common factor.

It will now be shown that *L has no greatest member and U has no least member.*

If possible, let k be the greatest number of L. We have

$$0 < k \text{ and } k^2 < 2.$$

Consider, now, the positive number $(4 + 3k)/(3 + 2k)$, we have

$$2-\left(\frac{4+3k}{3+2k}\right)^2 = \frac{2-k^2}{(3+2k)^2} > 0 \Rightarrow \left(\frac{4+3k}{3+2k}\right)^2 < 2;$$

$$\frac{4+3k}{3+2k} - k = \frac{2\left(2-k^2\right)}{3+2k} > 0 \Rightarrow \frac{4+3k}{3+2k} > k$$

Thus the positive number $(4 + 3k) / (3 + 2k) \in L$ and is greater than k. Thus k is not the greatest of L and we have a contradiction.

As above, it may also be shown that if k is the least member of U so that $k^2 > 2$, then $(4 + 3k)/(3 + 2k)$ is a still smaller member of U so that we again have a contradiction.

Conclusion. Thus we see that a section (L, U) may be such that

(*i*) L has no greatest member but U has a least;

(*ii*) L has a greatest member but U has no least;

(*iii*) L has no greatest member and U has no least.

In order to see that these are the only *three* possibilities, it is necessary to show that for no section (L, U) can L have a greatest member *and* also U a least. This fact is easily seen to be true, if we observe that in such a case the infinite number of numbers lying between the greatest member of L and the least member of U will neither belong to L nor to U and thus escape classification so that the sub-sets L, U will not constitute a section.

1.4.1. Modification in the Definition of a Section

It will be seen that to each rational number there correspond two sections according as the rational number in question is the greatest member of the lower class or the smallest member of the upper class. The presentation of the theory of real numbers is somewhat facilitated if we so modify the definition of the section that to each rational number there corresponds only one section. Accordingly we modify the definition by further insisting that the lower class does not have a *greatest member.*

Thus a pair of *non-empty* sub-sets L and U of Q is a section if

(*i*) $L \cup U = Q$

(*ii*) $x \in L, y \in U \Rightarrow x < y$

(*iii*) L has no greatest member.

Note. An equally suitable modification would have been that U has no smallest member.

1.4.2. Criteria for any Sub-set of Rational Numbers to be the Lower Class of a Section

The following simple statement whose truth can be easily established will prove very convenient for later developments.

A non-empty sub-set L of Q will form the lower class of a section if it is such that

(i) $L \neq Q$

(ii) $x \in L, y < x \Rightarrow y \in L$.

(iii) L has no greatest member.

If L satisfies these conditions then all those rational numbers which do not belong to L form the upper class U of the section (L, U).

Each of the sets L and U is the *Complement* of the other in Q.

EXAMPLES

1. *Show that the set $L = \{x \in Q : x > 0 \text{ or } x^2 < 2\}$ determines the lower class of a real number which is irrational.*

(*Gorakhpur 93; Poorv. 92, 95*)

Sol. Here we find that L is a proper non-empty sub-set of Q, since $1 \in L$ and $3 \notin L$.

Also, $x \in Q$ and $x < y$ for some $y \in L$

$\Rightarrow x \in L$, since $x > 0$ and $x < y \Rightarrow x^2 < y^2 < 2$

$\Rightarrow x^2 < 2$.

Suppose $x \in L$. Then if $x > 0$ and $x^2 < 2$, we can find a positive rational number $\dfrac{4+3x}{3+2x} \equiv b$, such that

$$2 - \left(\frac{4+3x}{3+2x}\right)^2 = \frac{2-x^2}{(3+2x)^2} > 0, \qquad \therefore x^2 < 2$$

i.e., $\quad 2 - b^2 > 0 \quad$ *i.e.,* $b^2 < 2$

and $$\left(\frac{4+3x}{3+2x}\right) - x = \frac{2(2-x^2)}{3+2x} > 0, \qquad \therefore x^2 < 2$$

i.e., $\quad b - x < 0 \quad$ *i.e.,* $b > x$.

Hence, if $x > 0$ and $x^2 < 2$, then there exists a positive rational number in L such that $b > x$, *i.e.*, there is no greatest rational number in L and so L determines a real number.

Also, we know that there is no rational number whose square is equal to 2 so the set U, *i.e.*, $Q - L$ has no least rational number in it.

Hence the given sub-set L determines an irrational number.

2. *Does the subset $L = \{x \in Q : x < (5/6)\}$, determine a real rational number.*

3. *State with reasons whether in each of the following the ordered pair (L, U) is a Dedekind's cut in the field Q of rational numbers :*

(*i*) $L = \{x : x \in Q, x < 5\}$ *and* $U = \{x : x \in Q, x > 5\}$

(*ii*) $L = \{x : x \in Q, x \notin U\}$ *and* $U = \{x : x \in Q, x^2 > 9\}$

(*iii*) $L = \{x : x \in Q, x \notin U\}$ *and* $U = \{x : x \in Q, x^2 > 2, x > 0\}$.

In (*i*) we observe that the rational number 5 neither belongs to L nor to U and hence according to the definition (L, U) is not a Dedekind's cut, etc.

Ex 4. *Specify the cut (or section) corresponding to* $\sqrt{3}$ *and prove that the cut so specified satisfies all the requirements.*

5. *Show that there exists an irrational number.*

If the Dedekind's cut or section (L, U) of Q represents a rational number then there is no greatest rational number in L and there is a least rational number in U and conversely when in a Dedekind's cut (L, U) of Q, L has no greatest rational number in it and U has a least rational number in it, then (L, U) represents a rational number.

Let Q be divided into two classes L and U such that L contains all negative rational numbers, the number zero and all the positive rational numbers whose square is less than 3 and U contains all positive rational numbers whose square is greater than 3.

Then symbolically

$$L = \{a \in Q : a \le 0\} \cup \{a \in Q : a > 0 \text{ and } a^2 < 3\}$$

and
$$U = \{b \in Q : b > 0 \text{ and } b^2 > 3\}.$$

Also, we can prove that (L, U) forms a cut of Q in which there is no greatest rational number in L and no least rational number in U.

Thus (L, U) does not represent a rational number and hence there exists an irrational number.

Ex 6. *Does the sub-set* $L = \left\{x \in Q : x < \frac{3}{4}\right\}$ *determines a real rational number ?* (*Gorakhpur 96*)

1.5. SET OF REAL NUMBERS

Def. Real numbers. *A section of rational numbers is called a real number.*

Thus the set of sections of rational numbers is the set of real numbers to be denoted as R.

Real rational numbers. Irrational numbers. *A section is said to be a real rational or an irrational number according as the upper class of the section does or does not have a smallest member.*

If a be the least member of the upper class of a real rational number (L, U), then we say that the rational number a corresponds to the real rational number (L, U).

Notation. *The real rational number corresponding to the rational number a will be denoted as a^*.*

Correspondence between rational and real rational numbers. In view of the modified definition of a section, we see that to each rational number, a, there corresponds a real rational number

$$a^* = (L, U)$$

where $$L = \{x : x < a\}, \quad U = \{x : x \geq a\}.$$

Also each real rational number (L, U) is the correspondent of only one rational number, *viz.*, the least member of the set U.

Thus there exists in a very natural way a one-one correspondence between the set of rational numbers and the set of real rational numbers. We shall denote the set of real rational numbers by Q^*.

Note. To a beginner it might appear strange to call a *section*, which is only a special type of a pair of sub-sets of rational numbers, a *real number*. There may be several reasons for this attitude on his part. One reason may be that the definition of a real number as a section is abstract and the idea of magnitude is foreign to it and the reader finds it difficult to disassociate number from any idea of magnitude which is the very basis of the manner in which he is introduced to the notion of number in Elementary Mathematics.

Another reason for his difficulties may be that he is at a loss to understand as to how the notion of order and the four operations of addition, etc. can be extended to the new set. This later difficulty is only temporary and, in the following sections, it is shown as to how this can be removed. The properties of the set of rational numbers given in § 1.3 will constitute the basis of all the developments that will be undertaken in this connection in the succeeding sections.

1.6. ORDER STRUCTURE OF THE SET OF REAL NUMBERS

Def. *A real number $\alpha_1 = (L_1, U_1)$ is said to be smaller than the real number $\alpha_2 = (L_2, U_2)$ if L_1 is a proper sub-set of L_2.*

Also if α_1 is smaller than α_2, we write

$$\alpha_1 < \alpha_2,$$

and also say α_2 is greater than α_1 and write

$$\alpha_2 > \alpha_1.$$

Thus $$\alpha_1 < \alpha_2 \Leftrightarrow L_1 \subset L_2 \ L_1 \neq l_2.$$

Thus in this case $L_2 \cup U_1$ is an infinite set.

Def. Zero, positive and negative real numbers. A real number is the real number zero, positive or negative according as

$$\alpha = 0^*, \alpha > 0^*, \alpha < 0^*$$

respectively. (*Gorakhpur 92*)

Ex. Show that

(*i*) **(L, U) is positive $\Leftrightarrow$ L contains an infinite number of positive members.**

(*ii*) **(L, U) is negative $\Leftrightarrow$ U contains an infinite number of negative members.**

1.6.1. Law of Order

Law of Trichotomy. *If α, β are two different real numbers, then*

$$\text{either } \alpha < \beta \quad \text{or} \quad \beta < \alpha$$

This property is also expressed by saying that the order relation in R, as defined here, is total. (*Poorvanchal, 94*)

The proof which is simple is left to the reader.

Law of Transitivity. If α, β, γ are three real numbers, then

$$\alpha < \beta, \beta < \gamma \Rightarrow \alpha < \gamma.$$

The proof which is simple is left to the reader.

1.6.2. Order-Completeness of the Set of Real Numbers

We shall now state and prove a fundamental theorem formulating what may be referred to by saying that the set of real numbers is *order-complete.* It will be seen that this property is not shared by the set of the rational numbers.

Before, however, stating and proving the order-completeness property of the set of real numbers, we shall derive three simple results.

I. *If $\alpha = (L, U)$, then*

$$a \in L \Rightarrow a^* < \alpha \text{ and } b \in U \ \alpha \le b^*.$$

Let $a^* = (L_1, U_1), b^* = (L_2, U_2)$.

Here $L_1 = \{x : x < a\}$ where $a \in L$.

Thus L_1 is a proper sub-set of L and as such $a^* < \alpha$.

Again if b be the least member of U, then L, L_2 are identical and as such $\alpha = b^*$.

Let b be not the least member of U. Since

$$L_2 = \{x : x < b\},$$

b does not belong to L so that L is a proper sub-set of L_2 and as such $\alpha < b^*$.

II. *Between two different real numbers, there lie an infinite number of real rational numbers.*

Let $$\alpha_1 = (L_1, U_1), \alpha_2 = (L_2, U_2)$$

be two different real numbers.

Suppose that $\alpha_1 < \alpha_2$.

There exist an infinite number of rational numbers which belong to L_2 as also to U_1.

If, a, be any one such rational number (other than the least member of U_1, if the same exists), we have, as shown above,

$$\alpha_1 < a^* < \alpha_2.$$

Hence the result.

III. *If a, b are two rational numbers, then* $a < b \Leftrightarrow a^* < b^*$.

The proof which is simple is left to the reader.

Note. The result proved above shows that the relationship of order between two real rational numbers satisfies the same basic properties as that between the corresponding rational numbers. We shall, however, in the following section state and prove a property of the set of real numbers not possessed by the set of rational numbers.

Ex. Show that the real number a^* is positive or negative according as the rational number a is positive or negative.

Order-completeness of the set of real numbers

Theorem. *Let L and U be two proper sub-sets of the set R of real numbers such that*

(*i*) $L \cup U = R$,

(*ii*) $\alpha \in L, \alpha \in U \Rightarrow \alpha < \alpha$,

i.e. every member of L is smaller than every member of U.

Then either the sub-set L has a greatest member or the sub-set U has a smallest member.

We consider the two sub-sets L_1, U_1 of Q such that L_1 consists of rational numbers corresponding to the real rational members of L and U_1 consists of rational numbers corresponding to the real rational members of U. Thus

$$L_1 = \{x : x^* \in L\}, \quad U_1 = \{x : x^* \in U\}.$$

Several possibilities will now have to be considered.

Case I. Let L_1 have a greatest member say, a.

We have $$a \in L_1 \Rightarrow a^* \in L.$$

It will be shown that a^* is the greatest member of L. If not, let α be a greater member of L.

Let b^* be any real rational number such that

$$a^* < b^* < a.$$

Now $$b^* < a \in L \Rightarrow b^* \in L \Rightarrow b \in L_1.$$

Also $$a^* < b^* \Rightarrow a < b.$$

Thus b, a member of L_1, is greater than its greatest member a so that we arrive at a contradiction.

Hence L has a greatest member, *viz.*, a^*.

Case II. Let R_1 have a smallest member, say, b. As in the preceding case, it may be shown that in this case R has the smallest member b^*.

Case III. Let neither L_1 have a greatest member nor U_1 have a least.

In this case (L_1, U_1) is an irrational number, say a, which must either $\in L$ or $\in U$.

Let $\alpha \in L$.

It will be shown that α is the greatest member of L. If not, let α be a greater member of L.

Let a^* be a real rational number such that

$$\alpha < a^* < \alpha.$$

Now $$a^* \in L \Rightarrow a \in L_1.$$

Thus there exists a member a of the class L_1 of a section

$$\alpha = (L_1, R_1)$$

such that $\alpha < a^*$, so that we arrive at a false statement.

Thus L has a greatest member.

It may similarly be shown that if $\alpha \in U$ then it is the smallest member of U.

We have thus proved that either L has a greatest member or U has a least member.

Note 1. The above theorem is sometimes stated in the following equivalent form :

If all the real numbers are divided into two non-empty sub-sets, L and U such that every member of L is less than every member of U, *then there exists a number α such that every real number less than α belongs to L and every real number greater than α belongs to U; α itself may belong to either class.*

The number, α, is the greatest member of L or the least member of U as the case may be. Also this member, α, is said to determine the section.

Note 2. The theorem discussed above indicates a fundamental difference between the sets of rational numbers and real numbers. This difference is generally described by saying that whereas the set R of real numbers is order complete, the set Q of rational numbers is not order complete.

1.7. ALGEBRAIC STRUCTURE OF THE SET R

We shall now define the compositions of addition and multiplication in R and also obtain their basic properties.

We shall first take up the case of Addition.

1.7.1. Addition Composition

Before defining the addition composition and establishing its various properties, we shall obtain a result which will be found useful in this discussion. *(Gorakhpur 92, 95; Poorv. 93)*

Let (L, U) be a section of rational numbers and $k > 0$ be any given rational number. Then there exist $x \in L$, $y \in U$ such that

$$y - x = k.$$

Let a, b be any two members of L and U respectively. The set of rational numbers being *Archimedean ordered*, there exists a positive integer n such that

$$nk > b - a \Rightarrow a + nk > b.$$

Consider the set of rational numbers

$$a,\ a + k,\ \dots,\ a + nk. \qquad \dots(1)$$

Now $\qquad a \in L,\ a + nk \in U.$

There exists, therefore, two consecutive members, say

$$a + rk,\ a + (r + 1)\,k.$$

of the set such that

$$a + rk \in L,\ a + (r + 1)\,k \in U.$$

These are then the required numbers x and y.

Def. Sum of two real numbers. *The sum of two real numbers*

$$\alpha_1 = (L_1, U_1),\ \alpha_2 = (L_2, U_2)$$

is the real number

$$(L, U)$$

where $\qquad L = \{x_1 + x_2 : x_1 \in L_1,\ x_2 \in L_2\}$

and U is the complement of L in **Q**.

As a matter of justification it has to be shown that L can be the lower class of a section and this fact can easily be proved.

We write, $\qquad (L, U) = (L_1, U_1) + (L_2, U_2) = \alpha_1 + \alpha_2.$

Def. The negative of a real number. Def. *The negative* $-\alpha$ *of a real number*

$$\alpha = (L, U)$$

is the real number

$$(L_1, U_2)$$

where $L_1 = \{-x : x \in U, x$ *is not the smallest member of* $U\}$ *and* U_1 *is the complement of* L_1 *in* **Q**.

It may be shown that L_1, as defined, satisfies the condition for being the lower class of a section.

We write $$-\alpha = (L_1, U_1).$$

The difference. Def. The difference, $\alpha - \beta$, of two real numbers α and β is defined by the equality

$$\alpha - \beta = \alpha + (-\beta),$$

so that to obtain $\alpha - \beta$ we add to α the negative $-\beta$ of β. (*Gorakhpur 95*)

Laws of Addition. In the following α, β, γ denote arbitrary real numbers.

A-1. *Addition composition is commutative, i.e.,*

$$\alpha + \beta = \beta + \alpha, \ \forall \alpha, \beta \in \mathbf{R}$$

A-2. *Addition composition is associative, i.e.,*

$$\alpha + (\beta + \gamma) = (\alpha + \beta) + \gamma, \ \forall \alpha, \beta, \gamma \in \mathbf{R}$$

The proofs, being very simple, are omitted.

A-3. *Addition composition possesses an additive identity.* We shall show that 0^* is an additive identity, *i.e.*, (*Gorakhpur 93, Agra, 93*)

$$\alpha + 0^* = \alpha \ \forall \ \alpha \in \mathbf{R}$$

Let $$\alpha = (L_1, U_1),\ 0^* = (L', U'),\ \alpha + 0^* = (L, U).$$

We have to show that L and L_1 are identical.

Let $x \in L$. There exists $x_1 \in L_1$ and there exists $x_2 \in L'$ such that

$$x = x_1 + x_2.$$

Also $$x_2 \in L' \Rightarrow x_2 < 0 \Rightarrow x_1 + x_2 < x_1 \Rightarrow x < x_1.$$

Also $$x < x_1 \in L_1 \Rightarrow x \in L_1.$$

Thus each member of L is a member of L_1.

Let $$y \in L_1$$

As L_1 has no greatest member, there exists $y + k \in L_1$ greater than y_1; $(k > 0)$.

Now $$y = (y + k) + (-k)$$

where $$y + k \in L_1,\ -k \in L'.$$

Thus $y \in L$

so that each member of L_1 is a member of L.

We thus see that L and L_1 are identical and as such

$$\alpha + 0^* = \alpha.$$

Note. Uniqueness of additive Identity. It may be seen that 0^* is the only number such that

$$\alpha + 0^* = \alpha \ \forall \ \alpha \in \mathbf{R}.$$

Suppose θ is also a number such that

$$\alpha + \theta = \alpha \ \forall \ \alpha \in \mathbf{R}.$$

Now $\theta + 0^* = \theta$, for 0^* is an identity and $\theta + 0^* = 0^* + \theta = 0^*$ for the addition is commutative and θ is an identity. Thus

$$\theta = 0^*.$$

A-4. Existence of additive inverse. We shall prove that

$$\alpha + (-\alpha) = 0^* \ \forall \ \alpha \in \mathbf{R}. \qquad \textit{(Gorakhpur 93, 95)}$$

Let $\alpha = (L_1, U_1), -\alpha = (L, U), \alpha + (-\alpha) = (L', U')$.

We have to show that 0, is the smallest number of U'.

Now each member of L', obtained as it is, on adding a number of L_1 to the negative of a member of U_1, is necessarily a negative rational number.

Again if k be any given negative rational member, $\exists \ a_1 \in L_1, b_1 \in U_1$ such that

$$b_1 - a_1 = -k \Rightarrow a_1 + (-b_1) = k$$

so that $k \in L'$.

Thus, every negative rational number is a member of L'.

Hence $(L', U') = 0^*$

Note. Uniqueness of additive inverse. It may be seen that $-\alpha$ is the only real number such that $\alpha + (-\alpha) = 0^*$. Let, if possible, α be also a number such that $\alpha + \beta = 0^*$. We have

$$\alpha + (-\alpha) = \alpha + \beta$$

$$\Rightarrow \quad -\alpha + [\alpha + (-\alpha)] = -\alpha + [\alpha + \beta]$$

$$\Rightarrow \quad [-\alpha + \alpha] + (-\alpha) = (-\alpha + \alpha) + \beta$$

$$\Rightarrow \quad 0^* + (-\alpha) = 0^* + \beta \Rightarrow -\alpha = \beta.$$

The reader may justify each of the above implications.

OA. Compatibility of order with addition. We show that

$$\alpha > \beta = \alpha + \gamma > \beta + \gamma.$$

Let $\alpha = (L_1, U_1), \beta = (L_2, U_2), \gamma = (L_3, U_3),$

and let $\alpha + \gamma = (L', U'), \beta + \gamma = (L'', U'').$

Since every member of L_2 is as well a member of L_1, it follows that every member of L'' is also a member of L'.

Let a_1 be a rational number which belongs to L_1 but *not* to L_3. Then $a_1 \in U_2$.

Let $a_1' > a_1$ be a member of L_1. Let

$$a_1' - a_1 = \varepsilon, \varepsilon > 0.$$

There exists $a_3 \in L_3$ and $b_3 \in U_3$ such that

$$b_3 - a_3 = \varepsilon.$$

We have $a_1' + a_3 = a_1 + \varepsilon + b_3 - \varepsilon = a_1 + b_3.$

Now $a_1' \in L_1, a_3 \in L_3 \Rightarrow a_1' + a_3 \in L'$

$$a_1 \in U_2, b_3 \in U_3 \Rightarrow a_1 + b_3 \in U''.$$

Thus there exists a member common to L' and U'', *i.e.*, there exists a rational member which belongs to L' but not to L''.

Hence L'' is a proper part of L and as such

$$\beta + \gamma < \alpha + \gamma.$$

Cor. *Between two different real numbers there lie an infinite number of irrational numbers.*

Let α, β be two real numbers and let $\alpha < \beta$. Let γ be any arbitrary irrational number. We have

$$\alpha < \beta \Rightarrow \alpha - \gamma < \beta - \gamma.$$

If a* be any one of the infinite number of real rational numbers between $\alpha - \gamma$ and $\beta - \gamma$, we have

$$\alpha - \gamma < a^* < \beta - \gamma \Rightarrow \alpha < \alpha^* + \gamma < \beta.$$

Now $\alpha^* + \gamma$ lies between α and β.

It may be seen that $a^* + \gamma$ is irrational.

1.7.2. Multiplication Composition

Before defining the product of two real numbers and developing the properties of multiplication, we obtain a result which will prove useful for the following discussion.

Let (L, U) be any positive real number and let k > 1 be a given rational number. Then there exists x ∈ L and y ∈ U such that

$$y/x = k$$

Let a, b be any positive members of L and U respectively. We write

$$k = 1 + l, \ l > 0.$$

There exists a positive integer n such that

$$n.al > b - a \Rightarrow a\,(1 + nl) > b.$$

Consider the set of numbers

$$a, ak, ak^2, ..., ak^n. \qquad ...(1)$$

Now $\quad ak^n = a\,(1 + l)^n > a\,(1 + nl) > b,$

$\therefore \quad ak^n \in U.$

There exist two consecutive members ak^r, ak^{r+1} of the set (1) such that

$$ak^r \in L, \ ak^{r+1} \in U$$

Then $x = ak^r$, $y = ak^{r+1}$ are the required numbers.

Def. The product of two real numbers.

(*Gorakhpur 91, 93; Poorv. 92, 94, 96*)

Let $\quad \alpha_1 = (L_1, U_1), \alpha_2 = (L_2, U_2).$

We firstly suppose that α_1, α_2 are positive real numbers.

Let a sub-set L of rational numbers be formed of

(*i*) all the negative rational numbers,

(*ii*) the rational number zero,

(*iii*) positive rational numbers which can be obtained on multiplying a positive member of L_1 with a positive member of L_2.

Let U be the complement of L in **Q**.

Then, by def., $\quad (L, U) = (L_1, U_1)\,(L_2, U_2) = \alpha_1\alpha_2.$

Of course, we have to verify, that L satisfies the conditions for being the lower class of a section. The verification is straight forward.

We now give definitions when α, β may not both be positive. We write

$$|\alpha| = \alpha \text{ if } \alpha \geq 0^* \text{ and } |\alpha| = -\alpha \text{ otherwise.}$$

By def.

$$\alpha\beta = \begin{cases} |\alpha||\beta| \text{ if } \alpha, \beta \text{ are both negative;} \\ -|\alpha||\beta| \text{ if one of } \alpha, \beta \text{ is positive and the other negative.} \end{cases}$$

If α, β be both zero, then, by def.,

$$\alpha\beta = 0^*.$$

Def. The reciprocal of a non-zero real number.

Let $\quad \alpha = (L_1, U_1)$

be a positive real number.

Let a sub-set L of $\mathbf{Q}$ be formed of

(*i*) the negative rational numbers,

(*ii*) zero,

(*iii*) the reciprocals of the members of the class U_1 excepting that of the least of U_1, if it exists.

Let U be the complement of L in $\mathbf{Q}$.

Then (L, U)

is called the reciprocal of $\alpha = (L_1, U_1)$ and denoted by $\alpha^\dagger$.

Of course we have to show that L can be the lower class of a section.

If α be negative, then, by def.

$$\alpha^\dagger = -(|\alpha|)^\dagger.$$

Division. If α, β be two given real numbers and $\beta \neq 0^*$, then we write

$$\alpha \div \beta = \alpha\beta^\dagger.$$

Laws of multiplication.

M.1. *Multiplication composition is commutative :*

$$\alpha\beta = \beta\alpha.$$

M.2. *Multiplication composition is associative :*

$$(\alpha\beta)\gamma = \alpha(\beta\gamma),$$

α, β, γ denote real numbers. (*Gorakhpur 92, 93, 95; Poorv. 92*)

M.3. $\alpha 1^* = \alpha, \ \forall \ \alpha \in \mathbf{R}.$ (*Gorakhpur 91, 94*)

The proofs of *M*.1, *M*.2, *M*.3 which are simple are omitted.

M.4. $\alpha.\alpha\dagger = 1^*, \alpha \neq 0^*.$

To start with, we suppose that α is positive.

Let $\alpha = (L_1, U_1), \alpha\dagger = (L, U), \alpha.\alpha\dagger = (L', U').$

Let a_1 be a positive member of L_1 and b_1 any, but not the least, member of U_1, so that $1/b_1$ is a positive member of L.

The positive members of L' are, therefore, of the form a_1/b_1.

Since $a_1/b_1 < 1$, we see that every member of L' is < 1.

Again, let k be a positive rational number < 1.

There exist positive members, a_2, b_2, of L_1, U_1 such that

$$a_2/b_2 = k.$$

Since $k = a_2(1/b_2)$, we see that *every rational number < 1, is a member of L'.*

Hence, 1, is the least member of U' and accordingly

$$(L', U') = 1^*.$$

Let, now, α be negative so that $\alpha\dagger$ is also negative.

We have, by the def. of the product of two negative numbers,

$$\alpha.\alpha^\dagger = |\alpha| |\alpha|^\dagger$$
$$= 1^*, \text{ as proved above.}$$

Note. This property is expressed by saying *that* $\alpha^\dagger$ *is the multiplicative inverse of* α. It may be shown that every non-zero real number admits of a *unique* multiplicative inverse.

Notation for inverse. In view of the property proved above, we shall write

$$\alpha^\dagger = \frac{1^*}{\alpha}.$$

Also since $\quad (\beta\alpha^\dagger)\,\alpha = \beta(\alpha^\dagger\,\alpha) = \beta\cdot(\alpha\alpha^\dagger) = \beta 1^* = \beta,$

we write $\quad \beta \div \alpha = \beta\alpha^\dagger = \dfrac{\beta}{\alpha}.$

A III. Distributive Law

$$\alpha\,(\beta + \gamma) = \alpha\beta + \alpha\gamma, \ \forall\ \alpha, \beta, \gamma \in \mathbf{R}.$$

(*Gorakhpur 94*)

Let $\quad \alpha = (L_1, U_1), \beta = (L_2, U_2), \gamma = (L_3, U_3).$

Let $\quad \alpha\,(\beta + \gamma) = (L, U)$ and $\alpha\beta + \alpha\gamma = (L', U').$

Firstly we consider the case, when α, β, γ are all positive.

All the negative rational numbers and the rational number zero are necessarily members of L as well as of L'.

The positive members of L are of the type

$$a_1\,(a_2 + a_3),$$

and the positive members of L' are of the type

$$a_1 a_2 + a_1'\,a_3,$$

where a_1, a_1' are positive members of L_1, and a_2, a_3 are positive members of L_2, L_3 respectively.

Since $\quad a_1\,(a_2 + a_3) = a_1 a_2 + a_1 a_3,$

we see on taking $a_1' = a_1$, that every positive member of L is also a member of L'.

Any member $a_1 a_2 + a_1' a_3$ of L' is clearly a member of L, if $a_1' = a_1$. In general, let $a_1 > a_1'$ so that $\alpha_1'/\alpha_1 < 1$.

We write $\quad a_1' a_3 = a_1\,[(a_1'/a_1)\,a_3] = a_1 a_3'$, say.

Now $\quad a_3' = (a_1'/a_1)\,a_3 < 1.\,a_3 = a_3,$

and therefore $a_3' \in L_3$.

Since $$a_1a_2 + a_1a_3 = a_1a_2 + a_1a_3' = a_1(a_2 + a_3'),$$
we see that every positive number of L' is also a number of L.

Thus $$\alpha(\beta + \gamma) = \alpha\beta + \alpha\gamma$$
when α, β, γ are all positive.

The proof of the law when α, β, γ may not all be positive is left to the reader.

OM. Compatibility of order with multiplication

$$\alpha < \beta, \gamma > 0^* \Rightarrow \alpha\gamma < \beta\gamma.$$

Firstly suppose that α, β are both positive.

Let $$\alpha = (L_1, U_1), \beta = (L_2, U_2), \gamma = (L_3, U_3);$$
$$\alpha\gamma = (L, U), \beta\gamma = (L', U').$$

All the negative rational numbers and the rational number zero are necessarily contained in L as well as L'.

Since every member of L_1 is a member of L_2, therefore, every member of L is also a member of L'.

Let a_2 be a positive number which is a member of L_2 but not of L_1. Since L_2 has no greatest member, there exists a member of L_2, say a_2' which is $> a_2$.

The numbers a_2, a_2' both belong to U_1.

Let $a_2'/a_2 = k$, which is greater than 1.

There exist positive members a_3, b_3 of L_3, U_3, respectively such that
$$b_3/a_3 = k.$$

We have $$a_2'a_3 = a_2ka_3 = a_2b_3.$$

Since a_2' is a positive member of L_2 and a_3 a positive member of L_3, therefore, $a_2'a_3$ *i.e.*, a_2b_3 belongs to L.

Also since a_2 is a member of U_1 and b_3 of U_3, therefore, a_2b_3 belongs to U and not to L.

Thus each number of L' is not a member of L. Therefore L is a proper part of L'. Hence the result.

The proofs of other cases are left to the reader.

1.8. THE SETS OF RATIONAL AND REAL RATIONAL NUMBERS

Isomorphism. Let **Q*** denote the set of real rational numbers; **Q** denoting the set of rational numbers. We have seen that there exists a one-one correspondence between **Q** and **Q*** which associates $a \in \mathbf{Q}$ to $a^* \in \mathbf{Q}^*$.

We shall see that this correspondence is

Order preserving, Sum preserving and Product preserving i.e., (*i*) $a < b \Rightarrow a^* < b^*$,

(*ii*) $a + b = c \Rightarrow a^* + b^* = c^*$

(*iii*) $ab = d \Rightarrow a^*b^* = d^*$.

Here a, b denote rational numbers.

The proofs of these results are very simple. We shall give only the proof for (*iii*).

If a, b be two rational numbers, then we shall show that

$$a^*.b^* = (ab)^*,$$

i.e., the product of two real rational numbers is also a real rational number which corresponds to the product of the corresponding rational numbers.

Let $a^* = (L_1, U_1)$, $b^* = (L_2, U_2)$, $(ab)^* = (L, U)$.

Let a, b be both positive. Let x, y be any positive members of L_1, L_2 respectively so that xy is a positive member of L.

Now $x < a$ and $y < b \Rightarrow xy < ab$.

Thus every member of L *is* $< ab$.

Again consider any positive rational number abk which is $< ab$ so that $k < 1$.

We write
$$k = \frac{1+k}{2} \cdot \frac{2k}{1+k}.$$

Now $k < 1 \Rightarrow (1 + k)/2 < 1$ and $2k/(1 + k) < 1$

We have
$$abk = \left(a.\frac{1+k}{2}\right) \cdot \left(b.\frac{2k}{1+k}\right).$$

Since $a.(1 + k)/2 < a$ and $b.[2k/(1 + k)] < b$, therefore $a.(1 + k)/2 \in L_1$ and $b.[2k/(k + 1)] \in L_2$ and accordingly their product $abk \in L$.

Thus every rational number $< ab \in L$.

Hence ab is the least member of U so that

$$a^*.b^* = (ab)^*$$

The case where a, b are both negative or only one of them is negative can now be easily disposed of.

The fact that there exists a one-one correspondence between **Q** and **Q*** which preserves, order, sum and product is expressed by saying that **Q** and **Q*** are **Isomorphic.**

Any true relation between rational numbers formulated in terms of order, addition and multiplication will remain true when each of these rational numbers is replaced by the corresponding real rational number.

Thus, for example, the proposition

$$2 + 3 = 5,$$

where 2, 3, 5 denote rational numbers, remains true even when they denote corresponding real rational numbers which have so far been denoted by the symbols 2^*, 3^*, 5^*.

In view of this, *we agree to denote, for future developments, a real rational number by the same symbol which denotes the corresponding rational number* so that if 'a' is a rational number, then the same symbol 'a' will also, now, be used to denote the corresponding real rational number which has so far been denoted by a^*.

The context in which the symbols may appear will fix the interpretation.

This use of a single symbol to denote two different concepts leads to no confusion, but is helpful, inasmuch as we have seen that a statement, which describes some relation between rational numbers, remains true when the symbols for rational numbers are interpreted as symbols for the corresponding real rational numbers.

An important note. In the following chapters, the word 'number' will always mean 'real number' and the word 'rational number' will mean 'real rational number'.

Complete ordered field of real numbers. The fact of the set of real numbers admitting of the two compositions of addition and multiplication satisfying the nine properties

$$A.1,\ A.2,\ A.3,\ A.4;$$

$$M.1.\ M.2,\ M.3,\ M.4;$$

$$A\,M$$

is referred to by saying that the set of real numbers is a **Field.**

Again the fact of the field of real numbers admitting an order relation which is transitive and compatible with addition and multiplication (defined by properties *OA* and *OM*) is referred to by saying that the set of real numbers is an ordered field.

Finally the ordered field being order-complete, we say that the set of real numbers is an **ordered complete field.**

While the set of rational numbers is an ordered field, it is *not* a complete ordered field.

1.9. RESTATEMENT OF THE BASIC PROPERTIES OF *R*

Having defined real numbers and established that the set of real numbers is a Complete Ordered Field, we restate below the basic properties of the set of real numbers as a complete ordered field.

FIELD STRUCTURE

Addition Composition. To each ordered pair a, b of real numbers, there corresponds a real number called their sum and denoted by $a + b$. The addition composition has the following properties :

I. $A.1.$ $a + b = b + a \ \forall \ a, b \in \mathbf{R}.$ [*Commutatively*]

II. $A.2.$ $(a + b) + c = a + (b + c) \ \forall \ a, b, c \in \mathbf{R}.$ [*Distributivity*]

III. $A.3.$ There exists a real number *viz.* '0' such that $\forall \ a \in \mathbf{R}.$

$$a + 0 = a. \qquad \text{[Existence of additive identity]}$$

IV. $A.4.$ To each real number a, there corresponds a real number *viz.*, $-a$, such that

$$a + (-a) = 0. \qquad \text{[Existence of additive Inverse]}$$

Multiplication Composition. To each ordered pair a, b of real numbers, there corresponds a real number called their product and denoted by ab. This multiplication composition has the following properties :

V. $M.1$ $ab = ba \ \forall \ a, b \in \mathbf{R}.$ [*Commutativity*]

VI. $M.2.$ $(ab)c = a(bc) \ \forall \ a, b, c \in \mathbf{R}.$ [*Associativity*]

VII. $M.3.$ There exists a real number, *viz.*, 1 such that $\forall \ a \in \mathbf{R}$

$$a1 = a. \qquad \text{[Existence of multiplicative identity]}$$

VIII. $M.4.$ To each real number $a \neq 0$, there corresponds another *viz.*, $1/a$ such that

$$a(1/a) = 1. \qquad \text{[Existence of multiplicative inverse]}$$

There is also a law which relates the two compositions, *viz.*,

IX. AM $a(b + c) = ab + ac, \ \forall \ a, b, c \in \mathbf{R}.$ [*Distributive law*]

The fact of the set of real numbers admitting the two compositions satisfying the Nine properties mentioned above is referred to as the set of real numbers having a field structure.

ORDERED FIELD STRUCTURE

'Greater than' relation. Given any two different real numbers a, b we have

$$\text{either } a > b \text{ or } b > a.$$

X. The 'Greater than' relation has the following property :

$$a > b \text{ and } b > c \Rightarrow a > c \qquad \text{[Transitivity]}$$

We also have a property relating to Addition Composition and the 'Greater than' relation referred to by saying that the Addition Composition is *compatible* with the 'Greater than' relation.

XI. AO $a > b \Rightarrow a + c > b + c.$

Also we have a property relating the Multiplication Composition and the 'Greater than' relation referred to by saying that the Multiplication Composition is compatible with the 'Greater than' relation.

XII. *MO* $a > b$ and $c > 0 \Rightarrow ac > bc.$

The twelve properties referred to above complete the description of the set of real numbers as an **ordered field.**

COMPLETE ORDERED FIELD STRUCTURE

XIII. Order-completeness of the set of real numbers.

If L and U be two proper sub-sets of the set R of real numbers such that

(*i*) $L \cup U = R,$

(*ii*) $\alpha \in L, \beta \in U \Rightarrow \alpha \subset \beta,$

i.e., every member of L is smaller than every member of U.

Then either the sub-set L has a greatest member or the sub-set U has a smallest member.

Note. The thirteen properties of the system of real numbers are basic and every other result pertaining to the system is deducible from them.

In the next chapter we shall also consider some concepts for which we shall need to utilize the completeness of this ordered field.

EXAMPLES

1. *Prove that* $a - b = 0^*$ *iff* $a = b, \ \forall a, b \in \mathbf{R}.$ (*Gorakhpur 95*)

Sol. If $a = b$, then

$$a - b = a - a = a + (-a) = 0^*, \text{ by the property of inverse}$$

and $$a - b = 0^* \Rightarrow a + (-b) = 0^*$$

$$\Rightarrow -b \text{ is the inverse of } a$$

But the inverse of a is equal to $-a$ and is unique.

$$\therefore \quad -b = -a, \quad \text{so} \quad -(-b) = -(-a)$$

where $-(-a)$ is the inverse of $-a$

$$\Rightarrow \quad b = a.$$

2. *If a, b, c are any three elements of* **R**, *then prove that*

$$a - (b - c) = (a - b) + c. \qquad \textit{(Gorakhpur 95)}$$

Sol.
$$\begin{aligned} a - (b - c) &= a - [b + (-c)] \\ &= a + [-\{b + (-c)\}] \\ &= a + (-b + c) \qquad \because -(a+b) = -a-b \\ &= \{a + (-b)\} + c \end{aligned}$$

by associative law of addition in **R**

$$= (a - b) + c.$$

3. *For any r, s ∈* **Q**, *prove that* $(r + s)^* = r^* + s^*$, *where the symbols have their usual meanings.* (*Gorakhpur 92, 95; Poorv. 93, 96*)

Sol. We know that there is an one-one correspondence $f : \mathbf{Q} \leftrightarrow \mathbf{Q}^*$ defined by $f(x) = x^*\ \forall\ x \in \mathbf{Q}$, where $\mathbf{Q}^*$ is the set of all real rational numbers.

$$\therefore \quad (r + s)^* = f(r + s) = f(r) + f(s) = r^* + s^*$$

4. *Prove that there exists an unique multiplicative inverse of a non-zero real number.* (*Gorakhpur 92, 94; Poorv. 95*)

Sol. (*i*) Let $a = (L_1, U_1) \in \mathbf{R}^+$ and $b = (L, U)$

where $L = \{x : x \leq 0 \text{ or } x = 1/y, \text{ where } y \in U_1 \text{ and } y \neq \text{least number of } U_1\}$

Suppose $\qquad a.b = (L, U).$

Let $0 < x_1 \in L_1$ and $0 < \left(\dfrac{1}{y_1}\right) \in L$. Then the positive elements of L' are of the form x_1/y_1.

Since $\qquad y_1 \in U_1, y_1 > x_1$, *i.e.*, $x_1/y_1 < 1$

$\Rightarrow$ every element of L' is less than 1.

Again, let z be any positive rational number such that $z < 1$ then there exists $0 < x_2 \in L_1$ and $0 < y_2 \in U_1$ such that $x_2/y_2 = z$.

Since $z = x_2/y_2 = x_2\,(1/y_2)$, so it follows that every rational number which is less than 1 belongs to L'.

Hence $L' = \{x : x < 1\}$ and 1 is the least member of U'.

$$\therefore \qquad (L', U') = 1^* \quad \text{or} \quad a.b = 1^*$$

(*ii*) If $a \in \mathbf{R}^-$, then $-a \in \mathbf{R}^+$, *i.e.*, a is positive and there exists a real number $c \in R^+$, such that $(-a)\,c = 1^*$.

$$\therefore \qquad a \in \mathbf{R}^- \text{ and } -c \in \mathbf{R}^-, \text{ so we have}$$

$$a.(-c) = (-a).c = 1^*$$

i.e., $-c$ is the multiplicative inverse of a.

Hence whether $a \in \mathbf{R}^+$ or $a \in \mathbf{R}^-$, there exists an element $b \in \mathbf{R}$ such that $a.b = 1^*$.

Now, we are to show that multiplicative inverse b is unique.

If possible let b_1 and b_2 be two multiplicative inverses of a non-zero element $a \in \mathbf{R}$.

Then, we have $a.b_1 = 1^*$ and $a.b_2 = 1^* \Rightarrow a.b_1 = a.b_2 \Rightarrow b_1 = b_2$, by left cancellation law, *i.e.*, the multiplicative inverse of a non-zero element in **R** is unique.

5. *If a, b ∈ R and a.b = 0*, then prove that either a = 0* or b = 0*.*

Sol. Let $a \neq 0^*$, then a^{-1} exists such that $a^{-1}.a = 1^*$.

Now $\mathbf{b} = 1^*.\mathbf{b}$, where 1^* is the multiplicative identity in **R**.

$$= (a^{-1}.a)b, \qquad (\because\ a^{-1}.a = 1^*)$$

$$= a^{-1}.(a.b), \qquad \text{(by associativity)}$$

$$= a^{-1}.0^*, \quad a.b = 0^* \qquad \text{(given)}$$

$$= 0^*$$

Similarly, if $b \neq 0^*$, we can show that $a = 0^*$.

Hence, if $a.b = 0^*$, then either $a = 0^*$ or $b = 0^*$.

6. *If a, b ∈* **R**, *prove that* $(a.b)^{-1} = a^{-1}.b^{-1}$, *where* $a \neq 0^*$, $b \neq 0^*$.

Sol.
$$(a.b)(b^{-1}.a^{-1}) = a.[b.(b^{-1}.a^{-1})], \qquad \text{(by associativity)}$$

$$= a[(b.b^{-1}).a^{-1}], \qquad \text{(by associativity)}$$

$$= a.[1^*.a^{-1}], \qquad (\because\ b.b^{-1} = 1^*)$$

$$= a.[a^{-1}], \qquad (\because\ 1^*a^{-1} = a^{-1})$$

$$= 1^* \qquad (\because\ a.a^{-1} = 1^*)$$

i.e., $(b^{-1}.a^{-1})$ is the inverse of $(a.b)$

i.e., $(a.b)^{-1} = b^{-1}.a^{-1} = a^{-1}.b^{-1}$ by commutative law.

7. *Show that the set of real numbers is dense.*

(Gorakhpur 94)

Sol. Let a and b be two distinct real numbers and let $a < b$.

Now, $a < b \Rightarrow L(a) \subset L(b)$

$\Rightarrow$ there exists $x \in L(b)$

such that $x \in U(a)$

$\Rightarrow x \in L(b)$ and $x \in U(a)$.

Now x can be so chosen that $x \in L(b)$ and $x \in U(a)$ but x is the least element in it. Then the real rational number x is such that $x > a$ and $x < b$, *i.e.*, $a < x < b$.

Therefore, between any two distinct real numbers a and b there exists a real rational number x such that $a < x < b$.

Proceeding in a similar way we can show that there exists another real rational number if such that $a < y < x$ and there exists a real rational number z such that $a < z < y$ and so on.

Hence between any two real numbers there is always an infinitely many rational numbers, *i.e.*, set of real numbers is dense.

8. *Show that the system of rational numbers is not complete with respect to Dedekind's cut.* (*Gorakhpur 95, 96; Poorv. 91*)

Sol. A system of numbers is said to be complete with respect to Dedekind's cut provided every cut of the system of numbers is produced by a member of the system.

Also, we know that if (L, U) represents a cut of Q, the set of rational numbers and if this cut is produced by a rational number x, say then the lower class L has no greatest in it and the upper class U has a least number x in it.

However, if (L, U) be a cut of $\mathbf{Q}$, such that

$$L = \{a \in \mathbf{Q} : a \le 0\} \cup \{a \in \mathbf{Q} : a > 0 \text{ and } a^2 < 3\}$$

and $$U = \{b \in \mathbf{Q} : b > 0 \text{ and } b^2 > 3\}$$

then we can prove that L has no greatest member in it and U has no least member in it.

Thus, we find that there exists a cut (L, U) in $\mathbf{Q}$ in which L has no greatest in it and U has no least in it.

Hence this cut of $\mathbf{Q}$ cannot be produced by any element of $\mathbf{Q}$, and so $\mathbf{Q}$ is not complete with respect to Dedekind's cut.

EXERCISES

1. Show that multiplicative identity is unique.

(*Gorakhpur 91, 94; Agra 93*)

2. For $a, b \in B$, prove that

(*i*) $a + (-b) = a - b$; (*ii*) $b - a = -(a - b)$.

3. If a is a real number, then prove that $-(-a) = a$. (*Agra 91*)

4. Prove that $(a - b) + (b - c) = a - c, \ \forall \ a, b, c \in \mathbf{R}$.

5. For any $r, s \in \mathbf{Q}$, prove that $(r - s)^* = r^* - s^*$, where the symbols have their usual meanings. (*Gorakhpur 95*)

6. Show that addition of real numbers is commutative and associative.

(*Gorakhpur 92, 95, 96; Poorv. 93, 95*)

7. Show that $a.(b.c) = (a.b).c, \ \forall \ a, b, c \in \mathbf{R}$.

(*Gorakhpur 92, 93, 95, Poorv. 92*)

8. Given a non-zero number a and two numbers b, c, show that the equation

$$ax + b = c$$

possesses one and only one solution.

9. Show that

$$-(a+b) = -a-b \ \forall \ a, b \in \mathbf{R}; \qquad (Gorakhpur\ 95)$$

$$\forall \ a, b \in \mathbf{R} \sim \{0\}$$

10. Show that $\forall \ a, b \ c \in R$

$$(a-b)-c = a-(b+c).$$

11. Show that

$$(a \div b) \div c = a \div (bc)$$

$$a \div (b \div c) = (ac) \div b,$$

$$\forall \ a \in \mathbf{R}, \forall \ b, c \equiv \mathbf{R} \sim \{0\}.$$

12. If a, b, c, d are four given real numbers and $b \neq 0$, $d \neq 0$, then

$$\frac{a}{b}+\frac{c}{d} = \frac{ad+bc}{bd}, \quad \frac{a}{b}\cdot\frac{c}{d} = \frac{ac}{bd}.$$

13. Show that

$$\frac{a}{b}+\frac{c}{d} = \frac{ad}{bc} \quad \text{if } b \neq 0, d \neq 0, c \neq 0.$$

14. Show that

$$a(-b) = -(ab), \ (-a)(-b) = ab \ \forall \ a, b \in \mathbf{R}.$$

(Agra 91)

15. Show that between any two distinct real numbers, there is always an irrational number and therefore infinitely many irrational numbers. *(Agra 91)*

16. Show that the sum of two positive real numbers is a positive real number. *(Gorakhpur 95)*

17. Prove that between two real numbers there lies at least one real number. *(Gorakhpur 91, 93)*

1.9.1. The Modulus of a Real Number

Let x be any real number. We have the following possibilities :

$$x > 0, x < 0, x = 0$$

We write

$$|x| = \begin{cases} x \text{ if } x \geq 0, \\ -x \text{ if } x < 0, \end{cases}$$

and call $|x|$ the *Modulus* or the *Absolute value* of the real number x.

It would be seen that $\forall \ x \in \mathbf{R}$, we have

$$|x| \geq 0$$

$$|x| \geq x \quad \text{and} \quad |x| \geq -x.$$

In fact

$$| x | = \max \{x, - x\},$$

where, $\max \{x, - x\}$ denotes the greater of the two numbers x and $- x$.

Some results involving moduli

(i) $| a + b | \leq | a | + | b |$, (*Agra 92*) (ii) $| ab | = | a | | b |$

Proof.

(i) $$a \leq | a |, \ b \leq | b |$$

$$\Rightarrow a + b \leq | a | + | b |$$

Similarly we may show that

$$- (a + b) \leq | a | + | b |.$$

It follows that

$$| a + b | \leq | a | + | b |$$

Thus we have proved the result (*i*).

(ii) $$a \geq 0 \wedge b \geq 0 \Rightarrow ab \geq 0$$

$$\Rightarrow | ab | = ab = | a | | b |$$

Again $$a \geq 0 \wedge b \leq 0 \Rightarrow ab \leq 0,$$

$$\Rightarrow | ab | = - (ab) = a(- b)$$

$$= | a | | b |$$

The two remaining cases may be similarly disposed of.

Theorem.

$$| a - b | < k \Leftrightarrow b - k < a < b + k.$$

Proof. We have

$$| a - b | = \max \{a - b, b - a\} < k.$$

$$\Leftrightarrow a - b < k \wedge b - a < k$$

$$\Leftrightarrow a < b + k \wedge b - k < a$$

$$\Leftrightarrow b - k < a < b + k.$$

Hence the result.

1.10. REPRESENTATION OF REAL NUMBERS BY POINTS ALONG A STRAIGHT LINE

Every thinking person possesses an intuitive idea of a straight line which, further, he conceives as composed of points. Even though this physical notion of a straight line and that of points on it has nothing to do with Analysis as such, yet it provides a very convenient and helpful

The symbol '$\wedge$' stand for 'and'.

picture of the set of real numbers and is often employed in the course of study of Analysis to provide suitable language and suggest ideas. One danger, which is inherent in this use should, however, be avoided; it may be that we accept a proposition suggested by this picture, obvious as it may seem, as obviously true and this obviousness may blind us to the necessity of a rigorous proof.

We now proceed to *see* how a straight line can be employed to provide a picture of the set of real numbers.

We consider a straight line and mark any two points O and A on it. The point O divides the line into two parts; the part containing the point A will be termed positive and the other negative.

According to the usual convention, the line in question is always drawn parallel to the printed lines of the page and the point A taken on the right of O. Representing the rational number 0 and 1 by the points O and A respectively, we find a point P of the line representing any rational number p/q, $(q > 0)$ by making from O, $|p|$ steps each equal to the qth part of OA to the right or to the left of O according as p is positive or negative.

It is easy to see that if a, b be two rational numbers and $a < b$, then the point representing b lies to the right of the point representing a.

If we call the points which represent rational numbers as rational points, we see that, since the set of rational numbers is dense, an infinite number of rational points lie between every two different rational points.

Insufficiency of Real Numbers to provide a picture of straight line. Even though as we have seen above, a line can be covered with rational points as closely as we like, there exist points of the line which are not rational. For example, a point P such that OP is equal to the diagonal of the square with side OA is one such point (§ 1.3 page 6). Also a point L on the line such that OL is a rational multiple p/q of OP cannot be a rational point. For, if possible, let L represent a rational number m/n, so that we have

$$\frac{p}{q}\cdot OP = OL = \frac{m}{n} \quad \text{or} \quad OP = \frac{mq}{np}$$

which shows that OP is rational *i.e.*, P is a rational point and this is a contradiction.

Thus we see that the set of rational numbers is not sufficient to provide us with a picture of complete straight line.

Real number. Let

$$\alpha = (L, U)$$

be a real number. The section (L, U) of rational numbers determines a section of the rational points of the line into two classes A and B such that A consists of rational points corresponding to the members of L and B of rational points corresponding to the members of U. Every point of the class A will lie to the left of every point of the class B.

From our intuitive picture of a straight line and its continuity, we can convince ourselves that there will exist a point P of the line separating the two classes in the sense that every point of the line lying to the left of P belongs to the class A and every point lying to the right of P belongs to the class B. *This point P*, we say, *denotes the real number* (L, U). Thus to every real number there corresponds a point of the line.

Conversely, let P be any point of the line. The point P divides the rational points of the line into two classes A and B such that the points lying to the left of P belong to A and those to the right of P belong to B; the point P, if rational, belongs to B. The classes A, B of rational points determine a section (L, U) of rational numbers, which corresponds to the point P.

Thus to every point there corresponds a real number.

The set of real numbers is called the Arithmetical Continuum and the set of points on a straight line is called the *Linear Geometric Continuum.* In view of what has been shown above, we see that there is a one-to-one correspondence between the two sets or continua and accordingly it may be found convenient to use the word 'point' for 'real number' as we shall often do.

OBJECTIVE QUESTIONS

Note. *For each of the following questions, four alternatives are given for the answer. Only one of them is correct. Choose the correct alternative.*

1. Product of two negative real numbers is :

(*a*) positive (*b*) negative

(*c*) zero (*d*) none of these

2. For real numbers the correct statement is :

(*a*) if $a > b$, then $a + c < b + c$ (*b*) if $a > b$, then $ac < bc$

(*c*) if $a > b$, then $-a < -b$ (*d*) none of these

3. In following statements which is true :

(*a*) $|a+b| \geq |a| + |b|$ (*b*) $|a+b| \leq |a| + |b|$

(*c*) $|a+b| = |a| + |b|$ (*d*) none of these

4. Choose the correct statement :

(a) $|a-b|=|a|-|b|$ (b) $|a-b|=|a|-|b|$

(c) $|a-b|=|a|-|b|$ (d) $|a-b|\geq|a|+|b|$

5. Between any two real numbers there exists irrational numbers :

(a) only one (b) at least one

(c) infinite (d) none of these

6. A real number $\alpha_1=(L_1, U_1)$ is said to be smaller that the real number $\alpha_2=(L_2, U_2)$ if

(a) $L_1 \subseteq L_2$ (b) $L_1 \supseteq L_2$

(c) $L_1 = L_2$ (d) $U_1 = U_2$

ANSWERS

1. (a) **2.** (c) **3.** (b) **4.** (c) **5.** (b)
6. (a)

2

Bounded Sets, Open and Closed Sets

Introduction. In this chapter we shall introduce some fundamental concepts of *Suprema* and *Infima* of bounded sets; *Limit points of sets, open* and *closed sets.* Two important theorems : *Bolzano-Weierestrass theorem* and *Heine-Borel theorem* will also be proved in this chapter.

2.1. CLOSED INTERVALS, OPEN INTERVALS, SEMI-CLOSED OR SEMI-OPEN INTERVALS

Let a, b be two given numbers such that $a < b$. We write

(*i*) $\{x : a \le x \le b\} = [\, a, b \,]$; (*ii*) $\{x : a < x < b\} =]\, a, b \,[$;
(*iii*) $\{x : a \le x < b\} = [\, a, b \,[$; (*iv*) $\{x : a < x \le b\} =]\, a, b \,]$

and refer to these sets of real numbers respectively as

(*i*) a closed interval
(*ii*) an open interval,
(*iii*) left closed-right open interval, and
(*iv*) left open-right closed interval.

Neighbourhood of a point. Let ξ be a number. Then any open interval $]\, a, b \,[$ of which ξ is a member is called a neighbourhood of ξ. In particular

$$]\, \xi - \varepsilon, \xi + \varepsilon \,[, \quad \varepsilon > 0$$

is a neighbourhood of ξ.

An open interval is a neighbourhood of each of its points and a closed interval is a neighbourhood of each of its points except the end points.

2.2. SETS BOUNDED ABOVE AND SETS BOUNDED BELOW

Bounded Sets. Let S be a given set of real numbers. We say that S is bounded above if there exists a number b such that

$$x \in S \Rightarrow x \le b,$$

so that no member x of S is greater than the number b.

The number b is called an *upper bound* of S. Of course, every number greater than an upper bound of a set is also an upper bound of the set.

Again a set S is bounded below if there exists a real number a such that

$$x \in S \Rightarrow a \leq x$$

So that no member x of S is smaller than the number a.

The number a is called a lower bound of S. Every number smaller than a lower bound of a set is also a lower bound of the set.

A set is said to be bounded if it is both bounded above and below. Thus a set S is bounded if and only if there exists an interval $[a, b]$ such that

$$S \subset [a, b].$$

Greatest member of a set bounded above. A number M is the greatest member of a set S, if

(*i*) $M \in S$ *i.e.*, M is a member of S,

(*ii*) M is an upper bound of S.

Existence of greatest members. A set may or may not possess a greatest member. Of course the greatest member of a set, if it exists, is unique. (Prove !)

Every finite set admits of a greatest member and no infinite set which is not bounded above can have a greatest member.

The cases of finite sets and of infinite sets which are not bounded above are thus trivial as regards the existence of greatest members. The case of importance which remains is that of infinite sets bounded above. Such a set may or may not have a greatest member. For example, the closed interval $[a, b]$ admits of the greatest member b but the open interval $]a, b[$ has no greatest member.

We may similarly deal with the notion of the smallest or least member of a set bounded below.

Theorem on upper bounds. *The set of upper bounds of a set of real numbers which is bounded above admits of a smallest member.*

Let S be a set bounded above. We consider two sub-sets L and U of R defined as follows :

$$L = \{x : x \text{ is not an upper bound of } S\},$$
$$U = \{x : x \text{ is an upper bound of } S\}.$$

Surely neither L nor U is an empty set.

Also $\quad L \cup U = R$

and $$x \in L, y \in U \Rightarrow x < y.$$

Then, because of the order completeness of the set of real numbers, either L has a greatest member or U has a least member.

We shall show that L cannot have a greatest member.

Let, if possible, L have the greatest member, say, ξ. Then ξ is *not* an upper bound of S. Thus there exists a member a of S such that $\xi < a$.

Let b be a real number such that

$$\xi < b < a.$$

Now b, which is greater than the greatest member ξ of L belongs to U.

Also b, being smaller than a member a of S, cannot be an upper bound of S. Thus $\xi \in L$ and not to U.

Thus we arrive at contradictory conclusions and as such we prove that L cannot have a greatest member.

It follows that U has a smallest member *i.e.*, the set of all the upper bounds of S has a smallest member.

Supremum of a set bounded above. Def. *The least of the upper bounds of a set bounded above (which of course exists) is called its supremum and is denoted as*

$$\sup S.$$

(*Gorakhpur 93, 94; Kanpur 93, 94, 95; Poorv. 92, 93*)

Illustration. The number b is the supremum of the set $[a, b]$ as well as of the set $]a, b[$.

Theorem on lower bounds. *The set of all the lower bounds of a set bounded below admits of a greatest member.*

The proof which is similar to that of the theorem on upper bounds, is left to the reader.

Infimum of a set bounded below. *The greatest lower bound of a set is also called its Infimum and is denoted by*

$$\inf S.$$

(*Gorakhpur 92, 94; Poorv. 96; Kanpur 93, 94, 95*)

Illustration. The number a is the infimum of the set $[a, b]$ as well as of the set $]a, b[$.

Oscillation of a bounded set. If ξ, η denote the supremum and infimum respectively of a bounded set, then the difference, $\xi - \eta$ is called the *oscillation of* the set.

EXAMPLES

1. *Find the infimum and supremum of the set* $S = \{ x \in \mathbf{I} : x^2 < 16\}$, *where* **I** *is the set of all integers.*

Sol. The given set can be rewritten as

$$S = \{-4, -3, -2, -1, 0, 1, 2, 3, 4\}$$

Evidently $-4 \in \mathbf{I}$ is a lower bound of the set S, because $-4 \leq x \ \forall \ x \in S$, hence inf. $S = -4$. Obviously sup. $S = 4$.

2. *Show that every non-empty finite sub-set of* **R** is bounded.

Sol. If S be a non-empty finite sub-set **R**, then only a finite number of elements are in S and out of these elements one element $a \in S$ shall be the smallest element of S and one element $b \in S$ shall be the greatest element of S due to order properties in **R**.

From the definition of upper and lower bounds, we find that a is a lower bound of S and b is an upper bound of S and as such the set S is bounded.

3. *Give an example each of a bounded set which contains its (i) infimum but does not contain supremum (ii) supremum but does not contain infimum.* *(Poorv. 93, 96)*

Sol. (*i*) The set $A = \{x : 1 \leq x < 3\}$ is such that its infimum $= 1 \in A$ and supremum $= 5 \notin A$.

(*ii*) The set $B = \{x : 3 < x \leq 7\}$ is such that its supremum $= 7 \in B$ and infimum $= 3 \notin B$.

4. *Every non-empty sub-set S of* **R** *which has a lower bound has the infimum.*

Sol. Let v be a lower bound of the given set $S \subseteq \mathbf{R}$.

Let S' be the set of negatives of element of S *i.e.*, $-x \in S'$ iff $x \in S$. Then $-v$ is an upper bound of S'.

In fact $x \geq v \ \forall \ x \in S$, $\therefore$ v is a lower bound of S.

$\Rightarrow -x - v \ \forall \ x \in S$

$\Rightarrow -x - v \ \forall \ -x \in S'$

$\Rightarrow$ the set S is bounded above

$\Rightarrow$ the set S' has supremum, say ω.

Now, let λ be any lower bound of S, then ω are to show that $-\omega$ is the infimum of S or $-\omega \geq \lambda$.

Now λ is a lower bound of S.

$\Rightarrow -\lambda$ is an upper bound of S

$\Rightarrow \omega \geq -\lambda \quad \therefore \omega$ is supremum of S'

$\Rightarrow -\omega \geq \lambda$.

5. *Show that supremum of a set is unique.*

(*Gorakhpur 93, 94; Poorv. 93, M.D.U. 1998*)

Sol. Let S be a set and s the supremum. If possible let there be two supremums s and s_1 of S. By definition, s and s_1 are both upper bounds of S.

Now if s is the supremum and s_1 is an upper bound of S, then we have

$$s \leq s_1 \qquad \text{...(i)}$$

And if s_1 is the supremum and s is an upper bound of S, then we have

$$s_1 \leq s \qquad \text{...(ii)}$$

From (*i*) and (*ii*), $s = s_1$. Hence the supremum s of S is unique.

EXERCISES

1. Find the supremum and infimum, if there exist, of the following sets :

(*i*) $x \in \mathbf{R}; x = (1/n), n \in N\}$

(*ii*) $x \in \mathbf{R} : -5 < x < 3\}$ (*Kanpur 93*)

2. Prove that every finite set is bounded.

3. Give examples to show that the supremum and infimum of a set may not belong to the set.

4. Show that

$$S \subset T \Rightarrow \inf T \leq \inf S \leq \sup S \leq \sup T.$$

5. Let S be any given set and let

$$T = \{x : -x \in S\}.$$

Show that

$$\sup T = -\inf S \quad \text{and} \quad \inf T = -\sup S.$$

6. Let S_1, S_2 be two given sets and let

$$S = \{x_1 + x_2 : x_1 \in S_1, x_2 \in S_2\}.$$

Show that

$$\sup S = \sup S_1 + \sup S_2, \inf S = \inf S_1 + \inf S_2.$$

7. Let S be any given set and let

$$T = \{| x - y | : x \in S, y \in S\}.$$

Show that

$$\sup T = \sup S - \inf S.$$

8. Show that every non-empty sub-set S of $\mathbf{R}$ which has an upper bound has the supremum.

9. Show that every sub-set of an unbounded set is not necessarily unbounded.

10. Show that infimum of a set is unique.

(*Gorakhpur 91, 92, Poorv. 96*)

2.3. LIMIT POINTS OF A SET

Def. *A number ξ is called a limit point of a set S if every neighbourhood of ξ contains a point of S other than ξ.*

(*Kanpur 92, 94, 95, 96*)

It may be shown that the definition of a limit point of a set, as given above, is equivalent to the following :

A number ξ is a limit point of a set S of real numbers. if every neighbourhood of ξ contains an infinite sub-set of S.

This equivalence is easy to establish.

A limit point of a set is also sometimes called an *Accumulation point* or *Condensation point* of the set.

A set may have no limit point, a unique limit point or any finite or infinite number of limit points.

Also a limit point of a set may or may not be a member of the set.

The derived set of a set. Def. *The set having as its members the limit points of a given set is called the derived set of the given set.*

The derived set of a set S will be denoted by S'. Thus

$$S' = \{x : x \text{ is a limit point of } S\}.$$

Illustrations

1. 0 is the only limit point of the set

$$\left\{\frac{1}{n} : n \in \mathbf{N}\right\}.$$

so that in this case the limit point of the set is not a member of the set.

2. The set **N** of natural numbers possesses no limit point so that in other words

$$N' = \phi$$

3. Every real number is a limit point of the set of rational numbers so that rational limit points are members of the set but irrational limit points are not. Here we have

$$\mathbf{Q}' = \mathbf{R}$$

4. Every real number is a limit point of the set of irrational numbers.

(*Refer cor. Page 16*)

5. The derived set of each of the intervals $[a, b]$, $]a, b[$, $[a, b[$, $]a, b]$ is the interval $[a, b]$.

6. $\pm$ 1 are the only two limit points of the set

$$\left\{-1-\frac{1}{n} : n \in \mathbf{N}\right\} \cup \left\{1+\frac{1}{n} : n \in \mathbf{N}\right\}$$

EXAMPLES

1. *Find the limiting point of the set of irrational numbers.*

Sol. Since every open interval containing a real number x_0 also contains irrational numbers distinct from x_0, so every number $x_0 \in \mathbf{R}$ is a limiting point of the set of irrational number.

2. *If the supremum of a set does not belong to the set, then it is a limiting point of the set.*

Sol. Let A be the set under consideration and u = supremum of A, where $u \notin A$.

We are to prove that u is a limiting point of A and for this we shall prove that every *nbd* of the point u contains a point of A different from u.

Let $(u - \varepsilon\ ,\ u + \varepsilon\ )$ be any *nbd* of the point u.

Now $\quad u - \varepsilon < u \Rightarrow u - \varepsilon$ is not an upper bound of the set A.

$\Rightarrow$ there exist some $x \in A$ such that

$$x > u - \varepsilon \qquad \text{...}(i)$$

Also, $x \neq u$ since $x \in A$ whereas $u \notin A$.

Again $u < u + \varepsilon$ and u is the supremum of the set A.

$\Rightarrow u + \varepsilon$ is an upper bound of A

$$\Rightarrow x < u + \varepsilon \qquad \text{...}(ii)$$

From (*i*) and (*ii*) we get $u - \varepsilon < x < u + \varepsilon$, where $x \neq u$.

$\Rightarrow (u - \varepsilon,\ u + \varepsilon)$ contains a point x of A different from u

$\Rightarrow u$ is a limiting point of the set A.

3. *If A and B are any sub-set of* $\mathbf{R}$, *then prove that* $A \subseteq B \Rightarrow A' \subseteq B'$.

Sol. Let $x \in \mathbf{R}$ and M be any *nbd* of the point x.

Then $\quad A \in B \Rightarrow (A \cap M) - \{x\} \subseteq (B \cap M) - \{x\}$

Now if $(A \cap M) - \{x\} \neq \phi$, for every *nbd* M of the point x,

i.e., if x is a limiting point of A

i.e., if $x \in A'$, by definition of A'.

Then $\quad (B \cap M) - \{x\} \neq \phi$, evidently as $A \leq B$.

i.e., $\quad x \in B'$, as above.

Therefore $\quad x \in A' \Rightarrow x \in B'$

i.e. $\quad A' \subseteq B'$

Hence $A \subseteq B \Rightarrow A' \subseteq B'$.

4. *If A and B are any sub-sets of R, then prove that* $(A \cap B)' \subseteq A' \cap B'$.

Sol. Since $(A \cap B)' \subseteq A$, so we have $(A \cap B)' \subseteq A$ and as $(A \cap B) \subseteq B$, so we have $(A \cap B)' \subseteq B'$.

Hence $(A \cap B)' \subseteq A' \cap B'$

5. *Find the derived set I of the integers.*

Sol. Let any number $x \in \mathbf{R}$.

If $x \in I$, then the *nbd* $\left(x - \frac{1}{2}, x + \frac{1}{2}\right)$ of the point x does not contain any point of I other than x.

$\therefore$ x is not a limiting point of set **I** of integers.

If $x \notin \mathbf{I}$, then taking $\varepsilon = \frac{1}{2}|x - n| > 0$, where n is an integer nearest to x, we find that the *nbd* $(x - \varepsilon, x + \varepsilon)$ of the point x does not contain any point of **I**.

$\therefore$ x is not a limiting point of the set **I** of integers.

Therefore there is no limiting point of the set **I** of integers. Hence $\mathbf{I}' = \phi$.

6. *Show that derived set of an infinite bounded set is bounded.*

Sol. Let A be an infinite bounded sub-set of **R**.

Let l = infimum and u = supremum of A.

Let $a > u$, then for a positive number $\varepsilon = \frac{1}{2}(a - u)$ we have

$$a - \varepsilon = a - \frac{1}{2}(a - u) = \frac{1}{2}(a + u) \qquad \text{...(i)}$$

Also
$$a > u \Rightarrow u < a$$
$$\Rightarrow u < \frac{1}{2}(u + a) < a$$
$$\Rightarrow u < a - \varepsilon < a, \qquad \text{from (i)}$$

Now as u = supremum of the set A, so

$$x \le u < a - \varepsilon \;\forall\; x \in A$$

$\Rightarrow (a - \varepsilon, a + \varepsilon) \cap A = \phi$

$\Rightarrow a$ is not a limiting point of A, where $a > u$

$\Rightarrow$ any number greater than u is not a limiting point of A'.

$\Rightarrow$ (every limiting point of A) u

$\Rightarrow$ (every element of A') $\le u$ ...(ii)

Similarly, let $b < l$, the infimum of A. Then for a positive number $\varepsilon = \frac{1}{2}(l - b)$, we have $b < b + \varepsilon < l$, as above.

Now, as l = infimum of the set A,

so $$b + \varepsilon < l < x \ \forall \ x \in A$$

$\Rightarrow (b - \varepsilon, b + \varepsilon) \cap A = \phi$

$\Rightarrow b$ is not a limiting point of A, where $b < l$.

$\Rightarrow$ any number greater than l is not a limiting point of A.

$\Rightarrow$ (every limiting point of A') $\geq l$

$\Rightarrow$ (every element of A') $\geq l$ ...(*iii*)

From (*ii*) and (*iii*) $l \leq y \in u, \ \forall \ y \in A'$

i.e., $$y \in [l, u] \ \forall \ y \in A'.$$

Hence $A' \subseteq [l, u]$, *i.e.*, A' is bounded.

EXERCISES

1. What do you understand by saying that ξ is *not* a limit point of a set S.

2. Show that

$$(S_1 \cup S_2)' = (S_1' \cup S_2').$$

3. *S is a set bounded above having no greatest member; show that*

$$sup \ S \in S'.$$

State and prove the corresponding theorem in respect of sets bounded below.

4. Find the derived set of the set

$$\left[\frac{1}{m} + \frac{1}{n} + \frac{1}{p} : m, n, p \in \mathbf{N}\right].$$

5. Give an example of a set which coincides with its derived set.

6. Find the limit points of the subset **N** of **R**.

7. Find the limiting point of the null set ϕ.

8. If the infimum of a set does not belong to the set, then prove that it is a limiting point of the set.

9. If ϕ be the empty set, then show that its derived set $\phi' = \phi$.

10. Find the derived set of the set **Q** of all rational numbers.

Existence of limit points. A finite set does not have a limit point and an infinite set may or may not have a limit point. We shall now have a theorem which establishes *Boundedness* of an infinite set as a sufficient condition for if to possess a limit point.

Bolzano-Weierstass Theorem. *Every infinite and bounded set has a limit point.*

(*Agra 93, 94; Kanpur 2005; Gorakhpur 92, 94; M.D.U. 2001, Kurkshetra*)

Let S be a given infinite and bounded set so that there exist an interval $[a, b]$ such that

$$S \subset [a, b]$$

We consider two sub-sets L, U of real numbers such that

$x \in L$ if it exceeds only a finite number of members of S,

$x \in U$ if x does not belong to L.

We may see that neither L nor U is empty. In fact

$$a \in L, b \in U.$$

Also $$L \cup U = \mathbf{R}.$$

Suppose that $$x \in L, y \in U.$$

Now y cannot be smaller than x, for, if it were so, it would exceed only a finite number of members of S and as such would belong to L.

Thus by the order completeness of $\mathbf{R}$, there exists a number, say g, such that

$$x < g \Rightarrow x \in L; x > g \Rightarrow x \in U.$$

We shall show that g is a limit point of S.

Consider any neighbourhood

$$]p, q[$$

of g.

Now $$p < g \Rightarrow p \in L$$

so that p exceeds only a finite number of members of S. Again

$$q > g \Rightarrow q \in U$$

so that q exceeds an infinite number of members of S.

Thus we conclude that an infinite number of members of S belong to the interval

$$]p, q[,$$

Thus g is a limit point of S.

Note. The above theorem establishes the existence of limit points of infinite bounded sets. It should, however, be understood, that the limit point g of S, as obtained, may not be the only limit point of S and that it may or may not be a member of S.

Ex. Show that boundedness of an infinite set is only a sufficient has not a necessary condition for it to possess a limit point.

Ex. Show that the derived set of an infinite and bounded set is not void.

Theorem. *A bounded set has greatest and least limit points.*

Let S be bounded set so that there exists an interval $[a, b]$ such that

$$S \subset [a, b]$$

Surely no limit point of S can be greater than b or smaller than a and as such every limit point of $S \in [a, b]$. Thus the derived set S' of S is bounded and in fact

$$S' \subset [a, b].$$

We write

$$G = \sup S', \; g = \inf S'$$

and show that G, g are themselves members of S *i.e.*, limit points of S.

Let $]\, p, q\, [$ be a neighbourhood of G. As G is the supremum of S', there exists $\xi \in S'$ such that

$$p < \xi < q.$$

Thus $]\, p, q\, [$ is a neighbourhood of ξ as well and as such $]\, p, q\, [$ contains an infinite number of members of S. Hence

$$G \in S'$$

We may similarly show that $g \in S'$.

Hence the result.

Upper and lower limits of a set. The greatest and least limit points of a bounded set are often referred to as the *upper and lower limits* respectively of the set.

Ex. 1. For a bounded set S, show that

(*i*) $\sup S' \le \sup S$; $\inf S' \ge \inf S$.

(*ii*) $\sup S$ does not belong to $S \Rightarrow \sup S$ belongs to S',

$\inf S$ does not belong to $S \Rightarrow \inf S$ belongs to S.

Ex. 2. Give an example of a set whereof no elements lie between its greatest and least limit points.

Ex. 3. Give the suprema, infima, upper limits and lower limits of the following sets :

(*i*) $\{-2, 2\} \cup \left\{1 + \dfrac{1}{n} : n \in \mathbf{N}\right\} \cup \left\{-1 - \dfrac{1}{n} : n \in \mathbf{N}\right\}.$

(ii) $\left\{\frac{1}{n} : n \in \mathbf{N}\right\} \cup \left\{1 + \frac{1}{n} : n \in \mathbf{N}\right\}$.

(iii) $\left\{(-1)^n \left(\frac{1}{4} - \frac{8}{n}\right) : n \in \mathbf{N}\right\}$.

(iv) $\left\{(-1)^n \left(1 + \frac{1}{n}\right) : n \in \mathbf{N}\right\}$.

(v) $\{3, 2\} \cup \{(2^{-n} + 1)/2^n : n \in \mathbf{N}\} \cup \{(2^n - 1)/2^n : n \in \mathbf{N}\}$.

2.4. CLOSED SETS.

Def. *A set S is said to be closed if every limit point of the set is a member of the set, i.e.,*

$$S \text{ is closed} \Rightarrow S' \subset S. \qquad \textit{(Agra 90; Kanpur 93, 95)}$$

Illustrations

1. A closed interval is a closed set but an open interval is not a closed set.
2. The set **Q** of rational numbers is not closed.
3. Every finite set is closed.

Note. While every limit point of a closed set S is a member of the set, it does not mean that every point of S is its limit point.

Points of a set which are not limit points of the set are called **Isolated points** of the set. There exists, as may easily be shown, a neighbourhood of an isolated point of a set which contains no point of the set other than itself.

Theorem. *The derived set of every set is a closed set.*

Let S be a set. We have to show that S' is a closed set.

Let ξ be a limit point of S' and $]\,p, q\,[$ a neighbourhood of ξ.

As $]\,p, q\,[$ is a neighbourhood of a limit point of S', it contains an infinite number of members of S'. Surely $]\,p, q\,[$ is a neighbourhood of each of these members of S' also and as such contains an infinite number of members of S.

Thus ξ is a limit point of S *i.e.*,

$$\xi \in S'$$

We have thus proved that the derived set S' is closed.

Open sets. Def. *A set is called* **open** *if some neighbourhood of each of its points is contained in the set.* (*Agra 93, Kanpur 92, 95, Patna 2003*)

It would be well to notice that the concepts of open and closed sets are neither mutually exclusive nor exhaustive.

Thus there may be a set which is both open and closed and a set which is neither open nor closed.

The set **R** of real numbers is an open as well as a closed set. The set **Q** of rational numbers is neither open nor closed.

It may be shown easily that while an open interval is an open set, a closed interval is not an open set.

Theorem. *A set is closed if and only if its complement is open.*

(*Agra 92*)

Let A be a closed set and let B denote its complement. It will be shown that B is an open set.

Let $$x \in B.$$

Since x does not belong to the closed set A, it is not a limit point of A. Thus $\exists$ a neighbourhood of x containing no point of A, *i.e.*, contained in B.

Thus B contains a neighbourhood of each of its points and is, therefore, open.

Let C be an open set and D its complement. It will be shown that D is a closed set.

Let x be a limit point of D. Suppose that $x \in C$. There exists therefore a neighbourhood of x contained in C and thus containing no point of D.

As such x cannot be a limit point of D.

Thus we prove that no limit point of D can belong to C and as such D is a closed set.

Hence the theorem.

2.5. COMPACT SETS (*Kurukshetra 94 S,*

Def. *A set of real numbers is called compact if it is both boundea and closed.*

Every infinite set is compact. A closed interval $[a, b]$ is also compact. The intervals

$$]\, a, b\,],\ [\, a, b\, [\ ,\ ]\, a, b\, [$$

are not compact.

The set $\{x : a \leq x\}$ though closed is not bounded and hence is not compact.

The set Q of rational numbers is neither closed nor bounded and is thus not compact.

Heine-Borel Theorem. In order to state the Heine-Borel theorem, we need to introduce the concept of an *Open cover of a set* which is given below.

Cover of a set. Let S be a set and let $\Im$ be a family of sets. We say that $\Im$ is a *cover* of S, if the union of members of $\Im$ contains S, as a sub-set *i.e.*, if every point of S belongs to some member of the family $\Im$.

We say that $\Im$ is an *open cover* if every member of $\Im$ is an open set.

We next prove an important theorem concerning open covers of compact sets.

Heine-Borel Theorem. *Every open cover of a compact set admits of a finite sub-cover.* [*Kurukshetra 94, 2001*]

Let S be a compact set and let $\Im$ be a open cover of S.

We shall show that there exists an open cover $\Im$ of S such that

(*i*) $\Im'$ contains only a finite number of members.

(*ii*) every members of $\Im'$ is as well a member of $\Im$.

As S is compact, it is bounded and closed and as such

$$a = \inf S \quad \text{and} \quad b = \sup S$$

are members of S.

We define a sub-set A of S as follows :

A member x of S belongs to A if and only if the sub-set of S contained in $[a, x]$ is covered by a finite sub-family of $\Im$.

We write

$$\Im = \sup A.$$

Proof. We shall firstly show that $\Im \in A$.

Suppose $\Im \in A$.

Now $\Im$ being *not* a member of A, it must be a limit point of A. As A is a sub-set of S, the limit point $\Im$ of A must also be a limit point of S. Again S being a closed set, $\Im \in S$.

There exists a member, say $]\,\alpha, \beta\,[$ of which covers $\Im$. Also since $\Im$ is sup A, there exists $\alpha' \in A$ such that

$$\alpha < \alpha' \leq \xi.$$

As $\alpha' \in A$, the sub-set of S contained in $[a, \alpha']$ is covered by a finite sub-family of $\Im$. It follows that the sub-set of S contained in $[\alpha, \beta]$ $(\beta > \xi)$ is covered by a finite sub-family of $\Im$ so that $\beta \in A$. We thus arrive at a contradiction. It follows that $\xi \in A$.

We shall now show that $\xi = b$.

Now let b be covered by a member $]\,\gamma, \delta\,[$ of $\Im$.

As $\gamma < b$, there exists a member γ' of A contained in $]\,\gamma, \delta\,[$.

It follows that the part of S contained in $[a, b]$ is S itself.

Hence the result.

EXAMPLES

1. *Show that the null set ϕ is closed.*

Sol. The null set ϕ is closed because

$$C(\phi) = \mathbf{R} - \mathbf{Q} = \mathbf{R}, \text{ which is open.}$$

2. *Prove that the union of two closed sets is a closed set. Prove a similar result for finite union.* (*Agra 91; Kanpur 93*)

Sol. Let A and B be two closed sets.

Then A and B are closed sets.

$\Rightarrow C(A)$ and $C(B)$ are open sets

$\Rightarrow C(A) \cap C(B)$ is an open set

$\Rightarrow C(A \cup B)$ is an open set,

as $C(A) \cap C(B) = C(A \cup B)$

$\Rightarrow A \cup B$ is a closed set.

Again, let $A_1, A_2, A_3, \ldots, A_n$ be n closed sets.

Then $C(A_1), C(A_2), C(A_3), \ldots, C(A_n)$ are n open sets.

$\Rightarrow \bigcap_{i=1}^{n} C(A_i)$ is an open set, because the intersection of a finite collection of open sets is an open set.

$\Rightarrow C\left(\bigcup_{i=1}^{n} A_i\right)$ is an open set, because we know from De Morgan's law

$$\bigcap_{i=1}^{n} C(A_i) = C\left(\bigcup_{i=1}^{n} A_i\right).$$

$\Rightarrow \left(\bigcup_{i=1}^{n} A_i\right)$ is a closed set.

3. *Show that the set* **Q** *of all rational numbers is not closed.*

Sol. Let x be any arbitrary element of $C(\mathbf{Q})$, where $C(\mathbf{Q}) = \mathbf{R} - \mathbf{Q}$, the set of irrational number.

Now $(x - \varepsilon, x + \varepsilon)$ contains an infinite number of irrational numbers for every $\varepsilon > 0$.

$\therefore$ There exists no open interval **I** such that $x \in \mathbf{I} \subset C(\mathbf{Q})$

$\Rightarrow C(\mathbf{Q})$ is not a *nbd* of x

$\Rightarrow C(\mathbf{Q})$ is not an open set

$\Rightarrow \mathbf{Q}$ is not a closed set.

4. *Is every finite set open ? Justify your answer by giving an example.*

Sol. No, the set **Q** of all rational numbers is infinite and is not open.

5. *Give an example of (i) a set which is neither an interval nor an open set, (ii) an interval which is not an open set.* (*Kanpur 92, 94, 95*)

Sol. (*i*) The set **N** of all natural numbers.

(*ii*) The closed interval $[a, b]$ as it is not an open set since it is not a *nbd* of a and b which belongs to it.

EXERCISES

1. Show that the intersection of any finite number of open (closed) sets is open (closed). (*Agra 93, Kerala 2001, Kanpur 2004*)

Prove a similar result for finite unions.

2. Show that the intersection (union) of any finite number of compact sets is compact.

3. Given a family

$$\{K_n : n \in \mathbf{N}\}$$

of non-empty compact sets such that

$$E_{n+1} \subset K_n \ \forall \ n \in \mathbf{N};$$

show that there exists at least one number ξ such that

$$\xi \in K_n \ \forall \ n.$$

4. Show that $S \cup S'$ is closed for every set S.

5. Show that $\dfrac{U}{n \in \mathbf{N}} \left] -\dfrac{1}{n}, \dfrac{1}{n} \right[= \{0\}$

and note that here is an example of an infinite number of open sets with non-open union.

6. Show that every finite set is a closed set.

7. Show that the set **I** of all integers is a closed set.

8. Show that no point of an open set can be its upper or a lower bound.

OBJECTIVE QUESTIONS

Note. *For each of the following questions, four alternatives are given for the answer. Only one of them is correct. Choose the correct alternative.*

1. If S be any subset of the set **R** of real numbers and if there exists a real number $\boldsymbol{u}$ such that $x \leq u \ \forall \ x \in S$, then number u is called :

(*a*) upper bound (*b*) lower bound

(*c*) *g. l. b.* (*d*) *l. u. b.*

2. Supremum of the set $\left\{1, \frac{1}{2}, \frac{1}{3}, \frac{1}{4}, ...\right\}$ is :

(*a*) 0 (*b*) 1

(*c*) $\frac{1}{2}$ (*d*) $\frac{1}{4}$

3. The set $A = \{ x : 1 \leq x < 3\}$ is such that it contains :

(*a*) *g. l. b.* and *l. u. b.* both

(*b*) *g. l. b.* but does not contain *l. u. b.*

(*c*) *l. u. b.* but does not contain *g. l. b.*

(*d*) neither *g. l. b.* nor *l. u. b.*

4. Every non-empty subset S of **R** which has a lower bound has :

(*a*) infimum (*b*) supremum

(*c*) no infimum (*d*) none of these

5. A real number m in the *g. l. b.* of a non-empty subset X of real numbers iff :

(*a*) $m = x \ \forall \ x \in X$ (*b*) $m \geq x \ \forall \ x \in X$

(*c*) $m \leq x \ \forall \ x \in X$ (*d*) none of these

6. The set $\{x : x \in \mathbf{R}, a < x < b\}$ is known as :

(*a*) closed interval (*b*) open interval

(*c*) right-half open interval (*d*) left-half open interval.

7. If in an interval, the *g. l. b.* and *l. u. b.* the sets of points corresponding to it belongs to the set, then interval is :

(*a*) closed (*b*) open

(*c*) infinite (*d*) semi-closed.

8. Subset of **R** which is a *nbd* of 3 is :

(*a*) [3, 6] (*b*) [3, 6)

(*c*) (2, 4) (*d*) (3, 6).

9. Set N of natural numbers is :

(*a*) a *nbd* of all its points (*b*) not a *nbd* of any of its points

(*c*) a *nbd* of *glb* only (*d*) none of these

10. A real number x_0 is such that the set A $\subseteq$ **R**, if every *nbd* of x_0 contains infinitely many points of A, then x_0 is called :

(*a*) g. l. b. (*b*) l. u. b.

(*c*) limiting point (*d*) none of these

11. Limiting point of the set A of rational numbers of the form $n/(n+1)$ where $n \in$ **N** is :

(*a*) 0 (*b*) 1

(*c*) n (*d*) $n+1$

12. If the supremum of a set is a limiting point of the set then it :

(*a*) does not belong to the set

(*b*) belongs to the set

(*c*) may or may not belong to the set

(*d*) none of these

13. Number of limiting points in every infinite bounded set is :

(*a*) zero (*b*) infinite (*c*) only one (*d*) at least one

14. The set of all the limiting points of a set A is known as

(*a*) limiting set (*b*) derived set (*c*) open set (*d*) closed set

15. If A and B be any subsets of **R** then value of $(A \cap B)'$ is :

(*a*) $A' \cap B'$ (*b*) $A' \cup B'$ (*c*) $\subseteq A' \cap B'$ (*d*) $\supseteq A' \cap B'$

16. Derived set of the set Q of all rational numbers is :

(*a*) **Q** (*b*) **N** (*c*) **R** (*d*) ϕ

17. Derived set of the set **N** of all natural numbers is

(*a*) ϕ (*b*) **N** (*c*) **R** (*d*) **Q**

18. Set **R** is

(*a*) open (*b*) closed (*c*) empty (*d*) none of these

19. A set which is neither an interval nor an open set is

(*a*) ϕ (*b*) **N** (*c*) **R** (*d*) **Q**

20. An interval which is not an open set is

(*a*) (a, b) (*b*) $[a, b)$ (*c*) $(a, b]$ (*d*) $[a, b]$

21. For any set A, interior of set Å is

(*a*) open (*b*) closed (*c*) null (*d*) infinite

22. If $i(A)$ is the interior of A, where A is any subset of **R**, then A is open iff :

(*a*) $i(A) \subseteq A$ (*b*) $i(A) \supseteq A$ (*c*) $i(A) = A$ (*d*) none of these

23. If A is any subset of **R**, and $i(A)$ is the interior of set A, then value of $i(i(A))$ is :

(*a*) A (*b*) $i(A)$ (*c*) ϕ (*d*) none of these

24. Singleton $\{a\}$, where a is a real number is :

(*a*) open set (*b*) closed set (*c*) derived set (*d*) limiting set

25. If A any subset of **R**, then derived set A' is :

(*a*) closed set (*b*) open set (*c*) limiting set (*d*) none of these

26. If a set is closed and bounded, then it is :

(*a*) covering set (*b*) compact set (*c*) derived set (*d*) cluster set.

ANSWERS

1. (*a*)	**2.** (*b*)	**3.** (*b*)	**4.** (*a*)	**5.** (*c*)
6. (*b*)	**7.** (*a*)	**8.** (*c*)	**9.** (*b*)	**10.** (*c*)
11. (*b*)	**12.** (*a*)	**13.** (*d*)	**14.** (*b*)	**15.** (*c*)
16. (*c*)	**17.** (*a*)	**18.** (*a*)	**19.** (*b*)	**20.** (*d*)
21. (*a*)	**22.** (*c*)	**23.** (*b*)	**24.** (*b*)	**25.** (*a*)
26. (*b*)				

3

Real Sequences

3.1. NOTATION OF A FUNCTION. DOMAIN AND RANGE OF A FUNCTION

Let S and T be two given sets. We suppose that there is given a rule which associates to each member x of S a member y of T. The rule is then called a function from the set S into the set T. The function is denoted by letters such as f, F, ϕ, etc.

If the function is denoted by f, then the member of T associated by the function f to $x \in S$ is denoted by $f(x)$ or f_x and we say that $f(x)$ or f_x is the value of the function f at x.

The set S is called the **Domain** of the function and the set with members $f(x)$ obtained when x varies over the domain S is called the **Range** of the function. Thus, the range of the function f with domain S is the set

$$\{f(x) : x \in S\}.$$

3.1.1. Composition of Functions

Let S be the domain and T the range of a function f and let ϕ be any other function whose domain is the range T of f. Thus, the range of f is the domain of ϕ.

Let $x \in S$. Surely then

$$\phi(f(x))$$

has a meaning in as much as $f(x)$ whose ϕ-value is in question belongs to the domain of ϕ.

Thus, here we have defined a function whose domain is the domain of f and range is contained in the range of ϕ on the condition that the range of f coincides with the domain of ϕ.

This function is known as the composite of ϕ and f and is also ofter written as $\phi \, o \, f$. Thus, we have

$$(\phi \, o \, f)(x) = \phi^{-}(f(x)), x \in S.$$

***Ex.** Exhibit the functions defined below as composites of functions :

(*i*) $\sin^2 x$; (*ii*) $\log \cos x$; (*iii*) $\log \sin x^2$.

Real Sequences. In this chapter, we shall be concerned with functions whose domain is the set **N** of natural numbers. A function whose domain is the set **N** of natural numbers is known as a **sequence.**

A sequence whose range is a sub-set of **R** may be called a **Real** *sequence.* (*Kanpur 93, Garhwal 99*)

Since in this chapter we shall be concerned with Real Sequences only, we shall refer to them as just sequences.

It has become customary to denote a natural number, *viz.*, the general member of **N** by n so that in our dealings with sequences we shall use n in place of x.

If $u_1, u_2, u_3, ..., u_n, ...$ denote the values of a sequence, then this sequence is also denoted as

$$\{u_n\}.$$

3.2. BOUNDED SEQUENCES

A sequence f is said to be bounded if there exist number K, k such that

$$k \leq f(n) \leq K \;\forall\; n \in \mathbf{N}.$$

Thus, a sequence is bounded if its range is bounded.

(*Ajmer 99, Agra 92*)

Ex. 1. *Examine the boundedness of the sequence* $f(n) = n(-1)^{n-1}$.

The given sequence is

$$1, -2, 3, -4, 5, ...$$

Here $\quad f(n) = n(-1)^{n-1} \;\forall\; n \in \mathbf{N}.$

There is no real number k for which $k \leq f(n)$.

Hence, $f(n) = n(-1)^{n-1}$ is not bounded below. Further there is no real number K for which $f(n) \leq K$.

Hence, $f(n) = n(-1)^{n-1}$ is not bounded above.

Therefore, the given sequence is an unbounded sequence.

Ex. 2. Examine which of the following sequences are bounded :

(*i*) $f(n) = (-1)^n$ (*ii*) $f(n) = 1 + \dfrac{(-1)^n}{n}$

(*iii*) $f(n) = (-1)^n n$ (*iv*) $f(n) = \left(1 + \dfrac{1}{n}\right)^n$

(*v*) $f(n) = 1 + \dfrac{1}{2} + ... + \dfrac{1}{n}$

(vi) $f(n) = 1 + \frac{1}{2} + \frac{1}{2^2} \ldots + \frac{1}{2^{n-1}}$.

(vii) $f(n)$ is the nth prime

(viii) $f(n) = n\,!$.

3.2.1. Bounds of a Bounded Sequence

The *supremum* and the *infimum* of the bounded range of a bounded sequence are called its supremum and infimum.

Ex. Give the characteristic properties of the supremum and infimum of a bounded sequence.

3.2.2. Limit points of a sequence

Whereas we have already introduced the notion of a limit point of an arbitrary set of real numbers, we shall now introduce that of a limit point of a sequence; the two notions have much in common.

Def. *A number ξ is said to be a limit point of a sequence f if given any neighbourhood of ξ, $f(n)$ belongs to it for an* **infinite** *number of values of n.* (*Kanpur 94, 98; Agra 92*)

This means that ξ is a limit point of a sequence f if and only if given any positive number ε,

$$f(n) \in \,]\xi - \varepsilon, \xi + \varepsilon[$$

for an infinite number of values of n.

It will be seen that if ξ is *not* a limit point of a sequence f, there exists a positive number ε such that

$$f(n) \in \,]\xi - \varepsilon, \xi + \varepsilon[$$

for *at the most a finite number* of values of n.

Ex 1. Show that ξ is a limit point of a sequence f if and only if given any positive number ε and a positive integer m, there exists an integer $m \geq m$ such that

$$f(m') \in \,]\xi - \varepsilon, \xi + \varepsilon[.$$

Ex 2. Show that every limit point of a range of a sequence is as well a limit point of the sequence but not necessarily conversely.

Illustrations.

1. '0' is the only limit point of the sequence f defined by $f(n) = \frac{1}{n}$.

2. -1 and 1 are two limit points of the sequence f defined by $f(n) = (-1)^n$.

3. If $f(n) = 2 \;\forall\; n$, 2 is the only limit point of f.

4. The sequence f where $f(n) = n$ has no limit point.

3.2.3. Existence of Limit Points

Theorem. *Every bounded sequence has a limit point.*

Let f be a given bounded sequence with A as its range. Then we distinguish between the following two possibilities :

(*i*) A is a finite set. (*ii*) A is an infinite set.

We first consider the case (*i*).

Of the different members of the finite set A, there must exist at least one, say ξ, such that

$$f(n) = \xi$$

for an infinite number of values of n.

This means that $]\xi - \varepsilon, \xi + \varepsilon[$ be a neighbourhood of ξ then

$$f(n) = \xi \in]\xi - \varepsilon, \xi + \varepsilon[$$

for an infinite number of values of n. Thus, ξ is a limit point of the sequence f.

We now consider the case (*ii*). The sequence being bounded, the range is bounded. The range A, being infinite and bounded, has, by Bolzano-Weierstrass theorem, at least one limit point, say η. Then an infinite number of members of the set A belong to $]\eta - \varepsilon, \eta + \varepsilon[$; ε being any given positive number. Then

$$f(n) \in]\eta - \varepsilon, \eta + \varepsilon[$$

for an infinite number of values of n and as such η is a limit point of the sequence f.

Hence the theorem.

3.2.4. Upper and Lower Limits of a Bounded Sequence

Let f be a bounded sequence and let E denote the set of its limit points. The sequence f being bounded, the set E of its limit points is also bounded.

We shall show that *the bounded set E admits of greatest and smallest members, i.e., a bounded sequence has greatest and smallest limit points.*

Let G, g be the supremum and infimum respectively of E so that we have to show that

$$G \in E, g \in E.$$

Let us first consider G. Consider a neighbourhood

$$]G - \varepsilon, G + \varepsilon[$$

of G. Now G being the supremum of E, there exists a member, say ξ, of E such that

$$G - \varepsilon < \xi \leq G < G + \varepsilon,$$

so that $]G - \varepsilon, G + \varepsilon[$ is also a neighbourhood of ξ.

Now ξ being a limit point of the sequence f,

$$f(n) \in \,]G - \varepsilon, G + \varepsilon[$$

for an infinite number of values of n. Thus, G is a limit point of the sequence f.

It may similarly be shown that g is a limit point of f.

Def. *The greatest and smallest limit points of a bounded sequence, exist as they do, are called the upper and lower limits of the sequence.*

The upper and lower limits of a bounded sequence are often denoted as

$$\overline{\lim_{n\to\infty}} f, \ \underline{\lim_{n\to\infty}} f \quad \text{or} \quad \overline{\lim_{n\to\infty}} f(n), \ \underline{\lim_{n\to\infty}} f(n)$$

respectively

The upper and lower limits of a bounded sequence are also often called its upper and lower limits of indetermination respectively.

3.2.5. Characteristic Properties of the Upper and Lower Limits of a Bounded Sequence

Theorem I. *A number G is the upper limit of a bounded sequence if and only if the following two conditions are satisfied :*

(i) If $\varepsilon > 0$ be a given positive number, then $f(n) > G + \varepsilon$ for at the most a finite number of values of n or equivalently $f(n) < G + \varepsilon$ for all except a finite number of values of n.

(ii) If $\varepsilon > 0$ be a given positive number, then $f(n) > G - \varepsilon$ for an infinite number of values of n.

First suppose that G is the greatest limit point of the sequence f. Let $\varepsilon > 0$ be a given positive number.

No limit point of the sequence f being greater than $G + \varepsilon$, only a finite number at the most of the members of the sequence can exceed $G + \varepsilon$. This proves (*i*).

As G is a limit point, $f(n) \in \,]G - \varepsilon, G + \varepsilon[$ for an infinite number of values of n and, in particular $f(n) \geq G - \varepsilon$ for an infinite number of n and thus we proves (*ii*).

Now suppose G satisfies the two properties above.

From (*i*) and (*ii*), we see that

$$f(n) \in \,]G - \varepsilon, G + \varepsilon[$$

for an infinite number of values of n, so that G is a limit point.

Again, no limit point can exceed G. If possible, let G' be a limit point greater than G. Let α be a number such that

$$G < \alpha < G'$$

and β be a number greater than G'. We have

$$G < \alpha < G' < \beta.$$

Because of (*i*) only a finite number of members of the sequence exceed α and as such only a finite number of values of the sequence, belong to $[\alpha, \beta]$ and as such G' is *not* a limit point of the sequence. Thus, we arrive at a contradiction.

Hence the theorem.

Theorem II. A number g is the lower limit of a bounded sequence f if and only if the following two conditions are satisfied :

(*i*) *If* $\varepsilon > 0$ *be a given positive number, then* $f(n) < g - \varepsilon$ *for at the most a finite number of values of n or that* $f(n) > g - \varepsilon$ *for all except a finite number of values of n.*

(*ii*) *If* $\varepsilon > 0$ *be a given positive number, then* $f(n) < g + \varepsilon$ *for an infinite number of values of n.*

This proof is left to the reader.

Note. The fact that $f(n) < G + \varepsilon$ for all except a finite number of values of n is also sometimes expressed by saying that

$f(n) < G + \varepsilon$ for all sufficiently large values of n or that $f(n) < G + \varepsilon$ for almost all values of n.

Again it may be seen that if $f(n) < G + \varepsilon$ for all except a finite number of values of n, then there exists a positive integer m such that

$$f(n) < G + \varepsilon \ \forall \ n \geq m$$

We may have similar alternative descriptions for the statement $f(n) > g - \varepsilon$ for all except a finite number of values of n.

Ex. 1. Point out the suprema, infima, the upper and lower limits whichever may exist, of the sequences defined as follows :

(*i*) $f(n) = (-1)^n / n$ (*ii*) $f(n) = 1 + (-1)^n$

(*iii*) $f(n) = (-1)^n \left(1 + \dfrac{1}{n}\right)$

(*iv*) $f(n) = \begin{cases} (n+1)/n & \text{when } n = 3m, \\ (n+2)/2n & \text{when } n = 3m+1, \\ 1/(n+1) & \text{when } n = 3m+2, \end{cases}$

(*v*) $f(n) = (4^n + 1)/4^n$ or $(1 - 4^n)/4^n$

according as n is even or odd.

(*vi*) $f(n)=\left[1+(-1)^n\right]^{n/n}$

Ex. 2. Construct a sequence with ± 3 as its supremum and infimum and ± 2 for its upper and lower limits such that no value of the sequence belongs to the interval [– 2, 2].

3.3. CONVERGENT SEQUENCES

Def. *A sequence is said to be convergent if it is bounded and has one and only one limit point.* (*Ajmer 98*)

Also if a sequence f is convergent with l as its only limit point, we say that l is the limit of the sequence f or that the sequence f converges to l and in symbols, we write

$\lim\limits_{n\to\infty} f(n)=l$ or simply $\lim f(n) = l$ or $\lim f = l$.

Moreover, if a sequence f is convergent, we say that $\lim f(n)$ exists.

It is very important to note that the statement

$$\lim f(n) = l$$

is equivalent to the following two assertions :

(*i*) The sequence f is convergent,

(*ii*) The limit of the convergent sequence f is l.

It is clear that a bounded sequence is convergent if and only if its upper and lower limits are equal and the common value of the two is the limit of the sequence.

Illustrations. The sequences f as defined below are convergent :

(*i*) $f(n)=\dfrac{1}{n}$ (*ii*) $f(n)=\dfrac{(-1)^n}{n}$ (*iii*) $f(n)=1+\dfrac{1}{n^2}$.

The sequences f as defined below are not convergent :

(*i*) $f(n) = (-1)^n$. (*ii*) $f(n) = 1 + (-1)^n$. (*iii*) $f(n) = n^2$.

Note. In the following, we shall have three theorems pertaining to convergence. The first of these will state necessary and sufficient conditions *for a given sequence to be convergent to a given limit* and the other two will relate to conditions for *just* convergence without any reference to the limit.

3.3.1. Conditions for Convergence to a given Limit

Theorem. *A necessary and sufficient condition for a sequence f to converge to l is that to each given* $\varepsilon > 0$ *there corresponds an integer m such that*

$$|f(n)-l| < \varepsilon \ \forall\ n \geq m.$$ (*Rohilkhand 91; Poorv. 93*)

The condition is necessary. Let f be convergent to l. Then f is bounded with l as its only limit point.

Let $\varepsilon > 0$ be a given positive number. Consider the interval

$$]\, l - \varepsilon,\ l + \varepsilon\, [.$$

Now l being the only limit point of the bounded sequence f, there can exist only a finite number of values of n such that the corresponding $f(n)$ does not belong to $]l - \varepsilon, l + \varepsilon[$. Let $m - 1$ be the greatest of such values of n. We then have

$$f(n) \in\ ]\, l - \varepsilon,\ l + \varepsilon\, [,\ \forall\, n \geq m$$

i.e.,

$$|f(n) - l| < \varepsilon\ \forall\, n \geq m.$$

Thus, the condition is necessary.

The condition is sufficient. We will first prove that under the given condition the sequence f is bounded.

We take 1 for ε so that $\exists\, p$ such that

$$|f(n) - l| < 1\ \forall\, n \geq p,$$

i.e.,

$$f(n) \in\ ]\, l - 1, l + 1\, [\ \forall\, n \geq p.$$

Let

$$k = \min\ \{l - 1,\ f(1),\ f(2), \ldots,\ f(p - 1)\}$$
$$K = \max\ \{l + 1,\ f(1),\ f(2), \ldots,\ f(p - 1)\}$$

Then, we have

$$k \leq f(n) \leq K\ \forall\, n \in \mathbf{N}$$

so that f is a bounded sequence.

The given condition shows that l is a limit point of the given sequence.

Let, if possible l' be another limit point of the sequence.

Suppose $l' > l$.

We take $\varepsilon = \frac{1}{3}(l' - l)$ so that

$$l' - \varepsilon = l + 2\varepsilon > l + \varepsilon.$$
$$\Rightarrow\ l' - \varepsilon > l + \varepsilon.$$

Now there exists m such that

$$f(n) \in\ ]\, l - \varepsilon,\ l + \varepsilon\, [\ \forall\, n \geq m,$$

and as such $f(n)$ is greater than $l + \varepsilon$ for at the most a finite number of values of n so that $f(n)$ can belong to $]\, l' - \varepsilon, l + \varepsilon\, [$ for at the most a finite number of values of n. It follows that l' is not a limit point of the sequence f and so we arrive at contradiction.

We have proved that, as consequence of the given condition, the sequence f is bounded with l and its only limit point, *i.e.*, the sequence f converges to l.

Hence the theorem.

3.3.2. Cauchy's General Principle of Convergence

The condition, as obtained in § 3.3.1, answers the question "*Is any given number, l, the limit of a sequence f ?*" If with the help of this condition, it be shown that a given number, l, is *not* the limit of f, then it will not follow that the sequence does not converge; there being another possibility also, *viz.*, that f may converge to a number *different* from l. Thus, this condition examines the question of '*Convergence to a number l*' and not that of '*convergence*' which question, as will be seen, is of far more frequent occurrence in the theoretical parts of the subject. To this purpose of examining the question of convergence are directed two tests which are developed in the following two sub-sections.

The principle of convergence due to Cauchy (1798–1857) which is the more general of the two tests to be considered is a development and refinement of the idea that *if a sequence f is convergent, then, for large values of n, the members of the sequence, being near the limit, are also near each other.*

Theorem. *A necessary and sufficient condition for a sequence f to be convergent is that to each given* $\varepsilon > 0$ *there corresponds an integer m such that*

$$|f(n+p) - f(n)| < \varepsilon, \ \forall\, n \geq m,\ p \geq 0.$$

(Ajmer 97; Garhwal 99)

The condition is necessary. Let the sequence f be convergent so that there exists a number l such that

$$\lim f = l.$$

Let $\varepsilon > 0$ be a given number. By § 3.3.1, there exists m such that

$$|f(n) - l| < \tfrac{1}{2}\varepsilon \ \forall\, n \geq m.$$

$$\therefore \quad |f(n+p) - l| < \tfrac{1}{2}\varepsilon \ \forall\, n \geq m \text{ and } p \geq 0$$

$$\therefore \quad \left|f(n+p) - f(n)\right| = \left|f(n+p) - l + l - f(n)\right|$$

$$\leq |f(n+p) - l| + |l - f(n)|$$

$$< \tfrac{1}{2}\varepsilon + \tfrac{1}{2}\varepsilon, \ \forall\, n \geq m \text{ and } p \geq 0$$

i.e., $$|f(n+p) - f(n)| < \varepsilon \ \forall\, n \geq m, \text{ and } p \geq 0.$$

Thus, the condition is necessary.

The condition is sufficient. Firstly we show that the boundedness of the sequence is a consequence of the given condition.

We take 1 for ε so that $\exists\, r$ such that

$$|f(n+p)-f(n)|<1 \;\forall\; n \geq r, p \geq 0.$$

Taking r for n, we have

$$|f(r+p)-f(r)|<1 \;\forall\; p \geq 0.$$

$$\therefore \qquad f(r)-1<f(r+p)<f(r)+1 \;\forall\; p \geq 0.$$

Let

$$k = \min\{f(r)-1,\ f(1),\ f(2),\ \ldots,\ f(r-1)\}$$
$$K = \max\{f(r)+1,\ f(1),\ f(2),\ \ldots,\ f(r-1)\}$$

Then, we have

$$k \leq f(n) \leq K \;\forall\; n$$

$\Rightarrow$ the sequence is bounded.

The sequence, being bounded, has at least one limit point. We shall show that as a consequence of the given condition, it has only one limit point. Let, if possible, l, l' be two limit points of the sequence.

Let $\varepsilon > 0$ be a given positive number.

Now $\exists\, m$ such that

$$|f(n+p)-f(n)|<\tfrac{1}{3}\varepsilon \;\forall\; n \geq m, p \geq 0.$$

As l is a limit point, $\exists\, m_1 \geq m$ such that

$$|f(m_1)-l|<\tfrac{1}{3}\varepsilon.$$

Again as l' is a limit point, $\exists\, m_2 \geq m$ such that

$$|f(m_2)-l'|<\tfrac{1}{3}\varepsilon.$$

As $m_1 \geq m,\ m_2 \geq m$, we have

$$|f(m_1)-f(m_2)|<\tfrac{1}{3}\varepsilon.$$

We have

$$|l-l'| = |l-f(m_1)+f(m_1)-f(m_2)+f(m_2)-l'|$$
$$\leq |l-f(m_1)|+f(m_1)-f(m_2)|+|f(m_2)-l'|$$

$$< \frac{1}{3}\varepsilon + \frac{1}{3}\varepsilon + \frac{1}{3}\varepsilon = \varepsilon$$

i.e.,

$$|\, l - l' \,| < \varepsilon$$

where ε is any given positive number. Thus, $|\, l - l' \,|$ is a non-negative number less than every positive number. This means that

$$|\, l - l' \,| = 0 \Rightarrow l = l'.$$

Hence, the sequence has just one limit point and is as such convergent.

3.3.3. Monotonic Sequences and their Convergence

A sequence f is said to be monotonically increasing, if

$$f(n+1) \geq f(n) \;\forall\; n$$

and monotonically decreasing if

$$f(n+1) \leq f(n) \;\forall\; n.$$

Thus, for example, f, defined by

$$f(n) = n$$

is monotonically increasing and f defined by

$$f(n) = 1/n$$

is monotonically decreasing.

Also f defined by $f(n) = (-1)^n$ is neither monotonically increasing nor decreasing.

A sequence which is either monotonically increasing or decreasing is called monotonic sequence.

We have, as given below, a very simple criterion for the convergence of monotonic sequences. (*Kanpur 92, Agra 92*)

Theorem. *A necessary and sufficient condition for a monotonic sequence to be convergent is that it is bounded.*

(*Garhwal 98, Ajmer 98, Poorv. 92*)

The condition is necessary, for every convergent sequence is bounded.

The condition is sufficient. Suppose first that f is a bounded monotonically increasing sequence.

Let l be the supremum of the sequence, *i.e.*, the supremum of the range of the sequence. We shall show that

$$\lim f = l.$$

Let $\varepsilon > 0$ be given.

Since l is the supremum of the sequence, there exists m such that

$$f(m) > l - e.$$

Also since l is monotonically increasing.

$$f(n) \geq f(m) \ \forall \ n \geq m$$

Thus

$$f(n) > l - \varepsilon \ \forall \ n \geq m.$$

Also

$$f(n) \leq l < l + \varepsilon \ \forall \ n.$$

It follows that

$$l - \varepsilon < f(n) < l + \varepsilon \ \forall \ n \geq m$$

$$\Rightarrow \quad |f(n) - l| < \varepsilon \ \forall \ n \geq m$$

$$\Rightarrow \quad \lim f(n) = l$$

Thus, the sequence is convergent.

We may similarly consider the case when f is a bounded monotonically decreasing sequence.

Cor. For a bounded monotonically decreasing sequence f.

$$\lim f = \sup f,$$

and for a bounded monotonically decreasing sequence f

$$\lim f = \inf f.$$

3.4. NON-CONVERGENT SEQUENCE

Bounded non-convergent sequences. A bounded sequence which *does not converge* is said to *oscillate finitely.*

A finitely oscillating sequence must necessarily have at least two limit points.

Unbounded sequences. In the case of sequences which are not-bounded, we distinguish following three types of behaviours :

(*i*) **Divergence to ∞.** A sequence f is said to be divergent to ∞, if to each given positive number Δ there corresponds an integer m such that

$$f(n) > \Delta \ \forall \ n \geq m$$

and in symbols, we write

$$\lim f = \infty \quad \text{or} \quad \lim f(n) = \infty.$$

(*ii*) **Divergence to $-\infty$.** A sequence f is said to be divergent to $-\infty$, if to each given positive number Δ there corresponds an integer m such that

$$f(n) < -\Delta \ \forall \ n \geq m.$$

(*iii*) **Infinite oscillation.** A sequence f is said to oscillate infinitely, if it is unbounded and is divergent neither to ∞ nor to $-\infty$.

Note. From above it follows that for a given sequence we distinguish between *five* types of different behaviours. For a monotonic sequence, however, there are only two types of possible behaviours in that a monotonically increasing sequence is either convergent or divergent to ∞ and a monotonically decreasing sequence is either convergent or divergent to $-\infty$.

Ex. 1. Show that the sequences f where

$$(i)\ f(n) = (-1)^n, \qquad (ii)\ f(n) = 1 + (-1)^n$$

oscillate finitely.

Ex. 2. Show that a divergent sequence can have no limit point.

Ex. 3. Given that f and g are two sequences such that g diverges to infinity and $g(n) \leq f(n)$ for all n; show that the sequence f also diverges to infinity.

Ex. 4. Show that a monotonic sequence which is not bounded diverges to ∞ or $-\infty$ according as the sequence is increasing or decreasing.

3.5. ALGEBRAIC OPERATIONS AND LIMITS

Let f, ϕ be two given sequences.

The sequences

$$f+\phi,\ f-\phi,\ f\phi,\ \frac{f}{\phi}$$

are defined as follows :

$$(f+\phi)(n) = f(n) + \phi(n),$$
$$(f-\phi)(n) = f(n) - \phi(n),$$
$$(f\phi)(n) = f(n)\,\phi(n),$$
$$(f/\phi)(n) = f(n)/\phi(n) \text{ where } \phi(n) \neq 0.$$

We say that the sequences $f+\phi,\ f-\phi,\ f\phi,\ \dfrac{f}{\phi}$ are defined by *point-wise* addition, subtraction, multiplication and division respectively.

We now state and prove a theorem concerning the convergence of the sequences $f \pm \phi, f\phi, \dfrac{f}{\phi}$.

3.5.1. Theorem

If f, ϕ be two sequences such that

$$\lim f = k,\ \lim \phi = l,$$

then

(*i*) $\lim (f+\phi) = k + l = \lim f + \lim \phi$,

(*Rohilkhand 94, Garhwal 92, 94*)

(*ii*) $\lim (f - \phi) = k - l = \lim f - \lim \phi$,

(*iii*) $\lim (f\phi) = kl = (\lim f)(\lim \phi)$,

(*iv*) $\lim (f/\phi) = k/l = (\lim f)/(\lim \phi)$ if $l \neq 0$.

(*i*) We have $\forall\, n$.

$$|[f(n)+\phi(n)]-[k+l]| = |[f(n)-k]+[\phi(n)-l]| \qquad ...(1)$$

$$\leq |f(n)-k| + |\phi(n)-l|.$$

Let $\varepsilon < 0$ be given. Now

$$\lim f = k$$

$\Rightarrow\ \exists\ m_1$ such that $|f(n)-k| < \frac{1}{2}\varepsilon\ \forall\ n \geq m_1$.

Also $$\lim \phi = l$$

$\Rightarrow\ \exists\ m_2$ such that $|\phi(n)-l| < \frac{1}{2}\varepsilon\ \forall\ n \geq m_2$.

Let $m = \max(m_1, m_2)$ so that we have $\forall\ n \geq m$

$$|f(n)-k| < \frac{1}{2}\varepsilon,\ |\phi(n)-l| < \frac{1}{2}\varepsilon \qquad ...(2)$$

From (1) and (2), we have $\forall\ n \geq m$

$$|[f(n)+\phi(n)]-(k+l)| < \varepsilon.$$

Hence, $\lim(f+\phi) = k + 1 = \lim f + \lim \phi$.

(*ii*) We have $\forall\ n$

$$|[f(n)-\phi(n)]-[k-l]| = |[f(n)-k]+[l-\phi(n)]|$$

$$\leq |f(n)-k| + |\phi(n)-l|.$$

The result now follows as in (*i*) above.

(*iii*) We have $\forall\ n$

$$f(n)\,\phi(n) - kl = \phi(n)[f(n)-k] + k[\phi(n)-l]$$

$$\Rightarrow \quad \left|f(n)\,\phi(n) - kl\right| \leq \left|\phi(n)\right|\left|f(n)-k\right| + \left|k\right|\left|\phi(n)-l\right| \forall\ n$$

The convergent sequence ϕ is bounded so that there exists $c > 0$ such that $\forall\ n$

$$|\phi(n)| \leq c.$$

$\therefore \quad |f(n)\,\phi(n) - kl| \leq c|f(n)-k| + |k|\ |\phi(n)-l|\ \forall\ n$

Let $\varepsilon > 0$ be given. Now

$$\lim f = k$$

$\Rightarrow\ \exists\, m_1$ such that $|f(n) - k| \le \varepsilon / 2c\ \forall\, n \ge m_1$.

Also $$\lim \phi = l$$

$\Rightarrow\ \exists\, m_2$ such that $|f(n) - l| \le \varepsilon / 2(|k| + 1)\ \forall\, n \ge m_2$.

Let $m = \max(m_1, m_2)$. Then $\forall\, n \ge m$, we have

$$|f(n)\,\phi(n) - kl| < c\frac{\varepsilon}{2c} + |k|\frac{\varepsilon}{2(|k|+1)} < \varepsilon.$$

$$\Rightarrow \qquad \lim f\phi = \text{kl} = (\lim f)\,(\lim \phi).$$

(*iv*) We have $\forall\, n$

$$\left|\frac{f(n)}{\phi(n)} - \frac{k}{l}\right| = \frac{l[f(n) - k] - k[\phi(n) - l]}{l\phi(n)}$$

$$\le \frac{|l|\,|f(n) - k| + |k|\,|\phi(n) - l|}{|l|\,|\phi(n)|} \qquad \text{...}(i)$$

Now $\lim \phi(n) = l \ne 0$.

$\Rightarrow\ \exists\, m_1$ such that $|\phi(n) - l| < \frac{1}{2}|l|\ \forall\, n \ge m_1$.

Thus, $\forall\, n \ge m_1$, we have

$$|l| - |\phi(n)| \le |\phi(n) - l| < \frac{1}{2}|l|.$$

$$\Rightarrow \qquad |\phi(n)| \ge \frac{1}{2}|l|. \qquad \text{...}(ii)$$

From (*i*) and (*ii*), we see that $\forall\, n \ge m_1$

$$\left|\frac{f(n)}{\phi(n)} - \frac{k}{l}\right| \le \frac{2}{|l|}|f(n)| - k| + \frac{2|k|}{|l|^2}|\phi(n) - l|.$$

Let $\varepsilon > 0$ be given. There exists m_2 and m_3 such that

$$|f(n) - k| < \frac{1}{2}|l|\ \forall\, n \ge m_2 \qquad \text{...}(iii)$$

$$|\phi(n) - l| < \frac{1}{4}\frac{|l|^2}{|k|+1}\varepsilon\ \forall\, n \ge m_2 \qquad \text{...}(iv)$$

Let $m = \max(m_1, m_2, m_3)$. Then $\forall\, n \ge m$, we have

$$\left|\frac{f(n)}{\phi(n)} - \frac{k}{l}\right| < \varepsilon.$$

$\Rightarrow \qquad \lim [f \div \phi] = k \div l$ where $l \neq 0$.

Note. As a particular case of the theorem proved above, we see that if the sequences f, ϕ be convergent, then the sequences

$$f + \phi,\ f - \phi,\ f\phi$$

are also convergent. Further if $\lim \phi \neq 0$, then the sequences $f \div \phi$ is also convergent.

The converse, however, may not be true as the following illustrations show :

(*i*) Let f and ϕ be defined by

$$f(n) = (-1)^n,\quad \phi(n) = (-1)^{n+1}$$

so that $$(f + \phi)(n) = 0\ \forall\, n.$$

As may be easily seen, the sequences f and ϕ are not convergent but the sequence $f + \phi$ is convergent.

(*ii*) Let f and ϕ be defined by

$$f(n) = (-1)^n,\quad \phi(n) = (-1)^n.$$

While neither f nor ϕ is convergent; it may be seen that each of the sequences

$$f - \phi,\ f\phi,\ f \div \phi$$

is convergent.

3.5.2. Theorem

f, g, h are three sequences such that

(*i*) $\quad f(n) \le g(n) \le h(n),\ \forall\, n$ ...(*i*)

(*ii*) $\quad \lim f = \lim h = l.$

Then

$$\lim g = l.$$

Let $\varepsilon > 0$ be given. Now

$$\lim f = l \quad \text{and} \quad \lim h = l,$$

$\Rightarrow \exists\, m$ such that $\forall\, n \ge m$ ...(*ii*)

$$l - \varepsilon < f(n) < l + \varepsilon,\ l - \varepsilon < h(n) < l + \varepsilon. \qquad ...(iii)$$

From (*i*), (*ii*) and (*iii*), we have

$$l - \varepsilon < g(n) < l + \varepsilon,\ \forall\, n \ge m.$$

$$\Rightarrow \qquad \lim g = l.$$

EXAMPLES

1. *If $f(n) > 0$ for all $n \in \mathbf{N}$, then show that $f(n) \to \infty$ as $n \to \infty \Leftrightarrow$ $\frac{1}{f(n)} \to 0$ as $n \to \infty$.*

Sol. Let $f(n) \to \infty$ as $n \to \infty$.

Let $\varepsilon > 0$ be given. Since $f(n) \to \infty$ as $n \to \infty$, hence for $1/\varepsilon > 0$ there exists $m \in \mathbf{N}$ such that

$$f(n) > 1/\varepsilon \text{ for all } n \geq m$$

$$\Rightarrow \quad \frac{1}{f(n)} < \varepsilon \ \forall\ n \geq m \ \Rightarrow \left|\frac{1}{f(n)}\right| < \varepsilon \ \forall\ n \geq m \quad [\because f(n) > 0]$$

$$\Rightarrow \quad \left|\frac{1}{f(n)} - 0\right| < \varepsilon \ \forall\ n \geq m$$

$$\Rightarrow \quad \frac{1}{f(n)} \to 0 \text{ as } n \to \infty.$$

Conversely, let $\frac{1}{f(n)} \to 0$ as $n \to \infty$.

Let $k > 0$ be arbitrary. Now $\frac{1}{f(n)} \to 0$ as $n \to \infty \Rightarrow$ for $1/k > 0$ there exists $m \in \mathbf{N}$ such that

$$\left|\frac{1}{f(n)} - 0\right| < \frac{1}{k} \text{ for all } n \geq m$$

$$\Rightarrow \quad \frac{1}{f(n)} < \frac{1}{k} \ \forall\ n \geq m \ \Rightarrow \ f(n) > k \ \forall\ n \geq m$$

$$\Rightarrow \quad f(n) \to \infty \text{ as } n \to \infty.$$

2. *If the sequences f, g diverge to infinity then prove that $f + g$ and fg diverge to infinity.*

Sol. The sequence f diverges to infinity $\Rightarrow$ for $k > 0$, there exists $m_1 \in \mathbf{N}$ such that $f(n) > k \ \forall\ n \geq m_1$. Again, the sequence g diverges to infinity $\Rightarrow$ for $l > 0$, there exists $m_2 \in \mathbf{N}$ such that $g(n) > l \ \forall\ n \geq m_2$.

Take $$m = \max.\ \{m_1, m_2\}.$$

$\therefore \qquad f(n)+g(n) > k+l > k \;\; \forall\, n \geq m$

and $\qquad fg > k \cdot l = k \;\; \forall\, n \geq m$

$\therefore$ Both $f + g$ and fg diverge to infinity.

$\lim\,[f \div \phi] = k \div l$ where $l \neq 0$.

3. *If f diverges to infinity and g is bounded then show that f + g diverges to infinity.*

Sol. The sequence g is bounded $\Rightarrow$ there exists $k_1 > 0$ such that $|g(n)| < k_1 \;\forall\, n \in \mathbf{N}$.

The sequence f diverges to infinity $\Rightarrow$ for $k > 0$ there exists $m \in \mathbf{N}$ such that $f(n) > k + k_1 \;\forall\, n \geq m$.

$\therefore$ For all $n \geq m$, $f(n)+g(n) > f(n) - |g(n)| > k + k_1 - k_1 = k$.

Thus, for $k > 0$ there exists $m \in \mathbf{N}$ such that

$$f(n) - g(n) > k \text{ for all } n \geq m.$$

Hence, $f + g$ diverges to infinity.

EXERCISES

1. If $\lim f = 0$ and g oscillates finitely, show that

$$\lim\,(fg) = 0.$$

2. If f is convergent and g is divergent, show that

$$\lim\,(f + g) = 0.$$

3. If f is convergent and g is divergent, show that

$$f + g$$

is divergent.

4. Show that

$$\lim f = l \;\Rightarrow\; \lim |f| = |l|.$$

Give an example to show that the reverse implication may not be true.

5. If $\lim f = \infty$ and $g(n) \geq f(n) \;\forall\, n$ show that

$$\lim g = \infty.$$

6. f and g are two convergent sequences such that

$$f(n) \geq g(n) \;\forall\, n$$

Show that $\lim f \leq \lim g$.

7. Given that f and g are two convergent sequences and two sequences F and G are defined as follows :

$F(n) = \max.\ \{f(n), g(n)\}$, $\;G(n) = \min.\ \{f(n), g(n)\}$.

Show that the sequences F and G are also convergent and that

$\lim F = \max\,(\lim f, \lim g)$, $\lim G = \min\,(\lim f, \lim g)$.

8. f is a convergent sequence and k is a number such that

$$f(n) \le k \ \forall \ n,$$

show that

$$\lim f \le k.$$

Give an example to show the possibility of equality.

3.6. SUB-SEQUENCES OF A SEQUENCE

Let f be a sequence with values

$$f(1),\ f(2),\ ...,\ f(n),\ ...$$

Then we are inclined to think of a sequence whose range of values forms a sub-set of the range of f as a sub-sequence of the sequence of f. Thus, for example, the sequence with values

$$f(2),\ f(5),\ f(9),\ ...$$

may be thought of as a sub-sequence of f. We now proceed to render the notion of sub-sequences precise.

Let ϕ be a sequence such that

(*i*) $\phi(n) \in \mathbf{N} \ \forall \ n \in \mathbf{N}$

(*ii*) $\phi(n+1) > \phi(n) \ \forall \ n \in \mathbf{N}$

(*iii*) ϕ is not bounded above.

As an implication of the above we may see that ϕ is a sequence which tends to infinity.

Now if f be any sequence, the composite sequence $f \, 0 \, \phi$ defined as

$$(f \, 0 \, \phi)\ (n) = f(\phi\ (n))$$

is called a sub-sequence of the sequence f.

We obtain different sub-sequences of a given sequence f for different choices of the sequences ϕ satisfying the conditions (*i*), (*ii*), (*iii*).

Illustrations. (*i*) Let $\phi\ (n) = n^2$.

Then the values of the sub-sequence $f \, 0 \, \phi$ are

$$f(\phi(1)),\ f(\phi(2)),\ f(\phi(3)),\ f(\phi(4)),\ ...$$

or equivalently

$$f(1),\ f(4),\ f(9),\ f(16),\ f(25),\ ...$$

(*ii*) Let $\phi\ (n) = n\,!$

Then the values of the sub-sequence $f \, 0 \, \phi$ are

$$f(1),\ f(2),\ f(6),\ f(24),\ ...$$

The following results may now be easily proved.

(i) Every sub-sequence of a convergent sequence is convergent and the limit of the sub-sequence is the same as that of the sequence.

(ii) There exists a convergent sub-sequence of a given sequence converging to a given limit point of the sequence.

3.7. THEOREM ON NESTED INTERVALS

Let

$$[a_1, b_1], \ [a_2, b_2], \ ..., \ [a_n, b_n], \ ..., \ n \in \mathbf{N}$$

be a set of closed intervals such that

$$[a_{n+1}, b_{n+1}] \subset [a_n, b_n], \forall n$$

and

$$\lim (b_n - a_n) = 0.$$

Then there exists one and only one point belonging to each member of the set of intervals.

Now

$$[a_{n+1}, b_{n+1}] \subset [a_n, b_n] \Rightarrow a_n \leq a_{n+1} < b_{n+1} \leq b_n,$$

so that we see that the sequence $\{a_n\}$ is monotonically increasing and bounded above and the sequence $\{b_n\}$ is monotonically decreasing and bounded below. Thus, the sequences $\{a_n\}$ and $\{b_n\}$ are convergent. Let

$$\lim a_n = \xi, \ \lim b_n = \eta.$$

We have

$$0 = \lim (b_n - a_n) = \lim b_n - \lim a_n = \eta - \xi$$

$$\Rightarrow \quad \xi = \eta.$$

We have

$$\xi = \sup \{a_n\}, \ \eta = \inf \{b_n\}$$

$$\Rightarrow \quad a_n \leq \xi \leq b_n \ \forall n.$$

Thus, we have a point ξ such that

$$\xi \in [a_n, b_n] \ \forall n.$$

To prove the uniqueness, suppose ξ' also belongs to each member of the set of intervals.

We have $$a_n \leq \xi, \ \xi' \leq b_n \ \ \forall n$$

$$\Rightarrow \quad |\xi' - \xi| \leq b_n - a_n \ \forall n.$$

Taking limits, we have

$$\xi' - \xi = 0 \Rightarrow \xi' = \xi.$$

Thus, we have proved the uniqueness also.

3.8. INFINITE SERIES : ITS CONVERGENCE AND SUM

Let f be a given sequence. Then a symbol of the form

$$f(1) + f(2) + ... + f(n) + ...$$

is called an *Infinite Series.* Another way of writing this infinite series is

$$\sum_{n=1}^{\infty} f(n) \text{ or simply } \Sigma f(n).$$

To the given sequence f which underlies the infinite series referred to above, we associate another sequence Σ defined as follows :

$$\Sigma(n) = f(1) + f(2) + ... + f(n)$$

Thus, $\Sigma(n)$ denotes the sum of the first n terms of the infinite series

$$\sum_{n=1}^{\infty} f(n).$$

The sequence Σ is said to be the sequence of *Partial Sums* of the given series.

The infinite series

$$\sum_{n=1}^{\infty} f(n)$$

is said to be convergent if the sequence Σ *is convergent. Moreover, in the event of the convergence of the infinite series,*

$$\lim \Sigma(n)$$

is said to be the **sum** *of the series.*

We shall now obtain some tests for the convergence of infinite series. As is to be expected, it will be seen that this discussion leans very heavily on the developments in respect of sequences.

A series which is not convergent may have any one of the following four types of behaviours. It may

(*i*) oscillate finitely;

(*ii*) oscillate infinitely;

(*iii*) diverge to $+\infty$;

(*iv*) diverge to $-\infty$;

depending on the behaviour of the sequence σ of partial sums.

Note. If we add the first two terms of an infinite series, and then add the sum so obtained to the third, and thus go on adding each term to the sum of the previous terms, we see that, as there is no last term of the series, the process will never be completed. In the case of a finite series, this process of addition will be completed at some stage, however large a number of terms the series may consist of. Thus, in the ordinary sense, the expression "*Sum of an infinite series*" has no meaning. The notion of limit has, therefore, been employed to give a meaning to this expression.

EXERCISES

1. Show that the infinite geometrical series

$$\sum_{n=0}^{\infty} x^n$$

is convergent, if and only if $|x| < 1$.

2. Show that the series Σa_n where $a_n = (-1)^n$ is not convergent.

3. Show that the following series are convergent :

(*i*) $\dfrac{1}{1.2}+\dfrac{1}{2.3}+\dfrac{1}{3.4}+\ldots+\dfrac{1}{n(n+1)}+\ldots$

(*ii*) $\dfrac{1}{1.2.3}+\dfrac{1}{2.3.4}+\dfrac{1}{3.4.5}+\ldots+\dfrac{1}{n(n+1)(n+2)}+\ldots$

(*iii*) $1+2x+3x^2+\ldots+nx^{n-1}+,\ldots,\ |x|<1.$

Also find the sum in each case.

4. Show that

$$\sum_{n=0}^{\infty} (-1)^n \frac{(2n+3)}{(n+1)(n+2)} = 1.$$

5. Show that

$$\sum_{n=1}^{\infty} \frac{n^2+9n+3}{(n+1)(2n+3)(2n+5)(n+4)} = \frac{5}{36}$$

6. Show that the arithmetic series

$$a + (a + d) + (a + 2d) + \ldots + (a + nd) + \ldots$$

is always divergent, except when a, d are both zero.

7. A series Σa_n is convergent and k is a constant; show that the series $\Sigma k a_n$ is also convergent.

8. A series Σa_n is given; a sequence $\{b_n\}$ is defined such that

$b_n = a_{m+n}$; m being a given positive integer;

show that the series Sa_n and Sb_n have the same behaviour in relation to convergence or otherwise.

9. Σa_n and Σb_n are two convergent series; S_1, S_2, being their sums; show that series $\Sigma\,(a_n + b_n)$ is also convergent with sum equal to $S_1 + S_2$.

10. Show that a sequence $\{a_n\}$ is convergent if, and only if, the series $\Sigma\,(a_{n+1} - a_n)$ is convergent.

3.8.1. General Test for the Convergence of Series with Positive Terms

Consider the positive term series

$$\sum_{n=1}^{\infty} f(n)$$

so that $$f(n) \geq 0 \ \forall \ n.$$

Now

$$f(n) \geq 0 \ \forall \ n$$

$$\Rightarrow \quad \sigma(n+1) - \sigma(n) = f(n+1) \geq 0,$$

$$\Rightarrow \quad \sigma(n+1) \geq \sigma(n) \ \forall \ n.$$

It follows that the sequence of partial terms of a series with positive terms is monotonically increasing.

Theorem. *A necessary and sufficient condition for the convergence of a positive term series*

$$\sum_{n=1}^{\infty} f(n), \ f(n) \geq 0 \ \forall \ n$$

is that the sequence Σ of partial sums of the series defined by

$$\sigma(n) = f(1) + f(2) + \ldots + f(n)$$

is bounded above.

As $f(n) \geq 0 \ \forall \ n$, the sequence σ is monotonically increasing.

A necessary and sufficient condition for the series to be convergent is that the sequence σ of its partial sums is convergent. Again a necessary and sufficient condition for the sequence σ which is monotonically increasing to be convergent is that it is bounded above. Hence the result.

Note 1. A series $\sum_{n=1}^{\infty} f(n)$ whose terms are not necessarily positive may fail to be convergent even if the sequence σ is bounded above. For example, consider

$$f(n) = (-1)^n$$

so that we have

$$\sigma(n) = \begin{cases} -1 \text{ if } n \text{ is odd,} \\ 0 \text{ if } n \text{ is even.} \end{cases}$$

The sequence σ even though bounded above is obviously not convergent and as such the series is not convergent.

It should be seen that, in general, boundedness of the sequence of partial sums is only a necessary but not a sufficient condition for the convergence of the series

$$\sum_{n=1}^{\infty} f(n),$$

and it is only for positive term series that this condition of boundedness if the partial sums is as well as sufficient condition.

Note 2. A positive term series which does not converge will diverge in plus infinity. In fact the sequence of partial sums of a positive term series which is monotonically increasing either tends to a finite limit or to plus infinity. Thus, whereas there are five possible behaviours for an arbitrary term series there are only *two* for a positive term series.

3.8.2. Cauchy's General Principles of Convergence of a Series

The necessary and sufficient condition for the convergence of an infinite series

$$\Sigma f(n)$$

is that given $\varepsilon > 0$*, there corresponds a positive integer m such that*

$$|f(n+1) + f(n+2) + \ldots + f(n+p)| < \varepsilon$$

$\forall\, n \geq m$ *and* $\forall\, p \geq 0$.

We write

$$\Sigma(n) = f(1) + f(2) + \ldots + f(n).$$

The necessary and sufficient condition for the convergence of the sequence Σ is that to $\varepsilon > 0$, there corresponds a positive integer m such that

$$|\sigma(n+p) - \sigma(n)| < \varepsilon \;\forall\, n \geq m, \text{ and } \forall\, p \geq 0$$

$$\Rightarrow \qquad |f(n+1) + f(n+2) + \ldots + f(n+p)| < \varepsilon$$

$\forall\, n \geq m$ and $\forall\, p \geq 0$.

EXAMPLES

1. *Show that the series*

$$1 - \frac{1}{2} + \frac{1}{3} - \frac{1}{4} + \frac{1}{5} - \frac{1}{6} + \ldots + \frac{(-1)^{n+1}}{n} + \ldots,$$

is convergent.

Sol. This is not a positive term series.

It is easy to see that

$$\frac{1}{n+1} - \frac{1}{n+2} + \frac{1}{n+3} - \frac{1}{n+4} + \ldots + \frac{(-1)^{p+1}}{n+p}$$

is positive and less than $1/(n+1)$,

Let ε be a positive number. We have

$$|\sigma(n+p)-\sigma(n)|=\left|\frac{1}{n+1}-\frac{1}{n+2}+\ldots+\frac{(-1)^{p+1}}{n+p}\right|$$

$$=\frac{1}{n+1}-\frac{1}{n+2}+\ldots+\frac{(-1)^{p+1}}{n+p}<\frac{1}{n+1}<\varepsilon,$$

if

$$n>(1/\varepsilon-1)$$

Let m be an integer greater than $(1/\varepsilon-1)$. Then, we have

$$|\sigma(n+p)-\sigma(n)|<\varepsilon\ \forall\ n\geq m,\ \text{and}\ \forall\ p\geq 0.$$

Hence the series converges.

2. *Show that the positive term series*

$$1+\frac{1}{2}+\frac{1}{3}+\frac{1}{4}+\frac{1}{5}+\ldots+\frac{1}{n}+\ldots$$

does not converge.

Sol. Suppose that it converges. There exists, therefore, a positive integer m such that $\forall\ n\geq m$ and $\forall\ p\geq 0$.

$$\left|\frac{1}{n+1}+\frac{1}{n+2}+\ldots+\frac{1}{n+p}\right|<\frac{1}{4}\cdot \qquad \left(\text{Taking } \varepsilon=\frac{1}{4}\right)$$

But taking $n=m$ and $p=m$, we see that

$$\frac{1}{m+1}+\frac{1}{m+2}+\ldots+\frac{1}{2m}>\frac{m}{2m}=\frac{1}{2}\cdot$$

Thus, we have a contradiction. Hence, the series does not converge.

3.8.3. Decimal Representation of Real Numbers

Firstly, we show that the *infinite series*

$$a+\frac{a_1}{10}+\frac{a_2}{10^2}+\ldots+\frac{a_n}{10^n}+\ldots \qquad \ldots(1)$$

is convergent, if, a, is any integer and $a_1, a_2, \ldots, a_n, \ldots$, *are integers between 0 and 9, i.e.,*

$$0\leq a_n\leq 9,\ \forall\ n\geq 1.$$

We write

$$a_n=a+\frac{a_1}{10}+\frac{a_2}{10^2}+\ldots+\frac{a_n}{10^n}$$

so that the sequence $\{a_n\}$ is monotonically increasing.

Also we have

$$a_n \le a + \frac{9}{10} + \frac{9}{10^2} + \ldots + \frac{9}{10^n}$$

$$= a + \frac{9}{10}\left[\frac{1-\left(\frac{1}{10}\right)^n}{1-\frac{1}{10}}\right]$$

$$= a + 1 - \frac{1}{10^n} < a+1, \ \forall\, n.$$

so that the sequence σ is bounded above.

Thus, the series is convergent.

This infinite series (1) is often written as

$$a.a_1\, a_2\, a_3\, a_4 \ldots\, a_n \ldots \qquad \ldots(2)$$

The symbol (2) represents the real number which is the sum of the infinite series (1).

We shall now show that every real number can be written in the form (2), *i.e.*, can be expressed as the sum of an infinite series of the form (1).

Consider a real number x. There exist two integers a, $a + 1$ such that

$$a \le x < a + 1.$$

Divide now the interval $[a, a + 1]$ into 10 equal parts. We then obtain an integer a_1 between 0 and 9 such that

$$a + \frac{a_1}{10} \le x < a + \frac{a_1+1}{10}$$

$$\Leftrightarrow \qquad a\,.\,a_1 \le x < a\,.\,(a_1 + 1).$$

Again divide the interval

$$\left[a + \frac{a_1}{10},\ a + \frac{a_1+1}{10}\right]$$

into 10 equal parts. We shall then obtain an integer, a_2, between 0 and 9, such that

$$a + \frac{a_1}{10} + \frac{a_2}{10^2} \le x < a + \frac{a_1}{10} + \frac{a_2+1}{10^2}$$

$$\Leftrightarrow \qquad a\,.\,a_1\, a_2 \le x < a\,.\,a_1\,(a_2 + 1).$$

Proceeding in this manner, we shall obtain a sequence $\{a_n\}$ of integers between 0 and 9 such that

$$a \,.\, a_1\, a_2 \,...\, a_n \le x < a \,.\, a_1\, a_2 \,...\, (a_n + 1).$$

We write

$$b_n = a \,.\, a_1\, a_2 \,...\, a_n,\ c_n = a \,.\, a_1\, a_2 \,...\, (a_n + 1).$$

As seen before, $\{b_n\}$ is convergent. Also since

$$c_n - b_n = \frac{1}{10^n} \Leftrightarrow c_n = b_n + \frac{1}{10^n},$$

we see that the sequence $\{c_n\}$ is convergent and

$$\lim c_n = \lim b_n,$$

From the inequalities

$$b_n \le x < c_n \ \forall\, n,$$

we conclude that

$$x = \lim b_n$$

$$\Leftrightarrow \qquad x = a \,.\, a_1\, a_2\, a_3 \,...\, a_n \,...$$

Note. In case x is a rational number the decimal representation will be terminating or recurring.

3.9. THE MEANING OF a^x, WHEN $a > 0$ AND x IS A RATIONAL NUMBER

When x is a positive integer, the symbol a^x denotes the product

$$a \,.\, a \,.\, a \,...\, a, \ (x \text{ times})$$

and when x is a negative integer so that, $-x$, is a positive integer, we have

$$a^x = 1/a^{-x}.$$

Thus, the concepts of multiplication and division (Chapter 1) are all that we require in order to define a^x, when x is an integer.

The theorem below is a fundamental for giving a meaning to the symbol, a^x, when x is a rational number, and a is positive.

Theorem. *If a is a given positive number, the equation*

$$x^m = a$$

in x has one and only one positive root.

We define two sets L and U of real numbers such that L consists of

(*i*) all the negative numbers,

(*ii*) zero,

(*iii*) positive real numbers x such that

$$a^m \le a$$

and U is the complement of L in **R**.

The sub-set U is not empty for every real number > max. $(a, 1)$ is a member of U.

Also we may see that

$$x \in L,\ y \in U \Rightarrow x < y.$$

Thus, there exists a number, say B, such that

$$x < B \Rightarrow x \in L,\ x > B \Rightarrow x \in U.$$

Clearly B cannot be negative.

Also $\qquad B \neq 0. \qquad$ If possible let $B = 0$.

Then

$$\frac{1}{n} \in U \ \forall\ n \in \mathbf{N}$$

$$\Rightarrow \qquad \left(\frac{1}{n}\right)^m > a \ \forall\ n \in \mathbf{N}$$

$$\Rightarrow \qquad \lim_{n\to\infty}\left(\frac{1}{n}\right)^m \geq a$$

$$\Rightarrow \qquad 0 \geq a$$

so that we have a contradiction in as much as

$$a > 0.$$

Thus, B is necessarily positive.

Now

$$B - \frac{1}{n} \in L, \quad B + \frac{1}{n} \in U$$

$$\Rightarrow \qquad \left(B - \frac{1}{n}\right)^m \leq a < \left(B + \frac{1}{n}\right)^m$$

$$\Rightarrow \qquad \lim_{n\to\infty}\left(B - \frac{1}{n}\right)^m \leq a < \lim_{n\to\infty}\left(B - \frac{1}{n}\right)^m$$

$$\Rightarrow \qquad B^m \leq a \leq B^m.$$

$$\Rightarrow \qquad B^m = a.$$

Thus, there exists a positive root, *viz.*, B of $x^m = a$.

If possible let B' be another positive root so that

$$B^m = a = B'^m.$$

Now

$$B' > B \;\Rightarrow\; B'^{\,m} > B'^{m}.$$
$$B' < B \;\Rightarrow\; B'^{\,m} < B'^{m},$$

so that we arrive at contradictory statements.

Thus, B is the only positive root of $x^m = a$.

Def. The unique positive root of the equation

$$x^m = a,\ a > 0,$$

m being a positive integer is written as

$$a^{1/m} \quad \text{or} \quad \sqrt[m]{a}$$

and called the mth root of a.

Note. If $a > 0$ and m is even, the equation $x^m = a$ also possesses a negative root, *viz.*, $-B$ but if m is odd, it cannot obviously have any negative root.

If $a < 0$ and m is even, the equation $x^m = a$ cannot obviously, have any root, positive or negative, but if m is odd, it has no positive root but has a negative root $-B$, where B is the positive root of $x^m = -a$.

To avoid this ambiguity and indefiniteness, we will always take the base, a, as positive, and the symbol $a^{1/m}$ will, then, always denote the unique positive root of $x^m = a$.

Def. *If n/m is a rational number where m is positive, then by def.,*

$$a^{n/m} = (\sqrt[m]{a})^n$$

a, being positive.

EXERCISES

1. Prove that

$$(\sqrt[m]{a})^n = (\sqrt[m]{a^n}).$$

2. x, y are rational numbers and a is a positive real number; show that

(*i*) $a^x \cdot a^y = a^{x+y}$. (*ii*) $(a^x)^y = a^{xy}$.

(Reduce x, y to a common denominator.)

3. a, b are two positive real numbers and x is a rational number; show that

$$(ab)^x = a^x b^x.$$

4. x, y are rational numbers such that $x > y$; show that

$$a^x \gtreqless a^y \text{ according as } a \gtreqless 1.$$

5. x is a positive rational number; show that

$$a^x \gtreqless 1, \text{ according as } a \gtreqless 1.$$

6. a, b are positive real numbers, and x is a positive rational number; show that

$$a > b \;\Rightarrow\; a^x > b^x,\ a < b \;\Rightarrow\; a^x < b^x.$$

7. Show that

$$\sqrt[n]{a} \to 1, \text{ when } n \to \infty, (a > 0).$$

For $a = 1$, the result is obvious.

Let $a > 1$. We write

$$\sqrt[n]{a} = 1 + h_n;\ h_n > 0.$$

We know that

$$(1 + h_n)^n > 1 + nh_n.$$

Now $\qquad \sqrt[n]{a} = 1 + h_n$

$\Rightarrow \qquad a = (+ h_n)^n > 1 + nh_n$

$\Rightarrow \qquad 0 < h_n < (a - 1)/n.$

Let ε be a positive number. There exists a positive integer m such that $(a-1)/n < \varepsilon \ \forall \ n \geq m.$

It follows that

$$-e < 0 < h_n < (a-1)/n < e \ \forall \ n \geq m,$$

$$\Rightarrow \qquad |\sqrt[n]{a} - 1| = |h_n| < e \ \forall \ n \geq m.$$

Hence, the result.

If $a < 1$, we write $a = 1/b$ so that $b > 1$,

$\therefore \quad \sqrt[n]{a} = 1/\sqrt[n]{b}$ and the result now follows.

8. Show that $\lim a^{-1/n} = 1$.

EXAMPLES

1. *Show that the sequence f defined by*

$$f(n) = \left(1 + \frac{1}{n}\right)^n \qquad \textit{(Ajmer 97, Agra 94)}$$

is convergent.

Sol. We have, by the binomial theorem,

$$\left(1+\frac{1}{n}\right)^n = 1 + n \cdot \frac{1}{n} + \frac{n(n-1)}{2!} \cdot \frac{1}{n^2} + \ldots + \frac{n(n-1)(n-2)\ldots 1}{n!} \cdot \frac{1}{n^n}$$

$$= 1 + 1 + \frac{1}{2!}\left(1 - \frac{1}{n}\right) + \ldots + \frac{1}{n!}\left(1 - \frac{1}{n}\right)\left(1 - \frac{2}{n}\right)\ldots\left(1 - \frac{n-1}{n}\right).$$

We may also similarly develop

$$\left(1+\frac{1}{n+1}\right)^{n+1}$$

by the binomial theorem.

Comparing these two developments, we may easily deduce that $\forall\, n$

$$\left(1+\frac{1}{n+1}\right)^{n+1} > \left(1+\frac{1}{n}\right)^{n}$$

$\Rightarrow$ $$f(n+1) > f(n)$$

$\Rightarrow$ the sequence f is monotonically increasing. Now

$$\left(1+\frac{1}{n}\right)^{n} \leq 1+1+\frac{1}{2!}+\frac{1}{3!}+\ldots+\frac{1}{n!}$$

$$< 1+1+\frac{1}{2}+\frac{1}{2^2}+\ldots+\frac{1}{2^{n-1}}$$

$$= 1+\frac{1-\frac{1}{2^n}}{1-\frac{1}{2}} = 3-\frac{1}{2^{n-1}} < 3\ \forall\, n.$$

Thus, the sequence f is bounded above.

The sequence f, being bounded and monotonically increasing, is convergent.

Hence the result.

2. *Show that the sequence f defined by*

$$f(1) = \sqrt{2}$$

$$f(n+1) = \sqrt{2f(n)}$$

converges to 2.

Sol. It may be seen that different members of the sequence f are

$$\sqrt{2},\ \sqrt{2\sqrt{2}},\ \sqrt{2\sqrt{2\sqrt{2}}},\ldots$$

We have

$$2\sqrt{2} > 2$$

$\Rightarrow$ $$\sqrt{2\sqrt{2}} > \sqrt{2} \Rightarrow f(2) > f(1).$$

Suppose that

$$f(n+1) > f(n).$$

It follows that $\sqrt{2f(n+1)} > \sqrt{2f(n)}$

$\Leftrightarrow$ $f(n+2) > f(n+1).$

Thus

$$f(n+1) > f(n) \Rightarrow f(n+2) > f(n+1).$$

By mathematical induction, we deduce that the sequence f is monotonically increasing.

Clearly, we have

$$f(1) < 2.$$

Suppose that $f(n) < 2$. Then

$$f(n+1) = \sqrt{2f(n)} < \sqrt{2.2} = 2.$$

Thus $f(n) < 2 \Rightarrow f(n+1) < 2.$

so that by mathematical induction, we deduce that

$$f(n) < 2 \; \forall \; n.$$

The sequence f, being bounded and monotonically increasing, is convergent.

Let $\lim f = l.$

We have

$$f(n+1) = \sqrt{2f(n)} \Rightarrow [f(n+1)]^2 = 2f(n).$$

Also $\lim f(n) = l \Rightarrow \lim f(n+1) = l.$

Thus we have

$$l^2 = 2l$$

$\Rightarrow$ $l(l-2) = 0 \Rightarrow l = 0$ or $l = 2.$

The limit, however, cannot be equal to zero. Thus

$$l = 2 \Rightarrow \lim f(n) = 2.$$

3. *If x_1, x_2 are positive and $x_{n+1} = \frac{1}{2}(x_n + x_{n-1})$, then of the two sequences with values*

$$x_1, x_3, x_5, \ldots, \text{ and } x_2, x_4, x_5, \ldots$$

one is decreasing and the other increasing and both converge to the same limit $\frac{1}{3}(x_1 + 2x_2)$.

Sol. Let $x_2 < x_1$. We have

$$x_2 < x_1 \Rightarrow x_2 < x_3 < x_1$$
$$x_2 < x_3 \Rightarrow x_2 < x_4 < x_3$$
$$x_4 < x_3 \Rightarrow x_4 < x_5 < x_3$$

Thus, we may see that

$$x_2 < x_4 < x_6 < \ldots < x_5 < x_3 < x_1$$

so that the sequence with values $x_1, x_3, x_5, \ldots$ is decreasing and that with values $x_2, x_4, x_6, \ldots$ increasing. Also both are bounded. It follows that they are both convergent.

We have

$$x_3 - x_2 = \frac{1}{2}(x_1 - x_2)$$

$$x_4 - x_2 = x_4 - x_3 + x_3 - x_2 = -\frac{1}{4}(x_1 - x_2) + \frac{1}{2}(x_1 - x_2)$$
$$= (x_1 - x_2)\left(-\frac{1}{4} + \frac{1}{2}\right)$$

$$x_5 - x_2 = x_5 - x_4 + x_4 - x_2 = (x_1 - x_2)\left(\frac{1}{5} - \frac{1}{4} + \frac{1}{2}\right)$$

$$x_6 - x_2 = x_6 - x_5 + x_5 - x_2 = (x_1 - x_2)\left(-\frac{1}{16} + \frac{1}{5} - \frac{1}{4} + \frac{1}{2}\right).$$

In general, we get

$$x_n - x_2 = (x_1 - x_2)\left(\frac{1}{2} - \frac{1}{2^2} + \frac{1}{2^3} - \frac{1}{2^4} + \ldots + \frac{(-1)^{n-1}}{2^{n-2}}\right)$$
$$= \frac{1}{3}(x_1 - x_2)\left[1 - \left(-\frac{1}{2}\right)^{n-2}\right]$$

Thus, $x_n - x_2 \to \frac{1}{3}(x_1 - x_2)$ so that $\{x_n\} \to \frac{1}{3}(x_1 + 2x_2)$ whether $n \to \infty$ through even or through odd integral values.

4. *If*

$$a_{n+1} = \sqrt{(k + a_n)}, \qquad \ldots(1)$$

where k, a_1 are positive, prove that the sequence $\{a_n\}$ is increasing or decreasing according as a_1 is less than or greater than the positive root of the equation

$$x^2 = x + k$$

and has, in either case, this root as its limit.

Sol. We have

$$a_{n+1}^2 - a_n^2 = (k + a_n) - (k + a_{n-1}) = a_n - a_{n-1}.$$

Thus, we see that

$$a_n > a_{n-1} \Rightarrow a_{n+1} > a_n \quad \text{and} \quad a_n < a_{n-1} \Rightarrow a_{n+1} < a_n$$

and thus $\{a_n\}$ is a monotonically sequence.

Also it is an increasing or a decreasing sequence according as

$$a_2 > a_1 \quad \text{or} \quad a_2 < a_1$$

The equation

$$x^2 - x - k = 0, \qquad k > 0$$

has one root positive and the other negative. Let α, $-\beta$ be these roots where α and β are positive. Now,

$$x^2 - x - k = (x - \alpha)(x + \beta) \ \forall\, x \qquad \text{...(2)}$$

$$\Rightarrow \quad a_1^2 - a_1 - k = (a_1 - \alpha)(a_2 + \beta). \qquad \text{...(3)}$$

Let $a_1 > \alpha$. Then from (3),

$$a_1^2 - a_1 - k > 0$$

$$\Rightarrow \quad a_2 = \sqrt{(a_1 + k)} < a_1.$$

Thus, $a_1 > \alpha \Rightarrow a_2 < a_1 \Rightarrow \{a_n\}$ is monotonically decreasing.

Now

$$a_n^2 = a_{n-1} + k > a_n + k,$$

$$\Rightarrow \quad a_n^2 - a_n - k > 0$$

From (2) it follows that

$$a_n > \alpha \ \forall\, n.$$

Thus, $\{a_n\}$ is monotonically decreasing sequence which is bounded below.

Hence, lim $\{a_n\}$ exists. Let it be l. We have $l \geq \alpha$.

Taking limits in (1), we get

$$l^2 = l + k$$

implying that l is equal to the positive root, α, of the equation

$$x^2 = x + k.$$

The case when $a_1 < \alpha$ may be similarly treated.

If $a_1 = \alpha$, then, as may be easily seen, that $a_n = \alpha \ \forall\, n$.

5. *If $\{a_n\}$ and $\{b_n\}$ are two bounded sequences, show that*

(*i*) $\overline{\lim}\, a_n + \overline{\lim}\, b_n \geq \overline{\lim}\,(a_n + b_n)$

(*ii*) $\underline{\lim}\, a_n + \underline{\lim}\, b_n \leq \underline{\lim}\,(a_n + b_n)$.

Sol. We write

$$\overline{\lim}\, a_n = G_1,\ \overline{\lim}\, b_n = G_2.\ \overline{\lim}\,(a_n + b_n) = G.$$

Let ε be any positive number.

There exists a positive integer m, such that $\forall\, n \geq m$

$$a_n < G_1 + \frac{1}{2}\varepsilon,\ b_n < G_2 + \frac{1}{2}\varepsilon,$$

$$\Rightarrow \qquad a_n + b_n < G_1 + G_2 + \varepsilon,\ \forall\, n \geq m$$

$$\Rightarrow \qquad G \leq G_1 + G_2 + \varepsilon \Rightarrow G \leq G_1 + G_2$$

as ε is an arbitrary positive number.

Thus, we have proved (*i*). We may similarly prove (*ii*).

6. *Show that the sequence $\{x^n\}$ is convergent if and only if $-1 \leq x < 1$.*

Sol. (*i*) *Let* $x > 1$. We write $x = 1 + h$ so that h is positive.

By mathematical induction, it may be easily shown that

$$x^n = (1 + h)^n > 1 + nh.$$

Let Δ be a positive number. We have

$$1 + nh > \Delta \quad \Leftrightarrow \quad n > (\Delta - 1)\, h.$$

Taking m as any positive integer $> (\Delta - 1)/h$, we see that

$$x^n > \Delta\ \forall\, n \geq m \Rightarrow \lim x^n = \infty.$$

(*ii*) *Let* $x = 1$. Clearly, in this case, $\lim x^n = 1$.

(*iii*) *Let* $0 < x < 1$. We write $x = 1/(1 + h)$ so that h is positive. We have

$$0 < x^n = 1/(1 + h)^n < 1/(1 + nh).$$

Let ε be a positive number. We have

$$1/(1 + nh) < \varepsilon \Leftrightarrow n > (1/\varepsilon - 1)/h.$$

Taking m as an integer $> (1/\varepsilon - 1)/h$, we see that

$$0 < x^n < \varepsilon\, |x^n| < \varepsilon\ \ \forall\ \ n \geq m.$$

$$\Rightarrow \qquad \lim x^n = 0$$

(*iv*) *Let* $x = 0$. Clearly, $\lim x^n = 0$.

(*v*) *Let* $-1 < x < 0$. We write $x = -\infty$ so that $0 < \alpha < 1$.

We have $\quad | x^n | = \alpha^n.$

It now follows from (*iii*) that $\lim x^n = 0$.

(*vi*) *Let* $x = -1$. Obviously x^n oscillates finitely.

(*vii*) *Let* $x < -1$. We write $x = -\alpha$ so that $x > 1$.

As $n \to \infty$, x^n takes values, both positive and negative greater in absolute value than any assigned positive number. Hence x^n oscillates infinitely.

7. *Show that lim* $\sqrt[n]{n} = 1$.

Sol. We write

$$a_n = \sqrt[n]{n}.$$

Let

$$a_n = 1 + h_n \quad \text{where} \quad h_n > 0.$$

We have

$$\begin{aligned} n &= a_n^{\ n} \\ &= (1 + h_n)^n \\ &= 1 + nh_n + \frac{1}{2}n(n-1)h_n^2 + \ldots + h_n^2 \\ &= > \frac{1}{2}n(n-1)h_n^{\ 2}. \end{aligned}$$

$$\Rightarrow \quad h_n^{\ 2} < 2/(n-1),$$

$$\Rightarrow \quad 0 < h_n < \sqrt{[2/(n-1)]}.$$

The result now follows.

To be rigorous, let ε be a positive number. Now

$$\sqrt{[2/(n-1)]} < \varepsilon \Leftarrow n > 1 + 2/\varepsilon^2$$

If m be an integer $> 1 + 2/\varepsilon^2$, we see that $\forall\, n \geq m$.

$$\Delta < h_n < \sqrt{[2/(n-1)]} < \varepsilon.$$

Hence, $h_n \to 0$.

8. $\{a_n\}$ *is a sequence such that*

$$\lim \{a_{n+1}/a_n\} = l$$

where $| l | < 1$; *show that* $\lim a_n = 0$.

Sol. Since $|l| < 1$, we can choose a positive number, ε, so small that

$$|l| + \varepsilon < 1.$$

There exists a positive integer m such that $\forall\, n \geq m$,

$$\left|\frac{a_{n+1}}{a^n} - l\right| < \varepsilon,$$

$$\Rightarrow \quad \left|\frac{a_{n+1}}{a_n}\right| - |l| \leq \left|\frac{a_{n+1}}{a_n} - l\right| < \varepsilon,$$

$$\Rightarrow \quad \left|\frac{a_{n+1}}{a_n}\right| < |l| + \varepsilon = k, \text{ say, where } k < 1$$

Changing n to m, $m + 1$, $m + 2$, ..., $(n - 1)$ and multiplying we get

$$\left|\frac{a_n}{a_m}\right| < k^{n-m} \Rightarrow |a_n| < k^n \cdot (|a_m|/k^m).$$

Since $k^n \to 0$, we have the required result.

Note. The general result obtained here enables us to prove the following particular but important results on limits :

(*i*) $\lim(x^n / n!) = 0, \ \forall\, x,$

(*ii*) $\lim(n^r / x^n) = 0, \ |x| > 1.$

(*iii*) $\lim \dfrac{m(m-1)...(m-n+1)x^n}{n!} = 0, \ |x| < 1.$

9. *If* $\{a_n\}$ *is a sequence such that* $a_n > 0\ \forall\, n$ *and*

$$\lim \{a_{n+1}/a_n\} = l > 1,$$

then, $\lim a_n = \infty$.

Sol. We choose a positive number ε such that $l - \varepsilon > 1$.

There exists a positive integer m such that $\forall\, n \geq m$

$$1 - \varepsilon < a_{n+1} / a_n < 1 + \varepsilon.$$

Thus, $\forall\, n \geq m$

$$a_{n+1}/a_n > 1 - \varepsilon = k, \text{ say, } (k > 1).$$

From this we deduce [As in Ex. 8 above] that

$$a_n > k^n \,.\, (a_m / k^m).$$

Since $k^n \to \infty$, we have the required result.

Note. Compare Ex. 8 and Ex. 9 above.

10. *If the sequences $\{a_n\}$, $\{b_n\}$ tend to 0 and if $\{b_n\}$ is a strictly monotonically decreasing sequence so that $b_n > b_{n+1}$ $\forall$ n, then*

$$lim \frac{a_n}{b_n} = lim \frac{a_n - a_{n+1}}{b_n - b_{n+1}}$$

provided that the limit on the right exists, whether finite or infinite.

Sol. Case I. Let $\lim \frac{a_n - a_{n+1}}{b_n - b_{n+1}} = l$ where l is finite. ...(*i*)

Let ε be a positive number. By virtue of (*i*), there exists a positive integer m such that $\forall n \geq m$,

$$l - \varepsilon < \frac{a_n - a_{n+1}}{b_n - b_{n+1}} < l + \varepsilon,$$

$$\Rightarrow \qquad (l - \varepsilon)(b_n - b_{n+1}) < (a_n - a_{n+1}) < (l + \varepsilon)(b_n - b_{n+1}),$$

for $(b_n - b_{n+1})$ is positive.

Changing n to n, $n + 1$, $n + 2$, ..., $(n + p - 1)$, in turn and adding, we see that $\forall n \geq m$ and $\forall p \geq 0$,

$$(l - \varepsilon)(b_n - b_{n+p}) < (a_n - a_{n+p}) < (l + \varepsilon)(b_n - b_{n+p}).$$

Keep n fixed and let $p \to \infty$. Since $a_{n+p} \to 0$, and $b_{n+p} \to 0$, we obtain

$$(l - \varepsilon)\, b_n \leq a_n \leq (1 + \varepsilon)\, b_n,$$

$$\Rightarrow \qquad (l - \varepsilon) \leq (a_n / b_n) \leq l + \varepsilon \ \forall n \geq m$$

$$\Rightarrow \qquad a_n / b_n \to l.$$

Case II. Let $\lim \frac{a_n - a_{n+1}}{b_n - b_{n+1}} = \infty.$ ...(*i*)

Let Δ be a positive number. By virtue of (*i*), there exists a positive integer m such that $\forall \geq m$

$$\frac{a_n - a_{n+1}}{b_n - b_{n+1}} > \Delta$$

$\Rightarrow$ $(a_n - a_{n+1}) > \Delta (b_n - b_{n+1})$ for $(b_n - b_{n+1})$ is positive.

As in Case I, we obtain

$$a_n - a_{n+p} > \Delta\, (b_n - b_{n+p})$$

$$\Rightarrow \qquad a_n \geq \Delta\, b_n, \ i.e., a_n / b_n \geq \Delta \ \forall n \geq m.$$

$$\Rightarrow \qquad a_n / b_n \to \infty.$$

11. *If $\{b_n\}$ is a strictly monotonically increasing sequence diverging to ∞ and $\{a_n\}$ be an arbitrary sequence, then*

$$lim\frac{a_n}{b_n}=lim\frac{a_{n+1}-a_n}{b_{n+1}-b_n},$$

provided that the limit on the right exists, whether finite or infinite.

Sol. Case I. Let $\lim\dfrac{a_{n+1}-a_n}{b_{n+1}-b_n}=l$, where l is finite. ...(*i*)

Let ε be a positive number. By virtue of (*i*), there exists a positive integer m_1 such that $\forall\, n>m_1$,

$$\left(l-\frac{1}{2}\varepsilon\right)(b_{n+1}-b_n)<(a_{n+1}-a_n)<\left(l+\frac{1}{2}\varepsilon\right)(b_{n+1}-b_n).$$

Changing n to n, $n+1$, $n+2$, ..., $n+p-1$, in turn and adding, we see that

$$\left(l-\frac{1}{2}\varepsilon\right)(b_{n+p}-b_n)<(a_{n+p}-a_n)<\left(l+\frac{1}{2}\varepsilon\right)(b_{n+p}-b_n).$$

Dividing by b_{n+p} and adding a_n/b_{n+p}, we obtain

$$\left(l-\frac{1}{2}\varepsilon\right)\left(1-\frac{b_n}{b_{n+p}}\right)+\frac{a_n}{b_{n+p}}<\frac{a_{n+p}}{b_{n+p}}<\left(l+\frac{1}{2}\varepsilon\right)\left(1-\frac{b_n}{b_{n+p}}\right)+\frac{a_n}{b_{n+p}} \quad ...(ii)$$

$\forall\, n>m_1$, and $\forall\, p\geq 0$.

We keep n fixed and let $p\to\infty$,

Since

$$\left(l-\frac{1}{2}\varepsilon\right)\left(1-\frac{b_n}{b_{n+p}}\right)+\frac{a_n}{b_{n+p}}\to l-\frac{1}{2}\varepsilon,$$

and

$$\left(l+\frac{1}{2}\varepsilon\right)\left(1-\frac{b_n}{b_{n+p}}\right)+\frac{a_n}{b_{n+p}}\to l+\frac{1}{2}\varepsilon,$$

we see that there exists a positive integer m_2, such that $\forall\, p\geq m_2$, we have

$$l-\frac{1}{2}\varepsilon-\frac{1}{2}\varepsilon<\left(l-\frac{1}{2}\varepsilon\right)\left(1-\frac{b_n}{b_{n+p}}\right)+\frac{a_n}{b_{n+p}}<l-\frac{1}{2}\varepsilon+\frac{1}{2}\varepsilon. \quad ...(iii)$$

and
$$l+\frac{1}{2}\varepsilon-\frac{1}{2}\varepsilon<\left(l+\frac{1}{2}\varepsilon\right)\left(1-\frac{b_n}{b_{n+p}}\right)+\frac{a_n}{b_{n+p}}<l+\frac{1}{2}\varepsilon+\frac{1}{2}\varepsilon. \quad ...(iv)$$

From (*ii*), (*iii*) and (*iv*), we obtain

$$l-\varepsilon<\frac{a_{n+p}}{b_{n+p}}<l+\varepsilon,$$

$\forall\, n > m_1$ and $\forall\, p \geq m_2$

$$\Rightarrow \qquad l - \varepsilon < \frac{a_n}{b_n} < l + \varepsilon \;\; \forall\, n \geq (m_1 + m_2).$$

$$\Rightarrow \qquad a_n / b_n \to l.$$

Case II. Let $\lim \dfrac{a_{n+1} - a_n}{b_{n+1} - b_n} = \infty.$ ...(*i*)

Let k be a positive number.

By virtue of (*i*), there exists a positive integer m_1 such that $\forall\, n \geq m_1$,

$$(a_{n+1} - a_n) > (k + 1)\,(b_{n+1} - b_n).$$

As in case I, we obtain

$$a_{n+p} - a_n > (k + 1)\,(b_{n+p} - b_n),$$

$$\Rightarrow \qquad \frac{a_{n+p}}{b_{n+p}} > (k+1)\left(1 - \frac{b_n}{b_{n+p}}\right) + \frac{a_n}{b_{n+p}} \qquad ...(vi)$$

Keeping n fixed and letting $p \to \infty$, we see that

$$(k+1)\left(1 - \frac{b_n}{b_{n+p}}\right) + \frac{a_n}{b_{n+p}} \to k + 1$$

so that there exists a positive integer m_2 such that $\forall\, p \geq m_2$,

$$(k+1) - 1 < (k+1)\left(1 - \frac{b_n}{b_{n+p}}\right) + \frac{a_n}{a_{n+p}} < (k+1) + 1 \qquad ...(vii)$$

From (*ii*) and (*iii*), we obtain

$$\frac{a_{n+p}}{b_{n+p}} > k \;\; \forall\, n \geq m_1 \text{ and } \forall\, p \geq, m_2,$$

$$\Rightarrow \qquad \frac{a_n}{b_n} > k \;\; \forall\, n \geq (m_1 + m_2).$$

$$\Rightarrow \qquad a_n / b_n \to \infty.$$

Note. Compare Ex. 10 and Ex. 11 above.

12. *If* $\lim a_n = l$, *then*

$$\lim \frac{a_1 + a_2 + a_3 + \ldots + a_n}{n} = l.$$

[*Kurukshetra 2000, M.D.U. 98*]

Sol. We write

$$a_n - l, = b_n,$$

so that the sequence $\{b_n\} \to 0$.

We have

$$\frac{a_1 + a_2 + \ldots + a_n}{n} = l + \frac{b_1 + b_2 + \ldots + b_n}{n}$$

so that we have to prove that $(b_1 + b_2 + \ldots + b_n)/n \to 0$ when $b_n \to 0$.

Let ε be a positive number so that there exists a positive integer p such that

$$|b_n| < \frac{1}{2}\varepsilon \ \forall\, n \geq p.$$

Also since $\{b_n\}$ is convergent, it is bounded and therefore there exists a positive number k such that

$$|b_n| < k, \ \forall\, n.$$

We write

$$\left|\frac{b_1 + b_2 + \ldots + b_n}{n}\right| = \left|\frac{b_1 + b_2 + \ldots + b_p}{n} + \frac{b_{p+1} + b_{p+2} + \ldots + b_n}{n}\right|$$

$$\leq \frac{|b_1| + |b_2| + \ldots + |b_p|}{n} + \frac{|b_{p+1}| + \ldots + |b_n|}{n}$$

$$\leq \frac{kp}{n} + \frac{\varepsilon\,(n-p)}{2n} < \frac{kp}{n} + \frac{\varepsilon}{2}.$$

Keep p fixed. We have

$$\frac{kp}{n} < \frac{\varepsilon}{2} \ \forall\, n > \frac{2kp}{\varepsilon}.$$

Let q be a positive integer greater than $2kp/\varepsilon$ so that for $n \geq q$ we have $kp/n < \varepsilon/2$.

Let $m = \max.\,(p, q)$.

Thus, $\forall\, n \geq m$, we have

$$\left|\frac{b_1 + b_2 + \ldots + b_n}{n}\right| < \varepsilon.$$

Hence the result.

Note. This result could be deduced from Ex. 11, Page 81, by putting $(a_1 + a_2 + \ldots + a_n)$ for a_n and n for b_n.

Note. The converse of the result obtained above is not true as may be seen on considering $a_n = (-1)^n$.

EXERCISES

1. From Cauchy's general principle of convergence, deduce that the sum, difference and product of two convergent sequences are convergent.
2. Show, with the help of Cauchy's general principle of convergence, that the sequence f where

$$f(n) = 1 + \frac{1}{2} + \frac{1}{3} + ... + \frac{1}{n}$$

is not convergent.
3. Show that the sequence f where

$$f(n) = \frac{1}{1!} + \frac{1}{2!} + ... + \frac{1}{n!}$$

is convergent.
4. Show that the sequence f where

$$f(n) = \frac{1}{n+1} + \frac{1}{n+2} + ... + \frac{1}{n+n}$$

(Agra 92; Kanpur 92, 94)

is convergent.
5. Show that when $n \to \infty$

(*i*) $$\lim \left[\frac{1}{\sqrt{(n^2+1)}} + \frac{1}{\sqrt{(n^2+2)}} + ... + \frac{1}{\sqrt{(n^2+n)}} \right] = 1$$

(*ii*) $$\lim \left[\frac{1}{n^2} + \frac{1}{(n+1)^2} + ... + \frac{1}{(2n)^2} \right] = 0.$$

(*iii*) $$\lim \left[\frac{1}{\sqrt{n}} + \frac{1}{\sqrt{(n+1)}} + ... + \frac{1}{\sqrt{(2n)}} \right] = \infty.$$

6. Prove that the sequence f defined by

$$f(1) = \sqrt{7}, \ f(n+1) = \sqrt{[7 + f(n)]}$$

converges to the positive square root of

$$x^2 - x - 7 = 0$$

7. If f is defined by

$$f(n) = a / [1 + f(n-1)],$$

where a, $f(1)$ are positive, show that the sequence f is convergent and converges to the positive root of the equation $x^2 + x = a$.
8. Given that f is a sequence such that

$$f(2) \le f(4) \le f(6) \le ... \le f(5) \le f(3) \le f(1)$$

and g defined by

$$g(n) = f(2n-1) - f(2n)$$

converges to 0, show that f is convergent.

9. A sequence $\{a_n\}$ of positive terms is defined by the conditions

$$a_1 = k > 0,\ a_{n+1}(2+a_n) = (3+2a_n),\ n \neq 1,$$

Show that the sequence converges to a limit independent of k and find the limit.

10. If $\{a_n\}$ is defined by

$$a_n \geq 0,\ 2a_{n+1}^2 = a_n^2 + a_n$$

show that $\{a_n\}$ tends to 0, 1 or infinity as n tends to infinity according to the value of a_1.

11. If $\{v_n\}$ is a sequence of positive numbers such that

$$v^2_{n+1} = 2v_n/(v_{n+1})$$

show that $v_n \to 1$, as $n \to \infty$.

12. A sequence $\{S_n\}$ is defined as follows :

$$S_1 = a > 0,\ S_{n+1} = \sqrt{\left(\frac{ab^2 + S_n^2}{a+1}\right)},\ b > a.$$

Show that $\{S_n\}$ is an increasing bounded sequence and $\lim S_n = b$.

13. If k is positive and α, $-\beta$ are the positive and negative roots of

$$x^2 - x - k = 0.$$

prove that if

$$v_n = \sqrt{(k - v_{n+1})} \quad \text{and} \quad v_1 < k,$$

then

$$v_n \to \beta,$$

14. If S_1, S_2 are positive and

$$S_{n+2} = \sqrt{(S_{n+1} S_n)},$$

prove that of the sequences with values

$$S_1,\ S_2,\ S_5,\ \ldots;\ S_2,\ S_4,\ S_6,\ \ldots$$

one is an increasing and the other a decreasing sequence and show that the common limit is $\sqrt[3]{(S_1 S_2^2)}$,

15. If $0 < S_1 < S_2$ and

$$S_n = \frac{2S_{n-1} S_{n-2}}{S_{n-1} + S_{n-2}},$$

so that S_n is the harmonic mean between S_{n-1} and S_{n-2}, show that

$$S_n \to 3S_1 S_2 / (2S_1 + S_2).$$

16. $a_{n+1} = \frac{1}{2}(a_n + b_n)$ and $b_{n+1} = \sqrt{a_n b_n}$,
show that the sequences $\{a_n\}$, $\{b_n\}$ converges to a common limit where

$$a > b > 0 \text{ and } a_1 = \frac{1}{2}(a+b) \text{ and } b_1 = \sqrt{ab}.$$

17. If x_1, y_1 are positive and if $\forall\, n \geq 1$,

$$x_{n+1} = \frac{x_n + y_n}{2}, \quad \frac{2}{y_{n+1}} = \frac{1}{x_n} + \frac{1}{y_n},$$

show that $\{x_n\}$ and $\{y_n\}$ are monotonic sequences which approach a common limit l where $l^2 = x_1 y_1$.

18. If $k \neq 0$ and $\{a_n\}$ is a sequence such that

$$(a_{n+1} - a_n) \to k,$$

then $a_n \to \infty$ or $-\infty$, according as k is positive or negative.

19. If λ, μ are two given numbers and $\{a_n\}$ is a given sequence such that $|\lambda| < 1$ and $(a_{n+1} + \lambda a_n + \mu) \to 0$,

then show that $a_n \to -\mu / (1+\lambda)$.

20. $\{a_n\}$ and $\{b_n\}$ are two sequences such that

(*i*) $b_n > 0$,

(*ii*) $S_n = (b_1 + b_2 + \dots + b_n)$ is divergent,

(*iii*) $a_n / b_n \to s$,

show that

$$\lim \frac{a_1 + a_2 + \dots + a_n}{b_1 + b_2 + \dots + b_n} = s.$$

21. Show that

$$\overline{\lim}\,(-a_n) = -\underline{\lim}\,(a_n).$$

when $\{a_n\}$ is a bounded sequence.

22. If $\{a_n\}$, $\{b_n\}$ are two bounded sequences of positive terms, show that

(*i*) $(\underline{\lim}\, a_n)(\underline{\lim}\, b_n) \leq (\underline{\lim}\,(a_n b_n)$, (*ii*) $(\overline{\lim}\, a_n)(\overline{\lim}\, b_n) \geq \overline{\lim}\,(a_n b_n)$.

OBJECTIVE QUESTIONS

Note. *For each of the following questions, four alternatives are given for the answer. Only one of them is correct. Choose the correct alternative.*

1. If in a sequence there exists a real number u such that $x_n \geq u \; \forall \; n \in N$, then the sequence is :

(*a*) bounded below (*b*) bounded above

(*c*) unbounded (*d*) none of these

2. If in a sequence $< x_n >$, $\lim_{n \to 0} x_n$ is finite, sequence is :

(*a*) convergent (*b*) divergent

(*c*) oscillatery (*d*) cauchy's

3. Every convergent sequence is :

(*a*) unbounded (*b*) bounded

(*c*) oscillatory (*d*) oscillates finitely

4. In a sequence $< x_n >$, $x_n + 1 \geq x_n \; \forall \; n \in W$, then sequence is

(*a*) monotonically increasing

(*b*) monotonically decreasing

(*c*) strictly monotonic increasing

(*d*) none of these

5. Every Cauchy sequence must be :

(*a*) monotonic (*b*) bounded above only

(*c*) bounded below only (*d*) bounded.

6. Sequence $< x_n >$, where $x_n = \left\{\sqrt{(n+1)} - \sqrt{n}\right\} \forall \; n \in N$, is :

(*a*) convergent (*b*) divergent

(*c*) oscillatory (*d*) monotonic

ANSWERS

1. (*b*) **2.** (*a*) **3.** (*b*) **4.** (*a*) **5.** (*d*)
6. (*a*)

4

Real Valued Functions of a Single Real Variable. Limit and Continuity

4.1. REAL VALUED FUNCTIONS OF A SINGLE REAL VARIABLE

In this chapter, we shall be concerned with functions whose domains and ranges are sub-sets of real numbers and with related important notions of *limit* and *continuity*. Such functions are referred to as *Real valued functions of a single real variable*. While these notions will be introduced in relation to functions with arbitrary domains, we shall obtain in this chapter a number of important properties of continuous functions with closed and bounded intervals, as their domains.

Illustrations

1. If $f(x) = 0$ when x is rational and $f(x) = 1$ when x is irrational, then f is a real valued function with the entire set **R** of real numbers as its domain. Its range is the two elementic set $\{0, 1\}$.

2. If $f(x) = [x]$ where $[x]$ denotes the greatest integer not greater than x, then f is a function with domain **R** and range as the set **I** of integers.

3. If $f(x) = a_0x^n + a_1x^{n-1} + ... + a_n$, where $a_0, ..., a_n$ are given real numbers, f is a function with domain **R**. What can be said about the range of this function ?

4. If $f(x) = (a_0x^n + a_1x^{n-1} + ... + a_n) / (b_0x^m + b_1x^{m-1} + ... + b_m)$, where $a_0, ... a_n, b_0, ..., b_m$ are given real numbers, then f is a function with domain **R** excluding the numbers wherefor

$$b_0x^m + b_1x^{m-1} + ... + b_m$$

assumes the value zero.

5. If $f(x) = x^{1/2}$, f is a function with domain $\{x : x \geq 0\}$. What is the range of f ?

***Ex.** Indicate the domains and ranges of the function f defined as follows :

(*i*) $f(x) = \sin x$. (*ii*) $f(x) = \tan x$. (*iii*) $f(x) = \sec x$. (*iv*) $f(x) = \cos^{-1} x$.

(*v*) $f(x) = \cot^{-1} x$. (*vi*) $f(x) = \text{cosec}^{-1} x$. (*vii*) $f(x) = e^x$. (*viii*) $f(x) = \log x$.

4.1.1. Algebra of Functions

It is possible to define, in a natural way, algebraic composition in the set of real valued functions defined in a given domain S.

We write $f = \text{£}$ if

$$f(x) = \text{£}(x) \ \forall \ (x)\ x \ \hat{\text{I}} \ S.$$

Let f, g be any two such functions. We give meanings to

$$f + g, \ f - g, \ fg, \ f \ . \ g$$

as below :

$$(f + g)(x) = f(x) + g(x)$$
$$(f - g)(x) = f(x) - g(x)$$
$$(fg)(x) = f(x)\, g(x)$$
$$(f \ . \ g)(x) = f(x) \ . \ g(x)$$

While the domain of each of the functions

$$f + g, \ f - g, \ fg$$

is S, *the domain of* f . g *is the set* S *excluding the set of points,* x *for which* $g(x) = 0$.

Constant functions. A function with one elementic range is called a *constant function.*

For example, let c denote a real number. Then the function f such that

$$f(x) = c \ \forall \ x \ \hat{\text{I}} \ S,$$

is a constant function.

When there is no cause for ambiguity, the constant function with range $\{c\}$ will itself be denoted as c.

Clearly to each real number there corresponds a constant function with domain S.

Zero function. Of the different constant functions, there is one which is specially important. This is the function whose value is the real number zero for every point of the domain of the function. This *zero function* will also be denoted by 0. It will be seen that this double use of the symbol will not cause any confusion inasmuch as the context will always help us to decide as to the particular meaning to be attached to the symbol.

Ex. Give example of functions f and g such that while $f \neq 0$, $g \neq 0$, but $fg = 0$.

4.1.2. Order Relation in the Set of Functions

Let f and g be two functions with domain $S \subset \mathbf{R}$. We shall write

$$f < g$$

to mean that $f(x) < g(x) \ \forall \ x \in S$.

It will be seen that this relation of order does not satisfy the Trichotomy law in that it is possible that if f and g be two different functions we may have

$$\text{neither } f < g \text{ nor } g < f.$$

For example, for the domain $S = [0, 1]$, of the two functions f and g defined as

$$f(x) = 3x, \ g(x) = \frac{3}{2}(1 - x),$$

$$\text{neither } f < g \text{ nor } g < f$$

This relationship of order is, however, transitive, *i.e.*,

$$f < g, \ g < h \Rightarrow f < h.$$

Also $$f < g \Rightarrow f + h < g + h$$

and $$f < g, \ h > 0 \Rightarrow fh < gh.$$

The proofs which are simple are left to reader.

4.1.3 Composites of Functions

Let S be the domain and T the range of a function f and let ϕ be any other function whose domain is the range T of f. Thus the range of f is the domain of ϕ.

Let x S so that

$$\phi(f(x))$$

has a meaning inasmuch as $f(x)$, whose ϕ-value is in question, belongs to the domain of ϕ.

Thus we have defined here a function whose domain is the domain of f and range is contained in the range of ϕ on the condition that the range of f coincides with the domain of ϕ.

This function is known as the *composite* of ϕ and f and is also often written as $\phi \ o \ f$. Thus we have

$$(\phi \ o \ f)(x) = \phi(f(x)), \ x \in S.$$

***Ex.** Exhibit the functions defined below as composites of functions :

(*i*) $\sin^2 x$ (*ii*) $\log \cos x$ (*iii*) $\log \sin x^2$.

4.1.4. Inverse of a Function

Let S be the domain and T the range of a function f so that to each $x \in S$ there corresponds $y \in T$ such that

$$y = f(x).$$

Suppose that y is a member of T. Now T being the range of the function f with domain S, there exists at *least one* $x \in S$ such that $y = f(x)$. There may, however, exist more than one such x since several different members of S can have the same f-value. If, however, we suppose that f is such that no two different members of S have the same f-value, we see that *associated with the function f with domain S and range T is a function say,* ϕ, *with domain T and range S such that*

$$y \in T \Rightarrow \phi(y) = x$$

and

$$\phi(y) = x \Rightarrow f(x) = y.$$

The fact that x exists and is unique is a consequence of the condition imposed on f. The function ϕ thus associated with f is called the **Inverse** of the function f.

It should be understood that not every function admits of an inverse. In §, 4.4, P. 120 we shall consider an important class of functions which admit of inverses.

Functions which admit of inverse functions will be referred to as **Inversible functions.**

4.1.5. Composite of a Function with its Inverse

Let f be a function with domain S and range T admitting of the inverse function ϕ with domain T and range S. Then the composite

$$\phi \, o \, f$$

is defined and is such that

$$(\phi \, o \, f)(x) = \phi(f(x)) = \phi(y) = x, \ \forall \ x \in S.$$

Thus $\phi \, o \, f$ is a function with domain S and range also S such that it is the **Identity** function *i.e.*, the $\phi \, o \, f$ image of every member of its domain coincides with itself.

Also it may be seen that the composite

$$f \, o \, \phi$$

is defined with domain T and is the Identity function on T.

Ex. 1. The function f defined by $f(x) = 3x + 4$ in **R** is inversible. Describe the inverse.

Ex. 2. The function f defined by $f(x) = x^2$ in **R** does not admit of an inverse.

Ex. 3. The function f with domain $[0, \infty[$ and defined by $f(x) = x^2$ has an inverse which may be described.

***Ex. 4.** The function given by $f(x) = \sin x$ in **R** does not possess any inverse.

***Ex. 5.** The function given by $f(x) = \sin x$ in $[-\pi/2, \pi/2]$ has an inverse.

***Ex. 6.** The function defined by $f(x) = \log x$ in $]0, \infty[$ admits of an inverse. Describe the inverse.

Bounded and unbounded functions. A function with a bounded range is said to be bounded. Also the bounds of the range of a bounded function are referred to as the bounds of the function.

Ex. **1.** *Determine whether the function f defined as follows is bounded or unbounded*

$$f(x) = \begin{cases} 0, & \text{if } x \text{ is irrational} \\ 1, & \text{if } x \text{ is rational.} \end{cases}$$

We know that a function f is said to be bounded or unbounded according as its range is bounded or unbounded. Here range of the given function $f = [0, 1]$ which is bounded as

$$0 \le f(x) \le 1.$$

Hence the given function is bounded.

Ex. 2. Show that the function f defined in $[0, 1]$ as below is not bounded :

$$f(x) = 1/x \text{ when } x \ne 0,\ f(0) = 0.$$

Ex. 3. Show that the function f in $[0, \infty[$ defined by

$$f(x) = x/(x+1)$$

is bounded. Find its bounds and show that while it attains its infimum, the supremum is not attained.

4.2. LIMITS

Let c be a limit point of the domain of definition S of a function f.

We say that the function f tends to the limit l as x tends to c or symbolically,

$$\lim_{x \to c} f(x) = l.$$

if, for every $\varepsilon > 0$, there exists $\delta > 0$ such that

$$|f(x) - l| < \varepsilon,$$

when x is a point of S satisfying the condition

$$0 < | x - c | < \delta.$$

This means that

$$f(x) \in]\, l - \varepsilon, l + \varepsilon\, [$$

when $\quad x \in S \cap]\, c - \delta, c + \delta\, [, x \neq c$

i.e. when $x \in]\, c - \delta, c\, [\cup]\, c, c + \delta\, [$. (*Agra 92, 94; Kanpur 92, 94, 96*)

Note. It may be seen that the inequality

$$0 < | x - c |$$

only excludes the possibility

$$0 = | x - c | \Leftrightarrow x = c.$$

In fact $\quad 0 < | x - c | \Leftrightarrow x \neq c.$

It should be noted that the existence of

$$\lim_{x \to c} f(x)$$

and the value of the limit in the event of its existence are quite independent of the value $f(c)$ of the function at c. In fact the function may not even be defined at c *i.e.*, c may not even belong to the domain of f.

Right-handed and Left-handed limits

$$\lim_{x \to (c+0)} f(x), \qquad \lim_{x \to (c-0)} f(x),$$

We say that the function f tends to l as x tends to c through values *greater* than c if for every $\varepsilon > 0$, there exists $\delta > 0$ such that

$$| f(x) - l | < \varepsilon$$

when x is any point of S satisfying the condition

$$c < x < c + \delta.$$

Also we then write

$$\lim_{x \to (c+0)} f(x) = l$$

The $\lim_{x \to (c+0)} f(x)$, if it exists, is also denoted by the symbol $f(c + 0)$.

Again we say that the function f tends to l as x tends to c through values smaller than c if to each $\varepsilon > 0$ there corresponds $\delta > 0$ such that

$$x \in S \cap]\, c - \delta, c\, [\Rightarrow f(x) \in]\, l - \varepsilon, l + \varepsilon\, [.$$

Also we then write

$$\lim_{x \to (c+0)} f(x) = l$$

The $\lim_{x \to (c-0)} f(x)$, if it exists, is also denoted by the symbol $f(c-0)$.

It may be easily seen that

$$\lim_{x \to c} f(x) = l \Rightarrow \lim_{x \to (c+0)} f(x) = l = \lim_{x \to (c-0)} f(x).$$

4.2.1. Difference between Value and Limit

If we put $x = a$ in $f(x)$, then we get the value of $f(x)$, *e.g.*, if $f(x) = \sec x$, $f(a) = \sec a$, if $f(x) = x - 7$, $f(7) = 0$.

If we are to find the limit, we shall put $x = a \pm h$ and will see to what number $f(a \pm h)$ tends as h approaches zero. The number A or the improper limit $+\infty$ or $-\infty$ to which $f(a \pm h)$ tends is known as limit of $f(x)$ at $x = 0$. Sometimes the limit and value turn out to be the same but it is not necessarily so.

4.2.2. Uniqueness of Limit

If a function $f : a \to R$, $A \subset R$, has a limit at a point a which belongs to A, then the limit is unique. (*Poorv. 92*)

Let l_1 and l_2 be two distinct limits of the function $f(x)$ at $x = a$. As $l_1 \neq l_2$, so $|l_1 - l_2| > 0$.

If we take $\varepsilon = \frac{1}{2}|l_1 - l_2| > 0$, then

$\lim_{x \to a} f(x) = l_2 \Rightarrow$ given $\varepsilon > 0$, there exists $\delta_1 > 0$ such that $|f(x) - l_1| < \varepsilon$, where $0 < |x - a| < \delta_1$.

And, $\lim_{x \to a} f(x) = l_2 \Rightarrow$ give $\varepsilon > 0$, there exists $\delta_2 > 0$ such that $|f(x) - l_2| < \varepsilon$,

when $$0 < |x - a| < \delta_2.$$

Let $$\delta = \min.\{\delta_1, \delta_2\}, \text{ then}$$

$$|f(x) - l_1| < \varepsilon \quad \text{and} \quad |f(x) - l_2| < \varepsilon,$$

when $$0 < |x - a| < \delta.$$

Now, $$|l_1 - l_2| = |l_1 + f(x) - f(x) - l_2|$$

$$\Rightarrow |l_1 - l_2| \leq |l_1 - f(x)| + |f(x) - l_2|$$

$$\Rightarrow |l_1 - l_2| \leq |f(x) - l_1| + |f(x) - l_2|$$

$$\Rightarrow |l_1 - l_2| \leq \varepsilon + \varepsilon, \quad \text{when } 0 < |x - a| < \delta$$

$$\Rightarrow |l_1 - l_2| \leq |l_1 - l_2| \text{ when } 0 < |x - a| < \delta.$$

This being absurd, hence our supposition is wrong and so $l_1 = l_2$, *i.e.*, $\lim\limits_{x \to 0} f(x)$ is unique if it exists. ...(*i*)

4.2.3. Cauchy's Existence Theorem for Functions

The necessary and sufficient condition that a function f (x) should tend to a definite limit as x tends to a is that for a given $\varepsilon > 0$, there exists a number $\delta > 0$, such that

$$|f(x_2) - f(x_1)| < \varepsilon$$

for every pair (x_1, x_2) in the neighbourhood $(a - \delta, a + \delta)$.

Necessary condition. Let the function $f(x)$ tends to a definite limit l when $x \to a$.

Then by the definition of the limit of a function, for a given $\varepsilon > 0$, there exists a number $\delta > 0$ such that

$$|x - a| \delta \Rightarrow |f(x) - l| < \frac{1}{2}\varepsilon$$

$\therefore$ x_1, x_2 be any two points in the neighbourhood $[a - \delta, a + \delta]$, then

$$|f(x_1) - l| < \frac{1}{2}\varepsilon \text{ and } |f(x_2) - l| < \frac{1}{2}\varepsilon \qquad ...(i)$$

$$\begin{aligned}\therefore \quad |f(x_2) - f(x_1)| &= |f(x_2) - l + l - f(x_1)| \\ &\le |f(x_2) - l| + |l - f(x_1)| \\ &\le |f(x_2) - l| + |f(x_1) - l| \\ &< \frac{1}{2}\varepsilon + \frac{1}{2}\varepsilon \qquad \text{from } (i) \\ &\Rightarrow |f(x_2) - f(x_1)| < \varepsilon\end{aligned}$$

Sufficient condition. Let for any given $\varepsilon > 0$, there exists a positive number δ such that

$$\begin{aligned}0 < |x_1 - a| < \delta \text{ and } 0 < |x_2 - a| < \delta \\ \Rightarrow |f(x_2) - f(x_1)| < \varepsilon \\ \Rightarrow f(x_2) - \varepsilon < f(x_1) < f(x_2) + \varepsilon\end{aligned}$$

Hence the set of values of the function $f(x)$ corresponding to the values of x in the neighbourhood $\mathbf{N}'(a, \delta)$ is bounded infinite set and so it has a maximum limit k and a minimum limit K, so that

$$f(x_2) - \varepsilon \le k \le K \le f(x_2) + \varepsilon$$

and
$$0 \le K - k \le 2\varepsilon,$$

But as $\delta \to 0$, $\mathbf{K} \to \mathbf{K}_0$ and $k \to k_0$,

where $K_0 \le K$ and $k_0 \ge k$ so that

$$0 \le K_0 - k_0 \le 2\varepsilon$$

i.e., $K_0 = k_0$, as ε is arbitrary.

Hence the function $f(x)$ will have an unique limit as $x \to a$.

EXAMPLES

1. *Does the* $\lim\limits_{x \to 0} \sin\left(\frac{1}{x}\right)$ *exist ? If yes, find it.*

(*Garhwal 90; Rohilkhand 92*)

Sol. Let
$$f(x) = \sin\left(\frac{1}{x}\right),$$

then
$$f(0+h) = \sin\left(\frac{1}{h}\right),$$

and
$$f(0-h) = \sin\left(\frac{1}{-h}\right) = -\sin\left(\frac{1}{h}\right).$$

$$\lim_{x \to 0} f(x) = \lim_{h \to 0} f(0+h)$$

$$= \lim_{h \to 0} \sin\left(\frac{1}{h}\right)$$

Now, as the sine of a very large angle oscillates between -1 and $+1$, so $\lim\limits_{h \to 0} \sin\left(\frac{1}{h}\right)$ is finite and oscillates between -1 and $+1$ but it does not tend to any unique and definite value. So the above limit (left hand limit) does not exist.

Similarly, we can show that the right hand limit also does not exist.

Hence $\lim\limits_{x \to 0} \sin\left(\frac{1}{x}\right)$ does not exist.

2. *Has the function* $e^{-1/(x-x_0)} \cos\left\{\frac{1}{(x-x_0)}\right\}$ *a limit when* $x \to x_0$ *? If so, find it.*

Sol. Let
$$f(x) = e^{-1/(x-x_0)} \cos\left\{\frac{1}{(x-x_0)}\right\}$$

$$\therefore \quad \lim_{x \to x_0 - 0} f(x) = \lim_{h \to 0} f(x_0 - h)$$

$$= \lim_{h \to 0} e^{-1/(x_0 - h - x_0)} \cos\left\{\frac{1}{x_0 - h - x_0}\right\}$$

$$= \lim_{h \to 0} e^{1/h} \cos(1/h)$$

$= \lim\limits_{z \to \infty} e^{z} \cos z$, putting $z = \dfrac{1}{h}$ and so $z \to \infty$ as $h \to 0$.

= ¥ × (something lying between – 1 and + 1), as the value of cos z lies between – 1 and + 1.

= – ¥ or ¥, according as cos z is – ve or + ve.

i.e., left hand limit = – ¥ or ¥

Again, $\lim\limits_{x \to x_0 + 0} f(x) = \lim\limits_{h \to 0} f(x+h)$

$= \lim\limits_{h \to 0} e^{-1/h} \cos(1/h)$

$= \lim\limits_{z \to \infty} e^{-z} \cos z$

= 0 ´ (something lying between – 1 and +1)

i.e., Right hand limit = 0.

Obviously, left and right hand limits exist but are not equal. Hence limit does not exist.

3. Let $f(x) = \begin{cases} x, & \text{if } x \text{ is rational} \\ -x, & \text{if } x \text{ is irrational} \end{cases}$

show that *$\lim\limits_{x \to 0} f(x)$ exists only when $a = 0$.*

Sol. Case I. *If a is a non-zero rational number*

$f(a - 0) = \lim\limits_{h \to 0} (a - h)$ or $\lim\limits_{h \to 0} -(a - h)$

according as $(a - h)$ is rational or irrational.

$= a$ or $-a$, *i.e.*, not unique.

$\therefore f(a - 0)$ does not exist Þ $\lim\limits_{x \to 0} f(x)$ does not exist.

Case II. *If $a = 0$.*

$f(0 - 0) = \lim\limits_{h \to 0} f(-h) = \lim h \to 0 (-h)$

$= \lim\limits_{h \to 0} (-h)$ or $\lim\limits_{h \to 0} h$, according as $-h$ is rational or irrational

= 0 in both the cases.

Again, $f(0+0) = \lim_{h \to 0} (0+h) = \lim_{h \to 0} h$, or $\lim_{h \to 0} (-h)$

according as h is rational or irrational.

$= 0$

Hence $\lim_{x \to 0} f(x)$ exists and is equal to zero.

Case III. *If a is an irrational number.*

In this case $f(a-0) = \lim_{h \to 0} f(a-h)$

$= \lim_{h \to 0} (a-h)$ or $\lim_{h \to 0} -(a-h)$

according as $(a-h)$ is rational or irrational.

$= a$ or $-a$, *i.e.*, is not unique.

$\therefore \lim_{x \to 0} f(x)$ does not exist.

Thus, $\lim_{x \to 0} f(x)$ exists only when $a = 0$.

4. *If*

$$f(x) = \frac{\sin [x]}{[x]}, \quad [x] \neq 0$$
$$= 0, \quad [x] = 0,$$

where $[x]$ *denotes the greatest integer less than or equal to x, then find* $\lim_{x \to 0} f(x)$.

(Agra 93; Rohilkhand 91)

Sol. Here, $f(0+0) = \lim_{h \to 0} f(0+h) = \lim_{h \to 0} f(h)$

$= \lim_{h \to 0} 0 \quad [\because [h] = 0]$

$= 0$

Also $f(0-0) = \lim_{h \to 0} f(0-h) = \lim_{h \to 0} f(-h)$

$$= \lim_{h \to 0} \frac{\sin [-h]}{[-h]} = \lim_{h \to 0} \frac{\sin(-1)}{(-1)} \quad [\because [-h] = -1 \neq 0]$$

$$= \frac{\sin(-1)}{(-1)} = \sin 1 \neq 0.$$

Since $f(0+0) \neq f(0-0)$, hence $\lim_{x \to 0} f(x)$ does not exist.

5. *If $\lim_{x \to 0} f_1(x)$ and $\lim_{x \to 0} [f_1(x).f_2(x)]$ both exist, then find whether $\lim_{x \to 0} f_2(x)$ exist or not.*

Sol. Let us consider, for this example, the following functions :

$$f(x) = x \text{ and } g(x) = \frac{|x|}{x}, \; x \neq 0$$

and also let us take $g(x) = 1$, if $x > 0$ and $g(x) = -1$ if $x < 0$.

Then $$f(x) \cdot g(x) = x \cdot \frac{|x|}{x} = |x|$$

$$\therefore \quad \lim_{x \to 0} f(x) = 0 \text{ and } \lim_{x \to 0} [f(x).g(x)] = 0$$

and both exist.

But $$\lim_{x \to 0+0} g(x) = 1 \quad \because g(x) = 1, \text{ if } x > 0$$

and $$\lim_{x \to 0-0} g(x) = -1, \quad \because g(x) = -1, \text{ if } x < 0$$

Thus $\lim_{x \to 0} g(x)$ does not exist.

Hence if $\lim_{x \to 0} f(x)$ and $\lim_{x \to 0} [f(x).g(x)]$ both exist it does not necessarily mean that $\lim_{x \to 0} g(x)$ also exists.

EXERCISES

1. Show that

(*i*) $\lim_{x \to 0} x \sin(1/x) = 0.$

(*ii*) $\lim_{x \to 0} \left(\frac{1}{x}\right) \sin x = 1$

2. If $\lim_{x \to 0} f_1(x)$ exists and $\lim_{x \to 0} f_2(x)$ does not exist, then show that

$$\lim_{x \to 0} [f_1(x) + f_2(x)]$$

does not exist.

3. If $f(x) = [x]$ where $[x]$ denotes the greatest integer not greater than x, show that

$$\lim_{x \to (2+0)} f(x) = 2, \quad \lim_{x \to (2-0)} f(x) = 1 \text{ and } \lim_{x \to 2} f(x) \text{ does not exist.}$$

4. If $\lim_{x \to c} f(x) = l$

and $\{x_n\}$ is a sequence with limit c, show that

$$\lim_{n\to\infty} \{f(x_n)\} = l.$$

Conversely, show that

$$\lim_{x\to c} f(c) = l,$$

if for every sequence $\{x_n\}$ with limit c,

$$\lim_{x\to\infty} \{f(x_n)\} = l.$$

5. Show that

(*i*) $$\lim_{x\to 1}\left(x^2 + 3x\right) = 4.$$

(*ii*) $$\lim_{x\to 0}\left[\frac{\left(2x^2+3\right)}{(x+1)}\right] = 3.$$ (*iii*) $$\lim_{x\to 1}\left[\frac{\left(x^2-1\right)}{(x-1)}\right] = 2.$$

4.2.4. Algebra of Limits

Let f and g be two functions with domain S and let c be a limit point of S. Let

$$\lim_{x\to c} f(x) = l, \quad \lim_{x\to c} g(x) = m.$$

Then

(*i*) $\lim (f \pm g)(x) = \lim f(x) \pm \lim g(x) = l \pm m$,

(*ii*) $\lim (fg)(x) = \lim f(x) \lim g(x) = lm$,

(*iii*) $\lim (f \div g)(x) = \lim f(x) \div \lim g(x) = l \div m$, if $m \neq 0$.

[The proofs are similar to those of the corresponding results on sequences, §3.51 P. 64].

(*i*) The proof is left to the reader.

(*ii*) Let $\varepsilon > 0$ be given.

We have, $\forall\ x \in S$

$$|f(x)g(x) - lm| = |g(x)(f(x) - l) + l(g(x) - m)|$$
$$\leq |g(x)||f(x) - l|$$
$$+ |l||g(x) - m)| \qquad ...(1)$$

There exists $\delta > 0$ such that when $0 < |x - c| < \delta$ and $x \in S$,

$$|f(x) - l| < \varepsilon', \ |g(x) - m| < \varepsilon' \qquad ...(2)$$

$\varepsilon' > 0$ being a given number.

Also $\forall \ x \in S$

$$|g(x)| \leq |m| + |g(x) - m|$$

so that when $\quad 0 < |x - c| < \delta$ and $x \in S$

$$|g(x)| < |m| + \varepsilon' \qquad ...(3)$$

From (1), (2) and (3), we obtain that when

$$0 < |x - c| < \delta \text{ and } x \in S,$$

then $$|f(x)\,g(x) - lm| < (|m| + \varepsilon')\,\varepsilon' + |l|\,\varepsilon'$$
$$= (|m| + |l| + \varepsilon')\,\varepsilon'.$$

We now suppose that ε' is a positive number smaller than 1 and smaller than $\varepsilon/(|m| + |l| + 1)$ where ε is the given positive number.

Thus we see that for $\varepsilon > 0$ other exists $\delta > 0$ such that

$$|f(x)\,g(x) - lm| < \varepsilon$$

when $$0 < |x - c| < \delta \text{ and } x \in S.$$

Hence the result *(ii)*

(iii) Let $\varepsilon \geq 0$ be given.

Since $\lim\limits_{x \to c} g(x) = m \neq 0$, there exists $\delta_1 > 0$ such that when $0 < |x - c| < \delta_1$ and $x \in S$.

$$|g(x) - m| < \frac{1}{2}|m|.$$

Also $$\forall x \in S$$

$$|m| \leq |g(x)| + |m - g(x)| \leq |g(x)| + \frac{1}{2}|m|.$$

Thus when $0 < |x - c| < \delta_1$ and $x \in S$,

$$|g(x)| > \frac{1}{2}|m|$$

so that there exists a neighbourhood of c such that for every point x of this neighbourhood which $\in S$, $f(x) \div g(x)$ has a meaning.

We have, $\forall \ x \in S$ such that $g(x) \neq 0$

$$\left|\frac{f(x)}{g(x)} - \frac{l}{m}\right| = \left|\frac{m[f(x) - l] - l[g(x) - m]}{m\,g(x)}\right|$$

$$\leq \frac{[|m| \, |f(x) - l| + |l| \, |g(x) - m|]}{|m| \, |g(x)|}$$

so that when $0 < |x - c| < \delta_1$ and $x \in S$,

$$\left|\frac{f(x)}{g(x)} - \frac{l}{m}\right| \leq \left|\frac{2}{|m|}\right| f(x) - l \left| + \frac{2|l|}{|m|^2}\right| g(x) - m| \quad ...(i)$$

Now there exists $\delta_2 > 0$ such that when $0 < |x - c| < \delta_2 . x \in S$

$$|f(x) - l| < \varepsilon', \; |g(x) - m| < \varepsilon' \quad ...(ii)$$

where ε' is a given positive number.

Let $\delta = \min(\delta_1, \delta_2)$.

We deduce, from (*i*) and (*ii*) that when

$$0 < |x - c| < \delta, \; x \in S,$$

$$\left|\frac{f(x)}{g(x)} - \frac{l}{m}\right| \leq \frac{2\varepsilon'}{|m|} + \frac{2|l|}{m^2}\varepsilon' = \frac{2}{m^2}(|m| + |l|)\,\varepsilon'.$$

We suppose that ε is chosen as a positive number smaller than

$$\varepsilon\, m^2/2\,(|l| + |m|).$$

Thus there exist $\delta > 0$ such that

$$\left|\frac{f(x)}{g(x)} - \frac{l}{m}\right| < \varepsilon$$

when $0 < |x - c| < \delta$ and $x \in S$.

Hence the result.

Ex. If $\lim f(x)$ and $\lim g(x)$ exist when $x \to c$, then

$$f < g \Rightarrow \lim f \leq \lim g.$$

Ex. If $|f|$ is defined by

$$|f|(x) = |f(x)| \; \forall \; x \in S,$$

show that when $x \to c$

$$\lim f(x) = 1 \Rightarrow \lim |f|(x) = |l|$$

but that the converse implication may not be true.

4.2.5. Limits at Infinity

Suppose that the domain S of a function f is not bounded above. In this case, if there exists a number l such that to every $\varepsilon > 0$ there corresponds $\Delta > 0$ such that

$$\left|f(x)-l\right| < \varepsilon$$

when $x \in S$ and $x < \Delta$, we write

$$\lim_{x\to\infty} f(x) = l.$$

Again if the domain S of f is not bounded below and there exists a number l such that to every $\varepsilon > 0$ there corresponds $\Delta > 0$ such that

$$\left|f(x)-l\right| < \varepsilon$$

when $x \in S$ and $x < -\Delta$, we write

$$\lim_{x\to-\infty} f(x) = l.$$

4.2.6. Infinite Limits

It may happen that the range of f is *not* bounded. We may then, on the lines of the definitions given in § 4.2.2 above give precise meanings to the statements :

$$\lim_{x\to a} f(x) = \pm\infty,\quad \lim_{x\to\infty} f(x) = \pm\infty,\quad \lim_{x\to-\infty} f(x) = \pm\infty.$$

4.3. CONTINUITY

Let f be a real valued function with domain S.

Let $$c \in S.$$

We say that the function f is continuous at c if to each $\varepsilon > 0$ there corresponds $\delta > 0$ such that when

$$|x - c| < \delta \text{ and } x \in S$$

we have $$|f(x) - f(c)| < \varepsilon |$$

i.e., we have

$$x \in\,]c - \delta, c + \delta[\, \cap S \Rightarrow f(x) \in\,]f(c) - \varepsilon, f(c) + \varepsilon[.$$

It will be seen that if c is *not* a limit point of S so that c is what is called and *Isolated point* of S, the function is necessarily continuous at c.

In fact, in this case there exists a neighbourhood of c containing no point of S other than c so that the only value x can take is c.

Also if c is a limit point of S, then f is continuous at c if and only if

$$\lim_{x\to c} f(x) \text{ exists and equals } f(c).$$

Continuity and Algebra of functions. The following results are immediate consequences of those established above in § 4.2.4, page 108.

Let f and ϕ be two functions defined in a domain S and continuous at a point $c \in S$. Then

$$f \pm \phi,\ f\phi,$$

are continuous at c. Again if $\phi(c) \neq 0$, then

$$f \div \phi$$

is also continuous at c.

Continuity in the domain. A function is said to be continuous in its domain if it is continuous at every point of the same.

Continuity of polynomials. Let f be defined by

$$f(\mathrm{x}) = a_0x^n + a_1x^{n-1} + \ldots + a_{n-1}x + a_n,$$

where $a_0, a_1, \ldots, a_{n-1}, a_n$ are given numbers.

The domain of f is $\mathbf{R}$. It may be seen that f is continuous in $\mathbf{R}$.

It is easy to show that the function ϕ defined by

$$\phi(x) = x \ \forall\ x \in \mathbf{R}$$

is continuous $\mathbf{R}$ and that every constant function *i.e.*, a function of the type

$$\psi(x) = c \ \forall\ x \in \mathbf{R}.$$

where c is a given number is also continuous in $\mathbf{R}$.

The fact that a polynomial arises as a result of a finite number of operations of addition, subtraction and multiplication performed on the continuous function ϕ defined by $\phi(x) = x$ and a number of constant functions which are also continuous shows that the function f is continuous in $\mathbf{R}$.

It also follows that a rational function is continuous in its domain of definition.

Continuity of composites of continuous functions. Let f be a function with domain S and range T let ϕ be a function with domain T so that the composite

$$\phi \circ f$$

is a function with domain S

Let $$u \in S$$

and let $$f(u) = v \in T.$$

We suppose that f is continuous at u and ϕ is continuous at $v = f(u)$ and show that the composite $\phi \circ f$ is also continuous at u. We have, by def.,

$$(\phi \, o \, f)(x) = \phi(f(x)); x \in S.$$

We write $\quad y = f(x)$ and $z = \phi(y)$

$\Rightarrow \quad z = (\phi \, o \, f)(x) = \phi(f(x))$.

Let $\varepsilon > 0$ be given. We write

$$v = f(u), w = \phi(v).$$

As ϕ is continuous at v, there exists $\delta_o > 0$ such that

$$|\phi(y) - \phi(v)| < \varepsilon \text{ when } |y - v| < \delta_o \text{ and } y \in T.$$

Also because f is continuous at u, there exists $\delta > 0$ such that

$$|y - v| = |f(x) - f(u)| < \delta_o$$

when $\quad |x - u| < \delta$ and $x \in S$.

We thus see that there exists $\delta > 0$ such that

$$|\phi(f(x)) - \phi(f(u))| < \varepsilon \quad \text{when } |x - u| < \delta \text{ and } x \in S$$

and as such $\phi \, o \, f$ is continuous at u. *(Garhwal 98)*

Discontinuities. A function f is said to be discontinuous at a point c of its domain if f is not continuous at c. The discontinuity of f at c can arise in either of the following two ways :

(*i*) $\lim\limits_{x \to c} f(x)$ exists and is different from $f(c)$,

(*ii*) $\lim\limits_{x \to c} f(x)$ does not exist.

We say that the discontinuity of f at c is of the *first kind or a simple discontinuity* if $\lim f(x)$ exists and is different from $f(c)$ and of the *second kind* if $\lim f(x)$ does not exist. *(Patna, 2003)*

Illustration

1. Let
$$f(x) = \begin{cases} 1 & \text{when } x \text{ is rational} \\ 0 & \text{when } x \text{ is irrational.} \end{cases}$$

The function is discontinuous at each point and each point is a point of discontinuity of the second kind.

***2.** If $f(x) = \sin 1/x$, $(x \neq 0)$ and $f(0) = 0$, $x = 0$ is a point of discontinuity of the second kind of the function f.

***3.** Let $f(x) = (x^2 - 1)/(x - 1)$, $x \neq 1$ and $f(1) = 0$.

The function f possesses discontinuity of the first kind at the point 1.

4.3.1. Theorem

A function f continuous in a closed interval [a, b] is bounded.

(Garhwal 98; Kanpur 96, Patna 2003, Osmania 2005)

We have to show that the range of a continuous function whose domain is a closed interval is bounded.

We suppose that f is *not* bounded.

We define a sequence with values in $[a, b]$ as follows :

To each natural number n there corresponds a point of $[a, b]$ which we denote by x_n such that

$$| f(x_n) | > n.$$

We have thus a sequence with values

$$x_1, x_2, ..., x_n, ...$$

in $[a, b]$ and the same is, as such, a bounded sequence and has, therefore, at least one limit point. Let ξ be a limit point of this sequence.

Now $\xi \in [a, b]$ and the function f is continuous at ξ. There exists therefore $\delta > 0$ such that

$$| f(x) - f(\xi) | < 1 \text{ when } | x - \xi | < \delta.$$

Now $$| f(x) | \leq | f(x) - f(\xi) | + | f(\xi) | \ \forall \ x \in [a, b].$$

This implies that

$$| f(x) | < 1 + | f(\xi) | \text{ when } | x - \xi | < \delta.$$

The interval $\xi - \delta, \xi + \delta$ [contains x_n for an infinite number of values of n and for this set of infinite values of n, we have

$$n < f(x_n) | < 1 + | f(\xi) |$$

implying that for an infinite number of values of n, *we have* $n < 1 + | f(\xi) |$ which is clearly not true.

Thus f is bounded function.

4.3.2. *A function f continuous in a closed interval* $[a, b]$ *has greatest and least values i.e., attains its supremum and infimum.*

(Kanpur 93; Garhwal 98; Kumaon 97; M.D.U. 2001, Osmania 2005)

We have to show that the range of a continuous function whose domain is a closed finite interval attains its bounds.

We know that f is bounded. Let M, m be the supremum and the infimum of f. It will be shown that there exists point α and β of $[a, b]$ such that

$$f(\alpha) = M, f(\beta) = m.$$

Let us consider the case of the supremum. Suppose that f does not have the value M for any $x \in [a, b]$, so that

$$M - f(x) \neq 0 \ \forall \ x \in [a, b].$$

Thus we deduce by § 4.2.3, P. 111 that the function with value

$$1/(M - f(x)); \; x \in [a,b]$$

is continuous in $[a, b]$ and is, as such, bounded.

Let k be a given number. Since M is the supremum of f, there exists $\xi \in [a, b]$ such that

$$f(\xi) > M - 1/k.$$

This implies that there exists $[a, b]$ such that

$$1/[M - f(\xi)] > k.$$

Since k is a given number, this shows that $1/[M - f(x)]$ is *not* bounded.

We thus arrive at contradictory statements.

Thus it is proved that the function f attains its supremum.

It may similarly be shown that f attains its infimum.

Note. As a consequence of the above theorem, it follows that if f is continuous in $[a, b]$, the range of f is a *sub-set* of $[m, M]$. It will be shown in cor. 2, p. 116 that $[m, M]$ is the range itself *i.e.*, every point of $[m, M]$ is the value of f for some point of $[a, b]$.

Intermediate value theorem

4.3.3. Theorem. *If f is continuous in $[a, b]$ and $f(a), f(b)$ are of opposite signs, there exists $\xi \in [a, b]$ such that $f(\xi) = 0$.*

(Garhwal 93, 98; Ajmer 98; Kurukshetra 2000)

Lemma I. If f is continuous at an interior point ξ of $]a, b[$ and $f(\xi) \neq 0$ there exists $\delta > 0$ such that $f(x)$ has the sign of $f(\xi)$ where $x \in [\xi - \delta, \xi + \delta]$.

If $\varepsilon > 0$ be given, there exists $\delta > 0$ such that

$$f(\xi) - \varepsilon < f(x) < f(\xi) + \varepsilon \text{ when } x \in [\xi - \delta, \xi + \delta].$$

Let $f(\xi) < 0$. Taking for ε a positive number less than $f(\xi)$, we see that $f(x) > 0$ when $x \in [\xi - \delta, \xi + \delta]$.

Let $f(\xi) < 0$. Taking for ε a positive number less than $-f(\xi)$, we see that $f(x) < 0$ when $x \in [\xi - \delta, \xi + \delta]$.

Hence the lemma I.

Lemma II. If f is continuous at an end point, a of $[a, b]$ and $f(a) \neq 0$, there exists an interval $[a, a + \delta]$ such that $f(x)$ has the sign of $f(a)$ when $x \in [a, a + \delta]$.

A similar result holds for continuity at b. The proofs are exactly similar to that of the result in Lemma I.

Proof of the main theorem

We suppose that

$$f(a) > 0 \text{ and } f(b) < 0$$

The set of points x of $[a, b]$ such that $f(x) > 0$ is finite. In fact, by Lemma II there exists an interval $[a, a + \delta]$, $\delta > 0$ for every point x of which $f(x) > 0$. We denote this set by S.

Let $\xi = \sup S$.

We shall show that $f(\xi) = 0$.

Let, if possible, $f(\xi) = 0$. Surely $\xi \neq a$.

As $f(\xi) < 0$, there exists $\delta > 0$ such that

$$f(x) < 0 \text{ when } x \in [\xi - \delta, \xi].$$

Also since ξ is sup S there exists $\eta \in S$ such that

$$\xi - \delta < \eta \leq \xi.$$

Now $$\eta \in S \Rightarrow f(\eta) > 0.$$

Also $$\eta \in [\xi - \delta, \xi] \Rightarrow f(\eta) < 0.$$

Thus we arrive at a contradiction and as such $f(\xi)$ is not < 0.

Now suppose $f(\xi) > 0$. Surely $\xi \neq b$, $\xi \neq a$. There exists $\delta > 0$ such that

$$f(x) > 0 \text{ where } x \in [\xi - \delta, \xi + \delta].$$

Thus $\xi + \delta \in S$ so that we arrive at a contradiction of the fact of ξ being sup S.

Thus $f(\xi) = 0$.

Cor. 1. *If f is continuous in $[a, b]$ and $f(a) \neq f(b)$, then f assumes every value between $f(a)$ and $f(b)$.* (*Garhwal 92, 94, 97; Ajmer 99*)

Let k be a number between $f(a)$ and $f(b)$.

Consider a function ϕ such that

$$\phi(x) = f(x) - k.$$

Now ϕ is continuous in $[a, b]$ and

$$\phi(a) = f(a) - k, \ \phi(b) = f(b) - k$$

are of opposite signs. Thus there exists $\xi \in [a, b]$ such that

$$\phi(\xi) = 0.$$

Also $$\phi(\xi) = 0 \Rightarrow f(\xi) = k$$

Hence the corollary.

Cor. 2. *If f is continuous in $[a, b]$, then f assumes every value between its infimum m and supremum M so that $[m, M]$ is the range of the function f.*

EXAMPLES

1. *Discuss the continuity of the function* $f(x) = \dfrac{\sin x}{x}$ *for* $x \neq 0$, $f(x) = 1$ *for* $x = 0$. *(Kanpur 93)*

Sol.
$$f(0+h) = \frac{\sin h}{h}, \quad f(+0) = 1$$
$$f(0-h) = \frac{\sin(-h)}{-h} = \frac{\sin h}{h}, \quad f(-0) = 1$$
$$f(0) = 1, \text{ by definition.}$$

so $f(x)$ is continuous at $x = 0$.

2. *Discuss the continuity at* $x = a$ *of the function* $f(x)$, *where*

Sol.
$$f(x) = \frac{x^2}{a} - a, \text{ for } 0 < x \leq a,$$
$$f(x) = a - \frac{a^3}{x^2}, \text{ for } x > a. \qquad \textit{(Poorv. 92, 93)}$$

$$f(a) = 0, \text{ Also, } f(a-h) = \frac{(a-h)^2}{a} - a$$
$$= \frac{h^2 - 2ah}{a} \qquad \text{(Here } 0 < x < a\text{)}$$
$$\therefore \quad f(a-0) = \lim_{h \to 0} \frac{h^2 - 2ah}{a} = 0$$

Similarly
$$f(a+h) = a - \frac{a^3}{(h+a)^2} = \frac{a\left(h^2 + 2ah\right)}{\left(h^2 + 2ah + a^2\right)}$$
$$\therefore \quad f(a+0) = \lim_{h \to 0} \frac{a\left(h^2 + 2ah\right)}{h^2 + 2ah + a^2} = 0$$
$$\therefore \quad f(a-0) = f(a+0) = f(a).$$

Hence $f(x)$ is continuous at $x = a$.

3. *Discuss the continuity of* $f(x)$ *at* $x = 0$, 1 *and* 2, *where*

Sol.
$$f(x) = -x^2, \text{ for } x \leq 0;$$
$$f(x) = 5x - 4, \text{ for } 0 < x \leq 1;$$
$$f(x) = 4x^2 - 3x, \text{ for } 1 < x \leq 2;$$
$$f(x) = 3x + 4, \text{ for } x \leq 2.$$

(Rohilkhand 94, Garhwal 97)

Sol. At $x = 0, f(0) = 0$

$$f(0-h) = -h^2 \Rightarrow f(-0) = \lim_{h\to 0}\left(-h^2\right) = 0$$

$$f(0+h) = 5h - 4 \Rightarrow f(+0) = \lim_{h\to 0}(5h-4) = -4$$

so, $f(x)$ has discontinuity at $x = 0$, as $f(-0) \neq f(+0)$.

At $x = 1$, $\quad f(1) = 5 \cdot 1 - 4 = 1.$

$$f(1-h) = 5 \times (1-h) - 4, \; (0 < x < 1)$$
$$= 1 - 5h$$

$$\therefore \quad f(1-0) = \lim_{h\to 0}(1-5h) = 1$$

$$f(1+h) = 4(h+1)^2 - 3(1+h) \qquad (1 < x < 2)$$
$$= 1 + 4h^2 + 5h$$

$$\therefore \quad f(1+0) = \lim_{h\to 0}\left(1 + 4h^2 + 5h\right) = 1.$$

Hence $f(x)$ is continuous at $x = 1$.

Similarly, it can be shown that $f(x)$ is continuous at $x = 2$.

4. *A function y is defined as follows :*

'y is the smallest positive number that makes x + y an integer'; for what values of x is the function discontinuous ?

Sol. Consider $x = \alpha$, where α is an integer.

If $\quad x = \alpha - h$, y is small $= \alpha - (\alpha - h) = h$

i.e., $\quad f(\alpha + 0) = 0,$

If $\quad x = \alpha + h,\; y = (\alpha + 1) - (\alpha - h) = 1 - h$

$\therefore \quad f(\alpha + 0) = 1,$

i.e., function is discontinuous at $x = \alpha$, where α is an integer.

Here it should be noted that if

$\alpha - 1 < x < \alpha$, then $x + y = \alpha$ and if $\alpha < x < \alpha + 1$, $x + y = \alpha + 1$.

5. *Discuss the continuity of the function* $f(x) = \dfrac{1}{x-a}\operatorname{cosec}\dfrac{1}{x-a}$, *for* $x \neq a$, $f(a) = 0$. *(Agra 91)*

Sol. We have

$$f(a+0) = \lim_{h\to 0} f(a+h) = \lim_{h\to 0}\frac{1}{h}\operatorname{cosec}\frac{1}{h}$$

$$= \lim_{h\to 0}\frac{1}{h} \times \lim_{h\to 0}\operatorname{cosec}\frac{1}{h}$$

Since cosec $\frac{1}{h}$ does not tend to a definite limit as h tends to zero but take values between the limits $+\infty$ and $-\infty$, we find that $\lim_{h\to 0} f(a+h)$ or $f(a+0)$ does not exist but oscillates between limits of which the greatest and the least are $+\infty$ and $-\infty$, so that $\overline{f(a+0)}=\infty$ and $\underline{f(a+0)}=-\infty$. Similarly, $\overline{f(a-0)}=\infty$ and $\underline{f(a-0)}=-\infty$.

Thus all the four limits $\underline{\overline{f(a\pm 0)}}$ are infinite with one sign or the other. Hence the function has infinite discontinuity at $x = a$. This discontinuity is, however, of the second kind as $f(a + 0)$ does not exist, nor does $f(a - 0)$ exist.

6. *Find the points of discontinuity of the function* $f(x) = \lim_{m\to\infty}\left\{\lim_{n\to\infty}(\cos m!\pi x)^{2n}\right\}$. *(Garhwal 92)*

Sol. Let $x = p/q$ (rational), where p and q are integers.

By taking m sufficiently large $m!\pi x$ is an integral multiple of π; so for such values and greater values of m,

$$\cos m!\pi x = \pm 1.$$

$\therefore$ For rational values of x,

$$f(x) = \lim_{n\to\infty}(\pm 1)^{2n} = 1.$$

If x is irrational, $\cos m!\pi x$ shall always lie between -1 and $+1$, whatever integral value m may have so that $(\cos m!\pi x)^{2n}$ (for a fixed value of m) $= (r_m)^{2n}$, where $r_m < 1$.

$$\therefore \quad f(x) = \lim_{m\to\infty}\lim_{n\to 0}(r_m)^{2n} = 0.$$

Since $f(x)$ is 1 for rational values of x, and $f(x)$ is zero for irrational values of x, it is a *totally discontinuous function, i.e., discontinuous for every value of x*. This is so because at any point $x = x_1$, $f(x_1 + 0)$ as also $f(x_1 - 0)$ shall not exist, there being infinite number of rational and infinite irrational values of x in any interval $(x_1 - h, x_1 + h)$, however small h may be, and for these the functional values widely differ. Thus the function has discontinuity of the second kind for every value of x.

EXERCISES

1. Discuss the continuity of the functions defined as follows :

(*i*) $f(x) = (x - a)\sin\frac{1}{x-a}$ for $x \neq a$, $f(x) = 0$ for $x = a$.

(*ii*) $f(x) = x\sin(1/x)$ for $x \neq 0$, $f(0) = 2$.

2. A function $f(x)$ is defined as follows in (0, 1)

$$f(x) = 0, \qquad \text{for } x = 0,$$

$$f(x) = \frac{1}{2} - x, \qquad \text{for } 0 < x < \frac{1}{2},$$

$$f(x) = \frac{1}{2}, \qquad \text{for } x = \frac{1}{2},$$

$$f(x) = \frac{3}{2} - x, \qquad \text{for } \frac{1}{2} < x < 1,$$

$$f(x) = 1, \qquad \text{for } x = 1.$$

Find the points of discontinuity. *(Rohilkhand 91, 93)*

3. Find whether the following functions are continuous for integral value of x :

(*a*) $f(x) = I(x)$, where $I(x)$ is the integral part of x.

(*b*) $f(x) = F_r(x)$, where $F_r(x)$ means fractional part of x.

(*c*) $f(x) = I(x) + F_r(x)$.

4. Show that the function $f(x) = (1 + 2x)^{1/x}$ for $x \neq 0$ and $f(x) = e^2$ for $x = 0$ is continuous. *(Rohilkhand 92)*

5. Let $f(x) = [x]$, where $[x]$ denotes the positive or negative excess of x over the nearest integer and let $[x] = 0$ when $x = n + \frac{1}{2}$, n being an integer. Discuss the continuity of $f(x)$. *(Garhwal 97)*

6. A function $f(x)$ is defined by

$$f(x) = \cos x, \qquad \text{for } x \geq 0$$
$$= -\cos x, \qquad \text{for } x < 0.$$

Is the function $f(x)$ continuous at $x = 0$? *(Agra 94)*

7. Discuss the point of continuity and discontinuity of the function

$$f(x) = \begin{cases} -x, & x \leq 0 \\ x, & 0 < x \leq 1 \\ 2 - x, & 1 < x \leq 2 \\ 1, & x > 2 \end{cases}$$

(Poorv. 91)

8. Show that the function $f(x) = |x|$ is continuous at $x = 0$.

9. Is the function $f(x) = \dfrac{e^{1/x}}{e^{1/x} + 1}$, when $x \neq 0$, $f(0) = 0$ continuous at $= 0$? *(Kanpur 94, 95)*

10. Discuss the continuity of the function $f(x) = \sin(1/x)$ when $x \neq 0$ and $f(x) = 0$ when $x = 0$. *(Garhwal 98, Agra 92, 93; Kanpur 92)*

11. If $f(x) = \dfrac{1}{x} \sin \dfrac{1}{x}$ for $x \neq 0$ and $f(0) = 0$, show that $f(x)$ is finite for every value of x in the interval $(-1, 1)$ but is not bounded. Determine the points of discontinuity of the function if any.

12. Given the function

$$f(x) = \lim_{n \to \infty} \left[\lim_{t \to 0} \frac{\sin^2 (n!\pi x)}{\sin^2 (n!\pi x) + t^2} \right]$$

show that $f(x)$ is totally discontinuous.

4.3.4. Uniform Continuity

We shall now introduce the notion of uniform continuity of a function. It should be seen that while the notion of continuity of a function at a point is essentially *Local* in character the same is *not* the case in respect of uniform continuity. The notion of uniform continuity of a function is *global* in character inasmuch as we can only talk of uniform continuity of a function over its domain rather than at a point of its domain.

Let f be a function with a domain S. Let $c \in S$. We know that f is continuous at c if to each $\varepsilon > 0$, there corresponds $\delta > 0$, such that

$$|f(x) - f(c)| < \varepsilon$$

for each $x \in S$ such that $|x - c| < \delta$.

Suppose that f is continuous at each point of S and ε is a given positive number. Then there exists δ corresponding to each point of the domain S. The question now arises, "*Does there exist a δ which holds uniformly for every point of the domain ?*" It may appear to a beginner that the infimum of the set consisting of all the values of δ would serve the purpose. The fallacy, in this approach however, lies in the fact that the infimum may be zero. In general, therefore, a choice of positive δ which may hold good uniformly for each point of the domain D may *not* be possible. We now define the notion of uniform continuity.

We say that f is uniformly continuous in ζ if these exists a 'δ' which holds uniformly for every point of the domain.

4.3.5. Theorem

If a function f is continuous in [a, b], it is uniformly continuous in [a, b].

[M.U. 2001]

Suppose that f is *not* uniformly continuous.

Let $\varepsilon > 0$ be given.

Consider an arbitrary sequence tending to zero with positive values

$$\delta_1, \delta_2, ..., \delta_n.$$

As f is assumed *not* to be uniformly continuous, there exists a pair of points x_n, y_n such that

$$|x_n - y_n| < \delta_n, |f(x_n) - f(y_n)| \geq \varepsilon.$$

Thus we have two bounded sequences with values

$$x_1, x_2, ..., x_n, ...; y_1, y_2, ..., y_n ...$$

Let ξ be a limit point of the sequence $\{x_n\}$.

There exists a convergent sub-sequence of the sequence $\{x_n\}$ with limit ξ. Let the member of this convergent sub-sequence be

$$x_{k_1,}\quad x_{k_2,}\quad x_{k_3,...}\quad k_{n,...}$$

where $\{k_n\}$ is a strictly monotonically increasing sequence of natural numbers tending to infinity.

It will then follow that ξ is also a limit of the sub-sequence of the sequence $\{y_n\}$ with members.

$$y_{k_1}, y_{k_2,}, y_{k_3,...} y_{k_n,...}$$

Now f is continuous at ξ. There exists therefore $\delta > 0$ such that

$$|f(x) - f(\xi)| < \frac{1}{2}\varepsilon, \text{ when } x \in [\xi - \delta, \ \xi + \delta].$$

Let u, v be arbitrary points in $[\xi - \delta, \xi + \delta]$. We have

$$|f(u) - f(\xi)| < \frac{1}{2}\varepsilon, \ |f(v) - f(\xi)| < \frac{1}{2}\varepsilon$$

$$\Rightarrow \qquad |f(u) - f(v)| < \varepsilon.$$

Thus if u, v be arbitrary points in $[\xi - \delta, \xi + \delta]$, we have

$$|f(u) - f(v)| < \varepsilon$$

Also an infinite number of pairs $x_{k_n}, y_{k_n} \in [\xi - \delta, \xi + \delta]$.

Thus we arrive at a contradiction and as such f is uniformly continuous in $[a, b]$.

Note. The properties of continuous functions referred to above may not be valid for functions continuous only in *open* intervals.

For example, consider f defined as follows in [0, 1] :

$$f(x) = 1/x \text{ when } x \neq 0, f(0) = 0.$$

It may be easily seen that f is *not* bounded in [0, 1]. The function f is not even uniformly continuous in [0, 1].

It will be seen that f is *not* continuous at 0.

Ex. 1. If $f(x) = x^2 + 3x$ in $[-1, 1]$, find δ so that

$$|x_2 - x_1| < \delta \Rightarrow |f(x_2) - f(x_1)| < \varepsilon,$$

where $\varepsilon > 0$ is given.

Ex. 2. Do as in Ex.1 above for

(*i*) $f(x) = x^3 + 3x^2 - 2x + 7$ in $[-3, 2]$;

(*ii*) $f(x) = (x + 2)/(2x + 3)$ in $[-1, 5]$.

4.4. MONOTONIC FUNCTIONS

A function f defined in S is said to be monotonically increasing if

$$x_1 \in S, x_2 \in S, x_2 > x_1 \Rightarrow f(x_2) \geq f(x_1).$$

Also f is said to be *strictly* monotonically increasing if

$$x_1 \in S, x_2 \in S, x_2 > x_1 \Rightarrow f(x_2) > f(x_1).$$

We may similarly give meanings to the statements :

(*i*) f is monotonically decreasing in S;

(*ii*) f is strictly monotonically decreasing in S.

Theorem. *Let f be monotonically increasing in $[a, b]$. Then if it is a point such that $a < c < b$* then

$$f(c-0) \text{ and } f(c+0)$$

both exist and

$$f(c-0) = \sup_{a<x<c} f(x), \quad f(c+0) \inf_{c<x<b} f(x).$$

Also

$$f(c-0) \leq f(c) \leq f(c+0)$$

The set

$$\{f(x) : a < x < c\}$$

is bounded above. In fact $f(c)$ is an upper bound of this set. Let the supremum of this set be denoted by ξ. We have

$$\xi \leq f(c).$$

Let $\varepsilon > 0$ be given. We have

$$f(x) \le \xi \ \forall \ x \in]a, c[\Rightarrow f(x) < \xi + \varepsilon$$

Also there exists $\delta > 0$ such that

$$a < c - \delta < c$$

and
$$f(c - \delta) > \xi - \varepsilon$$

As f is monotonically increasing, we have

$$f(x) > f(c - \delta) > \xi - \varepsilon \text{ when } x > c - \delta$$

Thus we see that there exists $\delta > 0$ such that

$$\xi - \varepsilon < f(x) < \xi + \varepsilon \text{ when } c - \delta < x < c$$

$$\Rightarrow \qquad f(c - 0) = \lim_{x \to (c-0)} f(x) = \xi = \sup_{a < x < c} f(x).$$

We may similarly prove that

$$f(c + 0) = \inf_{c < x < b} . f(x),$$

We may also show that

$$f(a + 0), \ f(b - 0)$$

exist and
$$f(a + 0) = \inf_{a < x < b} . f(x), \ f(b - 0) = \sup_{a < x < b} f(x).$$

Theorem. *Let f be monotonically decreasing in [a, b]. If c is any point such that a < c < b, then*

$$f(c - 0) \textit{ and } f(c + 0)$$

both exist and

$$f(c - 0) = \inf_{a < x < c} f(x), \ f(c + 0) = \sup_{c < x < b} f(x)$$

Also
$$f(c - 0) \ge f(c) \ge f(c + 0).$$

The proof is left to the reader.

Inverses of strictly monotonic functions.

Theorem. *A function f which is continuous and strictly monotonic with domain [a, b] admits of an inverse. Also the inverse is continuous.*

The proof of the existence of the inverse is left to the reader.

It will be seen that the domain of the inverse function ϕ is the interval

$$\left[f(a), f(b)\right] \text{ or } \left[f(b), f(a)\right]$$

according as the function f is increasing or decreasing.

We shall now prove that ϕ is continuous.

Let us suppose that f is increasing in $[a, b]$ so that the domain of the inverse ϕ is $[f(a), f(b)]$.

Let $$v \in \left[f(a), f(b)\right]$$

and left $\phi(v) = u \Rightarrow f(u) = v$.

Let $\varepsilon > 0$ be given. Let

$$f(u-\varepsilon) = v - \delta_1, \ f(u+\varepsilon) = v + \delta_2,$$

δ_1, δ_2 being necessarily positive.

As the function f is strictly monotonic,

$$y = f(x) \in]v - \delta_1, v + \delta_2 \text{ [when } x \in]u - \varepsilon, u + \varepsilon[.$$

Thus if $\delta = \min.(\delta_1, \delta_2)$, we see that

$$|x - u| < \varepsilon \Leftrightarrow |\phi(y) - \phi(v)| < \varepsilon \text{ when } |y - v| < \delta.$$

Hence ϕ is continuous at v.

As v is any point of $[f(a), f(b)]$, it follows that ϕ is continuous in its domain.

The continuity of ϕ may be similarly established if f is decreasing.

4.5. FUNCTIONS OF BOUNDED VARIATION

The section will be devoted to a discussion about the important concept of *the functions of bounded variation*. This concept is closely related to that of monotonic functions and plays an important part in various parts of Analysis.

A finite set of points

$$a = x_0, \ldots, x_{n-1}, x_n = 0$$

is said to be a **division** of $[a, b]$ if

$$a = x_0 < x_1 < x_2 < \ldots < x_{n-1} < x_n = b$$

The intervals $$\left[x_0, x_1\right], \left[x_1, x_2\right] \ldots, \left[x_{n-1}, x_n\right]$$

will be referred to as the sub-intervals of the division.

Let f be a function defined in an interval $[a, b]$ and let

$$D\left\{a = x_0, x_1, ..., x_{r-1}, x_r, ..., x_n = b\right\}$$

be a division of $[a, b]$.

Consider the sum

$$\sum_{r=1}^{r=n} \left| f(x_r) - f(x_{r-1}) \right| \qquad ...(1)$$

of the moduli of the differences of the values of f at the end points of the sub-intervals of the division D.

There corresponds a sum (1) to each division D of $[a, b]$. Consider the set of all such sums corresponding to the set of divisions of $[a, b]$. *Then f is said to be of bounded variation over $[a, b]$ if the set of these sums is bounded above and the supremum of this set is known as the total variation of the function f over $[a, b]$ and denoted by the symbol*

$$V_f(a, b)$$

Ex. What do you mean by saying that a function f is not of bounded variation over $[a, b]$.

Illustrations

1. *A bounded monotonic function is a function of bounded variation.*

If f is monotonically increasing in $[a, b]$ and

$$D\left\{a = x_0, ..., x_{r-1}, x_r, ..., x_n = b\right\}$$

is *any* division of $[a, b]$, we have

$$\sum_{r=1}^{r=n} \left| f(x_r) - f(x_{r-1}) \right| = \sum_{r=1}^{r=n} \left[f(x_r) - f(x_{r-1}) \right]$$

$$= f(b) - f(a)$$

Thus the monotonically increasing bounded function, f, is of bounded variation over $[a, b]$ and

$$V_f(a, b) = f(b) - f(a).$$

Similarly we may show that monotonically decreasing bounded function y is also of bounded variation and

$$V_g(a, b) = g(a) - g(b).$$

2. *A continuous function may* **not** *be a function of bounded variation.*

*Consider the function f defined in [0, 1] as follows :

$$f(x) = x \sin \frac{\pi}{x}, \text{ when } x \neq 0, f(0) = 0.$$

Clearly f is continuous in [0, 1].

Consider the division

$$D\left\{0, \frac{2}{2n+1}, \ldots, \frac{2}{7}, \frac{2}{5}, \frac{2}{3}, 1\right\}$$

n being a positive integer.

We have

$$\left|f(1) - f\left(\frac{2}{3}\right)\right| + \left|f\left(\frac{2}{3}\right) - f\left(\frac{2}{5}\right)\right| + \ldots + \left|f\left(\frac{2}{2n+1}\right) - f(0)\right|$$

$$= \frac{2}{3} + \left(\frac{2}{3} + \frac{2}{5}\right) + \left(\frac{2}{5} + \frac{2}{7}\right) + \ldots + \left(\frac{2}{2n+3} + \frac{2}{2n+1}\right) + \frac{2}{2n+1}$$

$$= 4\left[\frac{1}{3} + \frac{1}{5} + \frac{1}{7} + \ldots + \frac{1}{2n+1}\right]$$

The infinite series

$$\frac{1}{3} + \frac{1}{5} + \frac{1}{7} + \ldots$$

being divergent, the sequence σ where

$$\sigma(n) = \frac{1}{3} + \frac{1}{5} + \ldots + \frac{1}{2n+1}$$

is not bounded above. Thus there exists a value of n, such that for the corresponding division the sum of the moduli of the differences of the values of f for the end points of sub-intervals of the division D exceeds a given number.

Thus the function f in question, even though continuous, is not of bounded variation over [0, 1].

Also it may be seen that *a function of bounded variation is not necessarily continuous*. For example if $f(x) = [x]$ where $[x]$ denotes the greatest integer not greater than x, the function f is of bounded variation in [0, 2] but it is not continuous.

4.5.1. Some Properties of Functions of Bounded Variation

1. *A function of bounded variation is necessarily bounded.*

Sol. Let f be of bounded variation over $[a, b]$ and x be a point of $[a, b]$. Considering the division

$$D \{a, x, b]$$

of $[a, b]$ consisting of three points, a, x, b only, we have

$$|f(x)-f(a)|+|f(b)-f(x)| \leq V_f(a, b)$$

$$\Rightarrow \qquad |f(x)-f(a)| \leq V_t(a, b)$$

Thus we have

$$|f(x)| \leq f(a)|+|f(x)-f(a)| \leq |f(a)|+V_t(a, b).$$

$\Rightarrow$ f is bounded in $[a, b]$.

2. *The sum, difference and product of functions of bounded variation are also of bounded variation.*

Sol. Let f and g be two functions of bounded variation over $[a, b]$.

$$D\left\{a=x_0, x_1, \ldots, x_{r-1}, x_r \ldots, x_n=b\right\},$$

$$\sum_{r=1}^{n}\left|(f+g)(x_r)-(f+g)\left(x_{r-1}\right)\right|$$

$$=\sum_{r=1}^{n}\left|\left[f(x_r)+g(x_r)\right]-\left[f\left(x_{r-1}\right)+g\left(x_{r-1}\right)\right]\right|$$

$$\leq \sum_{r=1}^{n}\left|f(x_r)-f\left(x_{r-1}\right)\right|+\sum_{r=1}^{n}\left|g(x_r)-g\left(x_{r-1}\right)\right|$$

$$\leq V(a, b)+V_g(a, b).$$

$\Rightarrow f+g$ is of bounded variation over $[a, b]$ and

$$V_{f-g}(a, b) \leq V_f(a, b)+V_g(a, b).$$

Similarly it may be shown that $f-g$ is of bounded variation over $[a, b]$ and

$$V_{f-g}(a, b) \leq V_f(a, b)+V_g(a, b).$$

We consider now the case of product.

The functions f and g, being of bounded variation over $[a, b]$, are both bounded and accordingly there exists a number k such that $\forall\, x \in [a, b]$ and

$$|f(x)| \leq k, |g(x)| \leq k.$$

We have for any division

$$D\left\{a = x_0, \dots\dots x_{r-1}, x_r \dots, x_n = b\right\}.$$

$$\sum_{r=1}^{n} \left|(fg)(x_r) - (fg)\left(x_{r-1}\right)\right|$$

$$= \sum_{r=1}^{n} \left| f(x_r)\, g(x_r) - f\left(x_{r-1}\right) g\left(x_{r-1}\right)\right|$$

$$= \sum_{r=1}^{n} \left| f(x_r)\left[g(x_r) - g\left(x_{r-1}\right)\right] - g\left(x_{r-1}\right)\left[f\left(x_{r-1}\right) - f(x_r)\right]\right|$$

$$\le \sum_{r=1}^{n} \left\{\left| f(x_r)\right|\left| g(x_r) - g\left(x_{r-1}\right)\right| + \left| g\left(x_{r-1}\right)\right|\left| f\left(x_{r-1}\right) - f(x_r)\right|\right\}$$

$$\le k\left\{\sum_{r=1}^{n} \left| g(x_r) - g\left(x_{r-1}\right)\right| + \sum_{r=1}^{n}\left| f(x_r) - f\left(x_{r-1}\right)\right|\right\}$$

$$\le kV_g\,(a, b) + kV_f\,(a, b).$$

$\Rightarrow$ the function fg is of bounded variation over $[a, b]$.

3. *If f bounded variation over [a, b] and is such that there exists a number k such that* $|f(x)| \ge k\ \forall\ x \in [a, b]$, *then 1/f is also of bounded variation over* $[a, b]$.

Sol. For any division

$$D\left\{a = x_0, x_1, \dots, x_{r-1}, x_r, \dots, x_n = b\right\},$$

we have

$$\sum_{r=1}^{n} \left| \frac{1}{f(x_r)} - \frac{1}{f\left(x_{r-1}\right)}\right| = \sum_{r=1}^{n} \left| \frac{f\left(x_{r-1}\right) - f(x_r)}{f(x_r)\, f\left(x_{r-1}\right)}\right|$$

$$\le \frac{1}{k^2} \sum_{r=1}^{n} \left| f(x_r) - f\left(x_{r-1}\right)\right|$$

$$\le \frac{V_f\,(a, b)}{k^2}.$$

Thus, $1/f$, is of bounded variation over $[a, b]$

4. *If f is of bounded variation over [a, b] and c is a point of [a, b], then f is also of bounded variation over [a, b] and [c, b], and conversely, Also*

$$V_f(a, b) = V_f(a, c) + V_f(c, b)$$

Sol. Let f be of bounded variation over $[a, b]$. Let

$$D_1\left(a = x_0,\ x_1 \ldots,\ x_{p-1},\ x_p \ldots x_m = c\right)$$

$$D_2\left(c = y_0,\ y_1 \ldots,\ y_{q-1},\ y_q \ldots y_n = b\right)$$

by *any* two divisions of $[a, c]$, $[c, b]$ respectively.

They give rise to a division

$$D\left(a = x_0,\ \ldots,\ x_m,\ y_0,\ y_1 \ldots,\ y_{q-1},\ y_q, \ldots,\ y_n = b\right) \text{ of } [a, b].$$

We have

$$\sum_{p=1}^{m} \left| f(x_p) - f(x_{p-1}) \right| + \sum_{q=1}^{m} \left| f(y_q) - f(y_{q-1}) \right| \le V_{f(a,b)}$$

$$\Rightarrow \sum_{p=1}^{m} \left| f(x_p) - f(x_{p-1}) \right| \le V_f(a, b) \text{ and } \sum_{q=1}^{n} \left| f(y_q) - f(y_{q-1}) \right| \le V_f(a, b)$$

and accordingly f is of bounded variation over $[a, c]$ as also over $[c, b]$.

Now suppose that f is of bounded variation over $[a, c]$ and $[c, b]$.

Let $\left(a = z_0, \ldots, z_{r-1}, z_r, \ldots, z_t = b\right)$ be a division of $[a, b]$ and let $z_{s-1} < c < z_s$ where $s < t$.

We have

$$\sum_{r=1}^{t} \left| f(z_r) - f(z_{r-1}) \right| = \sum_{r=1}^{s-1} \left| f(z_r) - f(z_{r-1}) \right| + \left| f(z_s) - f(z_{s-1}) \right| + \sum_{r=s+1}^{t} \left| f(z_r) - f(z_{r-1}) \right|$$

$$\le \left\{ \sum_{r=1}^{s-1} \left| f(z_r) - f(z_{r-1}) \right| + \left| f(c) - f(z_{s-1}) \right| \right\} + \left\{ \left| f(z_s) - f(c) \right| + \sum_{r=s+1}^{t} \left| f(z_r) - f(z_{r-1}) \right| \right\}$$

$$\leq V_f(a, c) + V_f(c, b). \qquad ...(1)$$

Thus the function f is of bounded variation over $[a, b]$ if it is of bounded variation over $[a, c]$ and $[c, b]$. Also then, we deduce from (1)

$$V_f(a, b) \leq V_f(a, c) + V_f(c, b). \qquad ...(2)$$

Let now, ε be an arbitrary positive number.

Since $V_f(a, c)$ and $V_f(c, b)$ are the total variations of f over $[a, c]$ and (c, b) respectively, there exist divisions

$$D_1 \{a = x_0, x_1, ..., x_{p-1}, x_p, ..., x_m = c\}$$

$$D_2 \{c = y_0, y_1, ..., y_{q-1}, y_q, ..., y_n = b\}$$

of $[a, c]$ and $[c, b]$ respectively such that

$$\sum_{p=1}^{m} \left| f(x_p) - f(x_{p-1}) \right| > V_f(a, c) - \frac{1}{2}\varepsilon, \qquad ...(3)$$

$$\sum_{q=1}^{n} \left| f(x_q) - f(x_{q-1}) \right| > V_f(a, b) - \frac{1}{2}\varepsilon, \qquad ...(4)$$

From (3) and (4), we have

$$\sum_{p=1}^{m} \left| f(x_p) - f(x_{p-1}) \right| + \sum_{q=1}^{n} \left| f(y_q) - f(y_{q-1}) \right| > V_f(a, c)$$

$$+ V_f(c, b) - \varepsilon,$$

$$\Rightarrow \qquad V_f(a, b) \geq V_f(a, c) + V_f(c, b) - \varepsilon. \qquad ...(5)$$

As ε is an arbitrary positive numbers, we deduce from (5) that

$$V_f(a, b) \geq V_f(a, c) + V_f(c, b). \qquad ...(6)$$

Finally, from (2) and (6), we have

$$V_f(a, b) = V_f(a, c) + V_f(c, b),$$

as was to be proved.

4.5.3. Variation Function of a Function of Bounded Variation

We shall now define a function associated with a function of bounded variation.

Let f be a function of bounded variation over an interval $[a, b]$. If x be a point of $[a, b]$, then we denote by $V(x)$ the total variation of f over $[a, x]$ so that V is a function of x defined over $[a, b]$. This function will be referred to as the *Variation function* of f over $[a, b]$. We have, by def.,

$$V(x) = V_f(a, x).$$

4.5.4. Monotonically Increasing Character of the Variation Function

Let x_1, x_2 be two points of $[a, b]$ such that $x_2 > x_1$.

We have

$$0 \le \left| f(x_2) - f(x_1) \right| \le V_f(x_1, x_2) = V_f(a, x_2) - V_f(a, x_1)$$
$$= V(x_2) - V(x_1)$$
$$\Rightarrow \qquad V(x_2) \ge V(x_1)$$

and hence we have the monotonically increasing character of the Variation function V.

4.5.5. Characterisation of Functions of Bounded Variation

A function of bounded variation is expressible as the difference of two monotonically increasing functions.

We write

$$f(x) = \frac{1}{2}\left[V(x) + f(x)\right] - \frac{1}{2}\left[V(x) - f(x)\right]$$
$$= g(x) - h(x), \text{ say,}$$

and show that the function g and h are monotonically increasing in $[a, b]$.

Now if $x_2 > x_1$, we have

$$g(x_2) - g(x_1) = \frac{1}{2}\left[V(x_2) - V(x_1) + f(x_2) - f(x_1)\right]$$
$$= \frac{1}{2}\left\{V_f(x_1, x_2) - \left[f(x_1) - f(x_2)\right]\right\}.$$

Since $\quad V_f(x_1, x_2) \ge \left| f(x_1) - f(x_2) \right|,$

we deduce that

$$g(x_2) - g(x_1) \ge 0 \Leftrightarrow g(x_2) \ge g(x_1),$$

so that g is monotonically increasing in $[a, b]$.

Again, we have

$$h(x_2) - h(x_1) = \frac{1}{2}\left\{\left[V(x_2) - V(x_1)\right] - \left[f(x_2) - f(x_1)\right]\right\}$$

$$= \frac{1}{2}\left\{V_f(x_1, x_2) - [f(x_2) - f(x_1)]\right\}$$

so that as before $h(x_2) - h(x_1) \geq 0 \Leftrightarrow h(x_2) \geq h(x_1)$.

Hence the result.

With the help of the result proved here and in § 4.5.4 and & 4.5.5 we may state that *a function f is of bounded variation over [a, b]* **if and only if** *it can be expressed as the difference of two monotonically increasing functions.*

4.5.6. Variation Function of Continuous Function of Bounded Variation

We shall now show that the variation function of a continuous function of bounded variation is itself continuous.

Theorem. *If $c \in [a, b]$, then V is continuous at c, if and only if, f is continuous at c i.e., a point of continuity of f is also a point of continuity of V and conversely.*

Firstly suppose that V is continuous at c.

Let $\varepsilon > 0$ be given. Since V is continuous at c, there exist $\delta > 0$ such that

$$|V(x) - V(c)| < \varepsilon \text{ for } |x - c| < \delta. \qquad ...(1)$$

Also we have

$$\begin{cases} |f(x) - f(c)| \leq V(x) - V(c) & \text{if } x > c, \quad ...(2) \\ |f(x) - f(c)| \leq V(c) - V(x) & \text{if } x < c. \quad ...(3) \end{cases}$$

From (1), (2) and (3), we deduce that

$$|f(x) - f(c)| < \varepsilon \text{ for } |x - c| < \delta,$$

so that c is a point of continuity of the function f.

Now suppose that c is a point of continuity of f. Let $\varepsilon > 0$ be given. Now there exists $\delta > 0$ such that

$$|f(x) - f(c)| < \frac{1}{2}\varepsilon \text{ for } |x - c| < \delta.$$

Also there exists a division

$$D_1\,\{c = y_0, y_1, ..., y_{q-1}, y_q, ..., y_n = b\}$$

of $[c, b]$ such that

$$\sum_{q=1}^{n} \left| f(y_q) - f(y_{q-1}) \right| > V_f(c, b) - \frac{1}{2}\varepsilon. \qquad ...(4)$$

Since as a result of introducing additional points to the division D_1, the corresponding sum of the moduli of the differences of the function values at end points will not decrease, we may assume that $0 < (y_1 - c) < \delta$ so that

$$\left| f(y_1) - f(c) \right| < \frac{1}{2}\varepsilon. \qquad ...(5)$$

Thus (4) becomes

$$V_f(c, b) - \frac{1}{2}\varepsilon < \frac{1}{2}\varepsilon + \sum_{q=2}^{n} \left| f(y_q) - f(y_{q-1}) \right| \le \frac{1}{2}\varepsilon + V_f(y_1, b),$$

$$\Rightarrow \qquad V_f(c, b) - V_f(y_1, b) = \varepsilon,$$

$$\Rightarrow \qquad V(y_1) - V(c) = V_f(c, y_1) = \varepsilon.$$

Thus for $0 < (y_1 - c) < \delta$, we have

$$-\varepsilon < 0 < V(y_1) - V(c) < \varepsilon.$$

It follows that

$$\lim_{x \to (c+0)} V(x) = V(c)$$

Similarly $$\lim_{x \to (c-0)} V(x) = V(c).$$

Thus V is continuous at c.

From the preceding we deduce *that V is continuous in $[a, b]$ if and only if f is continuous in $[a, b]$*.

Cor. A continuous function is of bounded variation if and only if it can be expressed as the difference of two continuous monotonically increasing functions.

Ex. 1. Show that a polynomial function is of bounded variation over any finite interval.

Ex. 2. Show that sin x and cos x are of bounded variation over a finite interval.

EXAMPLES

1. *If f is a continuous function of x satisfying the functional equation.*

$$f(x+y) = f(x) + f(y),$$

show that $f(x) = ax$ where, a, is a constant. (*Poorv. 90*)

Sol. Taking $x = 0 = y$, we obtain

$$f(0) = 0 \qquad ...(1)$$

and taking $y = -x$, we obtain

$$f(-x) = -f(x). \qquad ...(2)$$

If, x, be a positive integer, we have

$$\begin{aligned} f(x) &= f(1+1+...+1) \\ &= f(1)+f(1)+...+f(1) \\ &= xf(1) = ax, \text{ say,} \end{aligned}$$

where $f(1) = a$.

Let x, now, be a negative integer. We write $x = -y$ so that y is a positive integer. We have

$$f(x) = f(-y) = -f(y) = -ay = ax$$

Again let $x = p/q$ be a rational number; q being positive. We have

$$f(p) = f\left(\frac{p}{q}q\right) = f\left(\frac{p}{q}\right) + f\left(\frac{p}{q}\right) + ...q \text{ times} = qf\left(\frac{p}{q}\right)$$

$$\Rightarrow \qquad ap = qf\left(\frac{p}{q}\right)$$

$$\Rightarrow \qquad f(x) = ax,$$

where x is a rational number p/q.

Suppose, now, that x is any real number.

Let $\{x_n\}$ be a sequence of rational numbers such that

$$\lim x_n = x.$$

We have, x_n, being rational,

$$f(x_n) = ax_n. \qquad ...(3)$$

Let $n \to \infty$. As f is a continuous function, we obtain from (3)

$$f(x) = ax \; \forall \; x.$$

Hence the result.

2. *Discuss the function defined by*

$$f(x) = \begin{cases} 1/q, & \textit{when } x \textit{ is a rational } p/q \neq \textit{ in its lowest terms,} \\ 0, & \textit{when } x \textit{ is irrational and } x = 0 \end{cases}$$

as regards its continuity.

Sol. It will be shown that f is continuous for irrational values of x and for $x = 0$ and discontinuous for *non-zero* rational values of x.

Let $p/q \neq 0$ be a rational number so that

$$f\left(\frac{p}{q}\right) = \frac{1}{q}.$$

We know that in every interval there lie an infinite number of irrational numbers. Accordingly in every neighbourhood of p/q, there exist irrational numbers, such that for these numbers x, we have

$$\left|f(x) - f\left(\frac{p}{q}\right)\right| = \left|0 - \frac{1}{q}\right| = \frac{1}{q}.$$

Thus, if, ε, be a positive number $< 1/q$, there cannot exist a neighbourhood of p/q for every point x of which

$$\left|f(x) - f\left(\frac{p}{q}\right)\right| < \varepsilon$$

Hence f is discontinuous for $x = p/q$.

Let, now, c, be an irrational number, so that $f(c) = 0$.

Suppose that we arrange the fractions $1/q$ in the order of increasing q.

Let, ε, be a positive number. Then there exist only a finite number of fractions $1/q$ such that $q < (1/\varepsilon)$. We can thus enclose, c, in an interval which does not enclose any rational number p/q for which $q < (1/\varepsilon)$. Then for irrationals x in this interval.

$$|f(x) - f(c)| = 0 < \varepsilon.$$

and for any rational $x = p/q$ in this interval

$$|f(x) - f(c)| = \frac{1}{q} < \varepsilon.$$

Thus f is continuous for $x = c$.

***3.** *Consider the continuity for $x = 0$ of f defined by*

$f(x) = 1/(1 - e^{1/x})$ when $x \neq 0$ and $f(0) = 0$.

Sol. Now $1/x \to \infty$ or $-\infty$ according as $x \to (0+0)$ or $(0-0)$.

$\Rightarrow$ $e^{1/x} \to \infty$ or 0 according as $x \to (0+0)$ or $(0-0)$.

Thus we have

$$\lim_{x \to (0+0)} f(x) = 0, \quad \lim_{x \to (0-0)} f(x) = 1.$$

Since these two limits are different, $\lim f(x)$ as $x \to 0$, does not exist. Thus f is discontinuous for $x = 0$ and the point of discontinuity is of the second kind.

Of course the function is continuous on the right and has a discontinuity of the second kind on the left of $x = 0$.

***4.** *Discuss the nature of discontinuity of the function f defined by*

$$f(x) = \lim_{n \to \infty} \frac{\log(2+x) - x^{2n} \sin x}{1 + x^{2n}},$$

as x = 1.

Show that f (0) and f (π/2) differ in sign, and explain still why f does not vanish in [0, π/2].

Sol. We know that

$$\lim_{n \to \infty} x^{2n} = \lim_{n \to \infty} \left(x^2\right)^n = \begin{cases} \infty \text{ if } x^2 > 1, \\ 1 \text{ if } x^2 = 1, \\ 0 \text{ if } x^2 < 1. \end{cases} \quad \text{(Ex. 6, page 64)}$$

Thus,

for $x > 1$,

$$f(x) = \lim_{n \to \infty} \frac{\left(1/x^{2n}\right) \log(2+x) - \sin x}{\left(1/x^{2n}\right) + 1} = -\sin x, \quad \text{...(i)}$$

for $-1 < x < 1$,

$$f(x) = \lim_{n \to \infty} \frac{\log(2+x) - x^{2n} \sin x}{1 + x^{2n}} = \log(2+x), \quad \text{...(ii)}$$

for $x = 1$,

$$f(x) = f(1) = \frac{\log 3 - \sin 1}{2}.$$

$$\therefore \quad \lim_{x \to (1+0)} f(x) = -\sin 1, \quad \lim_{x \to (1-0)} f(x) = \log 3.$$

As these two right and left hand limits $-\sin 1$ and $\log 3$ [being at least of opposite signs] are unequal, we see that

$$\lim_{x \to 1} f(x)$$

does not exist. Thus the function f is discontinuous at $x = 1$ and discontinuity is of the *second* kind.

Now, since $\pi/2 > 1$, we have from (*i*),

$$f(\pi/2) = -1 < 0.$$

Also since $-1 < 0 < 1$, we have from (*ii*),

$$f(0) = \log 2 > 0.$$

$\therefore f(0)$ and $f(\pi/2)$ are of opposite signs.

The point, 1, of discontinuity of f belongs to the interval $[0, \pi/2]$ so that the *function is not continuous in this closed interval.* This explains why f may not vanish anywhere in $[0, \pi/2]$ in spite of the fact that $f(0)$ and $f(\pi/2)$ are of opposite signs.

It will be seen that in the interval $[0, \pi/2]$, the function f is defined as follows :

$$f(x) = \begin{cases} \log(2+x) & \text{when } 0 \le x < 1, \\ \frac{1}{2}(\log 3 - \sin 1) & \text{when } x = 1. \\ -\sin x & \text{when } 1 < x \le \pi/2, \end{cases}$$

so that it is nowhere zero in $[0, \pi/2]$.

5. *Show that* $f(x) = \dfrac{1}{1+e^{1/(\sin n!\pi x)}}$ *can be made discontinuous at any rational point in the interval* (0, 1) *by the proper selection of n.*

(*Garhwal, 94*)

Sol. Let x be a rational number say p/q, where p and q are integers prime to each other. Fixing x to be q or greater than q, $n!\pi x = n!\pi \dfrac{p}{q}$ will be some integral multiple of π, and therefore $\sin(n!\pi x)$ will be zero, but $\cos(n!\pi x)$ will be 1 or -1, according as $n!x$ is an even or odd integer.

Now, $\{n!\pi(x \pm h)\}$

$$= \sin(n!\pi x)\cos(n!\pi h) \pm \cos(n!\pi x)\sin(n!\pi h)$$

$$= \pm\sin(n!\pi h) \quad \text{or} \quad \mp\sin(n!\pi h)$$

according as $n!x$ is even or odd.

When n!x is an even integer, then

$$f(x+0) = \lim_{h\to 0} \frac{1}{1+e^{1/\sin\{n!\pi(x+h)\}}} = \lim_{h\to 0} \frac{1}{1+e^{1/\sin(n!\pi h)}} = 0$$

$$f(x-0) = \lim_{h\to 0} \frac{1}{1+e^{1/\sin\{n!\pi(x-h)\}}}$$

$$= \lim_{h\to 0} \frac{1}{1+e^{-1/\sin(n!\pi h)}} = 1.$$

when $n!x$ *is an odd integer*

then
$$f(x+0) = \lim_{h\to 0} \frac{1}{1+e^{-1/\sin(n!\pi h)}} = 1$$

$$f(x-0) = \lim_{h\to 0} \frac{1}{1+e^{1/\sin(n!\pi h)}} = 0$$

Hence, $f(x)$ is discontinuous at any rational point x, where n has been suitably fixed in value.

6. *Let* $f(x) = x\left\{1+\frac{1}{3}\sin\log\left(x^2\right)\right\}$ *for* $x \neq 0$, $f(x) = 0$, *for* $x = 0$. *Show that* $f(x)$ *is continuous and monotone.* *(Patna 2003)*

Sol. $f(0) = 0$ (by definition)

$$f(+0) = \lim_{h\to 0} f(0+h) = \lim_{h\to 0} h\left[1+\frac{1}{3}\sin\log\left(h^2\right)\right]$$

$= 0$, since $\sin\left(\log h^2\right)$ lies between -1 and 1.

Also
$$f(-0) = \lim_{h\to 0} -h\left[1+\frac{1}{3}\sin\log\left(h^2\right)\right] = 0$$

so the function is continuous at $x = 0$.

We shall show that even when $x \neq 0$.

$$\lim_{h\to 0}\left[f(x\pm h) - f(x)\right] = 0$$

i.e., we shall prove $f(x)$ to be continuous for all non-zero values of x and also that $f(x)$ is monotone.

Now, $f(x+h) - f(x)$ (taking h as positive)

$$= (x+h)\left[1+\frac{1}{3}\sin\log(x+h)^2\right] - x\left[1+\frac{1}{3}\sin\log(x)^2\right]$$

$$= h+\frac{x}{3}\left[\sin\log(x+h)^2 - \sin\log x^2\right] + \frac{h}{3}\left[\sin\log(x+h)^2\right]$$

$$= h\left[1+\frac{1}{3}\sin\log(x+h)^2\right] + \frac{2x}{3}\left[\cos\log\{x(x+h)\} \times \sin\log\left(1+\frac{h}{2}\right)\right]$$

$$= h\left[1+\frac{1}{3}\sin\log(x+h)^2\right] + \frac{2x}{3}\left[\cos\left\{\log x^2 + \log\left(1+\frac{h}{x}\right)\right\}\right]$$

$$\times \sin\left(\frac{h}{x}-\frac{h^2}{2x^2}+\ldots\right)\Bigg]$$

$$= h\left[1+\frac{1}{3}\sin\log(x+h)^2\right]+\frac{2x}{3}\cos\left(\log x^2+\frac{h}{x}-\frac{h^2}{2x^2}+\ldots\right)$$

$$\left[\frac{h}{x}-\frac{h^2}{2x^2}+\ldots\right]$$

$$= h\left[1+\frac{1}{3}\sin\log(x+h)^2\right]+\frac{2x}{3}\left(\frac{h}{x}\right)\cos\left(\log x^2+\frac{h}{2}\right)$$

....

leaving out higher powers of h.

Taking – 1 as the least value of the sine and the cosine,

So $f(x)$ is an ever increasing function, *i.e.*, is monotonic and continuous.

7. *In the closed interval (– 1, 1), let f (x) defined as* $x^2 \sin\frac{1}{x^2}$ *for* $x \neq 0$, *and* $f(0) = 0$. *In the given interval, (a) Is the function bounded ? (b) Is it continuous ? (c) Is it uniformly continuous ? (d) Is it absolutely continuous ?*

Sol. Here $\qquad f(0) = 0,$

and $\qquad f(0+h) = h^2 \sin\frac{1}{h^2}, \quad \therefore\ f(+0) = 0$

Also, $\qquad f(-0) = 0.$

Hence $f(x)$ is continuous at $x = 0$.

Again $\qquad f(\alpha) = \alpha^2 \sin\frac{1}{\alpha^2}$

$\therefore \qquad f(\alpha+0) = \alpha^2 \sin\frac{1}{\alpha^2} = f(\alpha-0),$

where $-1 \leq \alpha < 0$ or $0 < \alpha \leq 1$.

So, the function is continuous at $x = \alpha \neq 0$.

Thus the function is *continuous* for all points in the interval (– 1, 1). Since it is continuous, it is *bounded* and also *uniformly continuous,* in the interval.

Now the fluctuation of $f(x)$ in the interval $\left[\frac{1}{\sqrt{\pi(r+1)}}, \frac{1}{\sqrt{\pi r}}\right]$ is greater than $\frac{2}{(r+1)\pi}$.

Therefore the sum of the absolute value of fluctuations in the sequence of intervals $\left[\dfrac{1}{\sqrt{(r+1)\pi}}, \dfrac{1}{\sqrt{r\pi}}\right]$ corresponding to $r = n, n + 1, \ldots$ is greater than

$$\frac{2}{\pi}\left[\frac{1}{n+1}+\frac{1}{n+2}+\ldots\right]$$

which exceeds any finite bound as the series within brackets is divergent, while, as is easily seen that the sum of the infinite set of the intervals tends to zero as r tends to infinite.

Hence, however large n may be, and so however, small the neighbourhood of the origin may be, the function is not of bounded total fluctuation, *i.e.,* it is not absolutely continuous although uniformly continuous in this neighbourhood.

8. *Consider the function $f(x) = x - [x]$ where x is a positive variable and $[x]$ denotes the greatest integer in x; and show that*

(i) $f(x)$ is discontinuous for all integral values of x, and continuous for all other values. (*Poorv. 90*)

(ii) In every interval which includes an integer, $f(x)$ is bounded between 0 and 1, and

(iii) the lower bound 0 is attained but the upper bound is not.

Sol. (*i*) We have $\quad f(x) = x - [x]$.

Let α be a positive integer.

Then
$$\begin{aligned} f(x) &= \alpha - [\alpha] = \alpha - \alpha = 0 \\ f(\alpha + h) &= (\alpha + h) - [\alpha + h] = \alpha + h - \alpha = h \\ f(\alpha + 0) &= 0 \end{aligned}$$

Also
$$\begin{aligned} f(\alpha - h) &= (\alpha - h) - [\alpha - h) \\ &= (\alpha - h) - (\alpha - 1) = 1 - h \end{aligned}$$

$\therefore \quad f(\alpha - 0) = 1$

Thus $\quad f(\alpha - 0) \neq f(\alpha + 0)$

$\therefore$ The function is discontinuous for integral values of x.

Again, if $x = \alpha + \beta$, where α is an integer and $0 < \beta < 1$, we have

$$\begin{aligned} f(\alpha + \beta) &= (\alpha + \beta) - [\alpha + \beta] = \alpha + \beta - \alpha = \beta \\ f(\alpha + \beta + h) &= (\alpha + \beta + h) - [\alpha + \beta + h] \\ &= \alpha + \beta + h - \alpha = \beta + h \end{aligned}$$

$\therefore \quad f(\alpha + \beta + 0) = \beta$

Again, $\quad f(\alpha + \beta - h) = (\alpha + \beta - h) - [\alpha + \beta - h]$
$$= \alpha + \beta - h - \alpha = \beta - h$$

$\therefore \quad f(\alpha + \beta - 0) = \beta.$

Thus, $\quad f(\alpha + \beta + 0) = f(\alpha + \beta) = f(\alpha + \beta - 0)$

Hence $f(x)$ is continuous for non-integral value of x.

Sol. (*ii*) $\quad f(x) = x - [x] = f_r(x),$

where $f_r(x)$ means the fractional part of x.

Let us consider the interval $(\alpha - H_1, \alpha + H_2)$ where each of H_1 and H_2

is positive and less than unity and α is a positive integer.

Since $\quad f_r(\alpha) = f(\alpha + 0) = 0$

and $\quad f_r(\alpha - 0) = 1.$

We observe that in an arbitrarily small neighbourhood $(\alpha - \varepsilon, \alpha + \varepsilon)$ of the point α these are infinite number of points of each of which the functional values $f_r(x)$ differ from 1 by an arbitrarily small quantity and also these are infinite number of points in the same neighbourhood at each of which the functional value $f_r(x)$ differ from zero also by an arbitrarily small quantity, and in the nature of things $f_r(x)$ is always less than 1 and can never be less than zero. Hence in any interval $(\alpha - H_1, \alpha + H_2)$ in which α and therefore $(\alpha - \varepsilon, \alpha + \varepsilon)$ is contained the upper bound of $x - [x] = f_r(x)$ is 1 and the lower bound is zero. Hence $f_r(x)$ is bounded between 0 and 1.

(*iii*) Since $f_r(x)$ is never equal to 1 hence $(\alpha - H_1, \alpha + H_2)$ for which $f_r(x) = 1$ and there is decidedly $x = \alpha$ in this interval for which $f_r(x) = 0$. Hence the upper bound of $f_r(x)$ namely 1 is not attained while the lower bound namely 0 is attained in any interval $(\alpha - H_1, \alpha + H_2)$ which includes an integral point $x = \alpha$.

EXERCISES

1. State with reasons for your conclusion which of the following sets of circumstances are sufficient and which of them are not to determine the value $f(0)$ for a function f, giving the value of $f(0)$ where possible :

(*i*) f is continuous at $x = 0$ and takes in any neighbourhood of $x = 0$ both positive and negative values.

(*ii*) For any positive ε, there is a δ such that

$$|f(x)| < \varepsilon \quad \text{for } 0 < |x| < \delta.$$

(*iii*) $[f(h) + f(-h) - 2f(0)]/h \to l$ and $f(h) \to a$ as $h \to 0$.

2. Show that the sum-function of the infinite saries

$$x^2 + x^2(1 - x^2) + x^2(1 - x^2)^2 + \dots + x^2(1 - x^2)^n + \dots.$$

is not continuous for $x = 0$ even though every term of series is continuous for $x = 0$.

3. Examine the continuity for $x = 0$ of the sum-functions of the following series :

(*i*) $$\frac{x}{x+1} + \frac{x}{(x+1)(2x+1)} + \frac{x}{(2x+1)(3x+1)} + \dots$$

(*ii*) $\displaystyle\sum_{n=0}^{\infty} (e^{-n|x|} - e^{-2n|x|})$ (*iii*) $\displaystyle\sum_{n=0}^{\infty} \frac{x^2}{\left[1 + nx^2\right]\left[1 + (n+1)x^2\right]}$

4. Let f be continuous in $[-1, 1]$ and assume rational values only and let $f(0) = 0$. Prove that $f(x) = 0$ everywhere.

5. A function f is continuous in the interval $[0, 1]$ and assumes only rational values in the entire interval. If $f(x) = \frac{1}{2}$ when $x = \frac{1}{2}$, prove that $f(x) = \frac{1}{2}$ everywhere.

6. Show that the function ϕ where $\phi(x) = x$ when x is rational and $\phi(x) = 1 - x$ when x is irrational assumes every value between 0 and 1 once and only once as x increases from 0 to 1, but is discontinuous for every value of x except $x = \dfrac{1}{2}$.

7. Discuss the continuity of f where

$$f(x) = \begin{cases} 0, & \text{when } x \text{ is irrational or zero.} \\ \dfrac{1}{q^3}, & \text{when } x = \dfrac{p}{q}, \text{ a fraction in its lowest terms.} \end{cases}$$

***8.** If a continuous function f satisfies the question

$$f(x + y) = f(x) f(y),$$

then either $f(x) = 0$ or else $f(x) = e^{ax} \ \forall \ x$.

9. Let S be an interval $]\, a, b\, [$. Show that

(*i*) there exists a continuous function on S which is not bounded.

(*ii*) there exists a continuous and bounded function on S not having greatest and least values.

(*iii*) there exists a continuous function on S which is not uniformly continuous.

***10.** Show that if

$$f(x) = (x-1)/[1 + e^{1/(x-1)}], \text{ if } x \neq 1; f(1) = 0.$$

f is continuous for every value of x.

***11.** Show that if

$$f(x) = (e^{1/x} - 1)/(e^{1/x} + 1), \text{ when } x \neq 0; f(0) = 0$$

is continuous for every value of x except $x = 0$. What is the nature of discontinuity at $x = 0$?

***12.** Consider the continuity of f where

$$f(x) = e^{1/x^2} (e^{1/x^2} - 1), \text{ when } x \neq 0; f(0) = 1.$$

13. Given the function $f(x) = \lim\limits_{t \to \infty} \dfrac{(1 + \sin \pi x)^t - 1}{(1 + \sin \pi x)^t + 1}$ is discontinuous at the points $x = 0, 1, 2, \ldots, n, \ldots$. (*Kumaon 99*)

14. Show that the function $f(x) = \dfrac{1}{x}$ is continuous in $(0, 1)$ when $x > 0$, but $f(x)$ is not uniformly continuous. (*Agra 92*)

OBJECTIVE QUESTIONS

Note. *For each of the following questions four alternatives are given for the answer. Only one of them is correct. Choose the correct alternative.*

1. Limit of a function $f(x)$ when $x \to a$ exists if :

(*a*) any one of $f(a-0)$ and $f(a+0)$ exists, (*b*) both $f(a-0)$ and $f(a+0)$ exist

(*c*) $f(a-0) = 0 = f(a+0)$ (*d*) $f(a-0) = f(a+0)$

2. A function $f(x)$ is said to be continuous at $x = a$ if :

(*a*) $\lim\limits_{x \to a} f(x)$ exists (*b*) $f(a)$ exists

(*c*) $\lim\limits_{x \to a} f(x) = f(a)$ (*d*) $\lim\limits_{x \to 0} f(x) \neq f(a)$

3. A polynomial function in R :
(a) is never continuous in **R** (b) may or may not be continuous in **R**
(c) is always continuous in **R** (d) is continuous in **R** except at $x = 0$.

4. If the function $f(x)$ is defined on **R** by :

$$f(x) = \begin{cases} 1, \text{ when } x \text{ is rational} \\ -1, \text{ when } x \text{ is irrational} \end{cases} \text{ then :}$$

(a) $f(x)$ is continuous at $x = 1$ only (b) $f(x)$ is continuous at $x = -1$ only
(c) $f(x)$ is continuous at $x = \pm 1$ only (d) $f(x)$ is not continuous at any point

5. Which two of the functions $\sin x$, e^x and $\log x$ are continuous on **R**?
(a) $\sin x$ and e^x (b) $\sin x$ and $\log x$
(c) e^x and $\log x$ (d) none of these

6. In the $(\in, \delta)$ definition of continuity of a function $f(x)$ at a point $x = a$, δ depends, in general, upon :
(a) $\in$ only (b) a only (b) both $\in$ and a (d) neither $\in$ nor a.

7. A function $f(x)$ is said to be continuous at a point $x = a$ if for each $\in > 0$ there exists a, $\delta > 0$, such that :
(a) $|f(x) - f(a)| < \in \Rightarrow |x - a| < \delta$
(b) $|f(x) - f(a)| < \in \Rightarrow |x - a| > \delta$
(c) $|x - a| < \delta \Rightarrow |f(x) - f(a)| > \in$
(d) $|x - a| < \delta \Rightarrow |f(x) - f(a)| < \in$

8. The implication $|x-a| < \delta \Rightarrow |f(x) - f(a)| < \in$ in the definition of continuity is equivalent to :
(a) $(f(a) - \in, f(a) + \in) \subseteq f(a - \delta, a + \delta)$
(b) $f(a - \delta, a + \delta) \subseteq (f(a) - \in, f(a) + \in)$
(c) $f([a - \delta, a + \delta]) \subseteq (f(a) - \in, f(a) + \in)$
(d) none of these

9. Which of the following is continuous at $x = 0$?
(a) $f(x) = 1/x$ (b) $f(x) = |x|/x$ (c) $f(x) = |x|$ (d) $f(x) = x/|x|$.

10. If $f(x) = \dfrac{\sin\{x\}}{\{x\}}$, $\{x\} \neq 0$ $= 0$, $\{x\} = 0$,
where $\{x\}$ denotes the greatest integer less than or equal to x, then $\lim\limits_{x \to 0} f(x)$ equals to :
(a) 1 (b) 0 (c) -1 (d) none of these

11. In order that function $f(x) = (x + 1)^{\cot x}$ is continuous at $x = 0$, $f(0)$ must be defined as :
(a) $f(0) = 0$ (b) $f(0) = e$ (c) $f(0) = 1/e$ (d) none of these

12. $$f(x) = \begin{cases} -1; & x < -1 \\ -x; & -1 \le x \le 1 \\ 1; & x > 1 \end{cases}$$
is continuous
(a) at $x = 1$ but not at $x = -1$ (b) at $x = -1$ but not at $x = 1$
(c) at both $x = 1$ and $x = -1$ (d) at none of $x = 1$ and $x = -1$.

ANSWERS

1. (d) 2. (c) 3. (c) 4. (d) 5. (a) 6. (b) 7. (d)
8. (b) 9. (c) 10. (d) 11. (b) 12. (d)

5

Real Valued Functions of a Single Real Variable. Derivability

5.1. DERIVABILITY

Derivative. Lef f be a real valued function will an interval $[a, b]$ as its domain.

Derivability at a point. Let c be a point of $[a, b]$.

Consider
$$\phi(x) = \frac{f(x) - f(c)}{x - c}, \quad x \neq c.$$

The domain of definition of the function ϕ is the interval $[a, b]$ excluding the point c, *i.e.*, the set

$$\{x : a \leq x < c, c < x \leq b\}.$$

The function f is said to be derivable at c, if

$$\lim_{x \to c} \frac{f(x) - f(c)}{x - c}$$

exists and the limit, called the derivative of f at c, is denoted by the symbol $f'(c)$.

[It may be seen that the fact or otherwise of the derivability of f at c depends only on the values of the function f in a neighbourhood of c; no particular neighbourhood, however, being material].

Right-hand derivative. Left-hand derivative *(Garhwal 90, 91)*

We are led to the consideration of the right hand and left hand derivatives

$$f'(c - 0), f'(c + 0)$$

on the basis of the right hand and left hand limits of $\phi(x)$.

It may be seen that

$$f'(c) \text{ exist} \Leftrightarrow f'(c + 0) \text{ and } f'(c - 0) \text{ exist and are equal.}$$

Derivatives at *a* and *b*. The function f is derivable at a if $f'(a+0)$ exists and by $f'(a)$ would then be meant $f'(a+0)$. Similarly f is derivable at b if $f'(b-0)$ exists and by $f(b)$ would then be meant $f'(b-0)$.

Derivability in [*a*, *b*]. The function f is said to be derivable in $[a, b]$ if it is derivable at every point thereof.

If f is derivable in $[a, b]$, there arises another function f' with domain $[a, b]$ such that the value of f' for any $x \in [a, b]$ is $f'(x)$.

The function f' is said to be the derivative of the function f.

Ex. Show that the functions f defined in **R** such that $\forall x$

(*i*) $f(x) = k$; k being a given number

(*ii*) $f(x) = x$ are derivable at every point.

5.1.1. Continuity and Derivability

Theorem. *Derivability at a point ⇒ Continuity at the point.*

(Kanpur 2003, 2004, Poorv. 91; Garhwal 90, 95; Rohilkhand 93, Ajmer 98)

Thus we shall show that a necessary condition for a function to be derivable at a point is that it is continuous at the point.

Let f be derivable at c so that

$$\lim_{x \to c} \frac{f(x) - f(c)}{x - c}$$

exists. We have

$$f(x) - f(c) = \left[\frac{f(x) - f(c)}{x - c}\right](x - c)$$

$$\Rightarrow \quad \lim_{x \to c} [f(x) - f(c)] = \lim_{x \to c} \frac{f(x) - f(c)}{x - c} \lim_{x \to c} (x - c)$$

$$= f(c).0 = 0$$

$$\Rightarrow \quad \lim_{x \to c} f(x) = f(c)$$

and as such f is continuous at c.

While continuity is a necessary condition for derivability, it is not sufficient as the following example will illustrate.

Let f be defined by

$$f(x) = |x|.$$

We shall show that while f is continuous at 0, it is not derivable at 0.

We have $$\frac{f(x) - f(0)}{x - 0} = \frac{|x|}{x} = \begin{cases} 1 \text{ if } x > 0, \\ -1 \text{ if } x < 0. \end{cases}$$

Thus $\lim_{x \to (0+0)} \dfrac{f(x) - f(0)}{x - 0} = 1 \neq -1 = \lim_{x \to (0-0)} \dfrac{f(x) - f(0)}{x - 0}$

$$\Rightarrow \qquad \lim_{x \to 0} \frac{f(x) - f(0)}{x - 0}$$

does not exist and as such f is not derivable at 0.

To example the continuity at 0, we take $\varepsilon > 0$.

We have $\left| f(x) - f(0) \right| = |x| < \varepsilon$ when $|x| < \delta$,

δ being any + *ve* number less than ε. Thus f is continuous at 0.

While f defined by $f(x) = |x|$

is not derivable for $x = 0$, it is derivable for every non-zero x. In fact it may be easily seen that

$$f(x) = \begin{cases} 1 & \text{if } x > 0, \\ -1 & \text{if } x < 0. \end{cases}$$

Thus there is only one point where this function ceases to be derivable while remaining continuous.

Weierstrass was the first mathematician who gave a function which is continuous for every value of x but not derivable for any value of x.

It is of course quite easy to construct functions having any finite number of points of continuity and non-derivability.

EXAMPLES

1. *If $f(x) = x^2 \sin(1/x)$, for $x \neq 0$ and $f(0) = 0$, then show that $f(x)$ is continuous and differentiable everywhere and the $f(0) = 0$. Also show that the function $f'(x)$ is discontinuous at $x = 0$.*

(Garhwal 93; Kanpur 95; Rohilkhand 90, 94, 96; Devi Ahilya 2001, Patna 2003)

Sol. $f(0+0) = \lim_{h \to 0} f(0+h) = \lim_{h \to 0} h^2 \sin(1/h) = 0,$

$f(0+0) = \lim_{h \to 0} f(0-h) = \lim_{h \to 0} \left(-h^2\right) \sin(-1/h) = 0.$

$\therefore \quad f(0+0) = f(0-0) = f(0),$

so the function is continuous at $x = 0$.

$$\mathbf{R} f'(0) = \lim_{h \to 0} \frac{f(0+h) - f(0)}{h}$$

$$= \lim_{h \to 0} \frac{h^2 \sin(1/h)}{h} = \lim_{h \to 0} h \sin(1/h) = 0$$

and $$Lf'(0) = \lim_{h \to 0} \frac{f(0-h) - f(0)}{h} = \lim_{h \to 0} h \sin(1/h) = 0$$

Thus $f(x)$ is differentiable at $x = 0$ and $f'(0) = 0$.

For all other values of x, $f(x)$ is easily seen to be continuous and differentiable.

Now, $$f'(x) = 2x \sin(1/x) - \cos(1/x) \text{ at}$$

$$x \neq 0 \text{ and } f'(0) = 0$$

$$\therefore \quad f'(0+0) = \lim_{h \to 0} f'(0+h) = \lim_{h \to 0} \left(2h \sin\frac{1}{h} - \cos\frac{1}{h}\right)$$

which does not exist.

Similarly it can be shown that $f'(0-0)$ does not exist. Hence f' is discontinuous at the origin.

2. *Show that the function defined by*

$$f(x) = x\left\{1 + \frac{1}{3} \sin \log (x)^2\right\}, \; f(0) = 0$$

is everywhere continuous and monotone but has no differential coefficient at the origin.

Sol. $$f(0) = 0 \quad \text{...(i)}$$

$$f(+0) = \lim_{h \to 0} f(0+h) = \lim_{h \to 0} h\left[1 + \frac{1}{3} \sin \log \left(h^2\right)\right]$$

$$= 0 \quad \text{...(ii)}$$

since $\sin\left(\log h^2\right)$ lies between -1 and 1.

Also, $$f(0-h) = -h\left[1 + \frac{1}{3} \sin\left(\log h^2\right)\right]$$

$$\therefore \quad f(-0) = 0 \quad \text{...(iii)}$$

So the function is continuous at $x = 0$.

We shall show that even when $x \neq 0$.

$$\lim_{h \to 0} \left[f(x \pm h) - f(x)\right] = 0,$$

i.e., we shall prove $f(x)$ to be continuous for all non-zero values of x and also that $f(x)$ is monotone.

Sol. Now, $f(x+h) - f(x)$

$$= (x+h)\left[1 + \frac{1}{3}\sin\log(x+h)^2\right] - x\left[1 + \frac{1}{3}\sin\log x^2\right]$$

$$= h + \frac{x}{3}\left[\sin\log(x+h)^2 - \sin\log x^2\right] + \frac{h}{3}\left[\sin\log(x+h)^2\right]$$

$$= h + \frac{x}{3}\left[2\cos\frac{\{2\log(x+h) + 2\log x\}}{2} \times \sin\frac{\{2\log(x+h) - 2\log x\}}{2}\right] + \frac{h}{3}\left[\sin\log(x+h)^2\right]$$

$$= h\left[1 + \frac{1}{3}\sin\log(x+h)^2\right] + \frac{2x}{3}\left[\cos\log\{x(x+h)\} \times \sin\log\left(1 + \frac{h}{2}\right)\right.$$

$$= h\left[1 + \frac{1}{3}\sin\log(x+h)^2\right] + \frac{2x}{3}\left[\cos\left\{\log x^2 + \left(\frac{h}{x} - \frac{h^2}{2x^2} + \ldots\right)\right\} \times \sin\left(\frac{h}{x} - \frac{h^2}{2x^2} + \ldots\right)\right]$$

$$= h\left[1 + \frac{1}{3}\sin\log(x+h)^2\right] + \frac{2x}{3}\cdot\frac{h}{2}\cos\left(\log x^2 + \frac{h}{x}\right) \qquad \ldots(iv)$$

leaving out higher power of h

$$\geq \frac{2h}{3} - \frac{2h}{3} = 0 \qquad \ldots(v)$$

Taking -1 as the least value of the sine and the cosine in (*iv*).

So, $f(x)$ is an ever-increasing function, *i.e.*, is monotonic from (*v*), and continuous from (*ii*), as the expression in (*i*) or (*iv*) tends to zero as $h \to 0$ even if it is negative.

Now about differentiability at $x = 0$,

$$\mathbf{R}f'(0) = \lim_{h \to 0} \frac{f(0+h) - f(0)}{h}$$

$$= \lim_{h \to 0} \frac{h\left\{1 + \frac{1}{3}\sin\log h^2\right\} - 0}{h}$$

$$= \lim_{h \to 0} \left\{1 + \frac{1}{3}\sin(2\log h)\right\}$$

which is indeterminate.

Again, $\mathbf{L}f'(0) = \lim_{h \to 0} \frac{f(0-h) - f(0)}{-h}$

$$= \lim_{h \to 0} \frac{-h\left\{1 + \frac{1}{3}\sin\log h^2 - 0\right\}}{-h}$$

$$= \lim_{h \to 0} \left\{1 + \frac{1}{3}\sin(2\log h)\right\}$$

which is indeterminate as before.

So the function is not differentiable at $x = 0$.

3. *Show that $f(x) = x\tan^{-1}(1/x)$ for $x \neq 0$ and $f(0) = 0$ is continuous but not differentiable at $x = 0$.* (*Kanpur 2002*)

Sol. Test the function for continuity.

$$\mathbf{R}f'(0) = \lim_{h \to 0} \frac{h\tan^{-1}(1/h) - 0}{h} = \tan^{-1}\infty = \pi/2.$$

and $\mathbf{L}f'(0) = \lim_{h\to 0} \frac{-h\tan^{-1}(-1/h) - 0}{-h} = \lim_{h\to 0} \tan^{-1}(-1/h)$

$$= -\pi/2.$$

Since $\mathbf{R}f'(0) \neq \mathbf{L}f'(0)$, f is not differentiable at $x = 0$.

4. *A function f is defined by $f(x) = x^p\cos(1/x), x \neq 0; f(0) = 0$. What conditions shoud be imposed on p so that f may be (i) continuous at $x = 0$, (ii) differentiable at $x = 0$?*

Sol. $$f(0+0) = \lim_{h\to 0}\left[(0+h)^p \cos\left(\frac{1}{0+h}\right)\right]$$

$$= \lim_{h\to 0} h^p \cos(1/h) \quad ...(i)$$

and $$f(0-0) = \lim_{h\to 0}\left[(0-h)^p \cos\left(\frac{1}{0-h}\right)\right]$$

$$= \lim_{h\to 0} (-h)^p \cos(1/h) \quad ...(ii)$$

Now the function $f(x)$ is to be continuous at $x = 0$, then

$$f(0+0) = f(0) = f(0-0)$$

i.e., the limits given in (*i*) and (*ii*) must both tend to zero.

This is possible only if $p > 0$, which is the required condition.

Now, $$\mathbf{R}f'(0) = \lim_{h\to 0} \frac{h^p \cos(1/h) - 0}{h}$$

$$= \lim_{h\to 0} h^{p-1} \cos(1/h) \quad ...(iii)$$

and $$\mathbf{L}f'(0) = \lim_{h\to 0} \frac{(-h)^p \cos(-1/h) - 0}{-h}$$

$$= \lim_{h\to 0} -(-1)^p\, h^{p-1} \cos(1/h) \quad ...(iv)$$

Now if $f'(x)$ exists at $x = 0$ then we must have $\mathbf{R}f'(0) = \mathbf{L}f'(0)$ and this is possible only if $p - 1 > 0$, *i.e.*, $p > 1$ which gives $\mathbf{R}f'(0) = 0 = \mathbf{L}f'(0)$. Hence in order that f is differentiable at $x = 0$, p must be greater than 1.

5. *Determine the set of all points where the function* $f(x) = \dfrac{x}{(1+|x|)}$ *is differentiable.*

Sol. Since $|x| = x,\ x > 0$, $|x| = -x,\ x < 0$, $|x| = 0,\ x = 0$.

$\therefore$ $$f(x) = \frac{x}{1-x},\ x<0;\ f(x)=0,\ x=0;$$

$$f(x) = \frac{x}{1+x},\ x>0.$$

we have $$f(0+0) = \lim_{h\to 0} \frac{h}{1+h} = 0$$

$$f(0-0) = \lim_{h\to 0} \frac{-h}{1+h} = 0.$$

Since $f(0+0) = f(0) = f(0-0) = 0$, so the function is continuous at $x = 0$.

Further
$$\mathbf{L}f'(0) = \lim_{h \to 0} \frac{f(0-h) - f(0)}{h}$$

$$= \lim_{h \to 0} \frac{\{(-h/1+h)\} - 0}{-h}$$

$$= \lim_{h \to 0} \frac{1}{1+h} = 1;$$

$$\mathbf{R}f'(0) = \lim_{h \to 0} \frac{(h/1+h) - 0}{h} = 1.$$

So, the function is differentiable at $x = 0$. It is obviously differentiable for all other real values of x. Hence it is differentiable in the open interval $(-\infty, \infty)$.

EXERCISES

1. If $\phi(x) = -x$ for $x \le 0$, $\phi(x) = +x$ for $x \ge 0$, test for its continuity and differentiability at $x = 0$.

2. $f(x)$ be an even function. If $f'(0)$ exists, find its value.

3. $f(x) = \begin{cases} -1, & -2 \le x \le 0, \\ x - 1, & 0 < x \le 2 \end{cases}$ and $g(x) = f(|x|) + |f(x)|$. Test the differentiability of $g(x)$ in open interval $(-2, 2)$.

4. $f(x) = x \dfrac{e^{1/x} - e^{-1/x}}{e^{1/x} + e^{-1/x}}$, $x \ne 0$; $f(0) = 0$. Show that $f(x)$ is continuous but not derivable at $x = 0$.

5. Prove that the function $f(x) = |x|$ is continuous at $x = 0$, but not differentiable at $x = 0$, where $|x|$ means the numerical value or the absolute value of x. *(Garhwal 91; Kanpur 92, 96)*

6. Suppose the function satisfies the conditions :

(*i*) $f(x + y) = f(x) f(y) \;\forall\; x, y$

(*ii*) $f(x) = 1 + xy(x)$ where $\lim_{h \to 0} g(x) = 1$.

Show that the derivative $f'(x)$ exists and $f'(x) = f(x)$ for all x.

7. Show that f defined by

$$f(x) = |x| + |x-1|$$

is continuous but not derivable for $x = 0$, $x = 1$.

8. Show that f defined by

$$f(x) = |x| + |x-1| + |x-2|$$

is continuous and not derivable for 0, 1, 2.

9. Show that f where

$$f(x) = \text{degree of } (u^{x^2} + u^2 + 2u - 3)$$

is continuous but not derivable at 2.

5.1.2. Derivability in Relation to Algebraic Compositions

Let f and ϕ be two real valued functions defined in an interval $[a, b]$. Let $c \in [a, b]$.

Let f and ϕ be derivable at c. Then

$$f \pm \phi,\ f\phi,$$

are also derivable at c and if $\phi(c) \neq 0$,

$$f \div \phi$$

is also derivable at c. Moreover

$$(f \pm \phi)'(c) = f'(c) \pm \phi'(c),$$

$$(f\phi)'(c) = f'(c)\phi(c) + f(c)\phi'(c),$$

$$(f \div \phi)'(c) = [\phi(c) f'(c) - f(c)\phi'(c)] \div \phi^2(c),$$

The proofs which are simple are left to the reader.

Note 1. It follows from above that if two functions are derivable at a point (or in an interval), (*i*) their sum, difference and product are also derivable at the point (or in that interval) (*ii*) their quotient is derivable at that point (or in that interval) provided that the denominator of the quotient is not zero at that point (or at any point of that interval).

Note 2. It may be easily deduced from above that every polynomial and every rational function are derivable at each point of their domain of definition.

5.1.3. Derivability of the Composites of Derivable Functions

Let f be a function with domain $[a, b]$ and ϕ a function with domain $[\alpha, \beta]$ such that the range of the function f is a sub-set of the domain of ϕ.

Let f be derivable at $c \in [a, b]$ and let ϕ be derivable at $u = f(c)$. Here $u \in [\alpha, \beta]$.

It will now be proved that the composite $\phi \, o \, f$ defined by

$$(\phi \, o \, f)(x) = \phi(f(x))$$

is derivable at c and

$$(\phi \, o \, f)'(c) = \phi'(u) f'(c) = \phi'(f(c)) f'(c).$$

We write, $\quad t = f(x), x \in [a, b]; \quad y = \phi(t), t \in [\alpha, \beta].$

Because of the derivability of f at c, we have a relation

$$t - u = f(x) - f(c) = (x - c) f'(c) + \varepsilon_1 (x - c) \qquad ...(1)$$

where ε_1 depends upon x such that

$$\lim_{x \to c} \varepsilon_1 = 0.$$

Also because of the derivability of ϕ at u, we have a relation

$$\phi(t) - \phi(u) = (t - u) \phi'(u) + \varepsilon_2 (t - u) \qquad ...(2)$$

such that $\quad \lim_{t \to u} \varepsilon_2 = 0.$

With the help of (1), we re-write (2) in the form

$$\phi(f(x)) - \phi(f(c)) = (x - c) \phi'(u) f'(c) + \varepsilon_1 (x - c) \phi'(u)$$

$$+ \varepsilon_2 (x - c) f'(c) + \varepsilon_1 \varepsilon_2 (x - c).$$

From this we obtain

$$\lim_{h \to 0} \frac{\phi(f(x)) - \phi(f(c))}{x - c} = \phi'(u) f'(c)$$

$$\Rightarrow \quad \lim_{h \to 0} \frac{(\phi \, o \, f)(x) - (\phi \, o \, f)(c)}{x - c} = \phi'(u) f'(c)$$

$$= \phi'(f(c) f'(c)$$

Hence the result.

Note. The proof given in elementary text books, which is based on the equality

$$\frac{\Delta y}{\Delta x} = \frac{\Delta y}{\Delta t} \cdot \frac{\Delta t}{\Delta x},$$

is incomplete inasmuch as no heed is paid to the case which arises when $\Delta t = 0$ for *some* point in *every* neighbourhood of the point x.

Thus, for example, this proof will not apply to the function f and ϕ where

$$t = f(x) = x^2 \sin 1/x, \ x \neq 0; \ f(0) = 0,$$

$$y = \phi(t) = t^2$$

5.1.4. Derivability of the Inverses of Strictly Monotonic Derivable Functions

Let f be a derivable strictly monotonic function with domain $[a, b]$ so that, being continuous and strictly monotonic, it is an inversible function. We denote the inverse of f by ϕ.

Let $c \in [a, b]$ and let f be derivable at c with non-zero derivative $f'(c)$. We shall show that the function ϕ is also derivable at $f(c)$ and that

$$\phi'(f(c)) f'(c) = 1.$$

We write

$$y = f(x), \ x = \phi(y); \ u = f(c), \ c = \phi(u).$$

We have

$$\frac{\phi(y) - \phi(u)}{y - u} = \frac{x - c}{f(x) - f(c)}$$

$$= 1 \div \left[\frac{f(x) - f(c)}{x - c}\right] \quad \text{...(1)}$$

We have seen that ϕ is a continuous function so that

$$\lim_{y \to u} \phi(y) = \phi(u) \Leftrightarrow \lim_{y \to c} x = c.$$

From (1), on taking limit when $y \to u$, we obtain

$$\phi'(u) = 1 \div f'(c).$$

Hence the result.

5.2. MEANING OF THE SIGN OF DERIVATIVE AT A POINT

Let c be an interior point of the domain $[a, b]$ of a function f. We suppose that $f'(c)$ exists and is positive. We have

$$\lim_{h \to 0} \frac{f(x) - f(c)}{x - c} = f'(c) > 0$$

Let $\varepsilon > 0$ be a positive number $< f'(c)$. Now there exists $\delta > 0$ such that

$$\left| \frac{f(x) - f(c)}{x - c} - f'(c) \right| < \varepsilon \text{ when } 0 < |x - c| < \delta,$$

$$\Rightarrow f'(c) - \varepsilon < \frac{f(x) - f(c)}{x - c} < f'(c) + \varepsilon \text{ when } x \in [c - \delta, c + \delta],\ x \neq c.$$

As ε was chosen to be smaller than $f'(c)$, we see that $\dfrac{f(x) - f(c)}{x - c} > 0$ when $x \in [c - \delta, c + \delta], x \neq c$ so that

if $\quad c < x \le c + \delta,$ then $f(x) - f(c) > 0 \Leftrightarrow f(x) > f(c),$

and if $\quad c - \delta \le x < c,$ then $f(x) - f(c) < 0 \Leftrightarrow f(x) < f(c).$

Thus we conclude that if $f'(c) > 0$, then there exists a neighbourhood $[c - \delta, c + \delta]$ of c such that

$$f(x) > f(c)\ \forall\ x \in]c, c + \delta],\ f(x) < f(c)\ \forall\ x \in [c - \delta, c[.$$

The conclusion is often stated in the following manner :

$$f'(c) > 0 \Rightarrow f \text{ is increasing at } c.$$

Now suppose that $f'(c) > 0$. We define a function ϕ as follows :

$$\phi(x) = -f(x)$$

$$\Rightarrow \qquad \phi'(c) = -f'(c) > 0.$$

We conclude that if $f'(c) < 0$, there exists a neighbourhood $[c - \delta, c + \delta]$ of c such that

$$\forall\ x \in]c, c + \delta],\ \phi(x) > \phi(c) \Leftrightarrow f(x) < f(c)$$

$$\forall\ x \in [c - \delta, c, [\ \phi(x) < \phi(c) \Leftrightarrow f(x) < f(c)$$

i.e., if $f'(c) < 0$, the function f is decreasing at c.

We now consider the end points. It may be shown that

(*a*) there exists an interval $\left[a, a+\delta\right]$ such that

$$f'(a)<0 \quad = \quad f(x)>f(a), \forall\, x \in]\,a,\ a+\delta\,].$$

$$f'(a)<0 \quad = \quad f(x)<f(a), \forall\, x \in]\,a,\ a+\delta\,].$$

(*b*) there exists an interval $[\,b-\delta,\ b\,[$ such that

$$f'(b)<0 \quad = \quad f(x)<f(b), \forall\, x \in [\,b-\delta,\ b\,[.$$

$$f'(b)<0 \quad = \quad f(x)>f(b), \forall\, x \in [\,b-\delta,\ b\,[.$$

5.3. DARBOUX'S THEOREM

If f is derivable in a closed interval [a, b] and $f'(a), f'(b)$ are of opposite signs, then there exists at least one point c of the interval [a, b] such that $f'(c)=0$. *(Garhwal 92, 95; Kanpur 2003; Dibrugarh 2004)*

For the sake of definiteness we suppose that $f'(a)$ is positive and $f'(b)$ negative. On this account there exist intervals $]\,a,\ a+h\,], [\,b-h,\ b\,[$, $(h>0)$, such that

$$\begin{cases} \forall\, x \in]\,a,\ a+h\,],\ f(x)>f(a); & \text{...}(i) \\ \forall\, x \in [\,b-h,\ b\,[,\ \ f(x)>f(b). & \text{...}(ii) \end{cases}$$

Again the derivable function *f* is continuous in $\left[a, b\right]$ (§ 5.1.1). Therefore it is bounded and attains its bounds (§ 4.3.1, § 4.3.2 pp. 113, 114) Thus *M* be the supremum of *f* in [*a*, *b*], there exists $c \in [a, b]$ such that

$$f(c) \ = \ M.$$

From (*i*) and (*ii*), we see that the supremum is not attained at the end points *a* and *b* so that *c* is an interior point of [*a*, *b*].

If $f'(c)$ be positive, then there exists an interval $]\,c,\ c+\eta\,], (\eta>0)$ such that for every point *x* of this interval $f(x)>f(c)=M$ and this is again, a contradiction.

If $f'(c)$ be negative, then there exists an interval $[\,c-\eta,\ c\,[, (\eta<0)$ such that for every point *x* of this interval $f(x)>f(c)=M$ and this is, again, a contradiction.

Hence $$f'(c) \ = \ 0.$$

Cor. If *f* is derivable in a closed interval [*a*, *b*] and $f'(a) \neq f'(b)$ and *k* is any number lying between $f'(a)$ and $f'(b)$, then there exists at least one point, $c \in \left[a, b\right]$ such that

$$f'(c) \ = \ k.$$

We write $$\varphi(x) \ = \ f(x)-kx.$$

The function φ is derivable in [*a*, *b*] and

$$\varphi'(a) = f'(a) - k, \ \varphi'(b) = f'(b) - k$$

are of opposite signs. Therefore there exists at least one point, *c*, of [*a*, *b*] such that $\varphi'(c) = 0$ equivalently $f'(c) - k = 0$.

Note. We have seen in § 4.3.3, P. 115, that if $\varphi(a), \varphi(b)$ are of opposite signs, then φ vanishes for at least one value, *c*, in [*a*, *b*], if φ is continuous in [*a*, *b*]. In this context, the importance of the Darboux's theorem lies in the fact that if *φ is the derivation of a function,* then the conclusion of the vanishing of φ remains valid even when *φ is not continuous.*

5.4. ROLLE'S THEOREM

If a function f with domain [*a*, *b*] *is*

(*i*) *continuous in* [*a*, *b*]

(*ii*) *derivable in*] *a*, *b* [,

(*iii*) $f(a) = f(b)$,

then there exists $a \in]a, b[$ *such that* $f'(c) = 0$.

(Kanpur 92; Poorv. 90, 93; Garhwal 99; Rohilkhand 93, Ajmer 96, 97; Kumaon 98; Patna 2003)

[It will be seen that while we have assumed continuity in a closed interval, we have assumed derivability only in the open *i.e.*, to say for the conclusion to be valid, we do not need derivability at the end points *a*, *b*].

Proof. The function *f*, being continuous in [*a*, *b*], is bounded and attains its bounds. Let *M*, *m* be the bounds of *f* and let *c*, *d* be such that

$$f(c) = M, f(d) = m.$$

Either $\quad M = m \quad$ or $\quad M \neq m.$

Now $\quad M = m \Rightarrow f(x) = M \ \forall \ x \in [a, b].$

$$\Rightarrow f'(x) = 0 \ \forall \ x \in [a, b].$$

Thus the theorem is true in this case.

Now suppose that $\quad M \neq m.$

As $f(a) = f(b)$ and $M \neq m$ at least one of *M* and *m* must be different from *f* (*a*) and *f* (*b*). Suppose

$$M \neq f(a), \ M \neq f(b).$$

Now $\quad f(c) = M \neq f(a) \Rightarrow c \neq a.$

and $\quad f(c) = M \neq f(b) \Rightarrow c \neq b.$

Thus $\quad a < c < b$

The function is derivable at c. We shall show that

$$f'(c) = 0.$$

Let $$f'(c) > 0.$$

Then there exists $\delta > 0$ such that $\forall\, x \in]c, c+\delta]$.

$$f(x) > f(c) = M.$$

But M, being the supremum, we have $\forall\, x \in [a, b]$.

$$f(x) \leq M.$$

Thus we have a contradiction, so that we cannot have

$$f'(c) > 0.$$

Now suppose $f'(c) < 0$.

Then there exists $\delta > 0$ such that $\forall\, x \in [c-\delta, c[$,

$$f(x) = f(c) = M.$$

This is, again, not possible. Thus we cannot have $f'(c) > 0$.

We thus conclude that $f'(c) = 0$.

5.4.2. Lagrange's Mean Value Theorem

If a function f with domain [a, b] is

(*i*) *continuous in [a, b]*

(*ii*) *derivable] a, b [, then there exists* $c \in]a, b[$ such that

$$\frac{f(b) - f(a)}{b - a} = f'(c).$$ (*Poorv. 91, 93; Garhwal 93, 97: Kanpur 2005; Ajmer 95, 99; Kumaon 98*)

We consider a function ϕ defined as follows :

$$\phi(x) = f(x) + Ax,$$

where A is such that

$$\phi(a) = \phi(b).$$

This requires $$A = -[f(b) - f(a)]/(b-a).$$

Now ϕ is continuous in $[a, b]$, derivable in $]a, b[$, $\phi(a) = \phi(b)$ and as such, by Rolle's theorem, there exists $c \in]a, b[$ such that

$$\phi'(c) = 0.$$

Also $$\phi'(x) = f'(x) + A$$

$$\Rightarrow \qquad 0 = \phi'(c) = f'(c) + A$$

$$\Rightarrow \qquad \frac{f(b) - f(a)}{b - a} = f'(c).$$

Hence the theorem.

Another form of statement. If a function f is (i) continuous in [a, a + h] and (ii) derivable in]a, a + h [, then there exists at least one number θ, between 0 and 1, such that

$$f(a+h)-f(a) = hf'(a+\theta h),\ \theta \in]0,1[.$$

Cor. Let f satisfy the two conditions of the Lagrange's mean value theorem. We may then show that

(*i*) $f'(x)=0\ \forall\ x \in [a,b] \Rightarrow f$ is constant in $[a, b]$,

(*ii*) $f'(x)>0\ \forall\ x \in [a,b] \Rightarrow f$ is strictly increasing in $[a, b]$,

(*iii*) $f'(x)<0\ \forall\ x \in [a,b] \Rightarrow f$ is strictly decreasing in $[a, b]$.

The proofs, being straightforward, are left to the reader.

It may be noticed that when we say that f is strictly increasing in $[a, b]$ we mean that

$$x_2 > x_1 \Rightarrow f(x_2) > f(x_1).$$

We may similarly give a meaning to a strictly decreasing function.

Cor. *If f is continuous at a point c, then*

$$\lim_{h\to 0} f'(x)=l \Rightarrow f'(c)=l.$$

The condition that $f'(x)\to l$ as $x\to(c+0)$ implies that there exists an interval $]c, c+h[, (h>0)$ for every point x of which f' exists and therefore, f is continuous. Since f is given to be continuous at c also, we see that f is continuous in the closed interval $[c, c + h]$. If, x, be a point of this interval there exists $\xi \in]c, x[$ such that we have

$$f(x)-f(c) = (x-c)\ f'(\xi) \Rightarrow \frac{f(x)-f(c)}{x-c} = f'(\xi).$$

Let $x\to(c+0)$. Then we have

$$f'(c+0) = \lim_{x\to(c+0)} f'(\xi) = \lim_{x\to(c+0)} f'(x) = l.$$

It may similarly be shown that

$$f'(c-0) = l.$$

Thus $f'(c)$ exists and is equal to l.

Note. The theorem above is equivalent to saying that a derived function f' cannot have discontinuity of the first kind.

EXAMPLES

1. *Verify Rolle's Theorem for* $f(x) = 2x^3 + x^2 - 4x - 2$.

(Garhwal 94; Rohilkhand 90)

$f(x) = 0$ gives $x = -\frac{1}{2}, \sqrt{2}, -\sqrt{2}$.

Sol. Now, $f\left(-\frac{1}{2}\right) = 0,\ f\left(\sqrt{2}\right) = 0,$ and $f\left(-\sqrt{2}\right) = 0.$

Thus $f\left(-\sqrt{2}\right) = 0 = f\left(\sqrt{2}\right)$.

Also, $$\text{R}f'(x) = \lim_{h \to 0} \frac{2(x+h)^3 + (x+h)^2 - 4(x+h) - 2 - \left(2x^3 + x^2 - 4x - 2\right)}{h}$$

$$= 6x^2 + 2x - 4$$

and $$\text{L}f'(x) = \lim_{h \to 0} \frac{2(x-h)^3 + (x-h)^2 - 4(x-h) - 2 - \left(2x^3 + x^2 - 4x - 2\right)}{-h}$$

$$= 6x^2 + 2x - 4.$$

$\therefore\ f'(x)$ exists for all values of x in the interval $\left(-\sqrt{2}, \sqrt{2}\right)$.

Also, $f(x)$ is continuous for all values of x in $\left(-\sqrt{2}, \sqrt{2}\right)$, since $f(x)$ is differentiable for all values of x in $\left(-\sqrt{2}, \sqrt{2}\right)$.

Hence all three conditions of Rolle's Theorem are satisfied.

$\therefore\ f'(x) = 0$ for atleast one value of x, where $-\sqrt{2} < x < \sqrt{2}$. Equating $f'(x)$ to zero, we get

$$6x^2 + 2x - 4 = 0 \Rightarrow (x+1)(3x-2) = 0$$

$$\Rightarrow x = -1,\ 2/3.$$

Evidently $x = \frac{2}{3}$ lies in $\left(-\sqrt{2}, \sqrt{2}\right)$.

2. *Discuss the applicability of Rolle's Theorem to the function* $f(x) = \log \frac{x^2 + ab}{(a+b)\,x}$ *in* (a, b), $0 \notin [a, b]$.

Sol. $f(a) = \log\left\{\frac{a^2 + ab}{(a+b)\,a}\right\} = \log 1 = 0$

$$f(b) = \log\left\{\frac{b^2+ab}{(a+b)\,b}\right\} = 0.$$

$$\therefore \qquad f(a) = f(b).$$

Also, $$\mathrm{R}f'(x) = \lim_{h\to 0}\frac{1}{h}\left[\log\left\{\frac{(x+h)^2+ab}{(a+b)(x+h)}\right\} - \log\left\{\frac{x^2+ab}{(a+b)\,x}\right\}\right]$$

$$= \lim_{h\to 0}\frac{1}{h}\left[\log\left\{1+\frac{2hx+h^2}{x^2+ab}\right\} - \log\left(1+\frac{h}{x}\right)\right]$$

$$= \lim_{h\to 0}\frac{1}{h}\left[\frac{2hx+h^2}{x^2+ab} - \frac{h}{x}\right] \qquad \text{...}(i)$$

$$= \frac{2x}{x^2+ab} - \frac{1}{x} \qquad \text{...}(ii)$$

Also $$\mathrm{L}f'(x) = \frac{2x}{x^2+ab} - \frac{1}{x} \qquad \text{...}(iii)$$

$$\therefore \qquad \mathrm{R}f'(x) = \mathrm{L}f'(x),$$

i.e., $f'(x)$ exists for all values of x in (a, b).

Also, $f(x)$ is continuous for all values of x in (a, b), since $f(x)$ is differentiable for all values of x in (a, b).

Hence all conditions of Rolle's Theorem are satisfied.

$\therefore\ f'(x) = 0$ for atleast one value of x, where $a < x < b$.

Equating $f'(x)$ to zero,

$$\frac{2x}{x^2+ab} - \frac{1}{x} = 0 \Rightarrow x^2 = ab$$

$$\Rightarrow x = \sqrt{(ab)},$$

which being the G.M. of a and b lies in (a, b).

3. *Let* $\dfrac{a_0}{n+1} + \dfrac{a_1}{n} + \dfrac{a_2}{n-1} + \ldots + \dfrac{a_{n-1}}{2} + a_n = 0.$ *Show that the function* $a_0x^n + a_1x^{n-1} + \ldots + a_n$ *vanishes at least once in (0, 1).*

Sol. Let $$f(x) = a_0 \frac{x^{n+1}}{n+1} + a_1 \frac{a^n}{n} + \ldots + a_{n-1} \frac{x^2}{2} + a_n x \qquad \ldots(i)$$

Here $f(x)$ is a polynomial

$\Rightarrow$ $f(x)$ is continuous and differentiable for all x,

$\Rightarrow$ $f(x)$ is continuous in the closed interval [0, 1] and differentiable in the open interval (0, 1).

Also, from (*i*), we find that $f(0) = 0$

and $$f(1) = \frac{a_0}{n+1} + \frac{a_1}{n} + \ldots + \frac{a_{n-1}}{2} + a_n = 0 \qquad \text{(given)}$$

i.e., $$f(0) = 0 = f(1).$$

Hence all the conditions of Rolle's theorem are satisfied and therefore $f'(x)$ for at least one value of x in (0, 1).

Now $$f'(x) = a_0 x^n + a_1 x^{n-1} + \ldots + a_{n-1} x + a_n.$$

Hence from Rolle's Theorem $a_0 x^n + a_1 x^{n-1} + \ldots + a_n$ must vanish atleast once in (0, 1).

4. *The function f' is defined in [0, 1] as follows :*

$$f(x) = 1 \text{ for } 0 < x < \frac{1}{2}, = 2 \text{ for } \frac{1}{2} \le x \le 1.$$

Show that f (x) satisfies none of the conditions of Rolle's Theorem, yet $f'(x) = 0$ for many points in [0, 1].

Sol. Here we find that

$$f\left(\frac{1}{2} - 0\right) = 1, \text{ whereas } f\left(\frac{1}{2} + 0\right) = 2.$$

i.e., $$f\left(\frac{1}{2} - 0\right) \ne f\left(\frac{1}{2} + 0\right),$$

so the function $f(x)$ is not continuous at $x = \frac{1}{2}$.

Also the continuity being the necessary condition for the existence of a finite derivative, the function $f'(c)$ does not exist for every point in $0 \le x \le 1$.

Also, $$f(0) = 1 \text{ and } f(1) = 2 \qquad \text{(given)}$$

so $$f(0) \ne f(1).$$

Hence all the three conditions of Rolle's theorem are not satisfied by $f(x)$ in [0, 1].

Here $f(x)$ being a constant function in [0, 1] there is possibility of the value of $f'(x)$ to be zero at many points in [0, 1].

5. *Considering the function $(x - 3) \log x$, show that the equation $x \log x = 3 - x$ is satisfied by at least one value of x lying between 1 and 3.*

Sol. Let $$f(x) = (x - 3) \log x \quad ...(i)$$

As functions $(x - 3)$ and $\log x$ are both continuous in the closed interval [1, 3] for all values of x.

$\therefore$ Their product is also continuous in [1, 3],

i.e., $$f(x) \text{ is continuous in } [1, 3]. \quad ...(ii)$$

Also, $$f'(x) = (x-3)\cdot\frac{1}{x} + 1.\log x$$

which exists in the open interval (1, 3) as it is neither indeterminate nor infinite for any value of x in (1, 3) ...(*iii*)

Also, from (*i*) $$f(3) = (3 - 3) \log 3 = 0$$

and $$f(4) = (1 - 3) \log 1 = 0$$

$$\therefore \quad f(3) = f(1) = 0 \quad ...(iv)$$

Hence from (*ii*), (*iii*) and (*iv*) we find that all the conditions of Rolle's Theorem are satisfied by the function $f(x)$.

Hence there exists atleast one value of x in (1, 3) for which $f'(x) = 0$.

i.e., $$(x-3)\cdot\frac{1}{3} + 1.\log x = 0, \text{ for atleast one value of } x \text{ in } (1, 3).$$

i.e., $$(x-3) + x \log x = 0$$

or $$x \log x = 3 - x, \text{ for at least one value of } x \text{ in } (1, 3).$$

6. *If two functions $f(x)$ and $g(x)$ have the same derivative for each x in (a, b), then show that they differ by a constant in (a, b).*

Sol. Given that $f'(x) = g'(x)$, for each $x \in (a, b)$.

So, $$f'(x) - g'(x) = 0,$$

i.e., derivative of

$$[f(x) - g(x)] = 0 \ \forall \ x \in (a, b).$$

$\therefore f(x) - g(x)$ reduces to a constant $\forall \ x \in (a, b)$.

i.e., $f(x)$ and $g(x)$ differ by a constant.

7. If $f(x + h) = f(x) + hf'(x + \theta h), 0 < \theta < 1$, *find the value of* θ *when* $f(x) = Ax^2 + Bx + C$.

Sol. We have $f(x) = Ax^2 + Bx + C$, so $f'(x) = 2Ax + B$ ...(*i*)

$$\therefore \quad f(x+h) = A(x+h)^2 + B(x+h) + C,$$
$$= \left(Ax^2 + Bx + C\right) + \left(2Axh + Ah^2 + Bh\right)$$

Also
$$f'(x+\theta h) = 2A(x+\theta h) + B$$
$$= 2Ax + 2A\theta h + B.$$

$$\therefore f(x+h) = f(x) + hf'(x+\theta h)$$
$$\Rightarrow \left(Ax^2 + Bx + C\right) + \left(2Axh + Ah^2 + Bh\right)$$
$$= \left(Ax^2 + Bx + C\right) + h(2Ax + 2A\theta h + B)$$
$$\Rightarrow Ah^2 = 2A\theta h^2 \Rightarrow \theta = \frac{1}{2}.$$

8. *Under suitable conditions, prove that in the interval (a, b) there exists atleast one point c, such that*

$$\begin{vmatrix} f(a) & f(b) \\ \phi(a) & \phi(b) \end{vmatrix} = (b-a)\begin{vmatrix} f(a) & f'(c) \\ \phi(a) & \phi'(c) \end{vmatrix}$$

(Garhwal 96)

Sol. From Mean Value Theorem, we know that if $f(x)$ and $\phi(x)$ be functions such that these are continuous in $a \le x \le b$ and $f'(x)$, $\phi'(x)$ exists in $a < x < b$, then there exists atleast one point in (a, b), such that

$$f(b) - f(a) = (b-a)\, f'(c) \qquad \text{...(i)}$$

and
$$\phi(b) - \phi(a) = (b-a)\phi'(c) \qquad \text{...(ii)}$$

Multiplying (*i*) by $\phi(a)$ and (*ii*) by $f(a)$, we get

$$\phi(b) f(a) - \phi(a) f(a) = (b-a)\phi'(c) f(a) \qquad \text{...(iii)}$$

$$f(b)\phi(a) - f(a)\phi(a) = (b-a)\phi'(c)\phi(a) \qquad \text{...(iv)}$$

Subtracting (*iv*) from (*iii*), we get

$$\phi(b) f(a) - f(b)\phi(a) = (b-a)\left[\phi'(c) f(a) - f'(c)\phi(c)\right]$$

or
$$\begin{vmatrix} f(a) & f(b) \\ \phi(a) & \phi(b) \end{vmatrix} = (b-a)\begin{vmatrix} f(a) & f'(c) \\ \phi(a) & \phi'(c) \end{vmatrix}$$

9. *A function f (x) is continuous in the closed interval $0 \le x \le 1$ and differentiable in the open interval $0 < x < 1$. Prove that*

$$f'(x_1) = f(1) - f(0), \text{ where } 0 < x_1 < 1.$$

As function $f(x)$ is continuous in $0 \le x \le 1$ and differentiable in $0 < x < 1, f(1) \ne f(0)$, hence the conditions of Mean Value Theorem are satisfied.

Hence applying the Mean Value Theorem in the interval (0, 1) to $f(x)$, we get

$$f(1) - f(0) = (1 - 0)\ f'(x_1), \quad \text{where } 0 < x_1 < 1.$$

Hence $$f'(x_1) = f(1) - f(0), \quad \text{where } 0 < x_1 < 1.$$

EXERCISES

1. Verify Rolle's theorem in $[a, b]$ when

(i) $f(x) = \log(x^2 + ab)/(a + b)\ x$.

(ii) $f(x) = (x - a)^m (x - b)^n$; m, n being positive integers.

2. Discuss the applicability of Rolle's theorem for

(i) $f(x) = |x|, x \in [-1, 1]$. *(Kanpur 2002)*

(ii) $f(x) = 2 + (x - 1)^{2/3}, x \in [0, 2]$.

3. Discuss applicability of Rolle's Theorem to the following functions :

(i) $f(x) = |x|$ in $(-1, 1)$.

(ii) $f(x) = x^2 + 1$, when $0 \le x < 1$, $= 3 - x$, when $1 < x \le 2$.

4. Any of the three conditions of Rolle's Theorem is not necessary for $f'(x)$ to vanish at same points in (a, b). Illustrate this by considering the function $f(x) = \sin(1/x)$ in $(-1, 1)$.

5. Verify Rolle's Theorem for $f(x) = x^2(1 - x^2)$ in $0 \le x < 1$. If $f(x)$ is defined in $[a, b]$ and $f'(c) > 0$ where $a < c < b$, prove that $f(x)$ is increasing at $x = c$.

6. Verify Lagrange's Mean Value Theorem for the function $f(x) = x(x - 1)(x - 2)$ in the interval $\left[0, \frac{1}{2}\right]$. *(Garhwal 92; Rohilkhand 94, 96)*

7. A function $f(x)$ is defined by $f(x) = x \sin(1/x)$ in $(-1, 1)$, $f(0) = 0$.

Are the conditions of first Mean Value Theorem of differential calculus satisfied in this case ?

8. Verify the Mean Value Theorem for the function $f(x) = \log x$ in the interval $[1, e]$. *[Rohilkhand 90]*

9. Verify Lagrange's mean value theorem for

(i) $f(x) = lx^2 + mx + n;\ x \in [a, b]$.

(ii) $f(x) = x(x - 1)(x - 2),\ x \in \left[0, \frac{1}{2}\right]$.

10. If, $f'(x) = k \ \forall \ x \in [a, b]$, k being a constant, show that

$$f(x) = kx + l,$$

whose l is a constant.

11. If f satisfies the conditions of Lagrange's theorem in $[a, b]$ and $f'(x) \geq 0 \ \forall \ x \in \,]a, b[$ and there exists no sub-interval of $]a, b[$ in which f' is 0, show that f is strictly increasing in $[a, b]$.

12. (*a*) If $f(x + h) = f(x) + hf'(x + \theta h)$, $0 < \theta < 1$, find the value of θ, when $f(x) = x^2$.

(*b*) With the help of Mean Value Theorem, show that if $x > 0$.

$$\log_{10}(x + 1) = (x \log_{10} e)/(1 + \theta x), \quad \text{where } 0 < \theta < 1.$$

(*Garhwal 97*)

5.4.3. Cauchy's Mean Value Theorem

(*i*) *continuous in* $[a, b]$

(*ii*) *derivable in* $]a, b[$ *and*

(*iii*) $F'(x) \neq 0$ *for any point* x *in* $]a, b[$.

then there exists at least one point, $c \in \,]a, b[$ such that

$$\frac{f(b) - f(a)}{F(b) - F(a)} = \frac{f'(c)}{F'(c)}.$$

(*Ajmer 98, Garhwal 96, 97, Kumaon 97; Osmania 2005, Kanpur 2002, 2005*)

Let a function φ be defined by

$$\varphi(x) = f(x) + AF(x).$$

where, A, is a constant to be determined such that

$$\varphi(a) = \varphi(b).$$

This requires

$$[F(b) - F(a)]\,A = -[f(b) - f(a)]. \qquad ...(1)$$

Now, $[F(b) - F(a)] \neq 0$, for, if it were 0, then F would satisfy all the conditions of Rolle's theorem, and its derivative would, therefore, vanish at least once in $]a, b[$ and the condition (*iii*) would be contradicted. On this account, we have from (1),

$$A = -[f(b) - f(a)]/[F(b) - F(a)].$$

The function φ is continuous in $[a, b]$, derivable in $]a, b[$, and $\varphi(a) = \varphi(b)$. Hence, by Rolle's theorem, there exists at least one point $c \in \,]a, b[$ such that $\varphi'(c) = 0$. But

$$\varphi'(x) = f'(x) + AF'(x).$$

$$\therefore \quad 0 = \varphi(c) = f'(c) + AF'(c).$$

$$\Rightarrow \quad \frac{f'(c)}{F'(c)} = -A = \frac{f(b)-f(a)}{F(b)-F(a)}, \qquad \because F'(c) \neq 0.$$

Another form of statement. If two functions f, F are continuous in $[a, a+h]$, *derivable in* $]a, a+h[$ *and* $F'(x) \neq 0$ *for any* $x \in]a, a+h[$, *then there exists at least one number* θ, *between 0 and 1 such that*

$$\frac{f(a+h)-f(a)}{F(a+h)-F(a)} = \frac{f'(a+\theta h)}{F'(a+\theta h)}, \; \theta \in]0,1[.$$

Note. Taking $F(x) = x$, we may see that Lagrange's mean value theorem is only a particular case of Cauchy's theorem.

5.5. HIGHER DERIVATIVES

Let f be derivable *i.e.*, let f' exist in a certain neighbourhood of c. This implies that f is defined and continuous in a neighbourhood of c. If the function f' has derivative at c, then this derivative is called the second derivative of f at c, and is denoted by $f''(c)$. In this case f' is necessarily continuous at c.

In general, if $f^{n-1}(x)$ exists in a neighbourhood of c, then the derivative of f^{n-1} at c, in case it exists, is called the nth derivative of f at c and is denoted by $f^n(c)$.

5.5.1. Generalised Mean Value Theorem : Taylor's Theorem

If a function f is such that (*Bhopal 2003*)

(*i*) *the* $(n-1)$*th derivative* f^{n-1} *is continuous in* $[a, a+h]$,

(*ii*) *the nth derivative* f^n *exists in* $]a, a+h[$,

and (*iii*) *p is a given positive integer,*

then there exists at least one number, θ, *between 0 and 1 such that*

$$f(a+h) = f(a) + hf'(a) + \frac{h^2}{2!} f''(a) + \frac{h^{n-1}}{(n-1)!} f^{n-1}(a)$$

$$+ \frac{h^n (1-\theta)^{n-p}}{(n-1)!\, p} f^n(a+\theta h). \quad ...(1)$$

The condition (*i*) implies the continuity of

$$f, f', f'', ..., f^{n-2} \text{ in } [a, a+h].$$

Let a function φ be defined by

$$\varphi(x) = f(x)+(a+h-x)f'(x)+\frac{(a+h-x)^2}{2!}f''(x)+\dots$$

$$+\frac{(a+h-x)^{n-1}}{(n-1)!}f^{n-1}(x)+A(a+h-x)^p,$$

where A is a constant to be determined such that

$$\varphi(a) = \varphi(a+h).$$

Thus A is given by

$$f(a+h) = f(a)+hf'(a)+\frac{h^2}{2!}f''(a)+\dots+\frac{h^{n-1}}{(n-1)!}f^{n-1}(a)+Ah^p \quad \dots(2)$$

The function φ is continuous in $[a, a+h]$, derivable in $]a, a+h[$ and $\varphi(a) = \varphi(a+h)$. Hence, by Rolle's theorem, there exists at least one number, θ, between 0 and 1 such that

$$\varphi'(a+\theta h) = 0$$

$$\text{But } \varphi'(x) = \frac{(a+h-x)^{n-1}}{(n-1)!}f^n(x) - pA(a+h-x)^{p-1}.$$

$$\therefore\ 0 = \varphi'(a+\theta h) = \frac{h^{n-1}(1-\theta)^{n-1}}{(n-1)!}f^n(a+\theta h) - pA(1-\theta)^{p-1}h^{p-1}$$

$$\Rightarrow \quad A = \frac{h^{n-p}(1-\theta)^{n-p}}{p.(n-1)!}f^n(a+\theta h), \text{ for } (1-\theta)\neq 0 \text{ and } h\neq 0.$$

Substituting the value of A in (2), we get the required result (1).

(*i*) *Remainder after n terms.* The term

$$\mathbf{R}_n = \frac{h^n(1-\theta)^{n-p}}{p(n-1)!}f^n(a+\theta h),$$

is known as the Taylor's remainder $\mathbf{R}_n$ after n terms and is due to *Schlomilch and Roche.*

(*ii*) Putting $p = 1$, we obtain

$$\mathbf{R}_n = \frac{h^n(1-\theta)^{n-1}}{(n-1)!}f^n(a+\theta h),$$

which form of remainder is due to *Cauchy.*

(*iii*) Putting $p = n$, we obtain

$$\mathbf{R}_n = \frac{h^n}{n!}f^n(a+\theta h),$$

which is due to *Lagrange.*

Cor. 1. Let x be a point of the interval $[a, a + h]$. Let f satisfy the conditions of Taylor's theorem in $[a, a + h]$ so that it satisfies the conditions for $[a, x]$ also.

Changing $a + h$ to x *i.e.*, h to $x - a$, in (1), we obtain

$$f(x) = f(a) + (x-a) f'(a) + \frac{(x-a)^2}{2!} f''(a) + \frac{(x-a)^3}{3!} f'''(a) + \dots +$$

$$\frac{(x-a)^{n-1}}{(n-1)!} f^{n-1}(a) + \frac{(x-a)^n (1-\theta)^{n-p}}{p.(n-1)!} f^n [a + \theta(x-a)],$$

$$0 < \theta < 1.$$

This result holds $\forall\, x \in [a, a+h]$. Of course, θ may be different for different points x.

Cor. 2. Maclaurin's theorem. Putting $a = 0$, we see that if $x \in [0, h]$, then

$$f(x) = f(0) + xf'(0) + \frac{x^2}{2!} f''(0) + \frac{x^3}{3!} f'''(0) + \dots +$$

$$\frac{x^{n-1}}{(n-1)!} f^{n-1}(0) + \frac{x^n (1-\theta)^{n-p}}{p(n-1)!} f^n(\theta x)$$

which holds when

(*i*) f^{n-1} is continuous in $[0, h]$, and (*ii*) f^n exists in $]0, h[$.

Putting $n = 1$ and $n = p$, respectively in the Schlomilch form of remainder

$$\frac{x^n (1-\theta)^{n-p} f^n(\theta x)}{p.(n-1)!}$$

we see that Cauchy's and Lagrange's forms are respectively

$$\frac{x^n (1-\theta)^{n-1}}{(n-1)!} f^n(\theta x) \text{ and } \frac{x^n}{n!} f^n(\theta x).$$

5.5.2. Power Series Representation of Functions

Taylor's infinite series. Suppose that *a given function f possesses a continuous derivative of* **every** *order in* $[a, a + h]$.

Then $\forall\, n \in \mathbf{N}$, there exists an equality of the form

$$f(a+h) = f(a) + hf'(a) + \dots + \frac{h^{n-1}}{(n-1)!} f^{n-1}(a) + R_n,$$

when R_n denotes Taylor's remainder after n terms.

We write $$S_n = f(a) + hf'(a) + \ldots + \frac{h^{n-1}}{(n-1)!} f^{n-1}(a),$$

$$\Rightarrow \qquad f(a+h) = S_n + R_n.$$

If $R_n \to 0$ as $n \to \infty$, we have

$$\lim S_n = f(a+h),$$

so that the infinite series

$$f(a) + hf'(a) + \ldots + \frac{h^{n-1}}{(n-1)!} f^{n-1}(a) + \frac{h^n}{n!} f^n(a) + \ldots$$

converges and its sum is equal to $f(a + h)$.

Thus we have proved that if

(i) f possesses continuous derivatives of every order in $[a, a + h]$

and (ii) Taylor's remainder $R_n \to 0$ *as* $n \to \infty$,

then $$f(a+h) = f(a) + hf'(a) + \ldots + \frac{h^n}{n!} f^n(a) \ldots$$

The infinite series

$$f(a) + hf'(a) + \ldots + \frac{h^n}{n!} f^n(a) \ldots$$

is known as *Taylor's series.*

It should be clearly understood that the mere convergence of this series does not mean that its sum is equal to $f(a + h)$.

5.5.3. Maclaurin's Infinite Series

From above, we deduce that if f possesses continuous derivatives of every order in $[0, h]$ and $R_n \to 0$ *as* $n \to \infty$, *then* $\forall\, x \in [0, h]$,

$$f(x) = f(0) + xf'(0) + \frac{x^2}{2!} + f''(0) + \ldots + \frac{x^n}{n!} f^n(0) + \ldots,$$

Note. The student should note that there is no fundamental distinction between the Taylor's and Maclaurin's infinite series each of which seeks to express the value of a function at any point in terms of the various derivatives of the function at any other point and the distance between the two points.

In this connection it is also interesting to see 'a priori' why the Taylor's or Maclaurin's expansions are not valid for an arbitrary function. Clearly the system of derivatives of a function at a point takes account of the values of the function in a neighbourhood of the point only and accordingly the value of any arbitrary function at a point cannot possibly be given by those in a neighbourhood of another point. The conditions imposed on the function for the validity of the Taylor's or Maclaurin's expansions provide, so to say, for the *linking up* of the values of the function at the two points.

5.6. YOUNG'S FORM OF TAYLOR'S THEOREM

If a function f be such that $f^n(a)$ exists and M is defined by the equation

$$f(a+h) = f(a) + hf'(a) + \frac{h^2}{2!} f''(a) + \ldots + \frac{h^{n-1}}{(n-1)!} f^{n-1}(a) + \frac{h^n}{n!} M,$$

then $M \to f^n(a)$ as $h \to 0$.

The fact that $f^n(a)$ exists implies that $f, f', f'' \ldots f^{n-1}$ all exist in a certain neighbourhood $]a-\delta, a+\delta[$ of a. Let $\varepsilon > 0$ be a given number.

Firstly we take $h \geq 0$. We define a function φ as follows :

$$\varphi(h) = f(a) + hf'(a) + \frac{h^2}{2!} f''(a) + \ldots + \frac{h^{n-1}}{(n-1)!} f^{n-1}(a) + \frac{h^n}{n!}\left[f''(a) + \varepsilon\right] - f(a+h).$$

We easily see that

$$\varphi(0) = 0,\ \varphi'(0) = 0, \ldots, \varphi^{n-1}(0) = 0 \text{ and } \varphi^n(0) = \varepsilon > 0.$$

Since $\varphi^n(0)$ is positive and $\varphi^{n-1}(0) = 0$, we see that there exists an interval $]0, \delta]$ such that for every point h of this interval $\varphi^{n-1}(h)$ is positive (§ 5.2, page 156).

Since $\varphi^{n-1}(h)$ is positive when $h \in]0, \delta_1]$ and $\varphi^{n-2}(0) = 0$, we see that $\varphi^{n-2}(h)$ is positive when $h \in]0, \delta_1]$ (§ 5.2, page 156).

Now successively applying the theorem of § 5.2, we deduce that $\varphi(h)$ is positive when $h \in]0, \delta_1]$.

Thus we see that there exists $\delta_1 > 0$ such that for $0 < h < \delta_1$

$$f(a) + hf'(a) + \frac{h^2}{2!} f''(a) + \ldots + \frac{h^{n-1}}{(n-1)!} f^{n-1}(a) + \frac{h^n}{n!}\left[f^n(a) - \varepsilon\right] - f(a+h) > 0.$$

Similarly we may prove that there exists a positive number δ_2 such that for

$$0 < h < \delta_2$$

$$f(a) + hf'(a) + \frac{h^2}{2!} f''(a) + \ldots + \frac{h^{n-1}}{(n-1)!} f^{n-1}(a) + \frac{h^n}{n!}\left[f^n(a) - \varepsilon\right] - f(a+h) > 0.$$

But, as given,

$$f(a) + hf'(a) + \frac{h^2}{2!} f''(a) + ... + \frac{h^{n-1}}{(n-1)!} f^{n-1}(a)$$

$$+ \frac{h^n}{n!} M - f(a+h) = 0.$$

Let $\eta = \min(\delta_1, \delta_2)$.

From above, we deduce that corresponding to every positive number ε, there exists $\eta > 0$ such that

$$f^n(a) - \varepsilon < M < f^n(a) + \varepsilon \;=\; \left|M - f^n(a)\right| < \varepsilon, \text{ when } 0 < h < \eta.$$

$$\therefore \qquad \lim_{h \to (0+0)} M = f^n(a).$$

Taking negative values of h, we may similarly show that

$$\lim_{h \to (0+0)} M = f^n(\text{a}).$$

Hence $\qquad M \to f^n(a)$ as $h \to 0$.

5.7. APPLICATION OF TAYLOR'S THEOREM

Applications of Taylor's theorem to the problem of *Extreme* values of a function and to the problem of the evaluation of certain special kinds of limits popularly known as *Indeterminate forms* is given in the following three sections.

5.7.1. Extreme Values of a Function : Maxima and Minima

If c be an *interior* point of the domain $[a, b]$ of a function f, then,

(*i*) $f(c)$ is said to be a *maximum* value of f, if there exists some neighbourhood $]\, c - \delta, c + \delta\, [$ of, c, such that $\forall\, x \in]c - \delta, c + \delta[$ other than c,

$$f(c) > f(x);$$

(*ii*) $f(c)$ is said to be a *minimum* value of f, if for every point x, other than c, of some neighbourhood $]\, c - \delta, c + \delta\, [$ of c,

$$f(c) < f(x);$$

(*iii*) $f(c)$ is said to be an *extreme* value of f, it is either a maximum or a minimum value.

For $f(c)$ to be an extreme value,

$$f(c) - f(x)$$

must keep the same sign for every point x, other than c, of some neighbourhood of c.

5.7.2. Theorem

If $f(c)$ be an extreme value of a function f, then $f'(c)$, in case it exists, is zero.

If $f'(c)$ be not 0, then in every neighbourhood of c there exists points x for which $f(x) > f(c)$ and points x for which $f(x) < f(c)_r$ (§ 5.2, page 156) so that $f(c)$ cannot be an extreme value.

Note 1. The theorem may also be stated a little differently as follows :

The necessary condition for $f(c)$ to be an extreme value is that $f'(c) = 0$, in case it exists.

To show that this condition is only necessary and not sufficient, we consider the function f where $f(x) = x^3$ when $x = 0$. Clearly $f'(0) = 0$.

Also, $x > 0 \Rightarrow f(x) > f(0)$, and $x < 0 \Rightarrow f(x) < f(0)$, so that $f(0)$ is not an extreme value even though $f'(0) = 0$.

Note 2. If $f(x) = |x|$, then $f(0)$ is a minimum value and $f'(0)$ does not exist. Remarks after § 5.11, page 155.

This example shows that $f(c)$ may be an extreme value even when f (c) does not eixst.

5.7.3. Sufficient Criteria for Extreme Values

Let, c, be an interior point of the domain $[a, b]$ of a function f. Let

I. $f^n(c)$ exist and be not zero,

II. $f'(c) = f''(c) = f'''(c) \ldots = f^{n+1}(c) = 0$;

then if n is odd, $f(c)$ is not an extreme value; and if n is even, $f(c)$ is a maximum or a minimum value according as $f^n(c)$ is negative or positive.

The condition (i) implies that $f, f'', \ldots, f^{n-1}$ all exist in a certain neighbourhood, $]c - \delta_1, c + \delta_1[$, of c.

As $f^n(c)$ exists and 0, there exists a neighbourhood $]c - \delta, c + \delta[$, of c, $(0 < \delta < \delta_1)$ such that

$$\left.\begin{aligned} f^{n-1}(x) < f^{n-1}(c) = 0, &\text{ when } c - \delta < x < c \\ f^{n-1}(x) > f^{n-1}(c) = 0, &\text{ when } c < x < c + \delta \end{aligned}\right]$$

(§ 5.2, page 156)... (ii)

in case $f^n(c)$ is positive;

$$\left.\begin{aligned} f^{n-1}(x) > f^{n-1}(c) = 0, &\text{ when } c - \delta < x < c \\ f^{n-1}(x) < f^{n-1}(c) = 0, &\text{ when } c < x < c + \delta \end{aligned}\right]$$

(§ 5.2, page 156)...(iii)

in case $f^n(c)$ is negative.

Because of (i), we have by Taylor's theorem, when $|h| < \delta$,

$$f(c+h) = f(c) + hf'(c) + \frac{h^2}{2!} f'(c) + \ldots + \frac{h^{n-1}}{(n-1)!} f^{n-1}(c+\theta h),$$

which, by virtue of (*ii*), gives

$$f(c+h) - f(c) = \frac{h^{n-1}}{n-1!} f^{n-1}(c+\theta h), \quad \text{...(iv)}$$

where, $c + \theta h$, $]\, c - \delta,\, c + \delta\, [$.

Let n be even. From (*ii*) and (*iv*), we deduce that if $f^n(c)$ be > 0, then for every point $x = c + h \in [c - \delta, c + \delta]$, other than c,

$$f(c+h) > f(c),$$

i.e., $f(c)$ is a minimum.

From (*iii*) and (*iv*), it may similarly be shown that $f(c)$ is a maximum if $f^n(c)$ be < 0.

Let n be odd. From (*ii*) and (*iv*), we deduce that if $f^n(c) > 0$, then

$$\begin{cases} f(c+h) > f(c), \text{ when } c < x = c + h \le c + \delta, \\ f(c+h) < f(c), \text{ when } c - \delta \le x = c + h < c \end{cases}$$

so that $f(c)$ is not an extreme value.

It may similarly be shown that $f(c)$ is not an extreme value when $f^n(c) < 0$.

Another Proof. We now give another proof which is dependent upon the Young's form of Taylor's theorem. We have

$$f(c+h) = f(c) + hf'(c) + \frac{h^2}{2!} f''(c) + \ldots + \frac{h^{n-1}}{(n-1)!} f^{n-1}(c) + \frac{h^n}{n!} M,$$

and $$M \to f^n(c) \text{ as } h \to 0.$$

With the help of *II*, we get

$$f(c+h) - f(c) = \frac{h^n}{n!} M. \quad \text{...(v)}$$

Since $M \to f^n(c)$ as $h \to 0$, there exists a number $\delta > 0$, such that for $0 < |h| \le \delta$, M has the same sign as $f^n(c)$. From (*v*), we now deduce that when n is even, $f(c+h) - f(c)$ has the same sign as $f^n(c)$ for $0 < |h|\ \delta$, so that $f(c)$ is a maximum or minimum according as $f^n(c)$ is negative or positive and when n is odd, $f(c+h) - f(c)$ changes sign with the change in the sign of h so that $f(c)$ is not an extreme value.

5.8. THE INDETERMINATE FORM (0/0)

5.8.1. Theorem

Let f, g be two functions such that

(*i*) $\lim_{x \to c} f(x) = 0 \qquad \lim_{x \to c} g(x) = 0,$ ...(*i*)

(*ii*) *f′ (c) g′ (c) both exist and g′ (c) ≠ 0,*

then $$\lim_{x \to c} \frac{f(x)}{g(x)} = \frac{f'(c)}{g'(c)}.$$

Since *f, g* are derivable at *c*, therefore they are continuous at *c* and accordingly, by (*i*), $f(c) = g(c) = 0$.

We have $$f'(c) = \lim_{h \to 0} \frac{f(c+h) - f(c)}{h} = \lim_{h \to 0} \frac{f(c+h)}{h}$$

$$g'(c) = \lim_{h \to 0} \frac{g(c+h) - g(c)}{h} = \lim_{h \to 0} \frac{g(c+h)}{h}.$$

$$\therefore \quad \frac{f'(c)}{g'(c)} = \lim_{h \to 0} \frac{[f(c+h)/h]}{[g(c+h)/h]} = \lim_{h \to 0} \frac{f(c+h)}{g(c+h)} = \lim_{h \to 0} \frac{f(x)}{g(x)}.$$

Note. This result may also be stated as follows :

If $f(c) = g(c) = 0$, and $f'(c)$, $g'(c)$ exist but $g'(c) \neq 0$, then

$$\lim [f(x) / g(x)] = f'(c) / g'(c) \quad \text{when } x \to c.$$

5.8.2. Theorem

Let f, g be two functions such that

(*i*) $\lim_{x \to c} f(x) = 0, \ \lim_{x \to c} g(x) = 0,$

(*ii*) *$\lim [f'(x)/g'(x)] = l$ when $x \to c$, then $\lim [f(x) / g(x)] = l$ when $x \to c$.*

The condition (*ii*) implies that $f'(x)$ and $g'(x)$ exist and $g'(x) \neq 0$ at every point $x \neq c$ of a certain neighbourhood $]\, c - \delta, c + \delta\, [$ of *c*.

We *suppose* that $f(c) = g(c) = 0$, for this change in the definitions of *f* and *g* influences neither the hypotheses nor the conclusion of the theorem.

If $x \in [c - \delta, c + \delta]$, we have, by Cauchy's theorem,

$$\frac{f(x)}{g(x)} = \frac{f(x) - f(c)}{g(x) - g(c)} = \frac{f'(\xi)}{g'(\xi)},$$

when ξ lies between *c* and *x* and also depends upon *x*.

By virtue of (*ii*), $f'(\xi)/g'(\xi) = l$ as $x \to c$.

$\therefore$ $$f(x)/g(x) \to l \text{ as } x \to c.$$

Note. The theorem may also be stated in another form as follows :

If $$\lim_{x \to c} f(x) = 0 = \lim_{x \to c} g(x)$$

then $$\lim_{x \to c} \frac{f(x)}{g(x)} = \lim_{x \to c} \frac{f'(x)}{g'(x)},$$

provided that the limit on the right exists.

It will later on be seen by means of an example that lim $[f(x)/g(x)]$ may exist without lim $[f'(x)/g'(x)]$ existing.

Note. The reader will find it useful to compare the hypotheses and the conclusions of the two preceding theorems. In § 5.8.1, the derivability of f and g is assumed at $x = c$ *only* whereas in § 5.8.2, while assuming the derivability of f, g in a neighbourhood of $x = c$, it is exactly at $x = c$ where we have not needed it.

5.8.3. Theorem

Let f, g be two functions such that when $x \to c$,

(*i*) $$\begin{cases} \lim f(x) = \lim f'(x) = \ldots = \lim f^{n-1}(x) = 0, \\ \lim g(x) = \lim g'(x) = \ldots = \lim g^{n-1}(x) = 0, \end{cases}$$

(*ii*) *lim* $[f^n(x)/g^n(x)] = l$,

then lim $[f(x)/g(x)] = l$ *when* $x \to c$.

This may be proved through repeated applications of the preceding theorem.

5.8.4. Theorem

Let f, g be two functions such that when $x \to c$,

(*i*) $$\begin{cases} \lim f(x) = \lim f'(x) = \ldots = \lim f^{n-1}(x) = 0, \\ \lim g(x) = \lim g'(x) = \ldots = \lim g^{n-1}(x) = 0, \end{cases}$$

(*ii*) $f^n(c)$, $g^n(c)$ *exist and* $g^n(c) \neq 0$,

then lim $[f(x)/g(x)] = f^n(c)/g^n(c)$ *when* $x \to c$.

By virtue of (*ii*), on applying the theorem of § 5.8.1, we see that

$$\lim \frac{f^{n-1}(x)}{g^{n-1}(x)} = \frac{f^n(c)}{g^n(c)}.$$

Also from the theorem of § 5.8.3, changing n to $(n-1)$, we see that

$$\lim \frac{f(x)}{g(x)} = \lim \frac{f^{n-1}(x)}{g^{n-1}(x)} = \frac{f^n(c)}{g^n(c)}.$$

5.8.5. Theorem

Let f, g be two functions such that

(i) $\lim_{x\to\infty} f(x) = 0,\ \lim_{x\to\infty} g(x) = 0,$

(ii) $\lim_{x\to\infty} [f'(x)/g'(x)] = l,$

then $\lim_{x\to\infty} [f(x)/g(x)] = l.$

We write $z = 1/x$ so that $x \to \infty \Rightarrow z \to (0+0)$.

Let $\quad F(z) = f(1/z),\ G(z) = g(1/z),$

so that $\quad \lim_{z\to(0+0)} F(z) = 0, \quad \lim_{z\to(0+0)} G(z) = 0. \quad$ ...(1)

We have $\quad F'(z) = -f'\left(\frac{1}{z}\right)\cdot\frac{1}{z^2},\ G'(z) = -g'\left(\frac{1}{z}\right)\frac{1}{z^2}.$

$$\frac{F'(z)}{G'(z)} = \frac{f'(1/z)}{g'(1/z)},$$

$$F'(z)/G'(z) \to l \text{ when } z \to (0+0) \text{ by } (iii). \quad ...(2)$$

Hence, from (1) and (2),

$$\lim_{x\to(0+0)} [F(z)/G(z)] = l = \lim_{x\to\infty} [f(x)/g(x)] = l.$$

5.9. THE INDETERMINATE FORM (∞/∞)

5.9.1. Theorem

If f, g be two functions such that when $x \to c$,

(i) $\lim |g(x)| = \infty,$ $\quad$ (ii) $\lim \frac{f'(x)}{g'(x)} = 0,$

then $\lim \frac{f(x)}{g(x)} = 0$, *when* $x \to c$.

The condition (*ii*) implies that there exists an interval $[c - \delta, c + \delta]$ around c such that for every point x, other than c, of this interval, $f'(x)$, $g'(x)$ exist and $g'(x) \neq 0$.

From Darboux's theorem of § 5.3, page 157, it follows that $g'(x)$ keeps the same sign, positive or negative $\forall\, x \in \,]c, c + \delta]$, and the same is true for $[c - \delta, c[$ also.

Firstly we consider $]c, c + \delta]$. Let $g'(x)$ remain positive in it.

Let $\varepsilon > 0$ be a number. There exists by virtue of (*ii*), a positive number $\delta_1 < \delta$ such that $\forall\, x \in \,]c, c + \delta_1[$.

$$\left|\frac{f'(x)}{g'(x)} - 0\right| < \frac{\varepsilon}{2},$$

$$|f'(x)| < \frac{1}{2}\varepsilon\,|g'(x)| = \frac{1}{2}\varepsilon g'(x),$$

$$-\frac{1}{2}\varepsilon g'(x) < f'(x) < \frac{1}{2}\varepsilon g'(x).$$

By virtue of the cor. to theorem of § 5.4.2, page 160, we have

$$-\frac{1}{2}\varepsilon\left[g(c+\delta) - g(x)\right] < \left[f(c+\delta) - f(x)\right] < \frac{1}{2}\varepsilon\left[g(c+\delta) - g(x)\right].$$

Thus $\forall\, x \in \,]c, c + \delta_1[$.

$$|f(c+\delta) - f(x)| < \frac{1}{2}\varepsilon\,|g(c+\delta) - g(x)|$$

$$\Rightarrow \quad |f(x)| - |f(c+\delta)| < \frac{1}{2}\varepsilon\,|g(c+\delta)| + \frac{1}{2}\varepsilon\,|g(x)|$$

$$\Rightarrow \quad |f(x)| < \frac{1}{2}\varepsilon\,|g(x)| + \frac{1}{2}\varepsilon\,|g(c+\delta)| + |f(c+\delta)|$$

$$\Rightarrow \quad |f(x)| < \frac{1}{2}\varepsilon\,|g(x)| + k \;\forall\, x \in \,]c, c+\delta_1[.$$

k being free from x,

$$\Rightarrow \quad \frac{|f(x)|}{|g(x)|} < \frac{1}{2}\varepsilon + \frac{k}{|g(x)|}.$$

There exists a positive number $\delta_2 < \delta_1$, such that

$$\forall\, x \in \,]c, c+\delta_2[, \quad \frac{k}{|g(x)|} < \frac{1}{2}\varepsilon.$$

Thus we see that to every $\varepsilon > 0$ there corresponds $\delta_2 \geq 0$ such that $\forall\, x \in]c, c+\delta_2[$,

$$\frac{|f(x)|}{|g(x)|} < \varepsilon \quad \Rightarrow \quad \lim_{x \to (c+0)} \frac{f(x)}{g(x)} = 0.$$

It may similarly be shown that

$$\lim_{x \to (c-0)} \frac{f(x)}{g(x)} = 0$$

Thus $\qquad f(x) / g(x) \to 0 \quad \text{as } x \to c.$

5.9.2. Theorem

If f, g be two functions such that when $x \to c$,

(i) $\lim |g(x)| = \infty$, $\qquad$ (ii) $\lim [f'(x) / g'(x)] = l$,

then $\lim f(x) / g(x) = l$ when $x \to c$.

We write $\qquad \varphi(x) = f(x) - lg(x).$

We have $$\lim_{x \to c} \frac{\phi'(x)}{g'(x)} = \lim_{x \to 0} \left[\frac{f'(x)}{g'(x)} - l\right] = 0, \text{ (by } ii).$$

$$\therefore \quad \lim_{x \to 0} \frac{\varphi(x)}{g(x)} = 0, \text{ (§ 5.8.2)}$$

$$\Rightarrow \quad \lim_{x \to c} \left[\frac{f(x)}{g(x)} - l\right] = 0 \quad \Rightarrow \quad \lim_{x \to c} \frac{f(x)}{g(x)} = l.$$

Note. As in § 5.8.5, it may be shown that the result remains true even when $x \to \infty$.

Note. It should be especially noted that in this preceding theorem nothing whatsoever has been said about the limit of $f(x)$ as $x \to c$ so that the result holds good independently of the behaviour of $f(x)$. In particular, therefore, the result holds good when $f(x) \to \infty$ as $x \to c$; this being the form in which the theorem is usually stated.

EXAMPLES

1. *If $a = -1$, $b \geq 1$ and $f(x) = 1/|x|$, show that the conditions of Lagrange's mean value theorem are not satisfied by the function f in [a, b] but that the conclusion is true if and only if $b > 1 + \sqrt{2}$.*

Sol. We have

$\forall\, x<0, \qquad f(x) = 1/|x| = -1/x \Rightarrow f'(x) = 1/x^2,$

and $\forall\, x>0, \qquad f(x) = 1/|x| = 1/x \Rightarrow f'(x) = -1/x^2$

For $x = 0$, f is not derivable.

Thus the conditions of the Lagrange's mean value theorem are not satisfied inasmuch as f is not derivable at the point, '0', of the open interval $]\,a, b\,[$.

Again
$$\frac{f(b)-f(a)}{b-a} = \frac{1/|b|-1/|a|}{b-a} = \frac{(1/b)-1}{b+1}$$

$$= \frac{1-b}{b(b+1)}, \text{ for } b \geq 1 \text{ and } a = -1.$$

As $[f(b)-f(a)]/(b-a)$ is negative, it cannot equal the derivative of the function f for any negative value of x; the derivative being necessarily positive for such values of x. Let, if possible, there exist $c>0$ such that

$$f'(c) = \frac{f(b)-f(a)}{b-a}.$$

This requires

$$-\frac{1}{c^2} = \frac{1-b}{b(b+1)} = -\frac{b-1}{b(b+1)} \Rightarrow c^2 = \frac{b(b+1)}{b-1}.$$

For the conclusion of the theorem to be true, we must have $c<b$. This requires

$$\frac{b(b+1)}{b-1} = < b^2 \Rightarrow b > 1+\sqrt{2}.$$

2. *A function f is twice derivable and satisfies $\forall\, x>a$ the inequalities*

$$|f(x)| < A, |f''(x)| < B,$$

where A and B are constants. Prove that $\forall\, x>a$,

$$|f'(x)| < 2\sqrt{(AB)}.$$

Sol. Let $x>a$ and h be any positive number. There exists a number, $\theta, (0<\theta<1)$ such that

$$f(x+h) = f(x) + hf'(x) + \frac{h^2}{2!}f''(x+\theta h)$$

$$\Rightarrow \qquad |hf'(x)| = \left|f(x+h) - f(x) - \frac{h^2}{2!}f''(x+\theta h)\right|$$

$$\leq \left|f(x+h)\right| + \left|f(x)\right| + \frac{h^2}{2!}\left|f''(x+\theta h)\right|$$

$$< \; 2A + \frac{Bh^2}{2}, \forall\, x > a$$

$$\Rightarrow \qquad \left|f'(x)\right| < \frac{2A}{h} + \frac{Bh}{2}, \forall\, x > a. \qquad \text{...}(i)$$

Now | $f'(x)$ | is independent of h and also, by (i) it is less than $(2A/h + Bh/2)\ \forall\ h > 0$. Thus | $f'(x)$ | must be less than the least value of $(2A/h + Bh/2)$. Also

$$\frac{2A}{h} + \frac{Bh}{2} = \left(\sqrt{\frac{2A}{h}} - \sqrt{\frac{Bh}{2}}\right)^2 + 2\sqrt{AB}$$

so that the least value of $\dfrac{2A}{h} + \dfrac{Bh}{2}$ is $2\sqrt{(AB)}$.

$$\therefore \qquad \left|f'(x)\right| < 2\sqrt{(AB)}.$$

3. *Show that θ which occurs in the Lagrange's form of remainder viz., $h^n/n!\,(f^n(a+\theta h))$, tends to the limit, $1\ (n+1)$, when $h \to 0$, provided that f^{n+1} is continuous at, a, and $f^{n+1}(a) \neq 0$.*

Sol. Since f^{n+1} is continuous at a, there exists an interval $[a - \delta, a + \delta]$ at every point of which $f^{n+1}(x)$ exists. Also, therefore, $f(x), f'(x), ..., f^n(x)$ are all continuous in $[a - \delta, a + \delta]$.

If $(a + h)$ be a point of this interval, we obtain the necessary conditions being satisfied,

$$f(a+h) = f(a) + hf'(a) + \ldots + \frac{h^n}{n!} f^n(a+\theta h),$$

$$f(a+h) = f(a) + hf(a) + \ldots + \frac{h^n}{n!} f^n(a) + \frac{h^{n+1}}{(n+1)!} f^{n+1}(a+\theta' h),$$

It follows that

$$f^n(a+\theta h) - f^n(a) = \frac{h}{n+1} f^{n+1}(a+\theta' h).$$

Again applying Lagrange's mean value theorem to the expression on the left, we see that

$$\theta h f^{n+1}(a + \theta\theta'' h) = \frac{h}{n+1} f^{n+1}(a+\theta' h)$$

$$\Rightarrow \quad \theta f^{n+1}(a+\theta\theta''h) = \frac{1}{n+1} f^{n+1}(a+\theta' h).$$

Let $h \to 0$. Then, we obtain

$$\lim \theta . f^{n+1}(a) = \frac{1}{n+1} f^{n+1}(a),$$

$\because f^{n+1}(x)$ is continuous.

$$\Rightarrow \quad \lim \theta = \frac{1}{n+1} \qquad \because f^{n+1}(a) \neq 0.$$

***4.** *If*

$$f(x) = e^{-1/x^2} \sin(1/x), \text{ for } x \neq 0 \text{ and } f(0) = 0,$$

show that

(i) the function f has at every point a differential coefficient and this is continuous at x = 0.

(ii) the differential coefficient vanishes at x = 0 and at an infinite number of points in every neighbourhood of x = 0.

Sol. We have

$$f'(x) = \frac{e^{-1/x^2}}{x^3}\left(2\sin\frac{1}{x} - x\cos\frac{1}{x}\right) \text{ when } x \neq 0$$

and
$$f'(0) = \lim_{x\to 0} \frac{e^{-1/x^2}\sin(1/x) - 0}{x}.$$

Since $e^{1/x^2} > 1/x^2 \; \forall \; x$, we see that

$$\left|\frac{e^{-1/x^2}\sin(1/x)}{x}\right| < \frac{x^2 \cdot 1}{|x|} = |x| \; \forall \; x$$

$$\Rightarrow \quad f'(0) = 0.$$

Also, when $x \neq 0$.

$$\left|f'(x) - f'(0)\right| = \left|\frac{e^{-1/x^2}}{x^3}\left(2\sin\frac{1}{x} - x\cos\frac{1}{x}\right)\right|$$

$$\leq \frac{e^{-1/x^2}}{|x|^3}(2+|x|).$$

Since $\quad e^{1/x^4} > 1/2x^4 \;\forall\; x,$

we see that

$$|f'(x) - f'(0)| \leq 2|x|(2 + |x|),$$

$$\Rightarrow \quad [f'(x) - f'(0)] \to 0 \text{ as } x \to 0$$

$$\Rightarrow \quad f' \text{ is continuous for } x = 0.$$

Also clearly f is continuous to every non-zero value of x.

We have now to show that f' vanishes at a point in every neighbourhood of $x = 0$.

Let $\delta > 0$ be a number. There surely exists a positive integer x such that

$$0 < \frac{2}{(2n+1)\pi} < \frac{1}{n\pi} < \delta.$$

It is easy to see that,

for $x = 1/n\pi$, $f'(x)$ is – ve or + ve according as n is even or odd;

for $x = 2/(2n + 1)\pi$, $f'(x)$ is +ve or –ve according as n is even or odd.

Therefore f', which is continuous, must vanish at least once between

$$2/(2n + 1)\pi \text{ and } 1/n\pi.$$

We may similarly dispose of left handed neighbourhood of 0.

EXERCISES

1. Examine the validity of the

(*i*) hypothesis, and

(*ii*) conclusion of the Lagrange's mean value theorem in the cases

$$(a)\; f(x) = 1/x, \qquad (b)\; f(x) = x^{1/3}$$

for the interval $[-1, 1]$.

2. Prove that, if $f'(c)$ exists and $\neq 0$, then, as $h \to 0$

$$\lim \frac{f(c+h) + f(c-h) - 2f(c)}{h} \quad \text{and} \quad \lim \frac{f(c+h) - f(c)}{f(c-h) - f(c)}$$

exist and find their values. State the theorems of limits you require in the proof.

Examine the existence of these limits if

$$f(x) = |x|.$$

3. If f, φ and ψ are continuous in $[a, b]$ and derivable in $]a, b[$ then show that there is a value, ξ, lying between a and b such that

$$\begin{vmatrix} f(a) & \varphi(a) & \psi(a) \\ f(b) & \varphi(b) & \psi(b) \\ f'(\xi) & \varphi'(\xi) & \psi'(\xi) \end{vmatrix} = 0.$$

Apply Rolle's theorem to the function

$$\begin{vmatrix} f(a) & \varphi(a) & \psi(a) \\ f(b) & \varphi(b) & \psi(b) \\ f(x) & \varphi(x) & \psi(x) \end{vmatrix}.$$ *(Poorv. 92; Garhwal 98)*

4. If $f(0) = 0$ and $f''(x)$ exists in $[0, \infty[$, show that

$$f'(x) - \frac{f(x)}{x} = \frac{1}{2} xf''(\xi),\ 0 < \xi < x$$

and deduce that if $f''(x)$ is positive for positive values of x, then $f(x)/x$ strictly increases as x increases.

5. Show by means of an example that the mere convergence of

$$f(x) + hf'(x) + \frac{h^2}{2!} f''(x) + \ldots$$

does not mean that it converges to $f(x + h)$.

6. Assuming f'' continuous in $[a, b]$, show that

$$f(c) - f(a)\frac{b-c}{b-a} - f(b)\frac{c-a}{b-a} = \frac{1}{2}(c-a)(c-b) f''(\xi),$$

where c and ξ both lie in $[a, b]$.

7. If $f^n(x) = 0$ for every x in $[a, b]$, then there are numbers

$$a_0, a_1, a_2 \ldots, a_{n-1}$$

such that $$f(x) = \sum_{r=0}^{n-1} a_r x^r,$$

in $[a, b]$.

8. Show that for the infinite series

$$\sum_{n=1}^{\infty} \left[\frac{nx}{1+n^2x^2} - \frac{(n-1)x}{1+(n-1)^2 x^2} \right],$$

the derivative of the sum is not equal to the sum of the derivatives for $x = 0$.

***9.** Applying Lagrange's mean value theorem in turn to the functions $\log x$ and e^x, determine the corresponding values of θ in terms of a and h.

Deduce that

(*i*) $0 < \left[\log(1+x)\right]^{-1} - x^{-1} < 1$. (*ii*) $0 < x^{-1} \log\left[\left(e^x - 1\right)/x\right] < 1$.

***10.** Show that

(*i*) $x^2 > (1+x)\left[\log(1+x)\right]^2$ for $x > 1$.

(*ii*) $x < \log\left[1/(1-x)\right] < x/(1-x)$, when $0 < x < 1$.

***11.** Prove that

$$\left|\log(1+x) - x + \frac{1}{2}x^2\right| \le \frac{1}{3}x^3.$$

if $0 \le x \le 1$.

***12.** Show that $a^x > x^a$ if $x > a \ge c$.

[Let $f(x) = x \log a - a \log x$. Show that $f(a) = 0$ and $f(x) > 0$].

***13.** If $\varphi''(x) \ge 0$ for every value of x, then

$$\varphi\left[\frac{1}{2}(x_1 + x_2)\right] \le \frac{1}{2}\left[\varphi(x_1) + \varphi(x_2)\right]$$

$$\varphi\left[\frac{1}{n}(x_1 + x_2 + \dots + x_n)\right] \le \frac{1}{n}\left[\varphi(x_1) + \dots + \varphi(x_n)\right]$$

where $x_1, x_2 \dots, x_n \in [a, b]$.

***14.** If

$$f(x) = x^2 \sin(1/x), \text{ when } x \ne 0, f(0) = 0$$

and $g(x) = x$,

show that $\lim_{x \to 0} [f'(x)/g'(x)]$ does not exist,

but $\lim_{x \to 0} [f(x)/g(x)]$ exists, *(Garhwal 99)*

and is equal to $f'(0)/g'(0)$.

***15.** If

$$f(x) = |x|, \; g(x) = 2|x|,$$

show that $f'(0)$ and $g'(0)$ do not exist but $\lim [f(x)/g(x)]$ exists and is equal to $\lim [f'(x)/g'(x)]$, when $x \to 0$.

16. If

$$f(x) = \sqrt{x}\,(1 + x \sin 1/x), \forall\, x > 0,$$

$$f(x) = -\sqrt{x}\,(1 + x \sin 1/x), \forall\, x < 0,$$

$$f(0) = 0,$$

show that f' exists everywhere and is finite except at $x = 0$ in the neighbourhood of which it oscillates between indefinitely great positive and negative values.

17. $$f(x) = 4 + 7x + x^2 (8 + x \sin 1/x),$$

where $x \sin(1/x)$ is zero for $x = 0$ has a first derivative but no second derivative at the origin.

18. If

$$f(x) = \sin x \sin(1/\sin x) \text{ when } 0 < x < \pi < x < 2\pi,$$

and $$f(x) = 0 \text{ when } x = 0, \pi, 2\pi;$$

show that f is continuous but not derivable for $x = 0, \pi, 2\pi$.

19. If $a_0, a_1, a_2, ..., a_n$ are real and

$$|a_0| + |a_1| + |a_2| + ... + |a_{n-1}| < a_n,$$

show that $$u(x) = a_0 + a_1 \cos x + a_2 \cos 2x + ... + a_n \cos nx$$

has at least $2n$ zeros in the interval $0 < x < 2\pi$. Show also that $u'(x)$ has at least $2n$ zeros in the interval $\alpha \leq x < 2\pi + \alpha$ for every real α.

[$u(0)$, $u(\pi/n)$, $u(2\pi/n)$..., $u(2n\pi/n)$ have positive and negative signs alternatively].

20. Determine the differential coefficient if any, of f defined by

$$f(x) = (x-a)\frac{Ae^{1/(x-a)} + Be^{-1/(x-a)}}{e^{1/(x-a)} + e^{-1/(x-a)}} \sin\frac{\pi}{2(x-a)} \text{ when } x \neq a$$

$f(a) = 0$, when $x = a$.

21. A function f is defined as follows :

$$f(x) = (x-w_1)(x-w_2)^2 (x-w_3)^2 \sin\frac{1}{(x-w_1)} \sin\frac{1}{(x-w_2)} \sin\frac{1}{(x-w_3)}$$

for all values of x except w_1, w_2, w_3 in the domain $[a, b]$ and $f(x) = 0$ when $x = w_1$ or w_2 or w_3. Show that

(*i*) df/dx does not exist at the point $x = w_1$,

(*ii*) df/dx exists but has a discontinuity of the second kind at $x = w_2$,

(*iii*) df/dx exists and is continuous at $x = w_3$.

OBJECTIVE QUESTIONS

Note. *Each of the following questions, four alternatives are given for the answer. Only are of them is correct. Choose the correct answer.*

1. Continuity is a condition for derivability which is :

(*a*) necessary and sufficient (*b*) necessary but not sufficient

(*c*) sufficient but not necessary (*d*) neither necessary nor sufficient

2. If $\phi(x) = \psi[f(x)]$, then $\phi'(x) =$

(*a*) $\psi'(x) + f'(x)$ (b) $\psi(x) f'(x) + f(x)\psi'(x)$

(c) $\psi'[f(x)] f'(x)$ (*d*) None of these

3. If f is finitely differentiable on $[a, b]$ and $f'(a) \neq f'(b)$ and k is any number lying between $f'(a)$ and $f'(b)$, then there exists a point $c \in (a, b)$ such that :

(*a*) $kf'(c) = 0$ (*b*) $\dfrac{f'(c)}{k} = 0$

(*c*) $f'(c) = k$ (*d*) $f'(c) \neq k$

4. Let $f'(x)$ be continuous at $x = 0$ and $f''(0) = 4$. The value of :

$\lim\limits_{x \to 0} \dfrac{2f(x) - 3f(2x) + f(4x)}{x^2}$ is

(*a*) 11 (*b*) 2

(*c*) 12 (*d*) none of these

5. The set of all points, where the function $\dfrac{f(x)}{(1+|x|)}$ is differentiable, is :

(*a*) $(-\infty, \infty)$ (*b*) $(0, \infty)$

(*c*) $(-\infty, 0) \cup (0, \infty)$ (*d*) does not exist.

6. If $f(x) = 1$ for $x < 0$ and $1 + \sin x$ for $0 \leq x < \pi/2$, then at $x = 0$, the derivative $f'(x)$ is :

(*a*) 1 (*b*) 0

(*c*) infinite (*d*) does not exist

7. If $f(x)$ is a differentiable function such that $f(x) \leq f(3)$ for $1 \leq x \leq 3$. Then

(*a*) $f'(3) = 0$ (*b*) $f'(3) = f(3)$

(*c*) $f'(3) = 3$ (*d*) $f'(3)$ does not exist.

8. If $f(x) = (x - x_0) g(x)$ where $g(x)$ is continuous at x_0, then $f'(x_0)$ is equal to

(*a*) 0 (*b*) x_0

(*c*) $g(x_0)$ (*d*) none of these

9. The value of c of the mean value theorem, if $f(x) = x(x-1)(x-2)$, $a = 0$, $b = 1/2$ is :

(a) 1 (b) $1+\sqrt{(21)}/6$

(c) $1-\sqrt{(21)}/6$ (d) none of these

10. The function $f(x)=x(x+3)e^{-x/2}$ satisfy all conditions of Rolle's theorem in the interval $[-3, 0]$. Then the value of c :

(a) 0 (b) 1

(c) 2 (d) -2

11. The value of c of the mean value theorem, if $f(x)=2x^2+3x+4$ in $[1, 2]$ is :

(a) 1 (b) 2

(c) 3/2 (d) 2/3

12. The function $f(x)=2x^3+x^2-4x-2$ satisfy all conditions of Rolle's theorem in the interval $\left[-\sqrt{2}, \sqrt{2}\right]$. Then the value of c is :

(a) -1 (b) $-2/3$

(c) 1 (d) 3/2

13. Which of the following is not an indeterminate form?

(a) $\infty+\infty$ (b) $\infty-\infty$

(c) ∞/∞ (d) $0\times\infty$

14. The formula L'Hospital's rule is :

(a) $\lim\limits_{x\to 0}\dfrac{f(x)}{g(x)}=\lim\limits_{x\to a}\left[\dfrac{g(x)}{f(x)}\right]$ (b) $\lim\limits_{x\to a}\dfrac{f(x)}{g(x)}=\lim\limits_{x\to a}\dfrac{f'(x)}{g'(x)}$

(c) $\lim\limits_{x\to 0}\dfrac{f(x)}{g(x)}=\dfrac{f(a)}{g(a)}$ (d) $\lim\limits_{x\to 0}\dfrac{f(x)}{g(x)}=\dfrac{f'(a)}{g'(a)}$

15. Consider

(i) $\lim\limits_{x\to 0}\dfrac{\sin x}{x}=1$, (ii) $\lim\limits_{x\to 0}\dfrac{\tan x}{x}=1$, then :

(a) Both (i), (ii) are true (b) (i) is false, (ii) is true

(c) (i) is true, (ii) is false (d) Both (i), (ii) are false

16. Value of $\lim\limits_{x\to\infty}\left(\dfrac{1}{x}\right)^{1/x}$ is :

(a) 0 (b) 1

(c) e (d) $1/e$

ANSWERS

1. (b)	**2.** (c)	**3.** (c)	**4.** (a)	**5.** (a)
6. (d)	**7.** (a)	**8.** (c)	**9.** (c)	**10.** (d)
11. (c)	**12.** (a)	**13.** (a)	**14.** (b)	**15.** (a)
16. (b)				

6

Real Valued Functions of a Single Real Variable. Riemann Integrability

6.1. INTRODUCTION

In elementary treatment of Integral Calculus the subject of integration is treated from the point of view of the inverse of differentiation so that a function φ is called an integral of a given function f if $\varphi'(x) = f(x)$ for all x belonging to the domain of the function f.

Historically, however, the subject arose in connection with the determination of areas of plane regions and was based on the notion of the limit of a type of sum when the number of terms in the sum tends to infinity each term tending to zero.

In fact the name Integral Calculus has its origin in this process of summation and the words '*To integrate*' literally mean '*To give the sum of*'. It was only afterwards that it was seen that the subject of integration can also be viewed from the inverse of differentiation point of view.

In elementary works the reference to integration from summation point of view is always associated with intuitively perceived geometrical concepts.

Consistent with our general approach, we shall, in this book, give a purely arithmetic treatment of the subject and the same will, moreover, basically be independent of the notion of differentiation. A function φ such that $\varphi'(x) = f(x)$ will be called a *Primitive* of f and the relation between *Integrals* and *Primitives* will be given later on by means of what is known as the Fundamental Theorem of Integral Calculus.

The first satisfactory rigorous aritnmetic treatment of Integration was given by Riemann (1826-1866) and we shall be following the definition given by him. Many refinements and generalisations of the subject have appeared since then. The most noteworthy of these is the theory of integration by Lebesgue (1902).

6.2. RIEMANN INTEGRABILITY AND INTEGRAL OF BOUNDED FUNCTIONS OVER BOUNDED INTERVALS

Let f be a *bounded* function whose domain is an interval $[a, b]$.

Let

$$\{a = x_0, x_1, x_2, x_3, ..., x_n = b\}$$

be a division of $[a, b]$.

Then

$$[x_0, x_1], [x_1, x_2], ..., [x_{r-1}, x_r], ..., [x_{n-1}, x_n],$$

are the sub-intervals in which the interval $[a, b]$ is divided.

Let the sub-interval $[x_{r-1}, x_r]$ as also its length, $x_r - x_{r-1}$, be both denoted by the same symbol, δ_r.

The function f which is bounded in $[a, b]$ is also necessarily bounded in each sub-interval δ_r. Let M_r, m_r be the* bounds of f in δ_r, *i.e.*, in $[x_{r-1}, x_r]$.

Set up the two sums S and s defined as follows :

$$S = M_1\delta_1 + ... + M_r\delta_r + ... + M_nS_n = \sum_{r=1}^{r=n} M_r\delta_r,$$

$$s = M_1\delta_1 + ... + m_r\delta_r + ... + m_nS_n = \sum_{r=1}^{r=n} m_r\delta_r,$$

The sums S and s are referred to as *upper* and *lower sums* respectively corresponding to the division D.

If M, m be the bounds of f in $[a, b]$ we have, for every value of r,

$$m \le m_r \le M_r \le M$$

$$m\delta_r \le m_r\delta_r \le M_r\delta_r \le M\delta_r.$$

Putting $r = 1, 2, ..., n$, and adding, we deduce that

$$m(b-a) \le s \le S \le M(b-a). \qquad ...(1)$$

Now, a pair of sums S, s corresponds to each division of $[a, b]$ and, from (1), we see that the sets of the sums S, s obtained by considering all possible divisions are bounded.

Upper and Lower Integrals. Def. The infimum *of the set of the upper sums, S, is called the* **upper integral** *of f over [a, b] and is denoted by*

$$U = \overline{\int_a^b} f(x)\,dx.$$

*The bounds are the suprema and the Infima.

The supremum *of the set of the lower sums, s is called the* **lower integral** *of f over [a, b] and is denoted by*

$$L = \underline{\int_a^b} f(x)\,dx.$$

[Clearly the supremum of the. sums, S, is $M(b-a)$ and the infimum of the sums, s, is $m(b-a)$ and they are attained.]

A bounded function f is said to be Riemann integrable, or simply integrable over [a, b], if its upper and lower integrals are equal; the common value of these integrals is called the **Riemann integral** *or simply the* **integral** *denoted by the symbol*

$$I = \int_a^b f(x)\,dx.$$

(Allahabad 98, Garhwal 91; Ajmer 98, Poorv. 92; Lucknow 94, 97)

Notes. 1. The numbers a, b are respectively called the lower and the upper limits of integration.

2. The definition of integrability as given above is based on the *notion of bounds.* Another equivalent approach based on the notion of limits is given in § 6.4.

3. It should be clearly understood that not *every bounded function is integrable, i.e.,* there do exist bounded functions f for which

$$\overline{\int_a^b} f(x)\,dx \neq \underline{\int_a^b} f(x)\,dx$$

[Refer Ex. 1, page 193].

The necessary and sufficient condition for the existence of

$$\int_a^b f(x)\,dx$$

i.e., for the integrability of f over $[a, b]$ is obtained in § 6.5.

4. The statement that, $\int_a^b f(x)\,dx$, exists, implies that the function f is bounded and integrable over $[a, b]$.

5. The concept of integrability of a function over an interval, as introduced here, is subject to the two following very important limitations :

(*i*) *the function is bounded*

(*ii*) *the interval is finite*

so that neither of the end points of the interval is infinite.

In chapter 9, we shall see how these limitations can be removed and the concept generalised so as to be applicable sometimes even to cases where the function is not bounded or where one or both the limits of integration are infinite.

6. If D_1, D_2 be two divisions such that $D_2 \supset D_1$, we shall say the division D_2 is *finer than* the division D_1.

7. Norm. The length of the greatest of all the sub-intervals $[x_{r-1}, x_r]$ of a division D will be called the **norm** of D and denoted by $|D|$.

8. The sums S, s corresponding to a division D will be denoted by the symbols $S(D)$ and $s(D)$ respectively. Clearly,

$$S(D) \geq s(D),$$

for all divisions D of $[a, b]$.

9. Oscillatory Sum. We have

$$S(D) - s(D) = \Sigma M_r \delta_r - \Sigma m_r \delta_r = \Sigma (M_r - m_r)\, \delta_r = \Sigma O_r \delta_r$$

where O_r denotes the oscillation of the function in the sub-interval δ_r. The sum $\Sigma O_r \delta_r$ which is called the *oscillatory* sum is denoted by $w(D)$.

As the oscillation O_r cannot be negative, it follows that each oscillatory sum consists of the sum of a finite number of non-negative terms.

EXAMPLES

1. *If f is defined on [0, 1] by $f(x) = x \;\forall\; x \in [0, 1]$, then prove that $f \in$ **R** [0, 1], and*

$$\int_0^1 f(x)\, dx = \frac{1}{2}. \qquad \textit{(Poorvanchal 91; Garhwal 97)}$$

Sol. Let any partition of [0, 1] be $D = \left\{0 = \frac{0}{n}, \frac{1}{n}, \frac{2}{n}, \ldots, \frac{r}{n}, \ldots, \frac{n}{n} = 1\right\}$.

Let the sub-intervals be $I_r = \left[\frac{r-1}{n}, \frac{r}{n}\right]$, for $r = 1, 2, \ldots, n$. If S_r be the length of this interval I_r, then

$$S_r = \frac{r}{n} - \frac{r-1}{n} = \frac{1}{n}$$

Also, if M_r and m_r be respectively the supremum and infimum of the function f in I_r, then $M_r = \frac{r}{n}$ and $m_r = \frac{r-1}{n}$, as $f(x) = x$.

$$\therefore \qquad S(D) = \sum_{r=1}^{n} M_r \delta_r = \sum_{r=1}^{n} \left(\frac{r}{n} \cdot \frac{1}{n}\right)$$

$$= \frac{1}{n^2} \sum_{r=1}^{n} r = \frac{1}{n^2}\left[\frac{1}{2} n(n+1)\right] = \frac{n+1}{2n}$$

$$= \frac{1}{2}\left(1 + \frac{1}{n}\right) \qquad \ldots(i)$$

Also, $$s(D) = \sum_{r=1}^{n} m_r \delta_r = \sum_{r=1}^{n} \left(\frac{r-1}{n}\right) \cdot \frac{1}{n}$$

$$= \frac{1}{n^2} \sum_{r=1}^{n} (r-1)$$

$$= \frac{1}{n^2}\left[\frac{1}{2}(n-1)(n-1+1)\right] = \frac{1}{2}\frac{(n-1)}{n}$$

$$= \frac{1}{2}\left(1-\frac{1}{n}\right) \qquad \text{...(ii)}$$

Again, $$\int_0^{\bar{1}} f(x)\,dx = \text{inf.}\left[s(D)\right]$$

$$= \lim_{n\to\infty} \frac{1}{2}\left(1+\frac{1}{n}\right) = \frac{1}{2} \qquad \text{...(iii)}$$

And $$\int_{\underline{0}}^{1} f(x)\,dx = \text{sup. } [s(D)]$$

$$= \lim_{n\to\infty} \frac{1}{2}\left(1-\frac{1}{n}\right) = \frac{1}{2} \qquad \text{...(iv)}$$

From (*iii*) and (*iv*), we find that

$$\int_{\underline{0}}^{1} f\,dx = \int_0^{\bar{1}} f\,dx = \frac{1}{2}$$

Hence $f \in \mathbf{R}\,[0, 1]$ and $\int_0^1 f(x)\,dx = \frac{1}{2}$.

2. *Show that a constant function is Reimann-integrable.*

(*Poorv.* 93)

Sol. Let a function f be defined on $[a, b]$ by $f(x) = c, \ \forall\, x \in [a, b]$, where c is a constant.

Let any partition of $[a, b]$ be $D = \{a = x_0, x_1, x_2, ..., x_r, ..., x_n = b\}$

Let its sub-intervals be $I_r = [x_{r-1}, x_r]$ for $r = 1, 2, ..., n$.

If δ_r be the length of this interval I_r, then

$$\delta_r = x_r - x_{r-1}.$$

Let M_r and m_r be respectively the supremum and infimum of the function f in I_r, then $M_r = c$, $m_r = c$, as $f(x) = c \;\forall\; x \in (a, b)$

$$\therefore \quad S(D) = \sum_{r=1}^{n} M_r \delta_r = \sum_{r=1}^{n} c\left(x_r - x_{r-1}\right)$$
$$= c\,[(x_1 - x_0) + (x_2 - x_1) + \ldots + (x_n - x_{n-1})]$$
$$= c\,(x_n - x_0) = c\,(b - a) = \text{constant.}$$

And
$$s(D) = \sum_{r=1}^{n} m_r \delta_r = \sum_{r=1}^{n} c\left(x_r - x_{r-1}\right)$$
$$= c\,(b - a) = \text{constant.}$$

$$\therefore \quad \overline{\int_a^b} f(x)\,dx = \text{infimum } S(D)$$
$$= c\,(b - a)$$

and
$$\underline{\int_a^b} f(x)\,dx = \sup.\left[s(D)\right]$$

Hence
$$\overline{\int_a^b} f(x)\,dx = \underline{\int_a^b} f(x)\,dx = c\,(b - a)$$

and so $f \in \mathbf{R}\,[a, b]$, *i.e.*, the function f is **R**-integrable and

$$\int_a^b f(x)\,dx = c\,(b - a).$$

3. *If $f(x)$ be defined on $[0, 1]$ as follows —*

$$f(x) = 1, \quad \text{when } x \text{ is rational}$$
$$= -1, \quad \text{when } x \text{ is irrational}$$

then prove that f is not Riemann integrable over $[0, 1]$.

Sol. Let any partition of $[0, 1]$ be $D = \{0 = x_0, x_1, x_2, \ldots, x_r, \ldots, x_n = 1\}$.

Let its sub-interval be $I_r = [x_{r-1}, x_r]$, for $r = 1, 2, \ldots, n$.

Clearly, $M_r = 1$ and $m_r = -1$.

$$\therefore \quad S(D) = \sum_{r=1}^{n} M_r \delta_r = \sum_{r=1}^{n} 1.\left(x_r - x_{r-1}\right)$$
$$= x_n - x_0 = 1 - 0 = 1.$$

And
$$s(D) = \sum_{r=1}^{n} m_r \delta_r = \sum_{r=1}^{n} (-1)\left(x_{r-1} - x_r\right)$$
$$= x_0 - x_n = -1.$$

$$\therefore \quad \overline{\int_0^1} f(x)\,dx = \text{inf.}\,\{s(D)\} = 1$$

and $\int_{\underline{0}}^{1} f(x)\,dx = \text{sup. } \{s(D)\} = -1$

$$\therefore \quad \int_{0}^{\overline{1}} f(x)\,dx \neq \int_{\underline{0}}^{1} f(x)\,dx$$

Hence $f \notin \mathbf{R}\,[0, 1]$, as the necessary and sufficient condition of Riemann-integrability is not satisfied.

EXERCISES

1. If

$$f(x) = \begin{cases} 0, \text{ where } x \text{ is rational,} \\ 1, \text{ where } x \text{ is irrational} \end{cases}$$

show that f is not integrable in any interval. *(Lucknow 95; Garhwal 90)*

2. Show that

$$\int_{a}^{\overline{b}} k\,dx = \int_{\underline{a}}^{b} k\,dx = k(b-a),$$

where k is a constant.

(This proves that every constant function is integrable.)

3. A function f is bounded in $[a, b]$; show that

$$(i) \quad \int_{a}^{\overline{b}} kf(x)\,dx = k\int_{a}^{\overline{b}} f(x)\,dx, \quad \int_{\underline{a}}^{b} kf(x)\,dx = k\int_{\underline{a}}^{b} f(x)\,dx,$$

where k is a positive constant.

$$(ii) \quad \int_{a}^{\overline{b}} kf(x)\,dx = k\int_{\underline{a}}^{b} f(x)\,dx, \quad \int_{\underline{a}}^{b} kf(x)\,dx = k\int_{a}^{\overline{b}} f(x)\,dx,$$

where k is a negative constant.

Deduce that if f is bounded and integrable over $[a, b]$ *then so is kf, where k is a constant, and that*

$$\int_{a}^{b} kf(x)\,dx = k\int_{a}^{b} f(x)\,dx.$$

[If M_r, m_r be the bounds of f in δ_r then kM_r km_r, (km_r, kM_r) are the bounds of kf in δ_r, where k is positive, (k is negative).

4. A bounded function f is integrable over $[a, b]$ and M, m are the bounds of f, show that

$$m(b-a) \leq \int_{a}^{b} f(x)\,dx \leq M(b-a).$$

6.3. DARBOUX'S THEOREM

To every $\varepsilon > 0$, *there corresponds* $\delta > 0$ *such that*

$$S(D) < \int_a^{\bar{b}} f(x)\,dx + \varepsilon \quad \textit{(Poorv. 91, 93; M.D.U. 97)}$$

$\forall$ *D with* $|D| \le \delta$.

Lemma. Let $|f(x)| \le k \ \forall\ x \in [a, b]$.

Let δ be a positive number and D_1 a division of $[a, b]$ such that

$$|D_1| \le \delta.$$

Let D_2 be a division of $[a, b]$ consisting of all the points of D_1 and at the most some p more.

Then we shall show that

$$S(D_1) - 2pk\delta \le S(D_2) \le S(D_1).$$

In particular, it will follow that

$$D_2 \supset D_1 \Rightarrow S(D_2) \le S(D_1).$$

Firstly suppose that $p = 1$ so that only one interval, say δ_r, of D_1 is divided into two intervals, say δ'_r, and δ''_r. Let M_r, M'_r, M''_r be the suprema of f in δ_r, δ'_r, δ''_r respectively.

We have

$$\begin{aligned} S(D_1) - S(D_2) &= M_r\delta_r - (M'_r\delta'_r + M_r''\delta_r'') \\ &= (M_r - M'_r)\,\delta'_r + (M_r - M''_r)\,\delta''_r, \end{aligned}$$

for $\delta_r = \delta'_r + \delta''_r$.

Now $\qquad |f(x)| \le k, \forall\ x \in [a, b]$

$\Rightarrow \qquad -k \le M'_r \le M_r \le k,$

$\Rightarrow \qquad 0 \le M_r - M'_r \le 2k.$

Similarly we have

$$0 \le M_r - M''_r \le 2k.$$

It follows that

$$0 \le S(D_1) - S(D_2) \le 2k(\delta'_r + \delta''_r) = 2k\delta_r \le 2k\delta.$$

Now supposing that each additional point is introduced one by one, we obtain the result.

We now prove the *main theorem.*

As f is bounded, there exists $k > 0$, such that

$$|f(x)| \le k \ \forall \ x \in [a, b].$$

Since $$\int_a^{\bar{b}} f(x)\, dx$$

is the infimum of the set of upper sums S, there exists a division

$$D_1 \{a = x_0, x_1, x_2, \ldots, x_{p-1}, x_p = b\}$$

such that $$S(D_1) < \int_a^b f(x)\, dx + \frac{\varepsilon}{2}.$$

The points of D_1 are $(p + 1)$ in number.

Let δ be the positive number such that

$$2k\,(p-1)\,\delta = \frac{1}{2}\varepsilon.$$

Let D be any division with norm less than or equal to δ.

Let D_2 be the division consisting of the points of D_1 as well as those of D. Applying the lemma to the divisions D and D_2, we have

$$S(D) - 2(p-1)k\delta \le S(D_2) \le S(D).$$

Also $$D_2 \supset D_1 \Rightarrow S(D_2) \le S(D_1).$$

Thus we obtain

$$S(D) - 2(p-1)\,k\delta \le S(D_1)$$

$$S(D) \le 2(p-1)\,k\delta + S(D_1)$$

$$< \frac{\varepsilon}{2} + \int_a^{\bar{b}} f(x)\, dx + \frac{\varepsilon}{2} = \int_a^{\bar{b}} f(x)\, dx + \varepsilon.$$

Hence the result.

Darboux's theorem II. *To every $\varepsilon > 0$ there corresponds $\delta > 0$, such that*

$$s(D) > \int_a^b f(x)\, dx - \varepsilon$$

for every division D, with

$$|D| \le \delta.$$

This proof is similar to that of the corresponding result on that upper integral proved above.

Note. Darboux's theorem may be symbolically exhibited as follows :

$$\lim s(D) = \int_{\underline{a}}^{b} f(x)\,dx, \quad \lim S(D) = \int_{a}^{\overline{b}} f(x)\,dx,$$

when the norm $|D|$ tends to zero.

Cor. I. *For every bounded function f*

$$\int_{a}^{\overline{b}} f(x)\,dx \geq \int_{\underline{a}}^{b} f(x)\,dx,$$

so that the upper integral ≥ the lower integral.

If possible, let

$$\int_{a}^{\overline{b}} f(x)\,dx < \int_{\underline{a}}^{b} f(x)\,dx.$$

Let, k be any number lying between the upper and lower integrals.

Now there exists by Darboux's theorem, a positive number δ_1 such that for every division whose norm is $\leq \delta_1$,

$$S < k.$$

Also, there exists a positive number δ_2 such that for every division whose norm is $\leq \delta_2$

$$s > k.$$

If, δ, be any positive number smaller than δ_1 as well as δ_2, then for every division whose norm is $\leq \delta$, we have

$$S < k < s \Rightarrow S < s,$$

which is not true.

Hence the result.

Cor. II.

$$S(D_1) \leq s(D_2)$$

even when D_1, D_2 are two *different* divisions.

This at once follows from the cor. 1 above.

6.4. ANOTHER EQUIVALENT DEFINITION OF INTEGRABILITY AND INTEGRAL

We shall firstly state and prove a theorem which will suggest an alternative definition of the integrability and integral of a function. Let f be a function defined in $[a, b]$. Let

$$D\{a = x_0 < x_1 < x_2 < ... < x_{r-1} < x_r < ... < x_n = b\}$$

be any division of $[a, b]$ and let ξ_r be an arbitrary point of

$$\delta_r = [x_{r-1}, x_r].$$

From the sum

$$\sum_{r=1}^{r=n} f(\xi_r)\,\delta_r = f(\xi_1)\,\delta_1 + ... + f(\xi_r)\,\delta_r + ... + f(\xi_n)\,\delta_n.$$

6.4.1. Theorem

If f is bounded and integrable over $[a, b]$, then to every $\varepsilon > 0$ there corresponds $\delta > 0$, such that for every division

$$D\left\{a = x_0, x_1 ..., x_{r-1}, x_r, ...x_n = b\right\}$$

of norm $\leq \delta$, and for every choice of $\xi_r \in \left[x_{r-1}, x_r\right]$,

$$\left|\sum_{r=0}^{r=n} f(\xi_r)\left(x_r - x_{r-1}\right) - \int_a^b f(x)\,dx\right| < \varepsilon.$$

In a more concise but less precise manner the result may be stated a little differently as follows — *If a function f is bounded and integrable, then*

$$\lim_{|D|\to 0} \sum_{r=1}^{r=n} f(\xi_r)\,\delta_r,$$

exists and is the integral of f over $[a, b]$.

As f is a bounded and integrable function, we have

$$\int_a^{\bar{b}} f(x)\,dx = \int_{\underline{a}}^{b} f(x)\,dx = \int_a^b f(x)\,dx.$$

Let ε be a positive number. By Darboux's theorem, there exists $\delta > 0$ such that for every division D whose norm is $\leq \delta$,

$$S(D) < \int_a^{\bar{b}} f(x)\,dx + \varepsilon = \int_a^b f(x)\,dx + \varepsilon, \qquad ...(1)$$

and

$$s(D) < \int_{\underline{a}}^{b} f(x)\,dx - \varepsilon = \int_a^b f(x)\,dx - \varepsilon \qquad ...(2)$$

If ξ_r be *any* point of the interval δ_r of D, we have

$$s(D) \leq \sum_{r=1}^{r=n} f(\xi_r)\,\delta_r \leq S(D) \qquad ...(3)$$

From (1), (2) and (3), we deduce that for every division D, whose norm is $\le \delta$,

$$\int_a^b f(x)\,dx - \varepsilon < \sum_{r=1}^{r=n} f(\xi_r)\,\delta_r < \int_a^b f(x)\,dx + \varepsilon$$

$$\Rightarrow \qquad \left| \sum_{r=1}^{r=n} f(\xi_r)\,\delta_r - \int_a^b f(x)\,dx \right| < \varepsilon.$$

Hence the theorem.

6.4.2. A Second Definition

We shall now prove that the definition of integrability as given in § 6.2 is equivalent to the following :

A function f is integrable over [a, b], if there exists a number 1 and to every $\varepsilon > 0$ there corresponds $\delta > 0$, such that for every division

$$D\ \{a = x_0, x_1, \ldots, x_{r-1}, x_r, \ldots, x_n = b\}$$

of norm $\le \delta$ and $\forall\ \xi_r \in [x_{r-1}, x_r]$,

$$\left| \sum_{r=1}^{r=n} f(\xi_r)(x_r - x_{r-1}) - 1 \right| < \varepsilon.$$

Also then the number I is said to be the integral of the function f over the interval [a, b].

In a more concise but less precise manner, this definition may be stated as follows :

A function f is integrable, if

$$\lim_{|D|\to 0} \sum_{r=1}^{r=n} f(\xi_r)\,\delta_r$$

exists and is independent of the choice of the interval δ_r and of the point ξ_r of δ_r; the limit I, if it exists is called the integral of f over [a, b].

The *equivalence* of the two definitions will now be established.

As a consequence of the theorem proved in § 6.4.1 above, we see that if a function f be integrable according to the former definition, then it is so according to the latter also.

Now, *let f be integrable according to the* **latter** *definition so that*

$$lim\ \Sigma\, f(\xi_r)\, d_r,$$

exists, as the norm $|D| \to 0$.

It will firstly be deduced that f is bounded in $[a, b]$.

If possible, let f be *not* bounded. There exists a division D such that for every choice of ξ_r in δ_r,

$$\left|\Sigma f(\xi_r)\delta_r - I\right| < 1 \Rightarrow \left|\Sigma f(\xi_r)\delta_r\right| < |I| + 1.$$

As f is not bounded in $[a, b]$, it must also be so in at least one sub-interval δ_r say in δ_m.

We take $\xi_r = x_r$ when $r \neq m$ so that every ξ_r, except ξ_m is fixed and, accordingly, every term of the sum $\Sigma f(\xi_r)\delta_r$ except the term $f(\xi_m)\delta_m$ is also fixed. Since f is not bounded in δ_m, we can choose a point ξ_m in δ_m such that

$$\left|\Sigma f(\xi_r)\delta_r\right| > |I| + I,$$

and thus we arrive at a contradiction. Hence f is bounded in $[a, b]$.

Now, let ε be any positive number. There exists a $\delta > 0$ such that for every division whose norm is $\leq \delta$.

$$\left|\Sigma f(\xi_r)\delta_r - I\right| < \frac{1}{2}\varepsilon,$$

$$\Leftrightarrow \qquad I - \frac{1}{2}\varepsilon < \Sigma f(\xi_r)\delta_r < I + \frac{1}{2}\varepsilon$$

for every-choice of point ξ_r, in δ_r.

If M_r, m_r be the bounds of f in δ_r, there exist points α_r, β_r of δ_r such that

$$f(\alpha_r) > M_r - \varepsilon/2(b-a), \quad f(\beta_r) < m_r + \varepsilon/2(b-a).$$

From these we deduce that

$$\Sigma f(\alpha_r)\delta_r > S - \frac{1}{2}\varepsilon \quad \Leftrightarrow \quad S < \Sigma f(\alpha_r)\delta_r + \frac{1}{2}\varepsilon. \qquad ...(2)$$

$$\Sigma f(\beta_r)\delta_r < s + \frac{1}{2}\varepsilon \quad \Leftrightarrow \quad S > \Sigma f(\beta_r)\delta_r - \frac{1}{2}\varepsilon. \qquad ...(3)$$

From (1), (2) and (3), we deduce, taking separately $\xi_r, = \alpha_r$ and β_r, that

$$I - \varepsilon < s \leq S < I + \varepsilon \qquad ...(4)$$

for every division whose norm is $\leq \delta$.

Also we know that

$$s \leq \underline{\int_a^b} f(x)\,dx \leq \overline{\int_a^b} f(x)\,dx \leq S. \qquad ...(5)$$

From (4) and (5), we have

$$I-\varepsilon<\int_{\underline{a}}^{b} f(x)\,dx \leq \int_{a}^{\bar{b}} f(x)\,dx<I+\varepsilon, \qquad \text{...(6)}$$

$$\Rightarrow \qquad 0 \leq \int_{a}^{\bar{b}} f(x)\,dx-\int_{\underline{a}}^{b} f(x)\,dx<2\varepsilon,$$

so that the non-negative number

$$\int_{a}^{\bar{b}} f(x)\,dx-\int_{\underline{a}}^{b} f(x)\,dx$$

is less than every positive number; ε being arbitrary.

It follows that

$$\int_{a}^{\bar{b}} f(x)\,dx-\int_{\underline{a}}^{b} f(x)\,dx=0. \qquad \text{...(7)}$$

so that the function is integrable according to the former definition also.

From (6) and (7), we have

$$I-\varepsilon<\int_{a}^{b} f(x)\,dx<I+\varepsilon,$$

and since ε is arbitrary, this gives

$$I=\int_{a}^{b} f(x)\,dx.$$

Thus the equivalence is completely established.

6.5. CONDITIONS FOR INTEGRABILITY

6.5.1. First Form

A necessary and sufficient condition for the integrability of a bounded function f is, that to every ε > 0, there corresponds a δ > 0 such that for every division D, whose norm is ≤ δ, the oscillatory sum w (D) is < ε.

(*Garhwal 92, 93, 95; Rohilkhand 90, 94; Dibrugarh 2003, 2005; Oshmania 2005*)

The condition is necessary. The bounded function *f* being integrable, we have

$$\int_{a}^{\bar{b}} f(x)\,dx=\int_{\underline{a}}^{b} f(x)\,dx=\int_{a}^{b} f(x)\,dx.$$

Let ε be any positive number. By Darboux's theorem, there exists $\delta > 0$ such that for every division D whose norm is $\leq \delta$,

$$\begin{cases} S(D) < \int_a^{\bar{b}} f(x)\,dx + \frac{\varepsilon}{2} = \int_a^b f(x)\,dx + \frac{\varepsilon}{2}, \\ s(D) > \int_{\underline{a}}^b f(x)\,dx - \frac{\varepsilon}{2} = \int_a^b f(x)\,dx - \frac{\varepsilon}{2}. \end{cases}$$

$$\Rightarrow \qquad \int_a^b f(x)\,dx - \frac{\varepsilon}{2} < s(D) \leq S(D) < \int_a^b f(x)\,dx + \frac{\varepsilon}{2}$$

$$\Rightarrow \qquad w(D) = S(D) - s(D) < \varepsilon$$

for every division D whose norm is $\leq \delta$.

The condition is sufficient. Let ε be any positive number. There exists a division D such that

$$S(D) - s(D) = \left[S(D) - \int_a^{\bar{b}} f(x)\,dx \right] + \left[\int_c^{\bar{b}} f(x)\,dx - \int_{\underline{a}}^b f(x)\,dx \right] + \left[\int_{\underline{a}}^b f(x)\,dx - s(D) \right] < \varepsilon.$$

Since each one of the three numbers

$$S(D) - \int_a^{\bar{b}} f(x)\,dx, \int_a^{\bar{b}} f(x)\,dx - \int_{\underline{a}}^b f(x)\,dx, \int_a^b f(x)\,dx - s(D)$$

is non-negative, we see that

$$0 \leq \int_a^{\bar{b}} f(x)\,dx - \int_a^b f(x)\,dx < \varepsilon.$$

As ε is an arbitrary positive number, we see that the non-negative number

$$\int_a^{\bar{b}} f(x)\,dx - \int_{\underline{a}}^b f(x)\,dx,$$

is less than every positive number, and hence

$$\int_a^{\bar{b}} f(x)\,dx - \int_{\underline{a}}^b f(x)\,dx = 0 \Rightarrow \int_a^{\bar{b}} f(x)\,dx = \int_{\underline{a}}^b f(x)\,dx,$$

so that f is integrable.

Note. This theorem is sometimes stated differently as follows :

A necessary and sufficient condition for a bounded function f to be integrable in [a, b] is that

$$\lim w(D) = 0,$$

when | D | the norm of the division D, tends to 0.

6.5.2. The Condition of Integrability

Second Form. *A necessary and sufficient condition for the integrability of a bounded function f is that to every* $\varepsilon > 0$ *there corresponds a division D such that the corresponding oscillatory sum* $w(D) < \varepsilon$.

(Ajmer 97)

That this condition is necessary follows at once from the first part of § 6.5.1.

The condition is sufficient. As in the second of § 6.5.1, we write

$$S(D) - s(D) = \left[S(D) - \overline{\int_a^b} f(x)\,dx\right] + \left[\overline{\int_a^b} f(x)\,dx - \underline{\int_a^b} f(x)\,dx\right] + \left[\underline{\int_a^b} f(x)\,dx - s(D)\right] < \varepsilon,$$

and see that

$$0 \le \overline{\int_a^b} f(x)\,dx - \underline{\int_a^b} f(x)\,dx < \varepsilon.$$

From this relation which holds for every positive ε, we deduce that

$$\overline{\int_a^b} f(x)\,dx - \underline{\int_a^b} f(x)\,dx = 0 \Rightarrow \overline{\int_a^b} f(x)\,dx = \underline{\int_a^b} f(x)\,dx.$$

Note. On comparing the two forms of conditions, the reader will easily see that, from the point of view of *necessity*, the first form is stronger than the second but, from the point of view of sufficiency, the second form is stronger than the first.

6.6. PARTICULAR CLASSES OF BOUNDED INTEGRABLE FUNCTIONS

6.6.1. Every Continuous Function is Integrable

(Lucknow 95, 97; Garhwal 93, 94; Kurukshetra 2000; Rohilkhand 92; Allahabad 99)

Suppose that f is continuous in $[a, b]$. Since f is continuous, it is bounded and also uniformly continuous.

Let ε be a positive number.

We divide $[a, b]$ into a finite number of sub-intervals, say, δ_r such that the oscillation of f in each of these sub-interval is

$$< \varepsilon / (b - a).$$

If $\omega(D)$ be the oscillatory sum for this division, we have

$$\omega(D) = \Sigma (M_r - m_r)\,\delta_r$$
$$< \Sigma [\varepsilon / b - a]\delta_r$$
$$= [\varepsilon / (b - a)]\,\Sigma\delta_r = [\varepsilon / (b - a)]\cdot(b - a) = \varepsilon,$$

$\Rightarrow$ $\omega(D) < \varepsilon.$

Hence f is integrable in $[a, b]$. [§ 6.5.2].

6.6.2. A Bounded Function f which has only a finite number of points of discontinuity in $[a, b]$ is integrable in $[a, b]$.

[*Poorv. 90, Kumaon 97; M.D.U. 96*]

Let

$$a_1, a_2, a_3, \ldots, a_p,$$

be the finite number of points of discontinuity of f in $[a, b]$.

Let ε be a positive number. We enclose the points $a_1, a_2, \ldots, a_p$, in p non-overlapping intervals.

$$[a_1', a_1''], [a_2', a_2''], \ldots, [a_p', a_p'']$$

such that the sum of their lengths is $< \varepsilon / 2(M - m)$; $M, m,$ being the bounds of f in $[a, b]$. The oscillation of f in each of these intervals is $\le (M - m)$ and accordingly the contribution to the oscillatory sum from these is

$$< [\varepsilon / 2(M - m)].[M - m] = \varepsilon / 2.$$

Now, f is continuous in the $(p + 1)$ sub-intervals

$$[a_1', a_1'], [a_2'', a_2'] \ldots, [a_p'' b].$$

As in § 6.6.1 above each of these $(p + 1)$ sub-intervals can be further sub-divided so that the part of the oscillatory sum arising from the sub-intervals of each of them separately is $< \varepsilon/2\,(p + 1)$.

Thus there exists a division of $[a, b]$ such that the corresponding oscillatory sum is

$$< \frac{\varepsilon}{2} + \frac{\varepsilon}{2(p+1)}(p+1) = \varepsilon.$$

Hence the function f is integrable in $[a, b]$. [§ 6.5.2].

6.6.3. *If a function f is bounded in* $[a, b]$ *and the set of its points of discontinuity has a* **finite number of limit points**, *then f is integrable in* $[a, b]$. *(Patna 2003; Dibrugarh 2003)*

Let
$$a_1, a_2, a_3, ..., a_p,$$
be the limit points of the set of points of discontinuity of f in $[a, b]$. We enclose them in, p, non-overlapping intervals

$$[a_1', a_1''], [a_2', a_2''] ... [a_p', a_p'']$$

such that the sum of their lengths is $< \varepsilon/2\,(M - m)$; M, m being the bounds of f in $[a, b]$. The contribution to the oscillatory sum by these intervals is $< \varepsilon/2$.

Only a *finite* number of points of discontinuity of f can lie in each of the $(p + 1)$ intervals

$$[a, a_1'], [a_1'', a_2'], ..., [a_p'', b],$$

so that, as in § 6.6.2, each of them can be so sub-divided that the part of the oscillatory sum arising from the sub-intervals of each of these $(p + 1)$ intervals is separately $< \varepsilon/2\,(p + 1)$.

Thus there exists a division of $[a, b]$ such that the corresponding oscillatory sum is

$$< \frac{\varepsilon}{2} + \frac{\varepsilon}{2\,(p+1)}(p+1) = \varepsilon.$$

Hence f is integrable in $[a, b]$. [§ 6.5.2]

6.6.4. *If f is monotonic in* $[a, b]$, *then it is integrable in* $[a, b]$.

(Garhwai 97, 99; M.D.U. 95; Kurukshetra 94; Ajmer 98)

Clearly f is bounded and $f(a)$, $f(b)$ are its two bounds.

Let ε be a positive number.

For the sake of definiteness, we suppose that f is monotonically increasing.

We divide $[a, b]$ so that the length of each sub-interval is

$$< \varepsilon / [f(b) - f(a) + 1].$$

Let
$$D\{a = x_0, x_1, x_2, ..., x_{r-1}, x_r, ..., x_n = b\}$$
be any one of such division. Let $\delta_r = x_r - x_{r-1}$.

Here $\qquad M_r = f(x_r),\ m_r = f(x_{r-1}).$

Thus we have $\quad w(D) = \Sigma (M_r - m_r)\,\delta_r$

$$= \Sigma\left[f(x_r) - f(x_{r-1})\right]\delta_r$$

$$< \frac{\varepsilon}{f(b) - f(a) + 1} \cdot \Sigma\left[f(x_r) - f(x_{r-1})\right]$$

$$= \frac{\varepsilon}{f(b) - f(a) + 1} \cdot \left[f(b) - f(a)\right] < \varepsilon.$$

Hence f is integrable in $[a, b]$.

Note. If we had taken $[f(b) - f(a)]$ instead of $[f(b) - f(a) + 1]$, the proof would not have been valid for the case when $f(b) - f(a) = 0$, *i.e.*, when f is a constant function. The artifice of taking, $[f(b) - f(a) + k]$, where k is positive, or in particular, $f(b) - f(a) + 1$, serves to make the proof applicable even to this case.

EXAMPLES

1. *If $f(x)$ be defined on $(0, 2)$ as follows :*

$$f(x) = x + x^2, \quad \text{when } x \text{ is rational}$$
$$= x^2 + x^3, \quad \text{when } x \text{ is irrational,}$$

then evaluate the upper and lower Riemann-integrals in $(0, 2)$.

(Lucknow 93, 96, 97)

Sol. We have $(x + x^2) - (x^2 + x^3) = x - x^3 = x(1 - x^2)$

Hence $\qquad (x + x^2) - (x^2 + x^3) > 0$ if $a < x < 1$

and < 0, if $1 < x < 2$.

Let M_r and m_r be the upper and lower bounds of the given function $f(x)$ in I_r, then for all value of r, we have

$$M_r = x + x^2, \quad \text{if } 0 < x < 1$$
$$= x^2 + x^3, \quad \text{if } 1 < x < 2$$

and m_r
$$= x^2 + x^3, \quad \text{if } 0 < x < 1$$
$$= x + x^2, \quad \text{if } 1 < x < 2.$$

$\therefore$ The upper Riemann-integral

$$= \int_0^1 \left(x + x^2\right) dx + \int_1^2 \left(x^2 + x^3\right) dx = \frac{83}{12}$$

and the lower Riemann-integral

$$= \int_0^1 \left(x^2 + x^3\right) dx + \int_1^2 \left(x + x^2\right) dx = \frac{53}{12}.$$

2. *If the function f (x) be defined on (a, b) as follows :*

$$f(x) = 1/q^2, \text{ when } x = p/q$$
$$= 1/q^3, \text{ when } x = \sqrt{(p/q)},$$

where p and q are relatively prime integer and f (x) = 0 for all other values of x, then show that the function f is Riemann-integrable in (a, b).

Sol. We can easily show that the function f is discontinuous for rational values and continuous for irrational values.

Also, we are given that if p/q is any rational number in (a, b), then

$$f(p/q) = 1/q^2$$

We know that in every interval, however small it may be, there are an infinite number of irrational numbers, so every interval around p/q contains irrational number x, for which

$$\left|f(x) - f(p/q)\right| = \left|0 - \frac{1}{q^2}\right| = \frac{1}{q^2}$$

Hence function f is discontinuous for all rational points in (a, b).

Again let y be an irrational point in (a, b), then we have $f(y) = 0$ or $1/q^2$ according as $y \neq \sqrt{(p/q)}$ or $y = \sqrt{(p/q)}$.

$\therefore$ If ε be any given positive number, then there will be only a finite number of fractions $1/q^2$ for which $1/q^2 > \varepsilon$ or $1/q > \sqrt{\varepsilon}$ or $q < \dfrac{1}{\sqrt{\varepsilon}}$. Thus an interval around y can be found which does not contain any rational number p/q for which $q < 1/\sqrt{\varepsilon}$.

$\therefore$ For any irrational point x in this interval, we get

$$\left|f(x) - f(y)\right| = 0 \text{ or } 1/q^3$$

$$\Rightarrow \quad \left|f(x) - f(y)\right| < \varepsilon.$$

Similarly, for any rational point $x = p/q$ in the same interval.

$$\left|f(x) - f(y)\right| = \left(\frac{1}{q^2}\right) - \left(\frac{1}{q^3}\right) \text{ or } \frac{1}{q^2}.$$

Thus in each case $f(x) - f(y) < \varepsilon$, hence $f(x)$ is continuous at $x = y$, where y is any rational point in (a, b).

Hence all the irrational points in (a, b) are points of continuity and all the rational points in (a, b) are points of discontinuity of the function f and so all points of discontinuity being rational points in (a, b) form a set of measure zero.

Hence the function f is **R**-integrable in (a, b).

3. *Show that the function defined as follows :*

$$f(x) = \frac{1}{2^n} \text{ when } \frac{1}{2^{n+1}} < x \le \frac{1}{2^n}, (n = 0, 1, 2, 3 ...)$$

$$f(0) = 0$$

is integrable in [0, 1], although it has an infinite number of point of discontinuity.

Sol. We have $f(0) = 0 \quad f(x) = 1$, when $\frac{1}{2} < x \le 1$,

$$f(x) = \frac{1}{2}, \text{ when } \left(\frac{1}{2}\right)^2 < x \le \frac{1}{2},$$

$$\cdots \quad \cdots \quad \cdots \quad \cdots$$

$$f(x) = \left(\frac{1}{2}\right)^{n-1}, \text{ when } \left(\frac{1}{2}\right)^n < x \le \left(\frac{1}{2}\right)^{n-1},$$

$$\cdots \quad \cdots \quad \cdots \quad \cdots \quad \cdots$$

$$\cdots \quad \cdots \quad \cdots \quad \cdots \quad \cdots$$

Since f is bounded and monotonically increasing in [0, 1], it is integrable (§ 6.6.4).

Alternatively we notice that f is continuous in [0, 1], except at the set of points

$$0, \frac{1}{2}, \left(\frac{1}{2}\right)^2, \left(\frac{1}{2}\right)^3, \ldots, \left(\frac{1}{2}\right)^n \ldots$$

which has only one limit point, *viz.*, 0 and hence f is integrable (§ 6.6.3).

4. *A function f is defined in [0, 1] as follows :*

$$f(x) = p/q$$

when x is any non-zero rational number p/q in its lowest terms and f (x) = 0 when x is irrational or 0.

Show that f is integrable in [0, 1] and the value of the integral is 0.

Sol. Let, ε, be a positive number.

There exists only a finite number of positive integers q such that

$$\frac{1}{q} > \frac{\varepsilon}{2} \Leftrightarrow q < \frac{2}{\varepsilon}.$$

We call the points p/q in [0, 1] for which $1/q > \varepsilon/2$ exceptional points. Surely they are finite in number.

Also we know that every interval encloses rational as well as irrational points.

Clearly the oscillation of the function f in any interval which does not include any exceptional point is $< \varepsilon/2$ and the oscillation of f in any interval which includes an exceptional point is at the most 1.

We now enclose the exceptional points in [0, 1] in intervals the sum of whose length is $< \varepsilon/2$ so that the part of the oscillatory sum arising from these $< \varepsilon/2$. Also, howsoever we may divide the remaining part of [0, 1] the part of the oscillatory sum arising from the same will be $< \varepsilon/2$.

Thus we have a division of [0, 1] such that the corresponding oscillatory sum is less than a given pre-assigned positive number ε. Hence by § 6.5.2, the function is integrable.

Also since for every division D, $s(D) = 0$, we have

$$\overline{\int_0^1} f(x)\,dx = \underline{\int_0^1} f(x)\,dx = 0.$$

Note. It has been seen in Ex. 2, page 208, that the function is discontinuous for every non-zero rational value of x so that the set of the limit points of the set of points of discontinuity of f is the entire interval [0, 1]. This shows that the condition obtained for integrability in § 6.6.3 is only sufficient and not necessary.

5. *A function f is integrable in $[a, b]$; show that*

(*i*) $$\lim \sum_{r=1}^{r=n} hf(a+rh) = \int_a^b f(x)\,dx,$$

when $h \to 0, n \to \infty, nh = b - a,$

(*ii*) $$\lim \sum_{p=1}^{p=n} f\left(ar^{p-1}\right)\left(ar^p - ar^{p-1}\right) = \int_a^b f(x)\,dx,$$

when $r \to 1, n \to \infty, r^n = b/a.$

6.7. PROPERTIES OF INTEGRABLE FUNCTIONS

6.7.1. *If a bounded function f is integrable in $[a, b]$, then it is also integrable in $[a, c]$ and $[c, b]$ where c is a point of $[a, b]$.*

Conversely, if f is bounded and integrable in [a, c], [c, b], then it is also integrable in [a, b].

Also in either case

$$\int_a^b f(x)\,dx = \int_a^c f(x)\,dx + \int_c^b f(x)\,dx,\ a < c < b.$$

(Garhwal 98; Kumaon 97, 98; Allahabad 98)

Suppose that f is bounded and integrable in $[a, b]$.

Let ε be a positive number.

There exists $\delta > 0$ such that for each division of $[a, b]$ whose norm is $\leq \delta$, the oscillatory sum is $< \varepsilon$. Let D be a division of $[a, b]$ such that $c \in D$ and $|D| \leq \delta$.

The oscillatory sum for the division D breaks itself into two parts, respectively consisting of the terms which arise from the sub-intervals $[a, c]$ and $[c, b]$. Since the terms of an oscillatory sum are all positive, each part must itself be $< \varepsilon$. Hence f is integrable both in $[a, c]$ and $[c, b]$.

Let now, f be obtained and integrable in $[a, c]$ and $[c, b]$.

Let ε be a positive number. There exist divisions of $[a, c]$ and $[c, b]$ such that the corresponding oscillatory sums are $< \varepsilon/2$. The divisions of $[a, c]$ and $[c, b]$ give rise to a division of $[a, b]$ for which the oscillatory sum is $< (\varepsilon/2 + \varepsilon/2) = \varepsilon$. Hence f is integrable in $[a, b]$.

The relationship of equality is to be proved now.

Let ε be a positive number.

As f is simultaneously integrable in $[a, c]$, $[c, b]$ and $[a, b]$, there exists $\delta > 0$ such that for divisions of norm $\leq \delta$, and of which c, is a point, we have

$$\left|\sum_{(a,c)} f(\xi_r)\,\delta_r - \int_a^c f(x)\,dx\right| < \frac{\varepsilon}{3}, \left|\sum_{(c,b)} f(\xi_r)\,\delta_r - \int_c^b f(x)\,dx\right| < \frac{\varepsilon}{3}$$

$$\left|\sum_{(a,b)} f(\xi_r)\,\delta_r - \int_a^b f(x)\,dx\right| < \frac{\varepsilon}{3};$$

where the meanings of the symbols $\sum_{(a,c)} f(\xi_r)\,\delta_r$, etc. are obvious.

Since $\quad \sum_{(a,c)} f(\xi_r)\,\delta_r + \sum_{(c,b)} f(\xi_r)\,\delta_r = \sum_{(a,b)} f(\xi_r)\,\delta_r,$

we deduce that

$$\left|\int_a^b f(x)\,dx - \int_a^c f(x)\,dx - \int_c^b f(x)\,dx\right| < \varepsilon,$$

$$\Rightarrow \qquad \int_a^b f(x)\,dx - \int_a^c f(x)\,dx - \int_c^b f(x)\,dx = 0;$$

ε being an positive number.

Cor. *If f is bounded and integrable in* $[a, b]$, *then it is also bounded and integrable in* $[\alpha, \beta]$ *where* $a < \alpha < \beta < b$.

f is integrable in $[a, b] \Rightarrow f$ is integrable in $[a, \beta]$

$\Rightarrow f$ is integrable in $[\alpha, \beta]$

6.7.2. Integrability of the Sum, Difference, Product and Quotient of Integrable Functions

Before taking up the main question, we state and prove a simple lemma.

Lemma. *The oscillation of a bounded function f in an interval* $[a, b]$ *is the supremum of the set of numbers*

$$\left\{\left|f(\alpha) - f(\beta)\right| : \alpha, \beta \in [a, b]\right\}.$$

Let m, M be the bounds of f in $[a, b]$. We have

$$m \le f(\alpha), f(\beta) \le M\,; \alpha, \beta \in [a, b]$$

$$\Rightarrow \qquad \left|f(\alpha) - f(\beta)\right| \le M - m; \qquad \text{...(1)}$$

and as such $M - m$ is an upper bound of the set in question.

Let $\varepsilon > 0$ be given.

Since M is the supremum of f, there exists a_1 $[a, b]$ such that

$$f(\alpha_1) > M - \frac{1}{2}\varepsilon. \qquad \text{...(2)}$$

Since m is the infimum of f, there exists $\beta_1 \in [a, b]$ such that

$$f(\beta_1) < m + \frac{1}{2}\varepsilon. \qquad \text{...(3)}$$

From (2) and (3), we have

$$f(\alpha_1) - f(\beta_1) > M - m - \varepsilon.$$

$$\Rightarrow \qquad \left|f(\alpha_1) - f(\beta_1)\right| \ge f(\alpha_1) - f(\beta_1) > M - m - \varepsilon.$$

There exist therefore a pair of numbers α_1, β_1 such that

$$\left|f(\alpha_1) - f(\beta_1)\right| > M - m - \varepsilon_1 \qquad ...(4)$$

where $\varepsilon > 0$ is arbitrary, so that no number less than $M - m$ is an upper bound of the set in question.

From (1) and (4), it follows that $M - m$ is the supremum of the set of numbers

$$\left\{\left|f(\alpha) - f(\beta)\right|;\ \alpha, \beta \in [a, b]\right\}$$

6.7.3. Integrability of the Sum and Difference

If f and g are two functions both bounded and integrable in $[a, b]$ then $f \pm g$ are also bounded and integrable in $[a, b]$, and

$$\int_a^b \left[f(x) \pm g(x)\right] dx = \int_a^b f(x)\, dx \pm \int_a^b g(x)\, dx.$$

(M.D.U. 97; Dibrugarh, 2005)

Let

$$D\left\{a = x_0, x_1, ..., x_{r-1}, x_r, ..., x_n = b\right\}$$

by *any* division of $[a, b]$:

Let

$$M_r', m_r'; M_r'', m_r''; M_r, m_r$$

be the bounds of f, g and $f + g$ in $\delta_r = [x_{r-1}, x_r]$. If α_1, α_2 be any two points of δ_r, we have

$$\left|\left[f(\alpha_2) + g(\alpha_2)\right] - \left[f(\alpha_1) + g(\alpha_1)\right]\right| \le \left|f(\alpha_2) - f(\alpha_1)\right| + \left|g(\alpha_2) - g(\alpha_1)\right|$$

$$\le (M_r' - m_r') + (M_r'' - m_r'')$$

$$\Rightarrow \quad M_r - m_r \le (M_r' - m_r') + (M_r'' - m_r''). \qquad ...(1)$$

Let $\varepsilon > 0$ be a given number.

Since f, g are integrable, there exists $\delta > 0$ such that for every division of norm $\le \delta$, the oscillatory sums of f and g are both less than $\frac{1}{2}\varepsilon$.

We now suppose that D is a division with norm $\le \delta$, so that for D, we have from (1),

$$\Sigma (M_r - m_r)\, \delta_r \le \Sigma (M_r' - m_r')\, \delta_r + \Sigma (M_r'' - m_r'')\, \delta_r < \frac{1}{2}\varepsilon + \frac{1}{2}\varepsilon = \varepsilon,$$

i.e., the oscillatory sum $\Sigma (M_r - m_r) \delta_r$ of $f + g$ for the division D is less than ε. Thus $f + g$ is integrable in $[a, b]$. Now to prove that

$$\int_a^b [f(x) + g(x)]\, dx = \int_a^b f(x)\, dx + \int_a^b g(x)\, dx.$$

Let ε be a positive number.

Since f, g are integrable, there exists $\delta > 0$ such that for every division of norm $\leq \delta$ and for every $\xi_r \in \delta_r$,

$$\left| \Sigma f(\xi_r) \delta_r - \int_a^b f(x)\, dx \right| < \frac{\varepsilon}{2}, \left| \Sigma g(\xi_r) \delta_r - \int_a^b f(x)\, dx \right| < \frac{\varepsilon}{2}.$$

It follows that

$$\left| \Sigma [f(\xi_r) + g(\xi_r)] \delta_r - \left[\int_a^b f(x)\, dx + \int_a^b g(x)\, dx \right] \right| < \varepsilon.$$

Thus we obtain

$$\int_a^b [f(x) + g(x)]\, dx = \int_a^b f(x)\, dx + \int_a^b g(x)\, dx \ (\S\ 6.4)$$

The case of difference may be similarly discussed.

6.7.4. Integrability of Product

If f, g are two functions, both bounded and integrable in $[a, b]$, then their product fg is also bounded and integrable in $[a, b]$.

(*Poorv. 92; Dibrugarh 2004*)

Since f, g are bounded, there exists k, such that

$$|f(x)| \leq k, |g(x)| \leq k, \forall\ x \in [a, b],$$

$$\Rightarrow |f(x) g(x)| \leq k^2, \forall\ x \in [a, b]$$

Thus fg is bounded.

Let

$$D \{a = x_0, x_1, x_2, \ldots, x_{r-1}, x_r, \ldots, x_n = b\}$$

be any division of $[a, b]$. Let

$$M_r', m_r'; M_r'', m_r''; M_r, m_r$$

be the bounds of f, g and fg in $\delta_r = [x_{r-1}, x_r]$. We have $\forall \alpha_1, \alpha_2 \in \delta_r$

$$f(\alpha_2)g(\alpha_2) - f(\alpha_1)g(\alpha_1) = g(\alpha_2)[f(\alpha_2) - f(\alpha_1)] + f(\alpha_1)[g(\alpha_2) - g(\alpha_1)]$$

$$|f(\alpha_2)g(\alpha_2) - f(\alpha_1)g(\alpha_1)| \le |g(\alpha_2) - f(\alpha_1)|$$

$$+ f|(\alpha_1)|\,|g(\alpha_2) - g(\alpha_1)|$$

$$\le k(M'_r - m'_r) + k(M''_r - m''_r).$$

$$\Rightarrow \qquad (M_r - m_r) \le k(M_r' - m_r') + k(M_r'' - m_r''). \qquad ...(1)$$

Now let ε be any positive number.

Since f, g are integrable, there exists $\delta > 0$, such that for every division of norm $\le \delta$, the oscillatory sums of f and g are both $< \varepsilon/2k$. We now suppose that D is a division of norm $\le \delta$, so that for D, we have, from (1).

$$\Sigma(M_r - m_r)\delta_r < k\,\Sigma(M_r' - m_r')\delta_r + k\Sigma(M_r'' - m_r'')\delta_r$$

$$< k(\varepsilon/2k) + k(\varepsilon/2k) = \varepsilon,$$

so that the oscillatory sum $\Sigma(M_r - m_r)\delta_r < \varepsilon$.

Hence fg is integrable in $[a, b]$.

Ex. Show by means of an example that the product of two non-integrable functions may be integrable.

6.7.5. Integrability of Quotient

If f, g are two functions, both bounded and integrable in $[a, b]$ and there exists a number, $t > 0$, such that $|g(x)| \ge t$ $\forall x \in [a, b]$, then f/g is bounded and integrable in $[a, b]$. (M.D.U. 96)

There exist positive numbers k and t such that $\forall x \in [a, b]$

$$|f(x)| \le k, \ |g(x)| \le k, |g(x)| \ge t.$$

Thus $\forall$ $x \in [a, b]$

$$|f(x)/g(x)| \le k/t \Rightarrow f/g \text{ is bounded.}$$

Let

$$D\{a = x_0, x_1, ..., x_{r-1}, x_r, ..., x_n = b\}$$

be a division of $[a, b]$ and let $M_r', m_r'; M_r'', m_r''; M_r, m_r$ be the bounds of f, g, f/g in $\delta_r = [x_{r-1}, x_r]$. Now $\forall\, \alpha_1, \alpha_2 \in \delta_r$, we have

$$\left|\frac{f(\alpha_2)}{g(\alpha_2)} - \frac{f(\alpha_1)}{g(\alpha_1)}\right| = \left|\frac{g(\alpha_1)[f(\alpha_2) - f(\alpha_1)] - f(\alpha_1)[g(\alpha_2) - g(\alpha_1)]}{g(\alpha_1)\, g(\alpha_2)}\right|$$

$$\leq (k/t^2)\left|f(\alpha_2) - f(\alpha_1)\right| + (k/t^2)\left|g(\alpha_2) - g(\alpha_1)\right|$$

$$\leq (k/t^2)(M_r' - m_r') + (k/t^2)(M_r'' - m_r'')$$

$$\Rightarrow \qquad (M_r - m_r) \leq (k/t^2)(M_r' - m_r') + (k/t^2)(M_r'' - m_r''). \quad \ldots(1)$$

Let, now ε be any positive number.

Since f, g are integrable, there exists a number $\delta > 0$ such that for every division D of norm $\leq \delta$, the oscillatory sums for f, g are both less than $t^2\varepsilon / 2k$. Thus for division D of norm $\leq \delta$, we have from (1)

$$\Sigma(M_r - m_r)\,\delta_r \leq (k/t^2)\,\Sigma(M_r' - m_r')\,\delta_r + (k/t^2)\,\Sigma(M_r'' - m_r'')\,\delta_r$$

$$< (k/t^2)(t^2\varepsilon/2k) + (k/t^2)(t^2\varepsilon/2k) = \varepsilon.$$

Hence f/g is bounded and integrable in $[a, b]$.

6.7.6. Integrability of the Modulus of an Integrable Function

If f is bounded and integrable in $[a, b]$, then $| f |$ is also bounded and integrable in $[a, b]$. *(Dibrugarh 2003)*

Since there exists a positive number k such that $\forall\, x \in [a, b]\, |f(x)| \leq k$, the function $| f |$ is bounded.

Let ε be any positive number.

Since f is integrable, there exists a division

$$D\{a = x_0, x_1, x_2, \ldots, x_{r-1}, x_r, \ldots, x_n = b\}$$

such that the corresponding oscillatory sum for f is $< \varepsilon$.

Let $M_r', m_r'; M_r, m_r$ be respectively the bounds of f and $| f |$ in $\delta_r = [x_{r-1}, x_r]$.

Now $\forall\, \alpha_1, \alpha_2 \in \delta_r$, we have

$$\left|\left[|f(\alpha_2)| - |f(\alpha_1)|\right]\right| \leq |f(\alpha_2) - f(\alpha_1)|$$

$$\leq M_r' - m_r'$$

$$\Rightarrow \qquad M_r - m_r \leq M_r' - m_r'$$

This gives

$$\Sigma (M_r - m_r)\,\delta_r \le \Sigma (M_r' - m_r')\,\delta_r < \varepsilon,$$

$$\Rightarrow \qquad \Sigma (M_r - m_r)\,\delta_r < \varepsilon.$$

Hence $| f |$ is integrable in $[a, b]$.

Remarks. The converse of this result is not true. If we take

$$f(x) = \begin{cases} 1 \text{ when } x \text{ is rational,} \\ -1, \text{ when } x \text{ is irrational,} \end{cases}$$

then $$\overline{\int_a^b} f(x)\,dx = (b - a), \quad \underline{\int_a^b} f(x)\,dx = -(b - a)$$

so that f is not integrable.

But since $| f(x) | = 1 \;\forall\; x$, therefore

$$\int_a^b |f(x)|\,dx \text{ exists and is equal to } (b - a).$$

6.7.7. Definition

The meaning of

$$\int_a^b f(x)\,dx,$$

where $b \le a$.

If f be bounded and integrable in $[b, a]$ where $a > b$, then, by def.,

$$\int_a^b f(x)\,dx = -\int_b^a f(x)\,dx.$$

Also by def., $$\int_a^a f(x)\,dx = 0.$$

It is easy to show that the results about integrals obtained in §§ 6.6, 6.7 hold true when the upper limit is less than or equal to the lower limit.

Note. The reader may carefully note that the statement :

$$\int_a^b f(x)\,dx \text{ exists}$$

means that f is bounded and integrable in $[a, b]$.

6.7.8. Inequalities for an Integral

Theorem. If f is bounded and integrable in $[a, b]$, and M, m are the bounds of f in $[a, b]$, then

$$m(b-a) \le \int_a^b f(x)\,dx \le M(b-a) \text{ if } b \ge a.$$

$$m(b-a) \ge \int_a^b f(x)\,dx \ge M(b-a) \text{ if } b \le a. \quad (Allahabad\ 99)$$

For $a = b$, the result is trivial.

If $b > a$, then for any division D, we have

$$m(b-a) \le \int_a^b f(x)\,dx \le S(D) \le M(b-a) \qquad (\S\ 6.2)$$

$$\Rightarrow \qquad m(b-a) \le \int_a^b f(x)\,dx \le M(b-a).$$

If $b < a$, *i.e.*, $a > b$, then, as proved above,

$$m(a-b) \le S(D) \le \int_b^a f(x)\,dx \le M(a-b)$$

$$\Rightarrow \qquad -m(a-b) \ge -\int_b^a f(x)\,dx \ge -M(a-b)$$

$$\Rightarrow \qquad m(b-a) \ge \int_a^b f(x)\,dx \ge M(b-a).$$

Hence the results.

Cor. 1. *If f is bounded and integrable in $[a, b]$, then there exists a number, μ, lying between the bounds of f such that*

$$\int_a^b f(x)\,dx = \mu(b-a). \qquad (Garhwal\ 93,\ Ajmer\ 99)$$

Cor. 2. *If f is continuous in $[a, b]$, then there exists a number, c, lying between a and b such that*

$$\int_a^b f(x)\,dx = (b-a)\,f(c).$$

(Allahabad 99; Lucknow 92; Garhwal 94, 97)

Cor. 3. *If f is bounded and integrable in $[a, b]$, and, k is a number such that $\forall\, x \in [a, b], |f(x)| \le k$,*

then

$$\left|\int_a^b f(x)\,dx\right| \le k\,|b-a|. \qquad (Garhwal\ 98)$$

For $a = b$, the result is trivial.

We have $\forall\, x \in [a, b]$,

$$-k \le f(x) \le k,$$

so that if M, m be the bounds of f in $[a, b]$,

$$-k \le m \le f(x) \le M \le k, \ \forall\, x \in [a, b]. \qquad ...(1)$$

Let $b > a$. Therefore, from 1,

$$-k(b-a) \le m(b-a) \le \int_a^b f(x)\,dx \le M(b-a) \le k(b-a).$$

$$\Rightarrow \qquad \left|\int_a^b f(x)\,dx\right| \le k\,|b-a|.$$

Let $b < a$. We have, from above

$$\left|\int_a^b f(x)\,dx\right| \le k\,|a-b| \Rightarrow \left|\int_a^b f(x)\,dx\right| \le k\,|b-a|.$$

Cor. 4. *If f is bounded and integrable in $[a, b]$ and*

$$\forall\, x \in [a, b],\ f(x) \ge 0,$$

then
$$\int_a^b f(x)\,dx \begin{cases} \ge 0, & \text{when } b \ge a; \\ \le 0, & \text{when } b \le a. \end{cases}$$

For $b = a$, the result is trivial.

Now $f(x) \ge 0\ \forall\, x \in [a, b] \Rightarrow m \ge 0.$

Let $b > a$. We have

$$\int_a^b f(x)\,dx \ge m(b-a) \ge 0. \qquad \because (b-a) \ge 0$$

Let $b < a$. We have, as proved above,

$$\int_b^a f(x)\,dx \ge 0.$$

$$\Rightarrow \qquad \int_a^b f(x)\,dx = -\int_b^a f(x)\,dx \le 0.$$

Cor. 5. *If*

$$\int_a^b f(x)\,dx,\ \int_a^b g(x)\,dx$$

both exist, then

$$f \geq g \Rightarrow \int_a^b f(x)\,dx \geq \int_a^b g(x)\,dx \ \text{ when } b \geq a,$$

$$f \geq g \Rightarrow \int_a^b f(x)\,dx \leq \int_a^b g(x)\,dx \ \text{ when } b \leq a.$$

Under the given condition $[f(x) - g(x)]$ is integrable and ≥ 0, $\forall\, x \in [a, b]$.

Therefore

$$\int_a^b [f(x) - g(x)]\,dx \geq 0 \ \text{ or } \leq 0,$$

according as $b \geq a$ or $b \leq a$

$$\left[\int_a^b f(x)\,dx - \int_a^b g(x)\,dx\right] \geq 0 \ \text{ or } \leq 0,$$

according as $b \geq a$ or $b \leq a$.

Hence the result.

Cor. 6. *If*

$$\int_a^b |f(x)|\,dx$$

exists, then

$$\left|\int_a^b f(x)\,dx\right| \leq \left|\int_a^b |f(x)|\,dx\right| \qquad \textit{(Patna 2003)}$$

It has been shown in § 6.7.6, page 217 that

$$\int_a^b |f(x)|\,dx$$

exists. We have $\forall\, x \in [a, b]$,

$$-|f(x)| \leq f(x) \leq |f(x)|.$$

If $b \geq a$, we have

$$-\int_a^b |f(x)|\,dx \leq \int_a^b f(x)\,dx \leq \int_a^b |f(x)|\,dx, \qquad \textit{(cor. 5)}$$

$$\left|\int_a^b f(x)\,dx\right| \leq \int_a^b |f(x)|\,dx = \left|\int_a^b |f(x)|\,dx\right|.$$

If $b \leq a$, we have, as proved above,

$$\left|\int_a^b f(x)\,dx\right| \le \left|\int_b^a |f(x)|\,dx\right|,$$

$$\Rightarrow \qquad \left|\int_a^b f(x)\,dx\right| \le \left|\int_a^b |f(x)|\,dx\right|.$$

6.8. FUNCTION DEFINED BY DEFINITE INTEGRALS

Let f be integrable in $[a, b]$. We write

$$\varphi(t) = \int_a^t f(x)\,dx,\ t \in [a, b].$$

It is, now, proposed to study the properties of the function φ with domain $[a, b]$, in relation to its continuity and derivability.

The function φ may be called the *Integral function* of the function f.

6.8.1. Continuity of the Integral Function

Theorem. *The integral function of an integrable function is continuous.* (*Garhwal 90. 92. 94; Ajmer 97; M.D.U. 2001*)

We have

$$\varphi(t) = \int_a^t f(x)\,dx.$$

Now there exists a number k such that $\forall\ x \in [a, b], |f(x)| \le k.$

Let c be a point of $[a, b]$. Let ε be *any* positive number. We have

$$\varphi(c) = \int_a^c f(x)\,dx,\ \varphi(c+h) = \int_c^{c+h} f(x)\,dx.$$

$$\therefore \qquad |\varphi(c+h) - \varphi(c)| = \left|\int_c^{c+h} f(x)\,dx\right|$$

$$\le |h|\,k. \qquad \text{(cor. 3 § 6.7.8, p. 218)}$$

$$< \varepsilon, \text{ if } |h| < \varepsilon / k.$$

Hence φ is continuous at any point, $c \in [a, b]$ and so in the interval $[a, b]$.

Ex. Prove that

$$\varphi(t) = \int_t^b f(x)\,dx,$$

is continuous in $[a, b]$.

6.8.2. Derivability of the Integral Function

The integral function φ, of a continuous function f is continuous and

$$\varphi' = f. \qquad \textit{(Osmania 2005)}$$

Let $c \in [a, b]$. We have

$$\varphi(c+h) - \varphi(c) = \int_c^{c+h} f(x)\,dx = hf(c+\theta h) \text{ where } (0 \le \theta \le 1).$$

(cor. 2, § 6.7.8)

Since f is continuous, we have

$$\lim f(c+\theta h) = f(c) \text{ when } h \to 0.$$

Hence
$$\lim_{h\to 0} \frac{\varphi(c+h) - \varphi(c)}{h} = \lim_{h\to 0} f(c+\theta h) = f(c)$$

$$\varphi'(c) = f(c).$$

As c, is any point of $[a, b]$, we have $\forall\, t \in (a, b)$,

$$\varphi'(t) = f(t) \Rightarrow \varphi' = f.$$

Primitive. Def. If there exists a derivable function φ such that φ′ is equal to a given function f in $[a, b]$, then we say that φ is a primitive of f. The theorem above shows that every continuous function f possesses a primitive *viz.*, the function

$$\int_a^t f(x)\,dx.$$

6.8.3. Fundamental Theorem of Integral Calculus

If

$$\int_a^b f(x)\,dx$$

exists and there exists a function φ such that $\forall\, x \in [a, b]$

$$\varphi'(x) = f(x),$$

hen
$$\int_a^b f(x)\,dx = \varphi(b) - \varphi(a). \qquad \textit{(Kurukshetra 2000)}$$

(Ajmer 96, 98; Garhwal 94, 97; Kumaon 98; Allahabad 95, 98)

Let ε be a positive number. Since $\varphi' = f$ is integrable in $[a, b]$, there exists a division

$$D\{a = x_0, x_1, x_2, \ldots, x_{r-1}, x_r, \ldots, x_n = b\}$$

such that

$$\left| \sum_{r=1}^{r=n} \varphi'(\xi_r)\,\delta_r - \int_a^b \varphi'(x)\,dx \right| < \varepsilon. \qquad ...(1)$$

We particularise the arbitrary point $\xi_r \in \delta_r = [x_{r-1}, x_r]$ in the following manner :

By the Lagrange's mean value theorem of Differential Calculus, there exists $\xi_r \in \delta_r$ such that

$$\varphi(x_r) - \varphi(x_{r-1}) = \varphi'(\xi_r)\,\delta_r$$

$$\Sigma\,\varphi'(\xi_r)\,\delta_r = \Sigma\left[\varphi(x_r) - \varphi(x_{r-1})\right] = \varphi(b) - \varphi(a). \qquad ...(2)$$

From (1) and (2) we have

$$\left| \varphi(b) - \varphi(a) - \int_a^b \varphi'(x)\,dx \right| < \varepsilon.$$

An ε is an arbitrary positive number, we have

$$\varphi(b) - \varphi(a) - \int_a^b \varphi'(x)\,dx = 0.$$

Hence the result.

6.9. MEAN VALUE THEOREMS OF INTEGRAL CALCULUS

6.9.1. First Mean Value Theorem

If

$$\int_a^b f(x)\,dx \text{ and } \int_a^b \varphi(x)\,dx,$$

both exist and $\varphi(x)$ *keeps the same sign, positive or negative, throughout the interval of integration, then there exists a number,* μ, *lying between* *he bounds of f such that* (Patna 2003, Dibrugarh 2005, Bhapal 2003

$$\int_a^b f(x)\,\varphi(x)\,dx = \mu \int_a^b \varphi(x)\,dx. \qquad ...(1)$$

First suppose that $\varphi(x)$ is positive $\forall\, x \in [a, b]$. If M, m be the bounds of f, we have $\forall\, x \in [a, b]$

$$m \le f(x) \le M$$

$$\Rightarrow \qquad m\varphi(x) \le f(x)\,\varphi(x) \le M\,\varphi(x), \text{ for } \varphi(x) \ge 0\ \forall\, x.$$

Thus

$$m \int_a^b \varphi(x)\,dx \le \int_a^b f(x)\,\varphi(x)\,dx \le M \int_a^b \varphi(x)\,dx \text{ if } b \ge a$$

$$m \int_a^b \varphi(x)\,dx \ge \int_a^b f(x)\,\varphi(x)\,dx \ge M \int_a^b \varphi(x)\,dx \text{ if } b \le a$$

In either case we see that there exists a number μ, lying between M and m, such that (1) is true. Hence the result.

The case when φ is negative may be similarly disposed off.

Cor. In addition to the conditions of the theorem, if f is continuous also, then there exists a number, ξ, belonging to the domain of integration such that

$$\int_a^b f(x)\,\phi(x)\,dx = f(\xi) \int_a^b \varphi(x)\,dx.$$

6.9.2. Second Mean Value Theorem

If

$$\int_a^b f(x)\,dx \;\; \textit{and} \;\; f(\xi) \int_a^b \varphi(x)\,dx$$

both exist and φ *is monotonic in* $[a, b]$, *then there exists* $\xi \in [a, b]$ *such that*

$$\int_a^b f(x)\,\phi(x)\,dx = \phi(a) \int_a^\xi f(x)\,dx + \phi(b) \int_\xi^b f(x)\,dx.$$

(This theorem is due to *Weierstrass*)

Abel's Lemma. The proof of the theorem depends upon a lemma which is due to Abel which we now state and prove.

If

(*i*) $a_1, a_2, \ldots, a_n$ *is a monotonically decreasing set of n positive numbers,*

(*ii*) $v_1, v_2, \ldots, v_n$ *is a set of any, n, numbers,*

and (*iii*) *k, K are two numbers such that*

$$k < v_1 + v_2 + \ldots + v_p < K, \qquad \textit{for } 1 \le p \le n,$$

then

$$a_1 k < \sum_{r=1}^{r=n} a_r v_r < a_1 K.$$

We write

$$S_p = v_1 + v_2 + \ldots + v_p.$$

We have

$$\sum_{r=1}^{r=n} a_r v_r = a_1 S_1 + a_2 (S_2 - S_1) + \ldots + a_r \left(S_r - S_{r-1}\right) + \ldots + a_n \left(S_n - S_{n-1}\right)$$

$$= (a_1 - a_2) S_1 + (a_2 - a_3) S_2 + \ldots + \left(a_{n-1} - a_n\right) S_{n-1} + a_n S_n.$$

Now by (*i*)

$$(a_1 - a_2), (a_2 - a_3), \ldots, \left(a_{n-1} - a_n\right), a_n$$

are all positive. Also, by (*iii*)

$$k < S_p < K \,\forall\, p \le n.$$

Therefore

$$\sum_{r=1}^{r=n} a_r v_r < (a_1 - a_2) K + (a_2 - a_3) K + \ldots + \left(a_{n-1} - a_n\right) K + a_n K = a_1 K,$$

$$\sum_{r=1}^{r=n} a_r v_r > (a_1 - a_2) k + (a_2 - a_3) k + \ldots + \left(a_{n-1} - a_n\right) k + a_n k = a_1 k.$$

Hence the lemma.

Proof of the theorem. Firstly, we prove the following :

If

$$\int_a^b f(x)\, dx \text{ and } \int_a^b \psi(x)\, dx$$

both exist. ψ *is monotonically* **decreasing** *and* **positive** *in* $[a, b]$, *then there exists a point,* $\xi \in [a, b]$ *such that*

$$\int_a^b f(x)\, \psi(x)\, dx = \psi(a) \int_a^{\xi} f(x)\, dx.$$

(This result is due to Bonnett).

Let

$$D\left\{a = x_0, x_1, \ldots, x_{r-1}, x_r, \ldots, x_n = b\right\}$$

be any division of $[a, b]$. Let M_r, m_r be the bounds of f in $\delta_r = \left[x_{r-1}, x_r\right]$. Let $\xi_1 = a$ and ξ_r, when $r \ne 1$, be any point of δ_r.

We have,

$$m_r \delta_r \le \int_{x_{r-1}}^{x_r} f(x)\, dx \le M_r \delta_r, \; m_r \delta_r \le f(\xi_r)\, \delta_r \le M_r \delta_r.$$

Putting $r = 1, 2, 3, ..., p$ where $p \le n$, and adding, we obtain

$$\sum_{r=1}^{r=p} m_r \delta_r \le \int_a^{x_p} f(x)\,dx \le \sum_{r=1}^{r=p} M_r \delta_r,$$

$$\sum_{r=1}^{r=p} m_r \delta_r \le \sum_{r=1}^{r=p} f(\xi_r)\,\delta_r \le \sum_{r=1}^{r=p} M_r \delta_r.$$

Thus we have

$$\left| \int_a^{x_p} f(x)\,dx - \sum_{r=1}^{r=p} f(\xi_r)\,\delta_r \right| \le \sum_{r=1}^{r=p} (M_r - m_r)\,\delta_r \le \sum_{r=1}^{r=n} (M_r - m_r)\,\delta_r,$$

$$\Rightarrow \int_a^{x_p} f(x)\,dx - \sum_{r=1}^{r=n} O_r \delta_r \le \sum_{r=1}^{r=p} f(\xi_r)\,\delta_r \le \int_a^{x_p} f(x)\,dx + \sum_{r=1}^{r=n} O_r \delta_r,$$

where $O_r = (M_r - m_r)$ is the oscillation of f in δ_r.

Now, $\int_a^t f(x)\,dx$, being a continuous function (§ 4.6.1, § 4.6.2) with t as variable, is bounded. Let C, D be its bounds. Therefore we have

$$C - \sum_{r=1}^{r=n} O_r \delta_r \le \sum_{r=1}^{r=p} f(\xi_r)\,\delta_r \le D + \sum_{r=1}^{r=n} O_r \delta_r.$$

In the statement of the Abel's lemma, we put, as is justifiable,

$$v_r = f(\xi_r)\,\delta_r, \quad a_r = \psi(\xi_r);$$

$$k = C - \Sigma\, O_r \delta_r, \; K = D + \Sigma\, O_r \delta_r,$$

and obtain

$$\psi(a)\left[C - \sum_{r=1}^{r=n} O_r \delta_r \right] \le \sum_{r=1}^{r=n} f(\xi_r)\,\psi(\xi_r)\,\delta_r$$

$$\le \psi(a)\left[D + \sum_{r=1}^{r=1} O_r \delta_r \right].$$

Let the norm of the division tend to 0. We then obtain, in the limit,

$$C\psi(a) \le \int_a^b f(x)\,\psi(x)\,dx = D\psi(a)\ .$$

$$\Rightarrow \qquad \int_a^b f(x)\,\psi(x)\,dx = \mu\,\psi(a),$$

where μ is some number between C and D.

The continuous function

$$\int_a^t f(x)\,dx$$

must assume, for some $\xi \in [a, b]$ the value μ which lies between its bounds C, D. (Cor. 2 to § 4.6.4). Thus we obtain

$$\int_a^b f(x)\,\psi(x)\,dx = \psi(a)\int_a^{\xi} f(x)\,dx.$$

We now turn to the theorem proper.

Let φ be monotonically decreasing so that the function ψ where

$$\psi(x) = \varphi(x) - \varphi(b)$$

is monotonically decreasing and positive.

There exists, therefore, a number, ξ, between a and b, such that

$$\int_a^b f(x)\left[\varphi(x) - \varphi(b)\right]dx = \left[\varphi(a) - \varphi(b)\right]\int_a^{\xi} f(x)\,dx$$

$$\Rightarrow \qquad \int_a^b f(x)\,\varphi(x)\,dx = \varphi(a)\int_a^{\xi} f(x)\,dx$$

$$+\,\varphi(b)\left\{\int_a^b f(x)\,dx - \int_a^{\xi} f(x)\,dx\right\}$$

$$= \varphi(a)\int_a^{\xi} f(x)\,dx + \varphi(b)\int_{\xi}^b f(x)\,dx.$$

Let φ be monotonically increasing so that, $-\varphi$, is monotonically decreasing.

There exists, therefore, by the preceding, a number ξ between a and b, such that

$$\int_a^b f(x)\left[-\varphi(x)\right]dx = -\,\varphi(a)\int_a^{\xi} f(x)\,dx - \varphi(b)\int_{\xi}^b f(x)\,dx,$$

$$\Rightarrow \qquad \int_a^b f(x)\,\varphi(x)\,dx = \varphi(a)\int_a^{\xi} f(x)\,dx + \varphi(b)\int_{\xi}^b f(x)\,dx.$$

Thus we have completely established the second mean value theorem.

Note. The reader may easily show that the theorem holds good even if $a > b$.

***Ex. 1.** Taking $f(x) = x$ and $\varphi(x) = e^x$, verify the two mean value theorems for the interval $[-1, 1]$.

Ex. 2. Show that the second mean value theorem does not hold good in the interval $[-1, 1)$ for $f(x) = \varphi(x) = x^2$. What about the validity of the first mean value theorem in this case ?

6.10. CHANGE OF VARIABLE IN AN INTEGRAL

If (i) $\int_a^b f(x)\,dx$ *exists.*

(ii) φ *is a derivable function with domain* $[\alpha, \beta]$ *such that*

$$\varphi(\alpha) = a,\ \varphi(\beta) = b,$$

(iii) φ' *is bounded and integrable and* $\varphi(t) \neq 0\ \forall\ t \in [\alpha, \beta]$,

(iv) $f \circ \varphi$ *is bounded and integrable in* $[\alpha, \beta]$.

then
$$\int_a^b f(x)\,dx = \int_\alpha^\beta f[\varphi(t)]\,\varphi'(t)\,dt.$$

Since $\varphi'(t) \neq 0$ for any $t \in [\alpha, \beta]$ it follows by Darboux's theorem (§ 5.3 page 177), that $\varphi'(t)$ must always have the same sign and therefore φ must be strictly monotonic in $[\alpha, \beta]$. Also therefore, $[a, b]$ is the range of the function f.

Let

$$D\{\alpha = x_0, x_1, x_2, \ldots, t_{r-1}, t_r, \ldots, t_{n-1}, t_n = \beta\}$$

be a division of $[\alpha, \beta]$ and let

$$D'\{\alpha = x_0, x_1, x_2, \ldots, x_{r-1}, x_r, \ldots, x_{n-1}, x_n = \beta\}$$

be the corresponding division of $[a, b]$; $\varphi(t_r)$ being equal to x_r.

By the Lagrange's mean value theorem, we have

$$x_r - x_{r-1} = \varphi(t_r) - \varphi(t_{r-1}) = (t_r - t_{r-1})\,\varphi'(\eta_r),$$

where η_r lies between t_{r-1} and t_r.

Let
$$\varphi(\eta_r) = \xi_r,$$

We have

$$\sum_{r=1}^{r=n} f(\xi_r)(x_r - x_{r-1}) = \sum_{r=1}^{r=n} f[\varphi(\eta_r)]\,\varphi'(\eta_r)(t_r - t_{r-1}) \quad \ldots($$

Now f is integrable in $[a, b]$.

Also $f \circ \varphi$, φ' are integrable in $[\alpha, \beta]$ so that $(f \circ \varphi)\,\varphi'$ is integrable $[\alpha, \beta]$.

As the norm of the division $D \to 0$, the norm of D' also $\to 0$. From (1), therefore we obtain in the limit

$$\int_a^b f(x)\,dx = \int_\alpha^\beta \left[(f \circ \varphi)\,\varphi'\right](t)\,dt = \int_\alpha^\beta f\left[\varphi(t)\right]\varphi'(t)\,dt.$$

Note. The theorem holds even if $\varphi'(t) = 0$ for a finite number of values of $t \in [\alpha, \beta]$. In this case we can divide the interval $[\alpha, \beta]$ into a finite number of sub-intervals in each of which φ is strictly increasing or decreasing and repeat the argument for each interval in turn and add the results.

.11. INTEGRATION BY PARTS

If

$$\int_a^b f(x)\,dx, \quad \int_a^b g(x)\,dx$$

oth exist and

$$F(x) = A + \int_a^x f(x)\,dx, \quad G(x) = B + \int_a^b g(x)\,dx$$

here A, B are two constants, then

$$\int_a^b F(x)\,g(x)\,dx = \left|F(x)\,G(x)\right|_a^b - \int_a^b G(x)\,f(x)\,dx.$$

[Here $\left|F(x)\,G(x)\right|_a^b$ denotes the difference $[F(b)\,G(b) - F(a)\,G(a)]$.

Proof. Let

$$D\left\{a = x_0, x_1, x_2, x_3, \ldots, x_{r-1}, x_r, \ldots, x_n = b\right\}$$

a division of $[a, b]$.

We have

$$\left|F(x)\,G(x)\right|_a^b = \sum_{r=1}^{r=n} \left[F(x_r)\,G(x_r) - F(x_{r-1})\,G(x_{r-1})\right].$$

$$= \Sigma\, F(x_r)\left[G(x_r) - G(x_{r-1})\right] + \Sigma\, G(x_{r-1})\left[F(x_r) - F(x_{r-1})\right]$$

$$= \Sigma\, F(x_r)\int_{x_{r-1}}^{x_r} g(x)\,dx + \Sigma\, G(x_{r-1})\int_{x_{r-1}}^{x_r} f(x)\,dx \qquad \ldots(1)$$

Let M_r, m_r, O_r denote the bounds and the oscillation of f and $_r', m_r', O_r'$ those of g in $\delta_r = [x_{r-1}, x_r]$. Now $\forall\, x \in \delta_r$, we have

$$\left|g(x) - g(x_r)\right| \le O_r', \left|f(x) - f(x_{r-1})\right| \le O_r,$$

$$\Rightarrow \quad \begin{cases} g(x_r) - O_r' \le g(x) \le g(x_r) + O_r'; \\ f(x_{r-1}) - O_r \le f(x) \le f(x_{r-1}) + O_r, \end{cases}$$

It follows that

$$\left.\begin{aligned} \left[g(x_r) - O_r'\right]\delta_r &\le \int_{x_{r-1}}^{x_r} g(x) \le \left[g(x_r) + O_r'\right]\delta_r; \\ \left[f(x_{r-1}) - O_r\right]\delta_r &\le \int_{x_{r-1}}^{x_r} f(x)\,dx \le \left[f(x_{r-1}) + O_r'\right]\delta_r \end{aligned}\right]$$

These give ...(2)

$$\left.\begin{aligned} \int_{x_{r-1}}^{x_r} g(x)\,dx &= \left[g(x_r) + \theta_r' O_r'\right]\delta_r \\ \int_{x_{r-1}}^{x_r} f(x)\,dx &= \left[f(x_{r-1}) + \theta_r O_r\right]\delta_r \end{aligned}\right] \quad \text{...(3)}$$

where $-1 \le \theta_r, \theta_r' \le 1$.

From (1), (2) and (3), we obtain

$$\left| F(x)\,G(x) \right|_a^b = \Sigma F(x_r)\,g(x_r)\,\delta_r + \Sigma G(x_{r-1})\,f(x_{r-1})\,\delta_r + \sigma \quad \text{...(4)}$$

where

$$\sigma = \Sigma\left[F(x_r)\theta_r' O_r' + G(x_{r-1})\theta_r O_r\right]\delta_r.$$

Since F, G are continuous, therefore, they are bounded. Let k be a number such that $\forall\, x \in [a, b]$,

$$|F(x)| \le k, |G(x)| \le k,$$

$$\therefore \quad |\sigma| \le k(\Sigma O_r + \Sigma O_r')\,\delta_r.$$

Let the norm of the division $D \to 0$. Then $\sigma \to 0$.

From (4) we now obtain

$$\left| F(x)\,G(x) \right|_a^b = \int_a^b F(x)\,g(x)\,dx + \int_a^b G(x)\,f(x)\,dx.$$

Hence the result.

Cor. If a function g is bounded and integrable in $[a, b]$ and a function f is derivable in $[a, b]$ and its derivative f' is bounded and integrable, then

$$\int_a^b f(x)\,g(x)\,dx = \left| f(x)\int_a^x g(x)\,dx \right|_a^x - \int_a^b \left\{ f'(x)\int_a^x g(x)\,dx \right\} dx$$

$$= f(b)\int_a^b g(x)\,dx - \int_a^b \left\{ f'(x)\int_a^x g(x)\,dx \right\} dx.$$

EXAMPLES

1. *If* $\quad G(x, \xi) = \begin{cases} x(\xi - 1), & \text{when } x \le \xi, \\ \xi(x-1) & \text{when } \xi < x, \end{cases}$

and if f is a continuous function of x in [0, 1] and if

$$g(x) = \int_0^1 f(\xi)\,G(x,\xi)\,d\xi,$$

show that $g''(x) = f(x), \forall\, x \in [0, 1]$ *and find g (0) and g (1).*

Sol. We have

$$g(x) = \int_0^x f(\xi)\,\xi\,(x-1)\,d\xi + \int_x^1 f(\xi)\,x\,(\xi - 1)\,d\xi$$

$$= (x-1)\int_0^x \xi f(\xi)\,d\xi + x\int_x^1 (\xi - 1)\,f(\xi)\,d\xi$$

$$= x\int_0^1 \xi f(\xi)\,d\xi - \int_0^x \xi f(\xi)\,d\xi - x\int_x^1 f(\xi)\,d\xi$$

$$g'(x) = \int_0^1 \xi f(\xi)\,d\xi - x\,f(x) + x\,f(x) - \int_x^1 f(\xi)\,d\xi.$$

$$g''(x) = f(x).$$

We may easily see that $g(0)$ and $g(1)$ are both zero.

2. *Prove that if the functions f and* ϕ *are bounded and integrable [a, b], then*

$$\left[\int_a^b f(x)\,\phi(x)\,dx\right]^2 \le \int_a^b [f(x)]^2\,dx \int_a^b \left[\phi(x)^2\right] dx.$$

Under what conditions does the sign of equality hold ?

Sol. We have

$$\left[\int_a^b f(x)\,\phi(x)\,dx\right]^2 = \left[\lim \Sigma\,(x_r - x_{r-1})\,f(\xi_r)\,\phi(\xi_r)\right]^2$$

$$\int_a^b [f(x)]^2\,dx = \lim \Sigma \left[\sqrt{(x_r - x_{r-1})}\,f(\xi_r)\right]^2$$

$$\int_a^b [\phi(x)]^2\,dx = \lim \left[\Sigma \sqrt{(x_r - x_{r-1})}\,\phi(\xi_r)\right]^2.$$

Now putting

$$a_r = \sqrt{(x_r - x_{r-1})}\, f(\xi_r),\ b_r = \sqrt{(x_r - x_{r-1})}\, \phi(\xi_r),$$

in the Cauchy's inequality

$$(\Sigma\, a_r b_r)^2 \le \Sigma\, a_r^{\,2}\, \Sigma\, b_r^{\,2}$$

we get the required result.

The sign of equality holds only when

$$\frac{a_1}{b_1} = \frac{a_2}{b_2} = \dots$$

$$\Leftrightarrow \qquad \frac{f(\xi_1)}{\phi(\xi_1)} = \frac{f(\xi_2)}{\phi(\xi_2)} = \dots$$

$\Leftrightarrow$ f, ϕ are both constant functions.

***3.** *If f is positive and monotonically decreasing in* $[1, \infty[$, *show that the sequence* $\{A_n\}$, *where*

$$A_n = \left\{ f(1) + f(2) + \dots + f(n) - \int_1^n f(x)\, dx \right\},$$

is convergent.

Deduce the convergence of $\left\{ 1 + \frac{1}{2} + \frac{1}{3} + \dots + \frac{1}{n} - \log n \right\}$.

Sol. We have

$$A_n = \left[f(1) - \int_1^2 f(x)\, dx \right] + \left[f(2) - \int_2^3 f(x)\, dx \right] + \dots$$

$$+ \dots + \left[f(n-1) - \int_{n-1}^n f(x)\, dx \right] + f(n)$$

Now, because of the monotonic character of the function f, each of the expressions within brackets is positive. Also $f(n)$ is positive. Thus A_n positive $\forall$ n. Again $\forall$ n

$$A_{n+1} - A_n = f(n+1) - \int_1^{n+1} f(x)\, dx + \int_1^n f(x)\, dx$$

$$= f(n+1) - \int_n^{n+1} f(x)\, dx < 0$$

$$\Rightarrow \qquad A_{n+1} < A_n\ \forall\ n.$$

Thus $\{A_n\}$ is monotonically decreasing. Also being positive, A_n bounded below. Hence $\{A_n\}$ is convergent.

Taking $f(x) = 1/x$, we can now deduce that

$$\lim_{n\to\infty}\left(1+\frac{1}{2}+\frac{1}{3}+\dots+\frac{1}{n}-\log n\right)$$

exists.

***4.** *Show that the function F defined in the interval [0, 1] by the condition that if r is a positive integer*

$$F(x) = 2rx \text{ when } 1/(r+1) < x < 1/r.$$

is integrable over [0, 1] and that

$$\int_0^1 F(x)\,dx = \frac{\pi^2}{6}. \qquad \textit{(Patna 2003)}$$

Sol. The function F, as given, is not defined at the points

$$0, 1, \frac{1}{2}, \frac{1}{3}, \dots \frac{1}{r} \qquad \dots(1)$$

We may, however, define F at these points in any manner we please provided F remains bounded.

Now, the only points of discontinuity of F are those given above in (1). The set formed by these points is infinite having only one limit point *viz.*, 0. Thus the function is integrable.

Consider

$$\psi(\varepsilon) = \int_\varepsilon^1 F(x)\,dx.$$

We know that ϕ is a continuous function so that

$$\psi(0) = \int_0^1 F(x)\,dx = \lim_{\varepsilon\to 0}\psi(\varepsilon).$$

We, now, find $\psi(\varepsilon)$. We take $\varepsilon = 1/n$ so that $\varepsilon \to 0$ as $n \to \infty$. We have

$$\int_{1/n}^1 F(x)\,dx = \int_{1/2}^1 F(x)\,dx + \int_{1/3}^{1/2} F(x)\,dx + \dots \int_{1/(r+1)}^{1/r} F(x)\,dx$$

$$+ \dots + \int_{1/n}^{1/(n-1)} F(x)\,dx.$$

Now $\displaystyle\int_{1/(r+1)}^{1/r} F(x)\,dx = \int_{1/(r+1)}^{1/r} 2r\,x\,dx = \int\frac{2r+1}{r(r+1)^2}.$

$$\therefore \quad \int_{1/n}^1 F(x)\,dx = \sum_{r=1}^{n-1}\frac{2r+1}{r(r+1)^2} = \sum_{r=1}^{n-1}\left(\frac{1}{r}-\frac{1}{r+1}+\frac{1}{(r+1)^2}\right)$$

$$= \sum_{r=1}^{n-1} \left(\frac{1}{r} - \frac{1}{r+1}\right) + \sum_{r=1}^{n-1} \frac{1}{(r+1)^2}$$

$$= 1 - \frac{1}{n} + \sum_{r=1}^{n-1} \frac{1}{(r+1)^2}.$$

*Now the series

$$\frac{1}{1^2} + \frac{1}{2^2} + \frac{1}{3^2} + \frac{1}{4^2} + \ldots$$

is convergent and its sum is $\pi^2/6$.

$$\therefore \qquad \lim_{n\to\infty} \int_{1/n}^{1} F(x)\,dx = 1 + \frac{\pi^2}{6} - 1 = \frac{\pi^2}{6}.$$

Thus $$\int_0^1 F(x)\,dx = \frac{\pi^2}{6}.$$

5. *Show that when* $-1 < x \le 1$,

$$\lim_{m\to\infty} \int_0^x \frac{t^m}{t+1}\,dt = 0.$$

Sol. Now $\forall\, x \in [0, 1]$, we have

$$0 \le \int_0^x \frac{t^m\,dt}{1+t} \le \int_0^x t^m\,dt = \frac{x^{m+1}}{m+1} < \frac{1}{m+1}$$

Let $-1 < x < 0$. Putting $t = -u$, we obtain

$$\left|\int_0^x \frac{t^m}{1+t}\,dt\right| = \left|\int_0^{-x} \frac{u^m}{1-u}\,du\right| < \frac{1}{1+x} \int_0^{-x} u^m\,du < \frac{1}{(m+1)(x+1)}.$$

Hence the result.

6. *Show that, when* $|x| < 1$,

$$\int_0^x \frac{dt}{1+t^4} = x - \frac{1}{5}x^5 + \frac{1}{9}x^9 - \frac{1}{13}x^{13} + \ldots$$

---

* We assume that $\frac{\pi^2}{6} = \frac{1}{1^2} + \frac{1}{2^2} + \frac{1}{3^2} + \frac{1}{4^2} + \ldots$

Sol. We have

$$\frac{1}{1+t^4} = 1 - t^4 + t^8 - t^{12} + \ldots + (-1)^{n-1}\, t^{4n-4} + \frac{(-1)^n t^{4n}}{1+t^4}$$

$$\Rightarrow \quad \int_0^x \frac{dt}{1+t^4} = x - \frac{1}{5}x^5 + \frac{1}{9}x^9 - \frac{1}{13}x^{13} + \ldots$$

$$+ \frac{(-1)^{n-1}\, x^{4n-3}}{4n-3} + (-1)^n \int_0^x \frac{t^{4n}}{1+t^4}\, dt.$$

Now, $\forall\, x \in [-1, 1]$, we have

$$0 \le \left| \int_0^x \frac{t^{4n}}{1+t^4}\, dt \right| < \left| \int_0^x t^{4n}\, dt \right| = \left| \frac{x^{4n+1}}{4n+1} \right| < \frac{1}{4n+1},$$

$$\lim_{n \to \infty} \int_0^x \frac{t^{4n}}{1+t^n}\, dt = 0.$$

Hence the result.

***7.** *If a function f is continuous in [0, 1], show that*

$$\lim_{n \to \infty} \int_0^1 \frac{nf(x)}{1+n^2x^2}\, dx = \frac{\pi}{2} f(0).$$

Sol. We write

$$\int_0^1 \frac{nf(x)}{1+n^2x^2}\, dx = \int_0^{1/\sqrt{n}} \frac{nf(x)}{1+n^2x^2}\, dx + \int_{1/\sqrt{n}}^1 \frac{nf(x)}{1+n^2x^2}\, dx.$$

By the first mean value theorem, we have

$$\int_0^{1/\sqrt{n}} \frac{nf(x)}{1+n^2x^2}\, dx = f(\alpha_n) \int_0^{1/\sqrt{n}} \frac{n\,dx}{1+n^2x^2}, \text{ where } 0 \le \alpha_n \le 1/\sqrt{n}$$

$$= f(\alpha_n) \tan^{-1}\sqrt{n}, \text{ which} \to f(0).\ \frac{1}{2}\pi \text{ as } n \to \infty.$$

Again $$\left| \int_{1/\sqrt{n}}^1 \frac{nf(x)}{1+n^2x^2}\, dx \right| = \left| f(\beta_n) \int_{1/\sqrt{n}}^1 \frac{n\,dx}{1+n^2x^2} \right|,$$

where $1/\sqrt{n} \le \beta_n \le 1$

$$= f(\beta_n)\left(\tan^{-1} n - \tan^{-1} \sqrt{n}\right)$$

$$\leq M\left(\tan^{-1} n - \tan^{-1} \sqrt{n}\right),$$

which $\to 0$ as $\to \infty$,

M, being the supremum of $|f(x)|$.

Hence the result.

***8.** *If f is bounded and integrable in the interval* $[a, b]$, *show that*

$$\lim_{n \to \infty} \int_a^b f(x) \cos nx \, dx = 0.$$

Sol. We write

$$I_n = \int_a^b f(x) \cos nx \, dx.$$

Let ε be a positive number. Since f is bounded and integrable in $[a, b]$, there exists a division

$$D\{a = x_0 < x_1 < x_2 < \ldots < x_{r-1} < x_r < \ldots < x_p = p\}$$

such that the corresponding oscillatory sum

$$\Sigma\left(x_r - x_{r-1}\right) O_r < \frac{1}{2}\varepsilon;$$

O_r being the oscillation of f in $[x_{r-1}, x_r]$.

We have

$$I_n = \Sigma \int_{x_{r-1}}^{x_r} f(x) \cos nx \, dx$$

$$= \Sigma f\left(x_{r-1}\right) \int_{x_{r-1}}^{x_r} \cos nx \, dx$$

$$+ \Sigma \int_{x_{r-1}}^{x_r} \left[f(x) - f\left(x_{r-1}\right)\right] \cos nx \, dx,$$

$$\Rightarrow \quad |I_n| \leq \Sigma \left|f\left(x_{r-1}\right)\right| \left|\int_{x_{r-1}}^{x_r} \cos nx \, dx\right|$$

$$+ \Sigma \left|\int_{x_{r-1}}^{x_r} \left\{f(x) - f\left(x_{r-1}\right)\right\} \cos nx \, dx\right|.$$

We have $\forall\, x \in \left[x_{r-1}, x_r\right]$

$$\left|f(x) - f\left(x_{r-1}\right)\right| \le O_r$$

$$\Rightarrow \qquad \left|\left[f(x) - f\left(x_{r-1}\right)\right] \cos nx\right| \le O_r.$$

Also

$$\left|\int_{x_{r-1}}^{x_r} \cos nx\, dx\right| \le \frac{1}{n}\left\{\left|\sin nx_r\right| + \left|\sin nx_{r-1}\right|\right\} < \frac{2}{n}.$$

It follows that

$$\left|I_n\right| \le \frac{2}{n} \Sigma \left|f\left(x_{r-1}\right)\right| + \Sigma\left(x_r - x_{r-1}\right) O_r$$

$$\le \frac{2}{n} \Sigma \left|f\left(x_{r-1}\right)\right| + \frac{\varepsilon}{2}.$$

Keeping the division D fixed, we see that $\Sigma\left|f\left(x_{r-1}\right)\right|$ is fixed. We now choose a positive integer m such that $\forall\, n \ge m$

$$\frac{2}{n} \Sigma \left|f\left(x_{r-1}\right)\right| < \frac{\varepsilon}{2}.$$

Thus $\forall\, n \ge m$ we have $\left|I_n\right| < \varepsilon.$

Hence the result.

It may similarly be shown that

$$\lim_{n \to \infty} \int_a^b f(x)\, nx\, dx = 0.$$

***9.** *Show that lim* $\{I_n\}$, *where*

$$I_n = \int_0^h \frac{\sin nx}{x}\, dx,\, n \in I$$

exists and that the limit is equal to $\pi/2$.

Sol. The integrand becomes continuous for every value of x, if we assign to it the value n for $x = 0$. The result will be proved in *three* steps.

I. Firstly, it will be proved that $\{I_n\}$ is convergent. Putting

$$nx = t,$$

we have

$$I_n = \int_0^{nh} \frac{\sin t}{t}\, dt.$$

$$\Rightarrow \qquad \left|I_{n+p} - I_n\right| = \left|\int_{nh}^{(n+p)h} \frac{\sin t}{t}\, dt\right|.$$

As $1/t$ is positive and monotonically decreasing when

$$t \in]\, nh, (n+p)\, h\, [,$$

we have, by Bonnett's form of the second mean value theorem,

$$\left|I_{n+p} - I_n\right| = \frac{1}{nh}\left|\int_{nh}^{\alpha} \sin t\, dt\right| \le \frac{2}{nh} < \varepsilon \;\forall\; n > 2/\varepsilon h.$$

Hence, by Cauchy's principle of convergence, $\{I_n\}$ converges (§ 3.3.2, page 60).

II. It will now be proved that, when $n \to \infty$,

$$\lim I_n = \lim \int_0^{\pi/2} \frac{\sin nx}{\sin x}\, dx.$$

We write

$$\int_0^{\pi/2} \frac{\sin nx}{x}\, dx = \int_0^{h} \frac{\sin nx}{x}\, dx + \int_h^{\pi/2} \frac{\sin nx}{x}\, dx.$$

As proved in the preceding example,

$$\int_0^{\pi/2} \frac{\sin nx}{x}\, dx \to 0 \text{ as } n \to \infty.$$

$$\lim I_n = \lim \int_0^{\pi/2} \frac{\sin nx}{x}\, dx.$$

Again, taking $f(x) = (1/x - 1/\sin x)$ in the preceding example,

$$\lim \int_0^{\pi/2} \left(\frac{1}{x} - \frac{1}{\sin x}\right) \sin nx\, dx = 0,$$

for f is continuous in $\left[0, \frac{1}{2}\pi\right]$, if we set $f(0) = 0$.

It follows that

$$\lim \int_0^{\pi/2} \frac{\sin nx}{x}\, dx = \lim \int_0^{\pi/2} \frac{\sin nx}{\sin x}\, dx$$

$$\lim I_n = \lim \int_0^{\pi/2} \frac{\sin nx}{x}\, dx = \lim \int_0^{\pi/2} \frac{\sin nx}{\sin x}\, dx.$$

To determine the actual value of the limit, we proceed by making $n \to \infty$ through odd integer values.

III. We have, as may be easily shown,

$$\frac{\sin(2n+1)x}{\sin x} = 2\left[\frac{1}{2} + \cos 2x + \cos 4x + \ldots + \cos 2nx\right]$$

so that

$$\int_0^{\pi/2} \frac{\sin(2\pi+1)x}{\sin x}\,dx = \frac{\pi}{2}.$$

Hence the result.

EXERCISES

1. If $f(y, x) = 1 + 2x$ for y rational and $f(y, x) = 0$ for y irrational, calculate

$$F(y) = \int_0^1 f(y, x)\,dx.$$

2. Integrate in [0, 2] the function f where $f(x) = x[x]$, and $[x]$ denotes the greatest integer not greater than x.

3. Evaluate

$$\int_0^2 f(x)\,dx$$

where $f(x) = 0$, when $x = n/(n+1), (n+1)/n, (n = 1, 2, 3 \ldots)$.

$f(x) = 1$, elsewhere.

Is f integrable in [0, 2]?

Examine for continuity the function f so defined at the point $x = 1$.

4. If the function f is defined in the interval (0, 1) as $f(x) = \sqrt{(1+x)^2}$, when x is irrational, $= (1 - x)$, when x is rational, then show that $f(x)$ is not Riemann-Integrable in (0, 1). *(Lucknow 94; Garhwal 93)*

5. The function $f(x)$ is defined in the interval (0, 1) by the condition that, for r is a positive integer,

$$f(x) = 2rx, \text{ when } \frac{1}{r+1} < x \le \frac{1}{r},$$

show that the function $f(x)$ is integrable in (0, 1) and evaluate $\int_0^1 f(x)\,dx$.

***6.** If a, b are positive and p is a positive integer, show that, as $n \to \infty$,

(*i*) $$\sum_{r=1}^{pn} \frac{1}{na+r} \to \log\left(1 + \frac{p}{a}\right).$$

(ii) $\displaystyle\sum_{r=1}^{pn} \frac{1}{na+rb} \to \frac{1}{b}\log\left(1+\frac{pb}{a}\right).$

7. A function f is defined, for $x \geq 0$, by

$$f(x) = \int_{-1}^{1} \frac{dt}{\sqrt{(1-2tx+x^2)}}.$$

Prove that if $0 \leq x \leq 1, f(x) = 2$. What is the value of f for $x > 1$? Has the function f a differential coefficient for $x = 1$?

[For $x > 1, f(x) = 2/x$; f is not derivable for $x = 1$ even though it is continuous there at].

8. If, for $x \geq 0$,

$$\phi(x) = \lim_{n\to\infty} \frac{x^n+2}{x^n+1} \quad \text{and} \quad f(x) = \int_0^x \phi(t)\, dt,$$

prove that f is continuous but not differentiable for $x = 1$.

9. If $f(x, y) = xy^2 e^{-xy} + x^2 y/(1+y)$ and a, b are positive, show that

$$\lim_{y\to\infty} \int_a^b f(x,y)\, dx = \int_a^b \left[\lim_{y\to\infty} f(x,y)\right] dx.$$

Also show that the equality does not hold for $a = 0$.

10. f is bounded and integrable in $[a, b]$, show that

$$\int_a^b [f(x)]^2\, dx = 0,$$

if, and only if, $f(c) = 0$ at every point, c, of continuity of f.

11. If $a > 0$, $n > 0$, show that

$$0 < n\int_0^1 \frac{x^{n-1}}{(1+x^2)^a}\, dx < 1.$$

12. The functions f and g are bounded and integrable in $[a, b]$. If, further

$$f(x) = \int_a^x f(t)\, dt \quad \text{and} \quad H(x) = \int_a^x f(t)\, g(t)\, dt, \quad \text{where } a \leq x \leq b$$

and if $F' = f$ and g is continuous, show that

$$H'(x) = f(x)\, g(x).$$

13. A function f is integrable in $[a - c, a + c]$ and

$$|f(x)| \leq M \;\forall\; x \in [a-c, a+c]$$

$$\int_{a-c}^{a+c} f(x)\,dx = 0 \text{ and } F(x) = \int_{a-c}^{x} f(t)\,dt.$$

Prove that

$$\left|\int_{a-c}^{a+c} F(x)\,dx\right| \le Mc^2.$$

***14.** Prove that

$$\lim_{x \to \infty} e^{-x^2} \int_0^x e^{t^2}\,dt = 0.$$

***15.** If f be defined in the interval [0, 1] by the condition that if, r is a positive integer, r = 1, 2, 3, ...

$$f(x) = (-1)^{r-1}, \text{ when } 1/(r+1) < x < 1/r,$$

prove that

$$\int_0^1 f(x)\,dx = \log 4 - 1.$$

16. Show that the function f defined in the interval [0, 1] by the conditions that

$$f(x) = \frac{1}{2^n} \text{ when } \frac{1}{2^{n+1}} < x \le \frac{1}{2n}, n = 0, 1, 2...$$

and $f(0) = 0$,

is integrable in the interval [0, 1] and find the value of

$$\int_0^1 f(x)\,dx.$$

(*Poorv.* 90; *Rohilkhand* 92)

17. A function f is defined in [0, 1] as follows :

$$f(x) = \frac{1}{a^{r-1}} \text{ where } \frac{1}{a^r} < x \le \frac{1}{a^{r-1}} \text{ for } r = 1, 2, 3, ..$$

where a is an integer greater than 2.

Show that $\int_0^1 f(x)\,dx$ exists and is equal to $\dfrac{a}{a+1}$.

18. If $f(x) = 0 \;\forall\; x \in [0, 1]$ except at the set of points

$$x_1, x_2, ..., x_n, ..., n \in \mathbf{N}$$

and $f(x_n) = 1/\sqrt{n}$; show that f is integrable in [0, 1].

***19.** By repeatedly employing the method of integration by parts to the integral

$$\int_0^x e^{-t}\,dt$$

show that

$$e^x = 1 + x + \frac{x^2}{2!} + \frac{x^3}{3!} + \ldots + \frac{x^n}{n!} + e^x \int_0^x \frac{t^n}{n!} e^{-t}\,dt,$$

and deduce the Maclaurin's infinite series for e^x.

***20.** Obtain by integration, from the identity

$$\frac{1}{1+t} = 1 - t + t^2 - t^3 + \ldots + (-1)^{n-1}\, t^{n-1} + (-1)^n \frac{t^n}{1+t},$$

the Maclaurin's infinite series for $\log(1+x)$ in $]-1, 1[$.

21. Show that if $a, b > 0$,

$$\frac{t^{a-1}}{1+t^b} = t^{a-1} - t^{a+b-1} + \ldots - (-1)^{n-1}\, t^{a+(n-1)(b-1)} + (-1)^n \frac{t^{a+nb-1}}{1+t^b},$$

and justify the equality

$$\int_0^1 \frac{t^{a-1}}{1+t^b}\,dt = \frac{1}{a} - \frac{1}{a+b} + \frac{1}{a+2b} - \frac{1}{a+3b} + \ldots$$

Deduce that $\dfrac{1}{1} - \dfrac{1}{4} + \dfrac{1}{7} - \dfrac{1}{10} + \ldots = \dfrac{1}{3}\left(\dfrac{\pi}{\sqrt{3}} + \log 2\right).$

22. Prove Bonnet's form of the second mean value theorem that if f' is continuous and of constant sign and $f(b)$ has the same sign as $f(b) - f(a)$, then

$$\int_a^b f(x)\,\varphi(x)\,dx = f(a)\int_a^{\xi} \varphi(x)\,dx$$

where ξ lies between a and b.

23. By applying the first mean value theorem of Integral Calculus, show that

$$\frac{\pi}{6} < \int_0^{1/2} \frac{dx}{\sqrt{\left[\left(1-x^2\right)\left(1-k^2x^2\right)\right]}} \le \frac{\pi}{6} \cdot \frac{1}{\sqrt{\left(1-\frac{1}{4}k^2\right)}}.$$

24. Prove that

$$\frac{1}{2}\left(1-\frac{1}{e}\right) < \int_0^1 e^{-x^2}\,dx < 1 \text{ and } 0 < \int_1^{\infty} e^{-x^2}\,dx < \left(\frac{1}{2c}\right)$$

and deduce that

$$\frac{1}{2}\left(1-\frac{1}{e}\right) < \int^{\infty} e^{-x^2}\,dx < 1 + \frac{1}{2c}.$$

25. Show that

(i) $$\frac{\pi}{6} > \int_0^1 \frac{dx}{\sqrt{(4 - x^2 + x^3)}} > \frac{1}{2}$$

(ii) $$\frac{\pi}{6} > \int_0^{1/2} \frac{dx}{\sqrt{(1 - x^{2n})}} > \frac{1}{2}, \forall\, n > 1.$$

***26.** By the use of the inequality $\sin x < x$, or otherwise prove that, if $0 < c < 1$, then

$$2 \int_0^1 \sqrt{(1 - c^2 \sin^2 x)}\, dx > \sqrt{(1 - c^2)} + 1.$$

***27.** Show that

(i) $$\left| \int_p^q \frac{\sin x}{x}\, dx \right| \le \frac{2}{p}, \text{ if } q > p > 0.$$

(ii) $$\left| \int_a^b \sin x^2\, dx \right| \le \frac{1}{a}.$$

***28.** If $f(t, x) = \pi t \sin \pi t x$ when $0 \le x \le 1/t$ and $f(t, x) = 0$ when $1/t < x \le 1$, show that

$$\int_0^1 \left\{ \lim_{t \to \infty} f(t, x) \right\} dx = 0 \ne 2 = \lim_{t \to \infty} \left\{ \int_0^1 f(t, x)\, dx \right\}.$$

***29.** If $f(t, x) = tx/(1 + t^2 x^4)$, show that

$$\int_0^1 \left\{ \lim_{t \to \infty} f(t, x) \right\} dx = 0 \ne \frac{\pi}{4} = \lim_{t \to \infty} \left\{ \int_0^1 f(t, x)\, dx \right\}.$$

30. If a is a fixed positive number, prove that

$$\lim_{h \to 0} \int_{-a}^{+a} \frac{h\,dx}{h^2 + x^2} = \pi.$$

31. If f is continuous in $[-1, 1]$, prove that

$$\lim_{h \to 0} \int_{-1}^{1} \frac{h f(x)}{1 + h^2 x^2}\, dx = \pi f(0).$$

[Split up the interval $[-1, 1]$ into three sub-intervals

$$\left[-1, -\sqrt{h}\right], \left[-\sqrt{h}, \sqrt{h}\right], \left[\sqrt{h}, 1\right].$$

***32.** If f is bounded and integrable in $[a, b]$, show that

$$\lim_{n \to \infty} \int_a^b f(x) \frac{\sin nx}{x} dx = 0, 0 < a < b.$$

***33.** Assuming that $|n| \neq |m|$, prove that

$$\lim_{y \to \infty} \frac{1}{y} \int_0^y \sin nx \sin mx \, dx = 0.$$

***34.** Show that for the validity of the second mean value theorem of Integral Calculus, φ must be necessarily monotonic by showing that the theorem does not hold if

$$\varphi(x) = \cos x, f(x) = x^2.$$

***35.** Illustrate the advantage of the second mean value theorem over the first by considering the integral

$$I = \int_1^{10} \frac{\sin x}{x} dx$$

and showing that

(*i*) the application of the first mean value theorem gives

$$I < 10 \log 10 < 24;$$

(*ii*) the application of the second mean value theorem gives

$$I < 2.$$

***36.** If

$$f(x) = \lim_{p \to \infty} \frac{\left(1 + \sin \frac{\pi}{x}\right)^p - 1}{\left(1 + \sin \frac{\pi}{x}\right)^p + 1}, \; x \in]0, 1[$$

draw the graph of the function f and show that f is integrable in $[0, 1]$.

OBJECTIVE QUESTIONS

Note. *For each of the following questions four alternatives are given for the answer. Only one of them is correct. Choose the correct alternative.*

1. If the partitions $P_1, P_2 \in P[a, b]$ them $P_1 \cup P_2$ is :

(*a*) family of partitions (*b*) common refinement

(*c*) normal (*d*) segment

2. If f be a real valued bounded function defined on $[a, b]$ and M, m be the supremum and infimum of f in (a, b) then :

(*a*) $m(b-a) \geq L(p, f) \leq \cup(p, f) \leq M(b-a)$

(*b*) $m(b-a) \leq L(p, f) \geq \cup(p, f) \leq M(b-a)$

(*c*) $m(b-a) \leq L(p, f) \leq \cup(p, f) \leq M(b-a)$

(*d*) $m(b-a) \leq L(p, f) \geq \cup(p, f) \leq M(b-a)$

3. If f is a bounded function defined on $[a, b]$, then corresponding to every $\in > 0$, there exists a $\delta > 0$, such that for all partitions P over $[a, b]$ for which $\| P \| < \delta$, we have

(*a*) $\cup(p, f) > \int_a^{\bar{b}} f(x)dx + \in$ (*b*) $\cup(p, f) < \int_a^{\bar{b}} f(x)dx + \in$

(*c*) $\cup(p, f) = \int_a^{\bar{b}} f(x)dx + c$ (*d*) $L(p, f) < \int_a^b f(x)dx - \in$

4. If $f(x)$ be defined on $[0, 1]$ and is such that $f(x) = 1$, when x is rational and $= -1$ when x is irrational, then for f:

(*a*) only upper Riemann integral exists

(*b*) only lower Riemann integral exists

(*c*) is Riemann integrable

(*d*) is not Riemann integrable

5. For $f(x)$ x for $x \in [0, 3]$ and let $P = [0, 1, 2, 3,]$ be the partition of $[0, 3]$, then value of $L(p, f)$ is :

(*a*) 0 (*b*) 3 (*c*) 6 (*d*) 9

6. A bounded function f is R-integrable in $[a, b]$ if the set of its points of discontinuity is :

(*a*) finite (*b*) infinite (*c*) oscillatory (*d*) none of these

7. Let $f \in \mathbf{R}[a, b]$. Then these exists a number μ, lying between the bounds m and M of f such that

(*a*) $\int_a^b f(x)dx < \mu(b-a)$ (*b*) $\int_a^b f(x)dx > \mu(b-a)$

(*c*) $\int_a^b f(x)dx = \mu(b-a)$ (*d*) none of these

8. If f is Riemann-integrable on $[a, b]$, then $|f|$ is :

(*a*) Riemann integrable

(*b*) not Riemann integrable

(*c*) neither integrable nor non-integrable

(*d*) none of these

9. If $f'(x) = f(x)\ \forall\ x \in [a, b]$ then :

(*a*) $f(x)$ is called a primitive of $f(x)$

(*b*) $f(x)$ is called a primitive of $f(x)$

(*c*) $f'(x)$ is called a primitive of $f(x)$

(*d*) $f'(x)$ is called a primitive of $f(x)$

10. If $f(x) = \int_a^x f(y)\, dy\ \forall\ x \in [a, b]$ then :

(*a*) $f(x)$ is integral function of $f(x)$

(*b*) $f(x)$ is integral function of $f(x)$

(*c*) $f(x)$ is primitive of $f(x)$

(*d*) $f(x)$ is primitive of $f(x)$.

ANSWERS

1. (*b*)	**2.** (*c*)	**3.** (*b*)	**4.** (*d*)	**5.** (*b*)
6. (*a*)	**7.** (*c*)	**8.** (*a*)	**9.** (*b*)	**10.** (*a*)

7

Sequences of Functions Point-wise and Uniform Convergence

7.1. INTRODUCTION

In this chapter we shall consider sequences whose members are real valued functions defined in a set S and distinguish between two types of convergence of such sequences.

Let f_n be a real valued function defined in a set S for each n.

Point-wise convergence. To each $c \in S$, there corresponds a real number sequence

$$\{f_n(c)\}$$

with values

$$f_1(c), f_2(c), \ldots, f_n(c) \ldots$$

We suppose that this sequence is convergent. In fact, we suppose that each of the sequences arising for different members of S is convergent. Thus we define in a natural way a real valued function say f, with domain S such that its value $f(c)$ for $c \in S$ is

$$\lim \{f_n(c)\}.$$

The function f, thus defined, is referred to as the *Pointwise* limit of the sequence $\{f_n\}$ of functions. Also in this case, we say that the sequence is point-wise convergent.

Thus if a function f is the point-wise limit of the point-wise convergent sequence $\{f_n\}$ of functions defined in S, to each $c \in S$ and to each $\varepsilon > 0$ there corresponds an integer m such that $\forall\, n \geq m$,

$$\left| f_n(c) - f(c) \right| < \varepsilon.$$

Of course, if we fix, ε, the choice of m may depend upon the choice of c.

Uniform convergence. We say that a sequence $\{f_n\}$ of real valued functions with domain S is uniformly convergent, if there exists a real valued function ϕ, with domain S and to each $\varepsilon > 0$ there corresponds an integer m such that $\forall\, x \in S$ and $\forall\, n \geq m$.

$$\left| f_n(x) - \phi(x) \right| < \varepsilon.$$

Also in this case we say that the function ϕ is the uniform limit of the sequence $\{f_n\}$.

It may be easily seen that

Uniform convergence $\Rightarrow$ *point-wise convergence*

and that in the event of uniform convergence

Uniform limit $=$ *Point-wise limit.*

As a result, we shall denote the uniform limit of $\{f_n\}$ by f instead of ϕ.

It should, however, be remembered that every point-wise convergent sequence is not uniformly convergent as is illustrated by the following counter-example. Let

$$f_n(x) = \frac{nx}{1+n^2x^2},\ x \in \mathbf{R}.$$

As may be easily seen, the sequence, $\{f_n\}$, is point-wise convergent with point-wise limit f such that

$$f(x) = 0\ \forall\ x \in \mathbf{R}.$$

We shall now show that the convergence is *not* uniform in any interval $[a, b]$ with, 0, as an interior point.

Suppose that $\{f_n\}$ is uniformly convergent in $[a, b]$ so that the point-wise limit f is also the uniform limit.

Let $\varepsilon > 0$ be given. Then there exists m such that $\forall\ x \in [a, b]$ and $\forall\ n \geq m$.

$$\left| \frac{nx}{1+n^2x^2} - 0 \right| < \varepsilon.$$

We take $\varepsilon = \frac{1}{4}$.

Now there exists an integer k such that

$$k \geq m \text{ and } 1/k \in [a, b].$$

Taking $n = k$ and $x = 1/k$, we have

$$\frac{nx}{1+n^2x^2} = \frac{1}{2}$$

which is not less than $\frac{1}{4}$.

Thus we arrive at a contradiction and as such see that the sequence is *not* uniformly convergent in any interval $[a, b]$ with, 0, as an interior point even though it is point-wise convergent there.

7.2. NECESSARY AND SUFFICIENT CONDITION FOR UNIFORM CONVERGENCE

A necessary and sufficient condition for a sequence $\{f_n\}$ of functions defined in a set S to be uniformly convergent is that to each $\varepsilon > 0$, there corresponds m such that $\forall\, n \geq m, \forall\, p \geq 0$ and $\forall\, x \in S$,

$$\left|f_{n+p}(x) - f_n(x)\right| < \varepsilon.$$

The condition is necessary. Let the sequence $\{f_n\}$ be uniformly convergent with f as its uniform limit.

Let $\varepsilon > 0$ be given. Then there exists m such that $\forall\, n \geq m$ and $\forall\, x \in S$,

$$\left|f_n(x) - f(x)\right| < \frac{1}{2}\varepsilon.$$

Also, since $\forall\, n \geq m, \forall\, p \geq 0$ and $\forall\, x \in S$,

$$\left|f_{n+1}(x) - f(x)\right| < \frac{1}{2}\varepsilon,$$

it follows that $\forall\, n \geq m, \forall\, p \geq 0$ and $\forall\, x \in S$

$$\left|f_{n+p}(x) - f_n(x)\right| = \left|f_{n+p}(x) - f(x) + f(x) - f_n(x)\right|$$

$$\leq \left|f_{n+p}(x) - f(x)\right| + \left|f(x) - f_n(x)\right|$$

$$< \varepsilon.$$

The condition is sufficient. From Cauchy's principle of convergence as proved in § 3.3.2, p. 46 it follows that the sequence is pointwise convergent. All that we have now to show is that the convergence is uniform. Let f be the point-wise limit of the sequence $\{f_n\}$. Let $\varepsilon > 0$ be given. Then there exists m such that $\forall\, n \geq m, \forall\, p \geq 0, \forall\, x \in S$,

$$\left|f_{n+p}(x) - f_n(x)\right| < \frac{1}{2}\varepsilon$$

$$\Rightarrow \qquad f_n(x) - \frac{1}{2}\varepsilon < f_{n+p}(x) < f_n(x) + \frac{1}{2}\varepsilon.$$

Keeping n fixed and letting p tend to infinity, we see that $\forall\, n \geq m$ and $\forall\, x \in S$

$$f_n(x) - \frac{1}{2}\varepsilon \leq f(x) \leq f_n(x) + \frac{1}{2}\varepsilon \Rightarrow \left|f_n(x) - f(x)\right| \leq \frac{1}{2}\varepsilon < \varepsilon.$$

Thus the convergence is uniform.

EXAMPLES

1. *Show that the sequence* $\{f_n\}$ *of functions where*

$$f_n(x) = n/(x+n),$$

is uniformly convergent in [0, k] whatever k may be, but not uniformly convergent in [0, ∞ [.

Sol. The sequence $\{f_n\}$ is point-wise convergent $\forall\, x \geq 0$ and the point-wise limit f is given by

$$f(x) = 1 \ \forall\, x \geq 0$$

Let $\varepsilon > 0$ be given. We have

$$|f_n(x) - f(x)| = \frac{x}{x+n} < \varepsilon \text{ if } n > x\left(\frac{1}{\varepsilon} - 1\right).$$

Let $m(\varepsilon, x)$ denote the integer just greater than $x\left(\frac{1}{\varepsilon} - 1\right)$. Obviously $m(\varepsilon, x)$ increases as x increases and $\to \infty$ as $x \to \infty$ so that it is not possible to choose any number m such that $\forall\, n \geq m$ and $\forall\, x \geq 0$,

$$|f_n(x) - f(x)| < \varepsilon$$

so that the convergence is *not* uniform in [0, ∞ [.

Now consider the interval $[0, k]$.

Let m be a integer greater than $k\,(1/\varepsilon - 1)$. We then see that $\forall\, n \geq m$ and $\forall\, x \in [0, k]$,

$$|f_n(x) - f(x)| < \varepsilon$$

so that the convergence is uniform in $[0, k]$.

***2.** *Show that the sequence* $\{f_n\}$ where

$$f_n(x) = x^n$$

is uniformly convergent in [0, k] where k is a number less than 1 and only point-wise convergent in [0, 1].

Sol. The given sequence is point-wise convergent in [0, 1] and the point-wise limit f is given by

$$f(x) = \begin{cases} 0, \text{ when } 0 \leq x < 1 \\ 1, \text{ when } x = 1. \end{cases}$$

Let $\varepsilon > 0$ be given. We suppose $\varepsilon < 1$.

We have when $x \neq 1$,

$$\left|f_n(x)-f(x)\right| = x^n < \varepsilon$$

$$(1/x)^n > (1/\varepsilon).$$

$$n \log(1/x) > \log(1/\varepsilon), x \neq 0$$

$$n > (\log 1/\varepsilon) \div \log(1/x).$$

Also if $x = 0$,

$$\left|f_n(x)-f(x)\right| = 0 < \varepsilon \text{ when } n > 1.$$

Let m (ε, x) denote the integer next greater than 1 and

$$\log(1/\varepsilon) \div \log(1/x).$$

Now

$$m\ (\varepsilon, x) \text{ increases and } \to \infty \text{ as } x \to 1,$$

so that there does not exist and m such that $\forall\ n \geq m$ and $\forall\ x \in [0, 1]$,

$$\left|f_n(x)-f(x)\right| < \varepsilon$$

so that the convergence is *not* uniform in [0, 1].

Now suppose that k is a number such that

$$0 \leq k < 1.$$

We see that in $[0, k]$, the greatest value of

$$\log(1/\varepsilon) \div \log(1/x) \text{ is } (\log(1/\varepsilon) \div \log(1/k)).$$

Let any integer greater than this value be denoted by m. Then we see that $\forall\ n \geq m$ and $\forall\ x \in [0, k]$,

$$\left|f_n(x)-f(x)\right| < \varepsilon$$

so that the convergence is uniform in $[0, k]$.

***3.** *Show that if*

$$f_n(x) = nxe^{-nx^2},$$

the sequence $\{f_n\}$ *is point-wise, but not uniformly, convergent in* $[0, \infty[$.

Sol. It may be easily seen that the sequence is point-wise convergent and that the point-wise limit is the function f such that $\forall\ x$

$$f(x) = 0.$$

If possible, let the sequence be uniformly convergent in $[0, \infty[$. So that, $\varepsilon > 0$ being given, there exists m such that $\forall\, n \geq m$ and $\forall\, x \geq 0$

$$\left| f_n(x) - f(x) \right| = nxe^{-nx^2} < \varepsilon. \qquad ...(1)$$

Let m_0 be an integer greater than m and $e^2\varepsilon^2$ and let $x = 1/m_0$. Then the inequality (1) holds for $x = 1/\sqrt{m_0}$ and $n = m_0$. These give

$$\sqrt{m_0}\,/\,e < \varepsilon \Leftrightarrow m_0 < e^2\varepsilon^2$$

so that we arrive at a contradiction.

Thus the convergence is *not* uniform in $[0, \infty[$.

***4.** *Show that the sequence f_n where*

$$f_n(x) = e^{-nx}$$

is point-wise but not uniformly convergent in $[0, \infty]$. *Also show that the convergence is uniform in* $[k, \infty[$; *k being any positive number.*

7.3. PROPERTIES OF UNIFORMLY CONVERGENT SEQUENCE OF FUNCTIONS

In the following, we shall suppose that each of the function f_n has $[a, b]$ as its domain.

7.3.1. Continuity of the Uniform Limit of a Uniformly Convergent Sequence of Continuous Functions

Consider a uniformly convergent sequence $\{f_n\}$ of continuous functions defined in $[a, b]$. Let the function f be the limit which is, of course, the uniform limit of the sequence. It will now be proved that f is a continuous function. Let c be any point of $[a, b]$.

We have $\forall\, x$ and $\forall\, n$,

$$\left| f(x) - f(c) \right| = \left| f(x) - f_n(x) + f_n(x) - f_n(c) - f(c) \right|$$

$$\leq \left| f(x) - f_n(x) \right| + \left| f_n(x) - f_n(c) \right| + \left| f_n(c) - f(c) \right|.$$

Let $\varepsilon > 0$ be given. As $\{f_n\}$ converges uniformly to f, there exists m such that $\forall\, x \in [a, b]$ and $\forall\, n \geq m$.

$$\left| f(x) - f_n(x) \right| < \frac{1}{3}\varepsilon.$$

In particular, we see that $\forall\, x \in [a, b]$,

$$\left| f(x) - f_m(x) \right| < \frac{1}{3}\varepsilon \Rightarrow \left| f(c) - f_m(c) \right| < \frac{1}{3}\varepsilon.$$

As f_m is continuous at c, there exists $\delta > 0$ such that $\forall\ x \in]c - \delta, c + \delta[$

$$\left|f_m(x) - f_m(c)\right| < \frac{1}{3}\varepsilon.$$

Thus we see that there exists $\delta > 0$ such that $\forall\ x \in]c - \delta, c + \delta[$

$$\left|f(x) - f(c)\right| < \varepsilon,$$

and as such f is continuous at c and, therefore, also at every point of the domain.

7.3.2. Integrability of the Uniform Limit of a Uniformly Convergent Sequence of Integrable Functions

Let $\{f_n\}$ be a uniformly convergent sequence with uniform limit f and let f_n be integrable $\forall\ n$. We shall show that the limit f is itself integrable and

$$\int_a^b f(x)\,dx = \lim_{n\to\infty} \int_a^b f_n(x)\,dx.$$

Let $\varepsilon > 0$ be given so that there exists m such that $\forall\ x \in [a, b]$ and $\forall\ n \geq m$,

$$\left|f_n(x) - f(x)\right| < \varepsilon / 4(b - a).$$

In particular, we have $\forall\ x \in [a, b]$,

$$\left|f_m(x) - f(x)\right| < \varepsilon / 4(b - a).$$

We write

$$f(x) - f_m(x) = R_m(x),$$

so that we have defined a new function R_m.

As the function f_m is integrable, there exists a division, say D, of $[a,\ b]$ such that the oscillatory sum $w_D(f_m)$ for the function f_m corresponding to the division D is $< \varepsilon/2$.

Also since $\left|R_m(x)\right| < \varepsilon / 4(b - a)\ \forall\ x \in [a, b]$, the oscillation of R_m in each sub-interval is

$$< \varepsilon / 2(b - a).$$

implying that

$$w_D(R_m) < \frac{\varepsilon}{2(b - a)}(b - a) = \frac{\varepsilon}{2}.$$

Also

$$f = f_m + R_m \Rightarrow w_D(f) \leq w_D(f_m) + w_D(R_m).$$

Thus if $\varepsilon > 0$ be given, there exists a division D of $[a, b]$ such that $w_D(f) < \varepsilon$, and accordingly f is integrable.

Now to prove

$$\int_a^b f(x)\,dx = \lim_{n\to\infty} \int_a^b f_n(x)\,dx.$$

Let $\varepsilon' > 0$ be given so that there exists m' such that $\forall\, x \in [a, b]$ and $\forall\, n \geq m'$

$$|f(x) - f_n(x)| < \varepsilon'/(b-a)$$

$$\Rightarrow \quad -\varepsilon'/(b-a) < f(x) - f_n(x) < \varepsilon'/(b-a)$$

$$\Rightarrow \quad -\varepsilon'/(b-a) < (f - f_n)(x) < \varepsilon'/(b-a).$$

Now f_n is integrable and f has also been proved integrable. Thus $f - f_n$ is integrable. We have

$$\left|\int_a^b (f - f_n)(x)\,dx\right| < \varepsilon'.$$

Thus if $\varepsilon' > 0$ is given, there exists m' such that $\forall\, n \geq m'$

$$\left|\int_a^b f(x)\,dx - \int_a^b f_n(x)\,dx\right| < \varepsilon' \Rightarrow \lim_{n\to\infty} \int_a^b f_n(x)\,dx = \int_a^b f(x)\,dx.$$

Cor. *It may be proved without difficulty that the sequence*

$$\left\{\int_a^t f_n(x)\,dx\right\}$$

of functions is uniformly convergent with limit

$$\int_0^t f(x)\,dx,\ t \in [a, b].$$

7.3.3. Derivability of the Point-wise Limit of a Sequence of Derivable Functions if the Derivatives are Continuous and the Sequence of Derivatives is Uniformly Convergent

Let $\{f_n\}$ be a sequence of derivable functions with point-wise limit f. Let f_n' be continuous $\forall\ n$ and let the sequence $\{f_n'\}$ be uniformly convergent with ϕ as uniform limit.

Then we shall show that ϕ is derivable and the derivative is equal to f, *i.e.*,

$$\phi' = \left(\lim_{n\to\infty} \{f_n\}\right)' = \lim_{n\to\infty} \{f_n'\}$$

i.e., the derivative of the limit of the sequence is equal to the limit of the sequence of derivatives.

As $\{f_n'\}$ is a uniformly convergent sequence of continuous functions, it follows that its uniform limit ϕ is a continuous function.

It follows that

$$\int_a^t \phi(x)\,dx = \lim \int_a^t f_n'(x)\,dx$$

$$= \text{iim}\left\{f_n(t) - f_n(a)\right\}$$

$$= \lim\left\{f_n(t)\right\} - \lim\left\{f_n(a)\right\} = f(t) - f(a).$$

Also ϕ being continuous, we have, by § 6.8.2

$$\phi'(t) = f(t)\ \forall\ t \in [a, b] \Rightarrow \phi' = f.$$

Note. It should be seen that the point-wise limit of a sequence of continuous functions may or may not be continuous. If, however, as has been shown in § 7.3 the convergence is uniform, the limit must necessarily be continuous. To illustrate this remark, we consider two examples.

1. Let f_n be defined by

$$f_n(x) = 1 - \left|1 - x^2\right|^n$$

in the domain

$$\left\{x : \left|1 - x^2\right| \le 1\right\} = \left[-\sqrt{2}, \sqrt{2}\right].$$

The given sequence $\{f_n\}$ is point-wise convergent in $\left[-\sqrt{2}, \sqrt{2}\right]$ with the point-wise limit f, where

$$f(x) = \begin{cases} 1 \text{ when } \left|1 - x^2\right| < 1 \\ 0 \text{ when } \left|1 - x^2\right| = 1 \; i.e., \text{ for } x = 0, \pm\sqrt{2}. \end{cases}$$

Thus we see that the point-wise limit is not continuous for $x = 0$ even though each f_n is continuous there at.

Thus we may conclude that the sequence cannot be uniformly convergent in $\left[-\sqrt{2}, \sqrt{2}\right]$. We may also see this fact directly as follows :

Suppose that the sequence is uniformly convergent in

$$\left[-\sqrt{2}, \sqrt{2}\right],$$

so that f is the uniform limit.

Take $\varepsilon = \frac{1}{2}$. Then there exists an integral m such that

$$\left|f_n(x) - f(x)\right| < \frac{1}{2}\ \forall\ n \ge m \text{ and } \forall\ x \in \left[-\sqrt{2}, \sqrt{2}\right].$$

In particular $\forall\, x \in \left[-\sqrt{2}, \sqrt{2}\right]$,

$$\left| f_m(x) - f(x) \right| < \frac{1}{2}.$$

Now $$\left| f_m(x) - f(x) \right| = \begin{cases} \left|1 - x^2\right|^m & \text{when } \left|1 - x^2\right| < 1 \\ 0 & \text{when } \left|1 - x^2\right| = 1 \end{cases}$$

Since $\lim\limits_{x \to 0} \left|1 - x^2\right|^m = 1$, there exists a neighbourhood of 0 for every point x of which $\left|1 - x^2\right|^m$ belongs to a neighbourhood

$$\left]1 - \frac{1}{4}, 1 + \frac{1}{4}\right] = \left]\frac{3}{4}, \frac{5}{4}\right[$$

of 1, and is as such greater than $\frac{1}{2}$.

Thus we arrive at a contradiction and as such we see directly that the convergence is *not* uniform in $\left[-\sqrt{2}, \sqrt{2}\right]$.

2. Let f_n be defined by

$$f_n(x) = \frac{nx}{1 + n^2 x^2},$$

in any domain $[a, b]$ with 0 as an interior point. It has been seen on p. 246 that the convergence is point-wise but not uniform. Also the point-wise limit f is given by

$$f(x) = 0 \;\forall\, x \in [a, b].$$

Thus we see that while convergence is not uniform, the point-wise limit itself is a continuous function.

***3.** Let f_n be defined by

$$f_n(x) = nxe^{-nx^2},\; x \in [0, 1].$$

The sequence $\{f_n\}$ is point-wise convergent with the point-wise limit f given by

$$f(x) = 0 \;\forall\, x \in [0, 1].$$

We have

$$\int_0^1 f_n(x)\, dx = -\frac{1}{2}\left| e^{-nx^2} \right|_0^1 = \frac{1}{2}\left(1 - e^{-n}\right)$$

$$\Rightarrow \quad \lim \int_0^1 f_n(x)\,dx = \frac{1}{2} \neq 0 = \int_0^1 f(x)\,dx = \int_0^1 \lim f_n(x)\,dx$$

and as such the integral of the point-wise limit is not equal to the limit of the sequence of integrals.

It follows from § 7.3.2 that the sequence is not uniformly convergent in [0, 1]. In fact, the fact of non-uniform convergence has also been directly established in Ex. on p. 252.

***4.** Let f_n be defined by

$$f_n(x) = \frac{nx}{1+n^2x^2},\ x \in [0, 1].$$

It has been shown in Ex. 2, p. 247 that while the sequence $\{f_n\}$ is point-wise convergent, it is uniformly convergent in [0, 1]. In spite of this, however, it will be seen that

$$\lim \int_0^1 f_n(x)\,dx = \int_0^1 \lim f_n(x)\,dx.$$

We have

$$\int_0^1 f_n(x)\,dx = \int_0^1 \frac{nx}{1+n^2x^2}\,dx = \frac{1}{2n}\log\left(1+n^2\right)$$

$$\lim \int_0^1 f_n(x)\,dx = 0.$$

Also since the point-wise limit f is such that $f(x) = 0\ \forall\ x$, we have

$$\int_0^1 \lim f_n(x) = 0.$$

Thus we have the equality, in spite of the absence of uniform convergence.

5. Let

$$f_n(x) = \frac{nx}{1+n^2x^2},\ x \in [0,1].$$

Now $\forall\ n$ and $\forall\ x$,

$$f_n'(x) = \frac{n\left(1-n^2x^2\right)}{\left(1+n^2x^2\right)^2}$$

$$\Rightarrow \quad f_n'(0) = n\ \forall\ n$$

$\Rightarrow$ the sequence $\{f_n'(0)\}$ is not convergent.

The given sequence is point-wise convergent with limit f such that

$$f(x) = 0 \; \forall \; x \Rightarrow f'(0) = 0.$$

Thus we see that in this case the limit of the sequence of derivatives does not equal the derivative of the limit of the sequence of functions.

It will be concluded that the sequence $\{f_n'(x)\}$ of derivatives cannot be uniformly convergent in this case.

7.4. INFINITE SERIES OF FUNCTIONS

We now consider series whose terms are functions defined in some set S. Let

$$f_1 + f_2 + \ldots + f_n + \ldots$$

be such a series. We write

$$\sigma_n = f_1 + f_2 + \ldots + f_n$$

so that $\{\sigma_n\}$ is a sequence of functions.

We say that the series is point-wise convergent if the sequence $\{\sigma_n\}$ is point-wise convergent. Also the series is said to be uniformly convergent, if the sequence $\{\sigma_n\}$ is uniformly convergent. Also the point-wise limit or the uniform limit of $\{\sigma_n\}$ as the case may be is said to be the point-wise sum or the uniform sum of the series.

It has been seen in Chapter 4, 5 and 6 that the sum of a *finite* number of continuous (derivable, integrable) functions is continuous (derivable, integrable). These results, however, may not be true in relation to the point-wise limits of infinite series. The following three results, however, are direct consequences of those proved in § 7.3, in relation to uniformly convergent sequences.

I. The sum of a uniformly convergent series of continuous functions is continuous.

II. The sum of a uniformly convergent series of integrable functions is integrable and the integral of the sum is equal to the sum of the series of integrals of the functions.

Thus if Σf_n be a uniformly convergent series of integrable functions in $[a, b]$, and σ denote its sum, then σ is integrable in $[a, b]$ and

$$\int_a^b \Sigma f_n(x)\,dx = \int_a^b \sigma(x)\,dx = \Sigma \int_a^b f_n(x)\,dx.$$

III. If Σf_n is a point-wise convergent series of derivable functions with continuous derivatives and the series $\Sigma f_n'$ of derivatives is uniformly convergent, then

$$\frac{d}{dx}(\Sigma f_n) = \Sigma f_n',$$

i.e., the derivative of the sum is equal to the sum of the derivatives.

7.5. TEST FOR THE UNIFORM CONVERGENCE OF A SERIES

7.5.1. General Principle of Convergence

The necessary and sufficient condition for the uniform convergence in $[a, b]$ *of a series* Σf_n *is that to every positive number,* ε, *there corresponds a positive integer m such that* $\forall\, n \geq m$ *and* $\forall\, x \in [a, b]$,

$$\left| f_{n+1}(x) + f_{n+2}(x) + \ldots + f_{n+p}(x) \right| < \varepsilon.$$

This result is an immediate consequence of the corresponding result for sequences proved in § 7.2, p 248.

7.5.2. Weierstrass's M-test for Uniform Convergence

A series Σf_n will converge uniformly in $[a, b]$, if there exists a convergent series ΣM_n of positive numbers such that $\forall\, n$ *and* $\forall\, x \in [a, b]$,

$$\left| f_n(x) \right| \leq M_n \qquad \ldots(i)$$

Let ε be a positive number. Since ΣM_n is convergent, there exists a positive integer m such that

$$\left| M_{n+1} + M_{n+1} + \ldots + M_{n+p} \right| < \varepsilon \qquad \ldots(ii)$$

$\forall\, n \geq m$ and $\forall\, p \geq 0$.

From (*i*) and (*ii*), we see that

$$\forall\, x \in [a, b], \forall\, n \geq m \text{ and } \forall\, p \geq 0,$$

$$\left| f_{n+1}(x) + f_{n+2}(x) + \ldots + f_{n+p}(x) \right|$$

$$\leq \left[M_{n+1} + M_{n+2} + \ldots + M_{n+p} \right] < \varepsilon.$$

Hence Σf_n is uniformly convergent in $[a, b]$.

EXAMPLES

1. *Prove that* $\Sigma a_n n^{-x}$ *is uniformly convergent on* $[0, 1]$ *if* Σa_n *converges uniformly in* $[0, 1]$.

Sol. Take $v_x(x) = n^{-x} = \dfrac{1}{n^x}$ and $u_n(x) = a_n$. The sequence $< n^{-x} >$ is monotonic decreasing on $[0, 1]$.

$\therefore \qquad \dfrac{1}{n^x} \leq \dfrac{1}{n^0} = 1 \ \forall \ n \in \mathbf{N}$ and $\forall \ x \in [0, 1]$.

$\therefore \qquad |u_n(x)| = |n^{-x}| \leq 1 \ \forall \ n \in \mathbf{N}$ and $\forall \ x \in [0, 1]$.

Thus $< v_n(x) >$ is uniformly bounded and monotonic decreasing sequence on [0, 1].

Also $\Sigma u_n(x) = \Sigma a_n$ is uniformly convergent on [0, 1].

Hence $\Sigma u_n(x) v_n(x) = \Sigma a_n n^{-x}$ is uniformly convergent on [0, 1].

2. *Show that the series*

$$\frac{x^2}{1+x} + \left(\frac{2x^2}{1+2x} - \frac{x^2}{1+x}\right) + \ldots + \left(\frac{nx^2}{1+nx} - \frac{(n-1)x^2}{1+(n-1)x}\right) + \ldots$$

converges uniformly on [0, 1]

Sol. Here, $\qquad u_1(x) = \dfrac{x^2}{1+x}$

$$u_2(x) = \frac{2x^2}{1+2x} - \frac{x^2}{1+x}$$

....

....

$$u_n(x) = \frac{nx^2}{1+nx} - \frac{(n-1)x^2}{1+(n-1)x}$$

Adding, $\qquad f_n(x) = \dfrac{nx^2}{1+nx}$

Hence $\qquad f(x) = \lim\limits_{n \to \infty} \dfrac{nx^2}{1+nx} = x.$

Let $\varepsilon > 0$ be given. Then

$$|f_n(x) - f(x)| < \varepsilon$$

$$\Rightarrow \quad \left|\frac{nx^2}{1+nx}\right| < \varepsilon$$

$$\Rightarrow \quad \frac{x}{1+nx} < \varepsilon \ \Rightarrow \ 1 + nx > \frac{x}{\varepsilon}$$

$$\Rightarrow \quad n > \frac{1}{\varepsilon} - \frac{1}{x}.$$

This inequality holds for all $n \geq m$ where m is positive integer just greater than $\frac{1}{\varepsilon}$. In particular, for $x = 0$, $m = 1$. Hence the series converges uniformly on [0, 1].

3. *If (x) denotes the positive or negative excess of x over the nearest integer and if x is midway between two integers, let x be zero. Test the uniform convergence of the series* $\Sigma \frac{(nx)}{n^2}$.

Sol. We have

$$(x) = \begin{cases} x, & \text{for } -\frac{1}{2} < x < \frac{1}{2} \\ 0, & \text{for } x = \frac{1}{2} \\ x - 1, & \text{for } \frac{1}{2} < x < 1\frac{1}{2} \\ 0, & \text{for } x = 1\frac{1}{2} \\ x - 2, & \text{for } 1\frac{1}{2} < x < 2\frac{1}{2} \text{ and so on} \end{cases}$$

Similarly,

$$(2x) = \begin{cases} 2x, & \text{for } -\frac{1}{4} < x < \frac{1}{4} \\ 0, & \text{for } x = \frac{1}{4} \\ 2x - 1, & \text{for } \frac{1}{4} < x < \frac{3}{4} \\ 0, & \text{for } x = \frac{3}{4} \text{ and so on.} \end{cases}$$

Similarly, we can write expressions for (3x), (4x), etc. It is obvious from above that

$$|(nx)| < \frac{1}{2}.$$

We then have

$$|u_n(x)| = \left|\frac{(nx)}{n^2}\right| < \frac{1}{2n^2} < \frac{1}{n^2}$$

But $\Sigma \frac{1}{n^2}$ is convergent. Hence the series $\Sigma \frac{(nx)}{n^2}$ is uniformly convergent.

4. *Examine for term by term integration the series for which $f_n(x) = nx.e^{-nx^2}$, indicating the interval over which your conclusion holds.*

Sol. Here $$f(x) = \lim_{n\to\infty} f_n(x) = \lim_{n\to\infty} \frac{nx}{e^{nx^2}}$$

$$= \lim_{n\to\infty} \frac{nx}{1 + nx^2 + \frac{n^2x^4}{2!} + \ldots}$$

$$= 0 \text{ for all finite values of } x.$$

Consider the interval $0 \le x \le 1$.

$$\int_0^1 f(x)\,dx = \int_0^1 0.dx = 0$$

And $$\int_0^1 f_n(x)\,dx = \int_0^1 nxe^{-nx^2}\,dx$$

$$= \left[-\frac{1}{2}e^{-nx^2}\right]_0^1 = \frac{1}{2}\left[1 - e^{-n}\right]$$

$$= \frac{1}{2} \text{ as } n \to \infty.$$

Hence term by term integration over the interval $0 \le x \le 1$ is not justified. In fact zero is a point of non-uniform convergence as we have already seen. However, term by term integration is justified over the interval $[c, 1]$ where $0 < c < 1$. For we have

$$\int_c^1 f_n(x)\,dx = \int_c^1 nxe^{-nx^2}\,dx$$

$$= \frac{1}{2}\left[e^{-nc^2} - e^{-n}\right] = 0 \text{ as } n \to 0.$$

Hence $$\int_c^1 f(x)\,dx = \lim_{n\to\infty} \int_c^1 f_n(x)\,dx.$$

Hence $[c, 1]$ is the interval in which term by term integration holds.

5. *Show that near $x = 0$, the series $u_1(x) + u_2(x) + \ldots$, where $u_1(x) = x, u_n(x) = x^{1/(2n-1)} - x^{1/(2n-3)}$ and real values of x are concerned, is discontinuous and non-uniformly convergent. Can the series be integrated term by term ?*

Sol. Here

$$u_1(x) = x$$

$$u_2(x) = x^{1/3} - x$$

$$u_3(x) = x^{1/5} - x^{1/3},$$

$$\ldots\ldots\ldots\ldots\ldots\ldots\ldots$$

$$u_n(x) = x^{1/(2n-1)} - x^{1/(2n-3)}$$

Hence $$f_n(x) = x^{1/(2n-1)}$$

$\therefore$ $$f(0) = 0$$

and $$f(x) = 1 \text{ for other values of } x.$$

Hence f is discontinuous at $x = 0$ and consequently zero is a point of non-uniform convergence of the series.

Now $0 \le x \le c < \infty$, we have

$$\int_0^c f(x)\,dx = \int_0^c dx = c$$

and $$\int_0^c f_n(x)\,dx = \int_0^c x^{1/(2n-1)}\,dx$$

$$= \frac{2n-1}{2n} \cdot c^{2n/(2n-1)}$$

$$\to c \text{ as } n \to \infty.$$

Hence the series is term by term integrable in the above interval though 0 is a point of non-uniform convergence of the series.

6. *Given the series $\Sigma\, u_n(x)$ for which*

$$f_n(x) = \frac{1}{2n^2} \log\left(1 + n^4 x^2\right).$$

Show that the series $\Sigma\, u_n'(x)$ does not converge uniformly, but that the given series can be differentiated term by term.

Sol. We have

$$f(x) = \lim_{n \to \infty} f_n(x) = \lim_{n \to \infty} \frac{\log\left(1 + n^4 x^2\right)}{2n^2} \quad \left[\text{from } \frac{\infty}{\infty}\right]$$

$$= \lim_{n \to \infty} \frac{\dfrac{4n^3 x^2}{1 + n^4 x^2}}{4n}$$

$$= \lim_{n \to \infty} \frac{n^2 x^2}{1 + n^4 x^2} = 0 \text{ for } 0 \le x \le 1$$

Hence $\quad f'(x) = 0.$

And $\quad \lim_{n \to \infty} f_n'(x) = \dfrac{n^2 x}{1 + n^4 x^2} = 0$ for $0 \le x \le 1.$

$$\therefore \quad f'(x) = \lim_{n \to \infty} f_n'(x),$$

so that term by term differentiation holds. The series $\Sigma\, u_n'(x)$ is no[t] uniformly convergent in $0 \le x \le 1$ since the sequence $\{f_n'(x)\}$ has 0 as [a] point of non-uniform convergence.

7. *Let*

$$u_n(x) = x^2 \left(x^{1/(2n-1)} - x^{1/(2n-3)}\right) \sin\frac{1}{x}$$

for $x \ge 0$, $u_n(0) = 0$, for any *positive integer greater than unity and*

$$u_1(x) = x^3 \sin\frac{1}{x} \quad \text{for } x \lesssim 0, u_1(0) = 0.$$

show that $\sum_1^\infty u_n(x)$ *converges for all values of x to f (x), wher[e]* $f(x) = x^2 \sin^{1}(1/x)$ for $x \lesssim 0$ and $f(0) = 0$.

Also show that f is discontinuous at x = 0, that $\sum_1^\infty u_n'(x)$ *is no[t] uniformly convergent in any interval including the origin, and tha[t]* $f'(x) = \sum_1^\infty u_n'(x)$ *for all values of x.*

Sol. We have, when $x \ne 0$

$$u_1(x) = x^3 \sin(1/x) = x^2 (x - 0) \sin 1/x$$

$$u_2(x) = x^2 \left[x^{1/3} - x\right] \sin(1/x),$$

$$u_3(x) = x^2 \left[x^{1/5} - x^{1/3}\right] \sin(1/x),$$

....

$$u_n(x) = x^2 \left[x^{1/(2n-1)} - x^{1/(2n-3)}\right] \sin(1/x),$$

then $$f_n(x) = x^2 \sin(1/x) \cdot x^{1/(2n-1)}.$$

$\therefore$ $$f(x) = \lim_{n \to \infty} f_n(x) = x^2 \sin(1/x) \text{ when } x \neq 0;$$

and $$f(x) = 0 \text{ when } x = 0.$$

Now, when $x \neq 0$, we have

$$f'(x) = 2x \sin(1/x) - \cos(1/x).$$

Since $\lim_{n \to 0} \cos(1/x)$ does not exist, f is discontinuous at $x = 0$,

$$\sum_1^{\infty} u_n'(x) = \lim_{n \to \infty} \sum_1^{n} u_n'(x)$$

$$= \lim_{n \to \infty} \left[\frac{d}{dx}\left\{x^2 \sin\frac{1}{x} - x^{1/(2n-1)}\right\}\right]$$

$$= \lim_{n \to \infty} \left[\frac{4n-1}{2n-1} x^{2n/(2n-1)} \sin\frac{1}{x} - x^{1/(2n-1)} \cos\frac{1}{x}\right]$$

$$= 2x \sin(1/x) - \cos(1/x)$$

and $$f'(x) = 2x \sin(1/x) - \cos(1/x).$$

Hence $$f'(x) = \sum_1^{\infty} u_n'(x)$$

when $x \neq 0$.

Obviously $f'(0) = \sum_1^{\infty} u_n'(0)$

Since the sum function $\{2x \sin(1/x) - \cos(1/x)\}$ of the series $\Sigma u_n'(x)$ is discontinuous at $x = 0$, the series $\Sigma u_n'(x)$ is non-uniformly convergent near $x = 0$.

EXERCISES

[In the following has been adopted the conventional method of designating a series of functions so that instead of denoting a term as f_n we have denoted it as $f_n(x)$.

1. Show that the series

$$\frac{x}{x+1}+\frac{x}{(x+1)(2x+1)}+\frac{x}{(2x+1)(3x+1)}+\ldots$$

is uniformly convergent in $]\,k,\infty\,[$, where k is a positive number. Show also that the series is non-uniformly convergent in $[\,0,\infty\,[$.

2. Show that the series

$$(1-x)^2+x(1-x)^2+x^2(1-x)^2+\ldots$$

is not uniformly convergent in $[0, 1]$

3. Show that the series

$$\Sigma\frac{x}{n(n+1)}$$

is uniformly convergent in $[0, k]$ where k is any positive number whatsoever but that it does converge uniformly in $]\,0,\infty\,[$.

4. Show that

$$\sum_{n=-\infty}^{\infty} e^{-(x-n)^2}$$

converges uniformly in any fixed interval $[a, b]$.

5. Show that the series

$$\sum_{n=1}^{\infty}\frac{1}{n^3+n^4x^2}$$

is uniformly convergent for all values of x; and that the derivative of the sum is given by term by term differentiation.

6. Show that

$$\sigma(x)=\sum_{n=0}^{\infty}\frac{1}{1+n^2+n^4x^2}$$

converges uniformly for all values of x; examine whether $\sigma'(0)$ can be found by term by term differentiation.

7. Show that the series $\Sigma\, n^{-x}$ is uniformly convergent in $]\,1+\delta,\infty\,[$ where δ is a positive number. Show also that term by term differentiation is valid in the same interval.

8. Show that the following series converge uniformly in the intervals indicated :

(*i*) $x - x^2 + x^3 - x^4 + x^5 - x^6 + \ldots, \left[-\frac{3}{4}, \frac{3}{4}\right]$

(*ii*) $e^x + e^{2x} + e^{3x} + e^{4x} + \ldots, \left[-2, -\frac{1}{2}\right].$

9. Show that

$$\sum_{n=1}^{\infty} \frac{x}{1+n^2x^2}$$

is uniformly convergent in [δ, 1] for any δ > 0 but is not uniformly convergent in [0, 1].

10. Show that the series

$$\sum_{n=1}^{\infty} \left[e^{-n|x|} - e^{-2n|x|}\right]$$

though convergent for all values of x, is not uniformity convergent in any interval which contains $x = 0$.

11. Show that, in the interval, [− 1, 1], each of the functions

$$\frac{nx}{1+nx^2}, \frac{nx}{1+n^2x^2}, \frac{nx}{1+n^3x^3}$$

tends to a limit as $n \to \infty$ and discuss the uniformly of convergence.

12. Discuss the uniform convergence of

$$1+\frac{e^{-2x}}{2^2-1}+\frac{e^{-4x}}{4^2-1}+\frac{e^{-6x}}{6^2-1}\ldots$$

in] 0, ∞ [.

The great German mathematician Weistrass stated and proved remarkable and interesting theorem, which we shall now state and prove.

Every continuous function, defined in an interval, can be approximated to uniformly, by a sequence of polynomials.

Let a real valued continuous function f be defined on $[a, b]$. We shall prove that there exists a sequence $\{P_n\}$ of polynomials which converges uniformly to f.

Without any loss of generality we suppose that $[a, b] = [0, 1]$ and that $(0) = 0 = f(1)$ and when $x \in R \sim [0, 1]$.

If the theorem is proved for this case, we consider.

$$g(x) = f(x) - f(0) - x[-f(1) - f(0)]$$

Clearly $g(0) = 0 = g(1)$. If g can be obtained as the limit of a uniformly convergent sequence of polynomials, then the same is true for $f, f - g$ being a polynomial.

I. We write

$$Q_n(x) = C_n\left(1 - x^2\right)^n; n \in \mathbf{N}, \text{ where } C_n \text{ is such that}$$

$$\int_{-1}^{+1} Q_n(x)\, dx = 1;\ n \in \mathbf{N}.$$

II. We now obtain an estimate of C_n. This result will be made to depend upon the equality

$$(1 - x^2)^n > \left(1 - nx^2\right).$$

To see the truth of this inequality, we consider the expression

$$\left(1 - x^2\right)^n - 1 + nx^2$$

which is 0 for $x = 0$, and whose derivative

$$2nx\left[1 - \left(1 - x^2\right)^n\right]$$

is positive for $x \in [0, 1]$.

Thus

$$\left(1 - x^2\right)^n - 1 + nx^2 > 0 \text{ for } x \in [0, 1].$$

We have

$$\int_{-1}^{1}\left(1 - x^2\right)^n dx = 2\int_0^1 \left(1 - x^2\right)^n dx$$

$$\geq \int_0^{1/\sqrt{n}} 2\left(1 - x^2\right)^n dx$$

$$\geq 2\int_0^{1/\sqrt{n}}\left(1 - nx^2\right) dx$$

$$= \frac{4}{3}\sqrt{n} > \frac{1}{\sqrt{n}}$$

Thus $$\int_{-1}^{1} C_n\left(1 - x^2\right)^n dx = 1 \Rightarrow C_n > \sqrt{n}.$$

It follows that

$$Q_n(x) \leq \sqrt{n}\left(1-\delta^2\right)^n \text{ where } \delta \leq |x| \leq 1$$

and as such $Q_n \to 0$ uniformly in $\delta \leq |x| \leq 1$.

We now write

$$P_n(x) = \int f(x+t)\, Q_n(t)\, dt.$$

We have

$$P_n(x) = \int_{-x}^{1-x} f(x+t)\, Q_n(t)\, dt.$$

on the basis of the fact that f is zero for all x outside [0, 1].

Also we have

$$P_n = \int_{-x}^{1-x} f(x+t)\, Q_n(t)\, dt = \int_0^1 f(t)\, Q_n(t-x)\, dt$$

so that P_n is a polynomial in x.

Now f being continuous, there exists $\delta > 0$ such that

$$\left|f(y) - f(x)\right| < \frac{\varepsilon}{2} \text{ where } \left|y - x\right| < \delta.$$

Let $\quad M = \sup. f(x)$:

we have

$$\left|P_n(x) - f(x)\right| \left|\int_{-1}^{+1} \left[f(x+t) - f(x)\right] Q_n(t)\right| dt$$

$$\leq \int_{-1}^{+1} \left|f(x+t) - f(x)\right| Q_n\, t \,\big|\, dt$$

$$\leq 2M \int_{-1}^{-\delta} Q_n(t)\, dt + \frac{\varepsilon}{2} \int_{-\delta}^{+\delta} Q_n(t)\, dt + 2M \int_{\delta}^{1} Q_n(t)\, dt$$

$$\leq 4M\sqrt{n}\left(1-\delta^2\right)^n + \frac{\varepsilon}{2} < \varepsilon.$$

for sufficiently large values of x.

Thus the theorem has been proved.

Note. For another proof, by an outstanding French mathematician Lebesgue, a reference may be made to a course on Mathematical analysis by Eduwart, Part I.

OBJECTIVE QUESTIONS

Note. *For each of the following questions, four alternatives are given for the answer. Only one of them is correct. Choose the correct alternative.*

1. For a sequence $u_1(x), u_2(x), ..., u_n(x)$ in an interval (a, b) if there is a positive number k, such that $|u_n(2)| < k$ for every value of x in (a, b) and for every integer n, then sequence is :

(*a*) convergent (*b*) uniformly convergent

(*c*) bounded (*d*) uniformly bounded.

2. Let $b_0 + b_1 + b_2 + ...$ be a convergent series of constants, then the series $\Sigma b_n e^{-b_n x}$ is :

(*a*) uniformly convergent in $x \geq 0$

(*b*) uniformly convergent in $x \leq 0$

(*c*) non-uniformly convergent

(*d*) uniformly bounded

3. If the sum function of a series each of whose terms is a continuous function of x to a certain interval, is a discontinuous function is that interval, then the series must be :

(*a*) uniformly convergent (*b*) non-uniformly convergent

(*c*) convergent only (*d*) none of these

4. A series of functions continuous in an interval (a, b), can be integrated term by term, if the series is

(*a*) convergent (*b*) divergent

(*c*) uniformly convergent (*d*) non-uniformly convergent

5. If a series is simply convergent and derived series is uniformly convergent then the series can be :

(*a*) term by term integrable (*b*) term by term differentiable

(*c*) whole series is continuous (*d*) none of the above

ANSWERS

1. (*a*) **2.** (*a*) **3.** (*b*) **4.** (*c*) **5.** (*b*)

8

Elementary Functions

8.1. INTRODUCTION

It is proposed in this chapter to define the elementary functions

$$\sin x, \cos x, \log x, e^x, a^x$$

and develop their basic properties. Of course acquaintance with these functions must have been already acquired by the reader earlier in his elementary courses. It is essential, therefore, that the difference in the manner in which we are introduced to these functions in the elementary course and the manner in which the introduction is proposed here is well understood.

The treatment now will be dependent on the set of real numbers as a complete ordered field and the properties of continuity, derivability and integrability of functions as based thereon. The treatment in the elementary courses, however, is a mixture of intuitive, geometrical and non-rigorously given considerations and the questions of existence are overlooked. Thus our introduction to sin x and cos x is based on the notions of angles and rotations as seen intuitively in the elementary courses.

Again while defining $\log_a x$, we say that this is the number y such that $a^y = x$ without caring to enquire as to whether or not y exists.

Thus it should be seen that the developments in this chapter constitute a basic necessity for those interested in Mathematical theory. Of course in previous chapters these functions have found a place but that is so merely in terms of illustrations and examples and never as a part of the text proper. This was considered necessary in order to illuminate the basic ideas. In order, however, that the reader may understand the reference to these functions in earlier chapters is not a part of the theory, we have, wherever these functions have made appearance, generally indicated the same by a star.

In this chapter we shall consciously accept an abuse of language inasmuch as notation for a function will not be distinguished from that for the functional value. This instead of denoting the function by sin we shall

8.2. TRIGONOMETRIC FUNCTIONS

The two functions

$$\sin x, \cos x$$

will be defined as the sum functions of two uniformly convergent power series.

Theorem. *The two series*

$$\sum_{n=0}^{\infty} \frac{(-1)^n\, x^{2n+1}}{(2n+1)!}, \sum_{n=0}^{\infty} \frac{(-1)^n\, x^{2n}}{(2n)!}$$

are uniformly convergent in every interval $[a, b]$.

Let M be any positive number greater than $|\, a \,|$ as well as $|\, b \,|$. Ther $\forall\, x \in [a, b]$, we have

$$\left|\frac{(-1)^n\, x^{2n+1}}{(2n+1)!}\right| \le \frac{M^{2n+1}}{(2n+1)!}, \left|\frac{-1^n\, x^{2n}}{(2n)!}\right| \le \frac{M^{2n}}{(2n)!}.$$

Now, as may be easily shown, the two series

$$\sum \frac{M^{2n+1}}{(2n+1)!}, \sum \frac{M^{2n}}{(2n)!}$$

are convergent.

Therefore, by Weierstrass's M-test, we prove that the two series ar uniformly convergent in $[a, b]$.

Definitions. This theorem justifies the following two definitions :

$$(A) \quad \sin x = x - \frac{x^3}{3!} + \frac{x^5}{5!} - \ldots + \frac{(-1)^n\, x^{2n+1}}{(2n+1)!} + \ldots \qquad (A)$$

$$(B) \quad \cos x = 1 - \frac{x^2}{2!} + \frac{x^4}{4!} - \ldots + \frac{(-1)^n\, x^{2n}}{(2n)!} + \ldots \qquad (B)$$

8.2.1. The functions sin *x*, cos *x* are defined and continuous $\forall\, x$.

This follows from § 7.4.

8.2.2. The functions sin *x*, cos *x* are derivable $\forall\, x$ and

$$(\sin x)' = \cos x \textit{ and } (\cos x)' = -\sin x.$$

If we differentiate term by term the convergent series (A), we get th uniformly convergent series (B) and hence, (§ 7.4), sin x is derivable a $(\sin x)' = \cos x$.

uniformly convergent series (*B*) and hence, (§ 7.4), sin *x* is derivable and (sin *x*)′ = cos *x*.

Again, on differentiating term by term the convergent series (*B*), we get the uniformly convergent series

$$-x+\frac{x^3}{3!}-\frac{x^5}{5!}+\ldots$$

and hence cos *x* is derivable and (cos *x*)′ = – sin *x*.

8.2.3. sin 0 = 0 and cos 0 = 1.

The proof is obvious.

8.2.4. sin (– *x*) = – sin *x* and cos (– *x*) = cos *x*, ∀ *x*.

The proof is obvious.

8.2.5. The Addition Theorem

$$\forall\, x,\, y$$

$$\sin(x+y)=\sin x\cos y+\cos x\sin y;$$

$$\cos(x+y)=\cos x\cos y-\sin x\sin y.$$

Giving any value to y and then keeping it fixed, we write

$$f(x)=\sin(x+y)-\sin x\cos y-\cos x\sin y;$$

$$g(x)=\cos(x+y)-\cos x\cos y+\sin x\sin y.$$

We have

$$f'(x)=\cos(x+y)-\cos x\cos y+\sin x\sin y=g(x),$$

$$g'(x)=-\sin(x+y)+\sin x\cos y+\cos x\sin y=-f(x),$$

$$\therefore\quad \frac{d\left[f^2(x)+g^2(x)\right]}{dx}=2f(x)f'(x)+2g(x)g'(x)$$

$$=2f(x)g(x)-2g(x)f(x)=0,\ \forall\, x$$

$$\Rightarrow\quad f^2(x)+g^2(x)$$

a constant.

Hence ∀ *x*

$$f^2(x)+g^2(x)=f^2(0)+g^2(0)=0$$

$$\Rightarrow\quad f(x)=0,\ g(x)=0.$$

Hence the theorem.

Cor. 1. $\cos^2 x+\sin^2 x=1\ \forall\, x$. We have

$$1 = \cos 0$$

$$= \cos(x-x)$$

Cor. 2. $|\sin x| \le 1, |\cos x| \le 1 \ \forall\ x.$

This follows from the above corollary.

Cor. 3. Changing y to x, we have

$$\sin 2x = 2 \sin x \cos x, \ \cos 2x = \cos^2 x - \sin^2 x.$$

Cor. 4. Changing y to $-y$, we have

$$\sin(x-y) = \sin x \cos y - \cos x \sin y,$$

$$\cos(x-y) = \cos x \cos y + \sin x \sin y.$$

8.3. THE NUMBER π.

The smallest positive root of the equation

$$\cos x = 0.$$

Theorem. *To prove that there exists a positive number, π, such tha cos (π/2) = 0 and cos x > 0, for 0 ≤ x < π/2.*

Consider the interval [0, 2].

We know that cos 0 (= 1), is positive and we will now show that co 2 is negative. We have

$$\cos 2 = 1 - \frac{2^2}{2!} + \frac{2^4}{4!} - \frac{2^6}{6!} + \ldots$$

$$= 1 - \frac{2^2}{2!}\left(1 - \frac{2^2}{3.4}\right) - \frac{2^6}{6!}\left(1 - \frac{2^2}{7.8}\right)\ldots$$

Since the brackets are all positive, we have

$$\cos 2 < 1 - \frac{2^2}{2!}\left(1 - \frac{2^2}{3.4}\right) = -\frac{1}{3},$$

so that cos 2 is negative.

Thus there exists at least, one number, α, between 0 and 2, such th cos α = 0. Also there cannot exist more than one such value : for, possible, let β be another so that

$$\cos \alpha = 0, \cos \beta = 0 \qquad 0 < \alpha, \beta < 2$$

By Rolle's theorem there exists at least, one number, l, between α ar β such that the derivative, $-\sin x$ of $\cos x$ vanishes for $x = l$, so that

$$\sin \lambda = 0, \text{ where } 0 < \lambda < 2.$$

But $$\sin \lambda = \frac{\lambda}{1!}\left(1 - \frac{\lambda^2}{2.3}\right) + \frac{\lambda^5}{5!}\left(1 - \frac{\lambda^2}{6.7}\right) + \ldots$$

which is clearly positive.

But $$\sin \lambda = \frac{\lambda}{1!}\left(1-\frac{\lambda^2}{2.3}\right)+\frac{\lambda^5}{5!}\left(1-\frac{\lambda^2}{6.7}\right)+...$$

which is clearly positive.

Thus there exists one and only one root of the equation $\cos x = 0$, lying between 0 and 2. Denoting twice this value by π, we see that $\pi/2$ is the least positive root of the equation $\cos x = 0$. Also, therefore, we have $\cos x > 0$ when $0 \le x < \pi/2$.

8.4. sin $x > 0$, WHEN $0 < x \le \frac{1}{2}\pi$.

Since the derivative of $\sin x$ is positive in $[0, \frac{1}{2}\pi]$ therefore $\sin x$ is strictly increasing. Also, since $\sin 0 = 0$, we see that $\sin x$ is positive when $0 < x \le \frac{1}{2}\pi$.

8.4.1. sin ($\pi/2$) = 1

Now $$\cos(\pi/2) = 0 \Rightarrow \sin^2(\pi/2) + \cos^2(\pi/2) = 1,$$

$$\Rightarrow \quad \sin^2(\pi/2) = 1 \Rightarrow \sin(\pi/2) = \pm 1.$$

But, by the Lagrange's mean value theorem $\exists\, \alpha \in]0, \pi/2[$ such that $\sin(\pi/2) = \sin(\pi/2) - \sin 0 = (\pi/2)\cos\alpha > 0,$

$$\Rightarrow \quad \sin(\pi/2) = 1.$$

8.4.2. cos $\pi = -1$, sin $\pi = 0$

We have

$$\cos\pi = 2\cos^2(\pi/2) - 1 = -1 \text{ and } \sin\pi = 2\sin\pi/2\cos\pi/2 = 0.$$

8.4.3. cos $2\pi = 1$, sin $2\pi = 0$

We have

$$\cos 2\pi = 2\cos^2\pi - 1 = 1 \text{ and } \sin 2\pi = 2\sin\pi\cos\pi = 0.$$

8.4.4. cos ($\pi/4$) = sin ($\pi/4$) = $1/\sqrt{2}$.

We have

$$0 = \cos(\pi/2) = 2\cos^2(\pi/4) - 1 \Rightarrow \cos(\pi/4) = 1/\sqrt{2},$$

rejecting the negative sign as $\cos \pi/4$ is necessarily positive. (§ 8.3)

Also

$$1 = \sin(\pi/2) = 2\sin(\pi/4)\cos(\pi/4)$$

$$\Rightarrow \quad \sin(\pi/4) = 1/\sqrt{2}.$$

8.4.5. $\sin\left(\frac{1}{2}\pi - x\right) = \cos x, \quad \cos\left(\frac{1}{2}\pi - x\right) = \sin x.$ ...(*i*)

$\sin\left(\frac{1}{2}\pi + x\right) = \cos x, \quad \cos\left(\frac{1}{2}\pi + x\right) = -\sin x.$ ...(*ii*)

$\sin(\pi + x) = -\sin x, \quad \cos(\pi + x) = -\cos x.$ ...(*iii*)

$\sin(\pi - x) = \sin x, \quad \cos(\pi - x) = -\cos x.$ ...(*iv*)

$\sin(2\pi + x) = \sin x, \quad \cos(2\pi + x) = \cos x.$ ...(*v*)

These are easily proved with the help of the addition formulae.

Note. Because of the formula (*v*), sin *x*, cos *x* are periodic functions with 2π as their period.

8.4.6. $\cos x > 0$, *when* $0 \le x < \frac{1}{2}\pi$ *or* $\frac{3}{2}\pi < x \le 2\pi$;

$\cos x < 0$, *when* $\frac{1}{2}\pi < x \le \pi$ *or* $\pi \le x < \frac{3}{2}\pi$.

When $0 \le x < \frac{1}{2}\pi$, we know from § 8.3 that $\cos x > 0$.

When $\frac{1}{2}\pi < x \le \pi$, we write $x = \frac{1}{2}\pi + y$ so that $0 < y \le \frac{1}{2}\pi$.

$$\therefore \quad \cos x = \cos\left(\frac{1}{2}\pi + y\right) = -\sin y < 0.$$

When $\pi \le x < \frac{3}{2}\pi$. We write $x = \pi + y$

so that $$\cos x = \cos(\pi + y) = -\cos y < 0.$$

When $\frac{3}{2}\pi < x \le 2\pi$, we write $x = \pi + y$

so that $$\frac{1}{2}\pi < y \le \pi.$$

$$\cos x = \cos(\pi + y) = -\cos y > 0.$$

8.4.7. $\sin x > 0$, **when** $0 < x < \pi$; $\sin x < 0$, **when** $\pi < x < 2\pi$

The proof is similar to that of the preceding theorem.

Ex. 1. Discuss how sin *x* and cos *x* vary (monotonically increase decrease), as *x* varies in the interval [0, 2].

Ex. 2. Show that when *n* is an integer

$$\sin n\pi = 0, \qquad \cos\frac{1}{2}(2n+1)\pi = 0;$$

and $$\sin\frac{1}{2}(2n+1)\pi = (-1)^n, \qquad \cos n\pi = (-1)^n.$$

8.5. THE FUNCTION TAN *X*

tan x is defined by the relation

$$\tan x = \frac{\sin x}{\cos x}.$$

Clearly tan x is defined, continuous and derivable for all values of x except those for which the denominator, cos x, vanishes which is the case for $x = \frac{1}{2}(2n+1)\pi$; n being any integer, positive, negative or zero.

From the formula (*iii*) of § 8.4.5, we have

$$\tan(\pi + x) = \tan x,$$

so that we see that, tan x, is a periodic function with period π.

Also we may easily show that when $x \neq \frac{1}{2}(2n+1)\pi$,

$$\frac{d(\tan x)}{dx} = \frac{d(\sin x / \cos x)}{dx} = \frac{1}{\cos^2 x}.$$

To show that

$$\lim_{x \to \left(\frac{1}{2}\pi - 0\right)} \tan x = \infty, \qquad \lim_{x \to \left(\frac{1}{2}\pi + 0\right)} \tan x = -\infty.$$

Let k be a positive number,

As $$\lim_{x \to \frac{1}{2}\pi} \sin x = 1,$$

$\exists \delta_2 > 0$ such that $\left(\text{taking } \varepsilon = \frac{1}{2}\right)$

$$\frac{1}{2} < \sin x \ \forall \ x \in \left[\frac{1}{2}\pi - \delta_1, \ \frac{1}{2}\pi + \delta_1\right]. \qquad ...(1)$$

Again since

$$\lim_{x \to \frac{1}{2}\pi} \cos x = 0,$$

$\exists \ \delta_2 > 0$ such that

$$-1/2\,k < \cos x < 0 \ \forall \ x \in \frac{1}{2}\pi, \frac{1}{2}\pi + \delta_2$$

As cos x is positive when $x \in \left[0, \frac{1}{2}\pi\right[$ and negative when $x \in \left]\frac{1}{2}\pi, \pi\right]$, we have

$$0 < \cos x < 1/2k \ \forall\ x \in \left[\frac{1}{2}\pi - \delta_2, \frac{1}{2}\pi\right] \qquad ...(2)$$

$$-1/2\,k < \cos x < 0 \ \forall\ x \in \frac{1}{2}\pi, \frac{1}{2}\pi + \delta_2$$

From (1) and (2), we have if $\delta = \min.(\delta_1, \delta_2)$

$$\tan x = \frac{\sin x}{\cos x} > k \ \forall\ x \in \left[\frac{1}{2}\pi - \delta, \frac{1}{2}\pi\right[$$

and from (1) and (3), we have

$$\tan x = \frac{\sin x}{\cos x} < -k \ \forall\ x \in \left]\frac{1}{2}\pi, \frac{1}{2}\pi + \delta\right].$$

Hence the result.

The inverse trigonometrical functions

$$\sin^{-1} y, \cos^{-1} y, \tan^{-1} y.$$

$\sin^{-1}$ y. Since as may be easily seen, $\sin x$ strictly increases from -1 to $+1$ as x increases from $-\frac{1}{2}\pi$ to $\frac{1}{2}\pi$, we see that $\sin x$ is inversible and its inverse is a function with domain $\left[-\frac{1}{2}\pi, \frac{1}{2}\pi\right]$ and range $[-1, 1]$. We write

$$y = \sin x \Leftrightarrow x = \sin^{-1} y.$$

$\sin^{-1} y$ is derivable in the open interval $]-1, 1[$ with $1/\sqrt{(1-y^2)}$ as the derivative. In fact, we have

$$(\sin x)'\left(\sin^{-1} y\right)' = 1$$

$$\left(\sin^{-1} y\right)' = \frac{1}{\cos x}, \ x \neq \pm\frac{\pi}{2}$$

$$= \frac{1}{\sqrt{\left(1-\sin^2 x\right)}} = \frac{1}{\sqrt{\left(1-y^2\right)}}, \ y \neq \pm 1.$$

$\cos^{-1}$y. Since $\cos x$ is a strictly decreasing function in $[0, \pi]$ with range $[-1, 1]$ we see that this is inversible. We write

$$y = \cos x \Leftrightarrow x = \cos^{-1} y$$

so that $\cos^{-1} y$ is a function with domain $[0, \pi]$ and range $[-1, 1]$.

It may be been that

$$\left(\cos^{-1} y\right)' = -1/\sqrt{\left(1-y^2\right)}, \ y \neq \pm 1.$$

$\tan^{-1} y$. Since $\tan x$ is strictly monotonic with domain $\left]-\frac{1}{2}\pi, \frac{1}{2}\pi\right[$ and range $]-\infty, \infty[$, the function is inversible. We write

$$y = \tan x \Leftrightarrow x = \tan^{-1} y$$

so that $\tan^{-1} y$ is a function with domain $]-\infty, \infty[$ and range $\left]-\frac{1}{2}\pi, \frac{1}{2}\pi\right[$.

It may be seen that

$$(\tan^{-1} y)' = 1/\left(1+y^2\right) \ \forall\ y.$$

8.6. LOGARITHMIC FUNCTION

The integral

$$\int_1^x \frac{dt}{t}$$

exists if and only if $x > 0$ as in that case the integrand is bounded. We write

$$\log x = \int_1^x \frac{dt}{t}$$

so that $\log x$ has a meaning if and only if $x > 0$.

Thus we have defined here what is called the logarithmic function with domain $]0, \infty[$. By virtue of the results obtained in § 6.8, p. 222, we see that $\forall\ x > 0$, *$\log x$ is continuous and derivable with derivative $1/x$.* We have $\log 1 = 0$.

Also the integrand being positive $\forall\ x > 0$, we have

$$x > 1 \Rightarrow \log x > 0 \text{ and } x < 1 \Rightarrow \log x < 0.$$

We shall now show that $\forall\ a, b > 0$,

$$\log ab = \log a + \log b \qquad ...(1)$$

$$\log(a \div b) = \log a - \log b. \qquad ...(2)$$

We have

$$\log(ab) = \int_1^{ab} \frac{dt}{t} = \int_1^a \frac{dt}{t} + \int_a^{ab} \frac{dt}{t}.$$

Taking $t = au$, we see that

$$\int_a^{ab} \frac{dt}{t} = \int_1^b \frac{du}{u} = \log b$$

so that $\qquad \log ab = \log a + \log b,\ \forall\ a, b > 0.$

Now consider

$$\log \frac{1}{b} = \int_1^{1/b} \frac{dt}{t}$$

and writing $u = bt$, we obtain

$$\log \frac{1}{b} = \int_1^{1/b} \frac{dt}{t} = \int_b^1 \frac{du}{u} = -\int_1^b \frac{du}{u} = -\log b.$$

$$\therefore \qquad \log \frac{a}{b} = \log a + \log \frac{1}{b} = \log a - \log b.$$

The function $\log x$ is strictly monotonically increasing with range $]-\infty, \infty[$.

We have

$$\log x = \int_1^x \frac{dt}{t}.$$

The integrand being positive we see that $\log x$ is strictly monotonically increasing. In fact, if $b > a$, we have

$$\log b - \log a = \int_1^b \frac{dt}{t} - \int_1^a \frac{dt}{t} = \int_1^b \frac{dt}{t} > 0.$$

Thus, the function is strictly monotonically increasing.

If n be a positive integer, we have, by (1)

$$\log 2^n = n \log 2.$$

Also $\qquad x > 2^n \Rightarrow \log x > \log 2^n = n \log 2.$

We thus see that while x increases, $\log x$ strictly increases and is not bounded above so that

$$\lim_{n \to \infty} \log x = \infty.$$

Again n being a positive integer, we have

$$\log 2^{-n} = -n \log 2.$$

Also $\qquad 0 < x < 2^{-n} \Rightarrow \log x < -n \log 2$

This shows that

$$\lim_{x \to (0+0)} \log x = -\infty.$$

Finally because of the continuity of $\log x$, it follows that it assumes every value in the range $]-\infty, \infty[$.

8.7. INVERSE OF THE LOGARITHMIC FUNCTION

Exponential Function. Since, as has been seen, log x is a strictly monotonically increasing continuous function with domain $]0, \infty[$ and range $]-\infty, \infty[$, it admits of continuous strictly, monotonically increasing inverse function with domain $]-\infty, \infty[$ and range $]0, \infty[$. This inverse function is called the Exponential function and we write

$$y = \log x \Leftrightarrow x = E(y)$$

The exponential function $E(y)$ has a meaning $\forall\, y$. The range of $E(y)$ being $]0, \infty[$, it does not assume any negative value or the value 0. Also $E(y)$ is derivable $\forall\, y$ and

$$E'(y) = E(y).$$

In fact, we have

$$(\log x)'\,(E(y))' = 1 \Leftrightarrow (E(y))' = x = E(y).$$

We also have

$$\lim_{y \to \infty} E(y) = 0, \quad \lim_{y \to \infty} E(y) = \infty.$$

We shall now show that $\forall\, u, v > 0, E(u+v) = E(u)\,E(v)$.

Let $\quad E(u) = a, \; E(v) = b,$

so that $\quad u = \log a, v = \log b,$

$\therefore \quad u + v = \log a + \log b = \log ab$

$\Rightarrow \quad E(u+v) = ab = E(u)\,E(v).$

Generalised power function. We have defined a^x when a is a positive real number and x is a rational number. We shall now give a meaning to a^x when x is any real number whatsoever. We write, by definition,

$$a^x = E(x \log a)\; \forall\, x, \; a > 0.$$

We shall verify that this definition is consistent with that already given when x is a rational number.

Let $x = n$; n being a positive integer. We have

$$\begin{aligned} E(n \log a) &= [E(\log a + \log a + \ldots n \text{ times})] \\ &= [E(\log a)][E(\log a)] \ldots n \text{ times} \\ &= a.a \ldots n \text{ times}. \end{aligned}$$

Let $x = -n$; n being a positive integer. We have

$$E(-n \log a) = E[(-\log a) + (-\log a) + \ldots n \text{ times}]$$

$$=\left[E\left(\log\frac{1}{a}\right)\right]\left[E\left(\log\frac{1}{a}\right)\right]\ldots n \text{ times}$$

$$=\frac{1}{a}\cdot\frac{1}{a}\ldots n \text{ times}.$$

Thus $E(x \log a)$ has the meaning already assigned to a^x when x is an integer.

Let $x=\dfrac{p}{q}$ where p, q are integers and q is a positive integer.

We have

$$E\left(q.\frac{p}{q}\log a\right)=\left[E\left(\frac{p}{q}\log a\right)\right]^q,$$

i.e.,
$$E(p\log a)=\left[E\left(\frac{p}{q}\log a\right)\right]^q$$

so that $E\left(\dfrac{p}{q}\log a\right)$ denotes the qth root of $E(p \log a)$. Hence the verification when x is a rational number.

The number *e*. We denote by e the number $E(1)$ so that e is a number such that

$$\log e = 1.$$

We have, by def.,

$$e^x = E(x \log e) = E(x)$$

so that what we have so far denoted by $E(x)$ can as is usual be denoted by e^x

8.8. POWER SERIES EXPANSION OF e^x, $\log(1 + x)$, $(1 + x)^m$

8.8.1. Let $f(x) = e^x$

The Function Possesses Continuous Derivatives of Every Order $\forall x$.

In fact, $\quad f^n(x)=e^x \;\forall x.$

If R_n denotes Lagrange's remainder after n terms, we have

$$R_n=\frac{x^n}{n!}e^{\theta x},\; 0<\theta<1.$$

Now $x>0\Rightarrow e^{\theta x}<e^x$ and $x<0 \Rightarrow e^{\theta x}<1.$

Thus
$$|R_n|=\left|\frac{x^n}{n!}\right|e^{\theta x} \Rightarrow \begin{cases}|x^n/n!|\,e^x & \text{if } x>0,\\ |x^n/n!|\,e^{-x} & \text{if } x<0.\end{cases}$$

Since $\forall\, x, \left|x^n / n!\right| \to 0$ as $n \to \infty$, we see that $\forall\, x, R_n \to 0$ as $n \to \infty$.

The validity of Maclaurin's infinite expansion for e^x has thus been established $\forall\, x$. Making substitutions, we obtain

$$e^x = 1 + x + \frac{x^2}{2!} + \frac{x^3}{3!} + \ldots + \frac{x^n}{n!} + \ldots$$

8.8.2. Let $f(x) = \log(1 + x)$.

Now log $(1 + x)$ possesses continuous derivatives of every order for every value of x such that $(1 + x)$ is positive *i.e.*, $x > -1$. In fact, when $x > -1$, we have

$$f^n(x) = \frac{(-1)^{n-1}(n-1)!}{(1+x)^n}.$$

If R_n denotes Lagrange's remainder after n terms, we have

$$R_n = \frac{x^n}{n!} f^n(\theta x) = (-1)^{n-1} \frac{1}{n} \left(\frac{x}{1+\theta x}\right)^n.$$

Let $0 \leq x \leq 1$ so that $x/(1+\theta x)$ is positive and < 1. We have

$$|R_n| < \frac{1}{n}.$$

Therefore $R_n \to 0$, when $0 \leq x \leq 1$.

Let $-1 < x < 0$. In this case $x/(1+\theta x)$ may not be numerically less than unity so that we fail to draw any conclusion from Lagrange's remainder R_n in this case. Taking Cauchy's remainder, we have

$$R_n = \frac{x^n}{(n-1)!}(1-\theta)^{n-1} f^n(\theta x) = (-1)^{n-1} x^n \cdot \left(\frac{1-\theta}{1+\theta x}\right)^{n-1} \frac{1}{1+\theta x}.$$

Since $|x| < 1, [(1-\theta)/(1+\theta x)]$ is positive and less than unity and hence

$$0 < \left(\frac{1-\theta}{1+\theta x}\right)^{n-1} < 1.$$

Also $$\frac{1}{1+\theta x} < \frac{1}{1-|x|}.$$

Finally $$x^n \to 0 \text{ as } n \to \infty.$$

Thus $$R_n \to 0 \text{ as } n \to \infty.$$

We have thus established that the Maclaurin's expansion for log $(1 + x)$ is valid when $-1 < x \le 1$. Making substitutions, we obtain

$$\log(1+x) = x - \frac{x^2}{2} + \frac{x^3}{3} - \frac{x^4}{4} \dots + (-1)^{n-1} \frac{x^n}{n} + \dots$$

8.8.3. Let $f(x) = (1 + x)^m$.

If m is a positive integer, then we know that $f^{m+1}(x)$ and the following successive derivatives are identically zero for every value of x, so that we may easily show that *when m is a positive integer*

$$(1+x)^m = 1 + mx + \frac{m(m-1)}{2!} x^2 + \dots + x^m,$$

for every value of x.

If m is a real number, not necessarily a positive integer, then $(1 + x)^m$ possesses continuous derivatives of every order for values of x such that $(1 + x)$ is positive, *i.e.*, $x > -1$.

We have $\quad f^n(x) = m(m-1)\dots(m-n+1)(1+x)^{m-n}.$

If R_n denotes Cauchy's remainder, we have

$$R_n = \frac{x^n}{(n-1)!}(1-\theta)^{n-1} f^n(\theta x)$$

$$= \frac{m(m-1)\dots(m-n+1)}{(n-1)!} x^n \left(\frac{1-\theta}{1+\theta x}\right)^{n-1} (1+\theta x)^{m-1}$$

If $|x| < 1$.

$$\frac{m(m-1)\dots(m-n+1)}{(n-1)!} x^n \to 0.$$

$$0 < \left(\frac{1-\theta}{1+\theta x}\right)^{n-1} < 1.$$

$$(1+\theta x)^{m-1} < \begin{cases} (1+1)^{m-1} = 2^{m-1}, \text{ if } (m-1) \text{ is positive} \\ (1-|x|)^{m-1}, \text{ if } (m-1) \text{ is negative} \end{cases}$$

Thus $R_n \to 0$ as $n \to \infty$, when $|x| < 1$.

Maclaurin's infinite expansion for $(1 + x)^m$ being thus valid, we see that

$$(1+x)^m = 1 + mx + \frac{m(m-1)}{2!} x^2 + \dots + \frac{m(m-1)\dots(m-n+1)}{n!} x^n + .$$

wnen $|x| < 1$.

Note. It is easy to see that we cannot directly prove that the Lagrange's remainder $R_n \to 0$, when $-1 < x < 0$.

8.8.4. Two Important Special Cases of Limits

8.8.5. Theorem

To prove that

$$\lim_{x \to \infty} \left(\frac{x^m}{e^{\alpha x}} \right) = 0,$$

where x, m are any positive numbers whatsoever.

Let $$f(x) = x^m, g(x) = e^{\alpha x}$$
so that $f(x) \to \infty$. and $g(x) \to \infty$, as $x \to \infty$.

We can write

$$m = n + p,$$

where n is some non-negative integer and p is a number such that $0 \le p < 1$.

It is easy to see that $f'(x), f''(x), ..., f^n(x)$ all tend to ∞ and $f^{n+1}(x) \to 0$ as $(x) \to \infty$.

Also $g'(x), g''(x), g'''(x)$... all tend to ∞.

Thus $$\lim_{n \to \infty} \{ f^{n+1}(x) / g^{n+1}(x) \} = 0$$

Therefore from the theorem of § 5.9.1, we deduce that

$$\lim [f(x) / g(x)] = 0, \text{ when } x \to \infty.$$

Note. The result of this theorem is roughly expressed by saying that $e^{\alpha x}$ tends to more rapidly than any positive power of x, when $x \to \infty$.

8.8.6. Theorem

To prove that

$$\lim_{n \to \infty} \frac{(\log x)^m}{x^\alpha} = 0,$$

where α, m are any positive numbers.

The independent proof is similar to that of the preceding result. Otherwise if we write $\log x = y$ we see that

$$\lim_{n \to \infty} \frac{(\log x)^m}{x^\alpha} = \lim_{y \to \infty} \frac{y^m}{e^{\alpha y}} = 0.$$

Note. *This shows that every positive power of x tends to infinity more rapidly than any positive power of log x as $x \to \infty$.*

Cor. $\lim_{x \to (0+0)} \left[x^{\alpha} (\log x)^m\right] = 0$; α, m being positive numbers.

Putting $x = 1/y$, we may easily obtain it.

8.8.7. Note on a Special Function

The function f defined as

$$f(x) = e^{-1/x^2} \text{ when } x \neq 0, f(0) = 0$$

possesses a remarkable property, *viz.*, $f^n(0) = 0 \; \forall \; n$.

We have

$$f'(0) = \lim_{x \to 0} \frac{f(x) - f(0)}{x} = \lim_{x \to 0} \frac{e^{-1/x^2}}{x} = \lim_{y \to \infty} \frac{y}{e^{y^2}} = 0.$$

$$f'(x) = \left(2/x^3\right) e^{-1/x^2}, \text{ when } x \neq 0.$$

$$\therefore \qquad f''(0) = \lim_{x \to 0} \frac{f'(x) - f'(0)}{x} = \lim_{x \to 0} \frac{2e^{-1/x^2}}{x^4}$$

$$= \lim_{y \to \infty} \frac{2y^4}{e^{y^2}} = 0.$$

$$f''(x) = \left(\frac{4}{x^6} - \frac{6}{x^4}\right) e^{-1/x^2}, \text{ when } x \neq 0,$$

$$\therefore \qquad f'''(0) = \lim_{x \to 0} \left(\frac{4}{x^6} - \frac{6}{x^4}\right) e^{-1/x^2} = \lim_{y \to \infty} \frac{4y^6 - 6y^4}{e^{y^2}} = 0$$

If we form the higher derivative from $x \neq 0$, we shall obviously always obtain the product of e^{-1/x^2} and a polynomial in $1/x$. Thus we see that all the higher derivatives will vanish at the point $x = 0$.

It is thus seen that this function possesses a continuous derivative $\forall \; x$. Maclaurin's infinite series for this function is

$$0 + x.0 + \frac{x^2}{2!}.0 + \frac{x^3}{3!}.0 + \ldots + \frac{x^n}{n!}.0 + \ldots$$

which is certainly not equal to $f(x)$ for any value of x, other than 0.

Thus we see that there exist functions which possess continuous derivatives $\forall \; x$ and yet cannot be expanded in terms of Maclaurin's series.

OBJECTIVE QUESTIONS

Note. *For each of the following questions, four alternatives are given for the answer. Only one of them is correct. Choose the correct alternative.*

1. $\cos x > 0$ when

(*a*) $x \geq 0$ (*b*) $0 \leq x < \pi/2$

(*c*) $0 < x \leq \pi/2$ (*d*) none of these

2. $\tan x$ is strictly monotonic with domain

(*a*) $\left[-\frac{1}{2}\pi, \frac{1}{2}\pi\right]$ (*b*) $\left[-\frac{1}{2}\pi, \frac{1}{2}\pi\right[$

(*c*) $\left]-\frac{1}{2}\pi, \frac{1}{2}\pi\right]$ (*d*) $\left]-\frac{1}{2}\pi, \frac{1}{2}\pi\right[$

3. The function $\log x$ is strictly monotonically increasing with range

(*a*) $]0, \infty[$ (*b*) $]-\infty, \infty[$

(*c*) $]-\infty, 0[$ (*d*) $[-\infty, \infty]$

4. $E(u+v) =$

(*a*) $E(u) + E(v)$ (*b*) $E(u) \,.\, E(v)$

(*c*) $E(u) - E(v)$ (*d*) $E(u)/E(v)$

5. $\sin x$, $\cos x$ are periodic functions whose period is

(*a*) π (*b*) $n\pi$

(*c*) 2π (*d*) $\pi/2$

ANSWERS

1. (*b*) **2.** (*d*) **3.** (*b*) **4.** (*b*) **5.** (*c*)

9

Improper Integrals

9.1. INTRODUCTION

The theory of Riemann integration, as developed in Chapter 6, expressly requires that the Domain of integration is finite and that the Integrand is bounded in the domain.

It is possible, however, so to extend the theory that the symbol

$$\int_a^b f(x)\, dx$$

may sometimes have a meaning (*i.e.,* denote a number), even when f is not bounded or when either a or b or both are infinite. In case f is unbounded or the limits a or b are infinite, the symbol

$$\int_a^b f(x)\, dx$$

is called an Improper or Generalised or Infinite integral. Thus

$$\int_0^1 \frac{dx}{x^3},\ \int_1^2 \frac{dx}{(1-x)(2-x)},\ \int_{-\infty}^{\infty} \frac{dx}{1+x^2},\ \int_0^{\infty} \frac{1}{\sqrt{x}}\, dx$$

are examples of improper integrals.

For the sake of distinction an integral which is not improper will be called a Proper integral.

It can be *proved that if a function f is not bounded in a finite interval $[a, b]$, then there exists *at least one point, c, of the interval such that in no neighbourhood of c, f is bounded.* Such a point is known as a point of **Infinite discontinuity** of the function f. It will always be assumed that the function f is such that the set of its points of infinite discontinuity in any interval, finite or infinite, are finite in number; the consideration of functions having an infinite number of points of infinite discontinuity being beyond the scope of the book.

---

* For the proof, use may be made of the completeness of order of the field of real numbers.

In a finite interval which encloses no point of infinite discontinuity, the function is always bounded and we assume, once for all, in order to avoid tedious repetition, that it is also integrable in such an interval.

9.2. DEFINITION

9.2.1. Convergence at the Left-End

Let, a, be the only point of infinite discontinuity of a function f in a finite interval $[a, b]$.

We write

$$\varphi(\varepsilon) = \int_{a+\varepsilon}^{b} f(x)\, dx, \quad 0 < \varepsilon \le b - a,$$

the integral on the right existing according to the assumption made in the last paragraph. Thus we have defined a function φ with domain $]\,0, b - a\,[$.

If, when $\varepsilon \to (0+0)$, the function φ tends to a finite limit, say I, we say that the improper integral

$$\int_{a}^{b} f(x)\, dx, \qquad \text{...(1)}$$

exists, or converges at a, and use the symbol (1) to denote the number I. Thus

$$\int_{a}^{b} f(x)\, dx = \lim_{\varepsilon \to 0} \int_{a+\varepsilon}^{b} f(x)\, dx,$$

provided that the limit on the right exists. In case the limit does not exist, we say that the improper integral $\int_{a}^{b} f(x)\, dx$ does not exist or that it does not converge.

Ex. Show that if $\int_{a}^{b} f(x)\, dx$ converges at a, then $\int_{a}^{c} f(x)\, dx$ $(a < c < b)$ also converges at a.

Ex. The improper integral $\int_{a}^{b} f(x)\, dx$ converges at a, and k is any constant; show that $\int_{a}^{b} kf(x)\, dx$ converges at a, and conversely.

9.2.2. Convergence at the Right-End

Let b be the only point of infinite discontinuity of f in a **finite** *interval* $[a, b]$. If then the proper integral

$$\int_{a}^{b-\varepsilon} f(x)\, dx, \quad 0 < \varepsilon < (b - a) \qquad \text{...(1)}$$

tends to a finite limit, as $\varepsilon \to (0+0)$, we say that the improper integral

$$\int_a^b f(x)\,dx \qquad \text{...(2)}$$

exists or converges at b and use the symbol (2) to denote the limit of (1).

Ex. Examine the convergence of the improper integrals :

(*i*) $\displaystyle\int_0^1 \frac{dx}{x^2}$, (*ii*) $\displaystyle\int_0^1 \frac{dx}{\sqrt{(1-x^2)}}$.

(*i*) Now 0 is the only point of infinite discontinuity of the integrand. We have, when $0 < \varepsilon \le 1$,

$$\varphi(\varepsilon) = \int_\varepsilon^1 \frac{dx}{x^2} = \left| -\frac{1}{x} \right|_\varepsilon^1 = \frac{1}{\varepsilon} - 1, \text{ which } \to +\infty \text{ as } \varepsilon \to (0+0).$$

Thus we see that the improper integral, in question, does not converge.

(*ii*) Now 1, is the only point of infinite discontinuity of the integrand in [0, 1]. We have

$$\varphi(\varepsilon) = \int_0^{1-\varepsilon} \frac{dx}{\sqrt{(1-x^2)}} = \sin^{-1}(1-\varepsilon),$$

which $\to \sin^{-1} 1 = \frac{1}{2}\pi$, as $\varepsilon \to 0$.

Thus the improper integral converges with value $\frac{1}{2}\pi$.

Note. The reader will note that the integrand in (*i*) is not defined for $x = 0$ and that in (*ii*) it is not defined for $x = 1$. We may assign to the integrand at such points any value we please without affecting the convergence and the value of the corresponding improper integral.

9.2.3. Convergence at both the End Point

Let the end points a and b be the only points of infinite discontinuity of f. We take any point, c, such that $a < c < b$.

If the improper integrals

$$\int_a^c f(x)\,dx \text{ and } \int_c^b f(x)\,dx$$

both converge at a and b respectively, we say that the improper integral

$$\int_a^b f(x)\,dx \text{ converges}$$

and write
$$\int_a^b f(x)\,dx = \int_a^c f(x)\,dx + \int_c^b f(x)\,dx.$$

It is easy to show that the fact of convergence and the value of the improper integral is independent of the choice of c.

If d be any point of $]\, a, b\, [$, we have

$$\int_{a+\varepsilon}^{d} f(x)\, dx = \int_{a+\varepsilon}^{c} f(x)\, dx + \int_{c}^{d} f(x)\, dx,$$

and see that

$$\lim_{\varepsilon \to 0} \int_{a+\varepsilon}^{d} f(x)\, dx \text{ exists finitely } \lim_{\varepsilon \to 0} \int_{a+\varepsilon}^{c} f(x)\, dx \text{ exists finitely.}$$

It follows that

$$\int_{a}^{d} f(x)\, dx \text{ converges} \Leftrightarrow \int_{a}^{c} f(x)\, dx \text{ converges.}$$

Also, in case they finitely, we have

$$\int_{a}^{d} f(x)\, dx = \int_{a}^{c} f(x)\, dx + \int_{c}^{d} f(x)\, dx. \quad \text{...(1)}$$

It may similarly be shown that

$$\int_{d}^{b} f(x)\, dx \text{ converges} \Leftrightarrow \int_{c}^{b} f(x)\, dx \text{ converges,}$$

and in case they converge, we have

$$\int_{d}^{b} f(x)\, dx = \int_{d}^{b} f(x)\, dx + \int_{c}^{b} f(x)\, dx. \quad \text{...(2)}$$

Adding (1) and (2), we get

$$\int_{a}^{d} f(x)\, dx + \int_{d}^{b} f(x)\, dx = \int_{a}^{c} f(x)\, dx + \int_{c}^{b} f(x)\, dx = \int_{a}^{b} f(x)\, dx.$$

Ex. Examine the convergence of the improper integrals :

(*i*) $\displaystyle\int_{0}^{1} \frac{dx}{\sqrt{(x - x^2)}}$, (*ii*) $\displaystyle\int_{0}^{2} \frac{dx}{x(2-x)}$, (*iii*) $\displaystyle\int_{0}^{\pi} \frac{dx}{\sin x}$.

9.2.4. General Case

Any finite number of points of infinite discontinuity. Let $c_1, c_2, c_3, \ldots, c_n$ be any finite number of points of discontinuity of f in $[a, b]$ where

$$a \le c_1 < c_2 < \ldots < c_{n-1} < c_n \le b.$$

If the improper integrals

$$\int_a^{c_1} f(x)\,dx,\ \int_{c_1}^{c_2} f(x)\,dx,\ \ldots,\ \int_{c_{n-1}}^{c_n} f(x)\,dx,\ \int_{c_n}^{b} f(x)\,dx$$

all exist in accordance with the definitions given above and we regard

$$\int_a^{c_1} f(x)\,dx = 0 \text{ if } c_1 = a \text{ and } \int_{c_n}^{b} f(x)\,dx = 0 \text{ if } b = c_n,$$

then we say that $\int_a^b f(x)\,dx$ exists and write

$$\int_a^b f(x)\,dx = \int_a^{c_1} f(x)\,dx + \int_{c_1}^{c_2} f(x)\,dx + \ldots + \int_{c_n}^{b} f(x)\,dx.$$

9.2.5. Infinite Range of Integration Convergence at ∞

Let f be bounded and integrable in $[a, t]$ where t is a number $\geq a$ so that the proper integral

$$\int_a^t f(x)\,dx$$

exists. We write

$$\varphi(t) = \int_a^t f(x)\,dx$$

so that φ is a function with domain $[a, \infty[$. If φ tends to a finite limit as $t \to \infty$, we say that the improper integral

$$\int_a^{\infty} f(x)\,dx \qquad \ldots(1)$$

exists or that it converges at ∞ and regard the symbol (1) as denoting the limit. Thus by definition

$$\int_a^{\infty} f(x)\,dx = \lim_{t \to \infty} \int_a^t f(x)\,dx,$$

provided the limit exists.

Ex. Examine the convergence of

(*i*) $\displaystyle\int_0^{\infty} \frac{dx}{1+x^2}$ (*ii*) $\displaystyle\int_0^{\infty} \frac{dx}{x^2}$

9.2.6. Convergence at – ∞

If f be bounded and integrable in $[t, b]$ where $t \leq b$, and

$$\int_t^b f(x)\,dx \qquad \ldots(1)$$

tends to a finite limit at $t \to -\infty$, we say that

$$\int_{-\infty}^{a} f(x)\,dx \qquad \text{...(2)}$$

converges or exists and regard the symbol (2) as denoting the limit of (1).

9.2.7. Let the Range of Integration be] – ∞, ∞ [.

If, c, is any number and

$$\int_{-\infty}^{c} f(x)\,dx, \int_{c}^{\infty} f(x)\,dx,$$

both exist in accordance with the definitions already given, then we say that

$$\int_{-\infty}^{\infty} f(x)\,dx$$

exists and write

$$\int_{-\infty}^{\infty} f(x)\,dx = \int_{-\infty}^{c} f(x)\,dx + \int_{c}^{\infty} f(x)\,dx.$$

It is easy to show that the existence and the value of

$$\int_{-\infty}^{\infty} f(x)\,dx$$

is independent of the choice of c.

9.2.8. General Case

If an infinite range of integration includes a finite number of points of infinite discontinuity, then we arbitrarily consider an interval $[a, b]$ which embraces all the points of infinite discontinuity and examine the existence of the three improper integrals

$$\int_{-\infty}^{a} f(x)\,dx, \int_{a}^{b} f(x)\,dx, \int_{b}^{\infty} f(x)\,dx,$$

in accordance with the definitions given above, and in case they all exist, we say that

$$\int_{-\infty}^{\infty} f(x)\,dx$$

exists and write

$$\int_{-\infty}^{\infty} f(x)\,dx = \int_{-\infty}^{a} f(x)\,dx + \int_{a}^{b} f(x)\,dx + \int_{b}^{\infty} f(x)\,dx.$$

Note. It will be seen that in order to examine the existence of a given improper integral, we have to examine the convergence of a system of improper integrals of the four types which have been considered in sub-sections 9.2.1, 9.2.2, 9.2.5, 9.2.6. Also we see that the case § 9.2.2 is analogous to that of § 9.2.1 and the case of § 9.2.6 is analogous to that of § 9.2.5.

9.2.9. From the foregoing, it will thus be seen that it is enough to consider Tests for the Convergence of

(i) $\int_a^b f(x)\,dx$ *at a, where f is bounded and integrable in*

$$[a+\varepsilon, b], 0 < \varepsilon \le (b-a).$$

(ii) $\int_a^\infty f(x)\,dx$ *at ∞, where f is bounded and integrable in*

$$[a, t],\ t \ge a.$$

Ex. Examine the existence of the following improper integrals and evaluate those which exist :

(i) $\int_1^\infty \frac{dx}{x(1+x)}$, (ii) $\int_0^\infty \frac{dx}{(1+x)\sqrt{x}}$,

(iii) $\int_0^\pi \frac{dx}{1+\cos x}$, (iv) $\int_{-1}^\infty \frac{dx}{\sqrt{[x(1+x)]}}$, (v) $\int_{-\infty}^\infty \frac{dx}{x(1+x^2)}$,

(vi) $\int_0^1 \log x\,dx$ (vii) $\int_{-2}^2 \frac{dx}{x^2(1-x)}$, (viii) $\int_0^\infty x^2 e^{-x}\,dx$.

Note. In the examples given above, the improper integrals are such that the integrands admit of primitives in terms of elementary functions. In such cases the examination of the existence is generally easy, but more advanced methods become necessary when the integrand does not possess a primitive in terms of elementary functions.

9.3. TEST FOR CONVERGENCE AT '*a*' POSITIVE INTEGRAND

Just as in the case of infinite series with positive terms it is comparatively easier to consider the convergence of integrals with positive integrands. We now proceed to consider the case of positive integrands. Let a be the only point of infinite discontinuity of f in $[a, b]$. The case where the integrand f is positive in a certain neighbourhood $]\,a, c\,[$ of, a, is particularly simple and important.

Since

$$\int_{a+\varepsilon}^b f(x)\,dx = \int_{a+\varepsilon}^c f(x)\,dx + \int_c^b f(x)\,dx,$$

it follows that

$$\int_a^c f(x)\,dx,\ \int_a^b f(x)\,dx$$

are either both convergent at, a or both non-convergent. It is, therefore, no loss of generality to suppose that f is positive in $[a, b]$.

The question of the existence of the integral is, in such a case, decided by comparison with another suitably chosen integral whose existence or otherwise is already known.

9.3.1. The Necessary and Sufficient Condition for the Convergence of the Improper Integral

$$\int_a^b f(x)\,dx$$

at a, where $f(x)$ is positive when $x \in]a, b]$, is that there exists k, > 0 such that

$$\int_{a+\varepsilon}^b f(x)\,dx < k \ \forall\ \varepsilon \in]0, b-a].$$

The proof follows from the fact that, since $f(x)$ is positive when $x \in]a, b]$, the integral

$$\int_{a+\varepsilon}^b f(x)\,dx$$

Monotonically increases as ε decreases and will, therefore, tend to a finite limit if, and only if, it is bounded above.

Note. In case $\varphi(\varepsilon)\int_{a+\varepsilon}^b f(x)\,dx$ is not bounded above, then $\varphi(\varepsilon) \to +\infty$ as $\varepsilon \to (0+0)$ and we say that the improper integral $\int_a^b f(x)\,dx$ diverges to ∞.

9.3.2. Comparison of Two Integrals

Let f and φ be two functions such that $\forall\ x \in]a, b]$,

$$f(x) \geq 0,\ \varphi(x) \geq 0.$$

Also let $f(x) \leq \varphi(x)$. Then

(i) $\int_a^b \varphi(x)\,dx$ *converges* $\Rightarrow \int_a^b f(x)\,dx$ *converges,*

(ii) $\int_a^b f(x)\,dx$ *does not converge* $\Rightarrow \int_a^b \varphi(x)\,dx$ *does not converges,*

(Lucknow 93, 95)

It is assumed that f and φ are both bounded and integrable in $[a+\varepsilon, b], 0 < \varepsilon \leq (b-a)$, and a, is the only point of infinite discontinuity $[a, b]$. We have $\forall\ \varepsilon \in]0, b-a]$.

$$\int_{a+\varepsilon}^{b} f(x)\,dx \le \int_{a+\varepsilon}^{b} \varphi(x)\,dx \qquad ...(1)$$

Let $\int_a^b \varphi(x)\,dx$ converge so that there exists a number k such that $\forall\ \varepsilon \in]0, b-a]$.

$$\int_{a+\varepsilon}^{b} \varphi(x)\,dx < k \qquad ...(2)$$

From (1) and (2), we have $\forall\ \varepsilon \in]0, b-a]$,

$$\int_{a+\varepsilon}^{b} f(x)\,dx < k$$

$$\Rightarrow \qquad \int_a^b f(x)\,dx \text{ converges at } a.$$

For the second part, we see that if $\int_a^b f(x)\,dx$ does not converge at a, then $\int_{a+\varepsilon}^{b} f(x)\,dx$ is not bounded above and consequently, from (1), $\int_{a+\varepsilon}^{b} \varphi(x)\,dx$ is also not bounded above so that $\int_a^b \varphi(x)\,dx$ does not converge.

9.3.3. Practical Comparison Test

If $f(x)/\varphi(x) \to l$ when $x \to a$, and, l, is neither 0 nor infinite, then the two integrals

$$\int_a^b f(x)\,dx \ \textit{and}\ \int_a^b \varphi(x)\,dx$$

either both converge or both do not converge.

Since $f(x)/\varphi(x)$ positive $\forall\ x$ limit $f(x)/\varphi(x) = l$ cannot be negative.

Let d be a positive number less than l so that there exists a number c, $(a < c < b)$, such that

$$l - \delta < f(x)/\varphi(x) < l + \delta$$

when $a < x \le c$, *i.e.*, $\forall\ x \in]a, c]$. We have $\forall\ x \in]a, c]$,

(*i*) $(l-\delta)\,\varphi(x) < f(x)$ and $f(x) < (l+\delta)\,\varphi(x)$. ...(*i*)

Now

$$\int_a^b f(x)\,dx \text{ converges}$$

$$\Rightarrow \qquad \int_a^c f(x)\,dx \text{ converges};$$

$$\Rightarrow \qquad (l-\delta)\int_a^c f(x)\,dx \text{ converges;}$$

$$\Rightarrow \qquad \int_a^c \varphi(x)\,dx \text{ converges;}$$

$$\Rightarrow \qquad \int_a^b \varphi(x)\,dx \text{ converges.}$$

It may similarly be shown that

$$\int_a^b \varphi(x)\,dx \text{ does not converge } \Rightarrow \int_a^b f(x)\,dx \text{ does not.}$$

Also, from (*ii*), we may prove that

$$\int_a^b \varphi(x)\,dx \text{ converges } \Rightarrow \int_a^b f(x)\,dx \text{ converges,}$$

$$\int_a^b f(x)\,dx \text{ does not converges } \Rightarrow \int_a^b \varphi(x)\,dx \text{ does not.}$$

Ex. Prove that if $f(x)/\varphi(x) \to 0$.

$$\int_a^b \varphi(x)\,dx \text{ converges } \Rightarrow \int_a^b f(x)\,dx \text{ converges}$$

and that if $f(x)/\varphi(x) \to \infty$,

$$\int_a^b f(x)\,dx \text{ converges } \Rightarrow \int_a^b \varphi(x)\,dx \text{ converge.}$$

9.3.4. A Useful Comparison Integral

$$\int_a^b \frac{dx}{(x-a)^n} \qquad \textit{(Dibrugarh 2005)}$$

The improper integral $\int_a^b \frac{dx}{(x-a)^n}$ *converges if and only if* $n < 1$.

We have, if $n \neq 1$,

$$\int_{a+\varepsilon}^b \frac{dx}{(x-a)^n} = \left|\frac{1}{(1-n)(x-a)^{n-1}}\right|_{a+\varepsilon}^b$$

$$= \frac{1}{(1-n)}\left[\frac{1}{(b-a)^{n-1}} - \frac{1}{\varepsilon^{n-1}}\right],$$

which tends to $1/(1-n)(b-a)^{n-1}$ or $+\infty$ according as $n < 1$ or > 1.

Again, if $n = 1$,

$$\int_{a+\varepsilon}^b \frac{dx}{x-a} = \log(b-a) - \log\varepsilon, \text{ which } \to +\infty \text{ as } \varepsilon \to (0+0).$$

Hence the result.

Note. The integral

$$\int_a^b \frac{dx}{(x-a)^n} \qquad ...(1)$$

is proper if $n \le 0$. The notion of the convergence of improper integrals has enabled us to give a meaning to the symbol (1) even for those values of n which are positive but < 1. For $n \ge 1$, the symbol does not represent any number.

9.4. With the help of § 9.3, we deduce two important practical tests for convergence at, a, of the integral of f by comparison with the integral of $\varphi(x) = \dfrac{1}{(x-a)^n}$.

I. *Let f be positive in* $[a, b]$. *Then the integral of f converges at a, if there exists a number,* $n > 0$ *and less than 1 and a fixed positive number k such that* $f(x) \le k/(x-a)^n \ \forall\, x \in \,]a, b]$.

Also, the integral of f does not converge, if, there exists a number $a \ge 1$ *and a fixed positive number k such that* $f(x) \ge k/(x-a)^n$ $\forall\, x \in \,]a, b]$.

II. *If* $\lim\limits_{x \to a} \left[f(x)(x-a)^n \right]$ *exists and is neither 0 nor infinite, then the integral of f converges if and only if* $n < 1$.

EXAMPLES

1. *Examine the convergence of*

(*i*) $\displaystyle\int_0^1 \frac{dx}{x^{\frac{1}{3}}\left(1+x^2\right)}$, (*ii*) $\displaystyle\int_0^1 \frac{dx}{x^2(1+x)^2}$, (*iii*) $\displaystyle\int_0^1 \frac{dx}{x^{\frac{1}{2}}(1-x)^{\frac{1}{2}}}$,

Sol. The integrands are all positive.

(*i*) Here, 0, is the only point of infinite discontinuity of the integrand. We have

$$f(x) = \frac{1}{x^{\frac{1}{3}}\left(1+x^2\right)}.$$

Take

$$\varphi(x) = \frac{1}{x^{1/3}}$$

Now

$$\lim_{x \to 0} \frac{f(x)}{\varphi(x)} = \lim_{x \to 0} \frac{1}{1+x^2} = 1.$$

$$\int_0^1 f(x)\, dx \quad \text{and} \quad \int_0^1 \varphi(x)\, dx$$

have identical behaviours. But $n=\frac{1}{3}$ being less than 1, the latter integral converges by § 9.3.4. Hence the given integral also converges.

(*ii*) Here, 0, is the only point of infinite discontinuity of the given integrand. We have

$$f(x)=\frac{1}{x^2(1+x)^2}$$

Take $$\varphi(x)=\frac{1}{x^2}$$

Now $$\lim_{x\to 0}\frac{f(x)}{\varphi(x)}=1.$$

$$\int_0^1 f(x)\,dx \text{ and } \int_0^1 \varphi(x)\,dx$$

have identical behaviours.

But $n = 2$ being greater than 1, the latter integral does not converge by § 9.3.4. Hence the given integral also does not converge.

(*ii*) Here, 0 and 1, are the two points of infinite discontinuity of the integrand. We have

$$f(x)=\frac{1}{x^{\frac{1}{2}}(1-x)^{\frac{1}{3}}}.$$

We take any number between 0 and 1, say, 1/2, and examine the convergence of the two improper integrals.

$$\int_0^{\frac{1}{2}} f(x)\,dx \text{ and } \int_{1/2}^1 f(x)$$

at 0, and, 1, respectively.

To examine the convergence of the former integral, we take

$$\varphi(x)=\frac{1}{\sqrt{x}}$$

so that $f(x)/\varphi(x)\to 1$ as $x\to 0$.

$\therefore$ By § 9.3.4, $\int_0^{\frac{1}{2}} f(x)\,dx$ converges.

For the latter, we take

$$\psi(x)=\frac{1}{(1-x)^{\frac{1}{3}}}$$

so that $f(x)\psi(x)\to 1$ as $x\to 1$.

$\therefore$ by § 9.3.4, $\int_{\frac{1}{2}}^{1} f(x)\,dx$ converges.

Hence

$$\int_0^1 f(x)\,dx \text{ converges.}$$

2. Beta Function. *Show that*

$$\int_0^1 x^{m-1}(1-x)^{n-1}\,dx$$

exists if and only if m and n, are both positive.

(*Garhwal 93; Ajmer 97, 99, M.D.U. 97*)

Sol. The integral is proper if $m \geq 1$ and $n \geq 1$.

The number, 0 is a point of infinite discontinuity if $m < 1$ and the number, 1, is a point of infinite discontinuity if $n < 1$.

Let $m < 1$ and $n < 1$.

We take any number, say, 1/2 between 0 and 1 and examine the convergence of the two improper integrals

$$\int_0^{\frac{1}{2}} x^{m-1}(1-x)^{n-1}\,dx, \quad \int_{\frac{1}{2}}^{1} x^{m-1}(1-x)^{n-1}\,dx$$

at, 0, 1 respectively.

Convergence at 0. We write

$$f(x) = x^{m-1}(1-x)^{n-1} = (1-x)^{n-1}/x^{1-m}.$$

and take $\varphi(x) = 1/x^{1-m}$, so that

$$f(x)/\varphi(x) \to 1 \text{ as } x \to 0.$$

As

$$\int_0^{\frac{1}{2}} \varphi(x)\,dx = \int_0^{\frac{1}{2}} \frac{1}{x^{1-m}}\,dx$$

is convergent at, 0, if and only if, $1-m<1 \Leftrightarrow 0<m$, we deduce *that the integral*

$$\int_0^{\frac{1}{2}} x^{m-1}(1-x)^{n-1}\,dx$$

is convergent at 0, if and only if, $m > 0$.

Convergence at 1. We write

$$f(x) = x^{m-1}(1-x)^{n-1} = x^{m-1}/(1-x)^{1-n}$$

and take $\varphi(x) = 1/(1-x)^{1-n}$, so that

$$f(x)/\varphi(x) \to 1 \text{ as } x \to 1.$$

As $$\int_{\frac{1}{2}}^{1} \varphi(x)\,dx = \int_{\frac{1}{2}}^{1} \frac{1}{(1-x)^{1-n}}\,dx$$

is convergent, if and only if, $1-n<1 \Leftrightarrow 0<n$, we deduce that the integral

$$\int_{\frac{1}{2}}^{1} x^{m-1}(1-x)^{n-1}\,dx$$

converges if, and only if, $n > 0$.

Thus $\int_0^1 x^{m-1}(1-x)^{n-1}\,dx$ exists for positive values of m, n only. It is a function of two variables called Beta function denoted by $B(m, n)$ and defined for all positive values of m and n.

3. *Show that*

$$\int_0^{\frac{1}{2}\pi} x^m \operatorname{cosec}^n x\,dx \qquad \ldots(i)$$

exists if and only if $n < (m+1)$. *(Garhwal 96)*

Sol. Writing

$$f(x) = \frac{x^m}{(\sin x)^n} = \left(\frac{x}{\sin x}\right)^n \cdot x^{m-n} = \left(\frac{x}{\sin x}\right)^n \frac{1}{x^{n-m}},$$

we see that

$$\lim_{x\to 0} f(x) = \begin{cases} 0 & \text{if } m-n>0 \\ 1 & \text{if } m-n=0 \\ \infty & \text{if } m-n<0 \end{cases}$$

Thus (i) is a proper integral if $(m-n) \geq 0$; and improper if $(m-n) < 0$; 0 being the only point of infinite discontinuity of the integrand in this case.

Let $(m-n) < 0 \; (n-m) > 0$

Take $\varphi(x) = 1/x^{n-m}$.

Then $f(x)/\varphi(x) \to 1$ as $x \to 0$.

Also $$\int_0^{\frac{1}{2}} \varphi(x)\, dx = \int_0^{\frac{1}{2}\pi} \frac{1}{x^{n-m}}\, dx$$

converges if and only if,

$$(n-m) < 1 \Leftrightarrow n < (m-1).$$

Therefore the integral converges if and only if $n < (m+1)$, which also includes the case $n \le m$ when the integral is proper.

4. *Examine the convergence of*

$$\int_0^1 x^{n-1} \log x\, dx. \qquad \text{(Garhwal 93)}$$

The integrand is negative in the interval] 0, 1] and we therefore, consider

$$\int_0^1 -x^{n-1} \log x\, dx = \int_0^1 x^{n-1} \log\left(\frac{1}{x}\right) dx.$$

The integrand is proper if $(n-1) > 0$, inasmuch as the integrand, in that case $\to 0$ as $x \to 0$ and accordingly, 0, is not a point of infinite discontinuity in this case.

Let $(n-1) \le 0$ so that we have now to examine the convergence at 0. Let m be a positive number such that

$$(m+n-1) > 0$$

We have

$$\lim_{x+0} \left[-x^{m+n-1} \log x\right] = 0,$$

so that ε being a given positive number,

$$-x^{m+n-1} \log x < \varepsilon \Rightarrow -x^{n-1} \log x < \varepsilon / x^m$$

for values of x, sufficiently near 0.

Now the integral of ε / x^m converges at 0 if and only if $m < 1$.

It is possible to choose a number $m < 1$ such that

$$(m+n-1) = [(m-1)+n] > 0$$

if and only if $n > 0$. Thus the integral converges if $n > 0$.

When $n = 0$, the integrand becomes $x^{-1} \log x$. We have

$$\int_\varepsilon^1 x^{-1} \log x\, dx = -\frac{(\log \varepsilon)^2}{2} \text{ which} \to -\infty \text{ as} \to 0.$$

When $n < 0$, we have

$$x^{-1} < x^{n-1} \; \forall \; x \in [0, 1],$$

so that in this case also the integral does not converge.

Thus the given integral converges if and only if n > 0.

5. *Show that*

$$\int_0^{\frac{1}{2}\pi} \sin x \log \sin x \, dx$$

is convergent with value log (2/e).

Integrating by parts,

$$\int_\varepsilon^{\frac{1}{2}\pi} \sin x \log \sin x \, dx = \left| -\cos x \log \sin x + \log \tan \frac{1}{2} x + \cos x \right|_\varepsilon^{\frac{1}{2}\pi}$$

$$= -(1 - \cos \varepsilon) \log \sin \frac{1}{2} \varepsilon + \cos \varepsilon \log 2 \cos \frac{1}{2} \varepsilon - \cos \varepsilon + \log \cos \frac{1}{2} \varepsilon.$$

Now, when $\varepsilon \to 0$.

$$\lim (1 - \cos \varepsilon) \log \sin \frac{1}{2} \varepsilon = \lim 2t^2 \log t; \text{ when } t \to 0, \left(t = \sin \frac{1}{2} \varepsilon. \right)$$

$$\therefore \qquad \int_0^{\frac{1}{2}\pi} \sin x \log \sin x \, dx = \log 2 - 1 = \log (2/e).$$

EXERCISES

1. Test the convergence of the following infinite integrals;

(i) $\displaystyle\int_0^{\pi/2} \frac{\log x}{x^n} dx$ (ii) $\displaystyle\int_0^{\pi/2} \frac{\sin x}{x^n}$ (iii) $\displaystyle\int_0^1 \frac{x^{a-1}}{1+x} dx.$

(*Dibrugarh 2005*)

(iv) $\displaystyle\int_0^{\pi/2} \frac{x^{a-1}}{1-x} dx$ (v) $\displaystyle\int_0^2 \frac{\log x}{\sqrt{(2-x)}} dx$ (vi) $\displaystyle\int_0^{\pi/4} \frac{dx}{\sqrt{(\tan x)}}$

(vii) $\displaystyle\int_0^{\pi/2} \sin^{m-1} x \cos^{n-1} x \, dx$ (viii) $\displaystyle\int_0^1 x^{a-1} e^{-x} \, dx$

(ix) $\displaystyle\int_0^3 \frac{dx}{\left[(x-1)^2 (x-2)\right]^{1/5}}$ (x) $\displaystyle\int_0^1 \frac{x^p \log x \, dx}{(1+x)^2}$

(xi) $\displaystyle\int_0^\pi \frac{\sqrt{x}}{\sin x} dx$ (xii) $\displaystyle\int_0^\pi \frac{\sqrt{x}}{\sin x} dx$

2. Discuss the convergence of the integral

$$\int_0^1 x^{m-1} (1-x)^{n-1} \log \left(\frac{1}{x} \right) dx.$$

3. Show that

$$\int_0^{\frac{1}{2}\pi} \log \sin x \, dx.$$ *(Dibrugarh 2005)*

converges.

9.5. *f* (*x*), NOT NECESSARILY POSITIVE

General Test for Convergence. We now obtain a general test for convergence at, *a*, of the infinite integral

$$\int_a^b f(x)\, dx. \qquad ...(1)$$

The necessary and sufficient condition for the convergence of the improper integral (1) at, a, is that to every given $\psi > 0$ *there corresponds* $\delta > 0$ *such that*

$$\left| \int_{a+\varepsilon_1}^{a+\varepsilon_2} f(x)\, dx \right| < \eta,$$

$\forall$ *positive number* $\varepsilon_1, \varepsilon_2$ *less than or equal to* δ.

We write

$$\varphi(\varepsilon) = \int_{a+\varepsilon}^{b} f(x)\, dx.$$

The necessary and sufficient condition for $\lim \varphi(\varepsilon)$ to exist finitely is that to every $\eta > 0$ there corresponds $\delta > 0$ such that

$\forall\, 0 < \varepsilon_1, \varepsilon_2 < \delta$

$$\Rightarrow \quad \left| \varphi(\varepsilon_1) - \varphi(\varepsilon_2) \right| < \eta$$

$$\Leftrightarrow \quad \left| \int_{a+\varepsilon_1}^{b} f(x)\, dx - \int_{a+\varepsilon_2}^{b} f(x)\, dx \right| < \eta$$

$$\Leftrightarrow \quad \left| \int_{a+\varepsilon_1}^{a+\varepsilon_2} f(x)\, dx \right| < \eta.$$

9.5.1. From above, we deduce an Important Sufficient Test for Convergence *viz.*,

$$\int_a^b |f(x)|\, dx \textit{ exists} \Rightarrow \int_a^b f(x)\, dx \textit{ exists.}$$

(Here it is assumed that the proper integral $\int_{a+\varepsilon}^{b} f(x)\, dx$ *exists.*)

The result follows from the general condition of convergence above and the inequality

$$\left| \int_{a+\varepsilon_1}^{a+\varepsilon_2} f(x)\,dx \right| \le \left| \int_{a+\varepsilon_1}^{a+\varepsilon_2} |f(x)|\,dx \right|. \quad (Cor.\ 6\ to\ \S\ 6.7.8,\ page\ 218)$$

Def. Absolute convergence. *The improper integral* $\int_a^b |f(x)|\,dx$ *is said to be absolutely convergent if* $\int_a^b |f(x)|\,dx$ *is convergent.*

From the result proved above, it follows that *Every absolutely convergent integral is convergent.*

Ex. 1. *Test the convergence of* $\int_0^1 \frac{\sin(1/x)}{\sqrt{x}}\,dx.$

Let $f(x) = \sin(1/x)/\sqrt{x}$. Here there is no neighbourhood of the point, 0, in which $f(x)$ constantly keeps the same sign.

Now $\forall\, x \in]0, 1]$, we have

$$\left| \frac{\sin 1/x}{\sqrt{x}} \right| = \frac{|\sin 1/x|}{\sqrt{x}} \le \frac{1}{\sqrt{x}}.$$

Also
$$\int_0^1 \frac{1}{\sqrt{x}}\,dx$$

is convergent. It follows that

$$\int_0^1 \left| \frac{\sin 1/x}{\sqrt{x}} \right| dx \text{ is convergent} \Rightarrow \int_0^1 \frac{\sin 1/x}{\sqrt{x}}\,dx \text{ is absolutely convergent.}$$

9.6. CONVERGENCE AT ∞

Convergence of

$$\int_a^\infty f(x)\,dx,$$

where f is bounded and integrable in $[a, t]$, $\forall\, t \ge a$.

Positive integrand. Let $f(x)$ be positive in $[a, t]$. The necessary and sufficient condition for $\int_a^\infty f(x)\,dx$ to be convergent is that there exists a positive number; k, such that

$$\int_a^t f(x)\,dx \le k\ \forall\, t \ge a.$$

The proof follows immediately from the fact that $f(x)$ being positive $\forall\, x$, the integral $\int_a^t f(x)\,dx$ monotonically increases as t increases and will, therefore, tend to a finite limit or to ∞ according as it is bounded above or not.

From this we immediately deduce that if $f(x)$ and $\varphi(x)$ are both positive and $\varphi(x) \leq f(x) \ \forall x \in [a, \infty[$, then

$$\int_a^{\infty} \varphi(x)\, dx \textit{ converges if } \int_a^{\infty} f(x)\, dx \textit{ converges.}$$

Also it may be shown that *if $f(x)$ and $\varphi(x)$ are positive and when $x \to \infty$, $\lim f(x)/\varphi(x)$ exists and is equal to l, and, l, is neither 0 nor ∞, then*

$$\int_a^{\infty} f(x)\, dx \text{ and } \int_a^{\infty} \varphi(x)\, dx$$

either both converge or both do not converge.

Ex. 1. What conclusion can be drawn if, l, is zero or infinite.

9.6.1. A Useful Comparison Integral

To prove that

$$\int_a^{\infty} \frac{dx}{x^n}, \ (a > 0) \qquad ...(1)$$

converges if and only if $n > 1$.

We have, if $n \neq 1$.

$$\int_a^t \frac{dx}{x^n} = \frac{1}{1-n}\left[\frac{1}{t^{n-1}} - \frac{1}{a^{n-1}}\right]$$

so that when $t \to \infty$, the integral tends to $1/(n-1)a^{n-1}$ or ∞ according as $n > 1$ or $n < 1$.

For $n = 1$, we have

$$\int_a^t \frac{dx}{x} = \log\frac{t}{a} \text{ which} \to \infty \text{ as } t \to \infty.$$

Hence the result.

Adopting (1) as the comparison integral and employing the test of § 9.6, we may now easily obtain the following practical tests for convergence at ∞.

9.6.2. If $f(x)$ is Positive in $[a, \infty[$ then the Integral Converges, if there exists a Positive Number n greater than 1 and a Fixed Positive Number k such that

$$f(x) \leq k/x^n \ \forall x \geq a.$$

Again, the integral does not converge, if there exists a positive number $n \leq 1$ and a fixed positive number k such that

$$f(x) \geq k/x^n \ \forall x \geq a.$$

9.6.3. If, when $x \to \infty$, $\lim \left[f(x) x^n \right]$ exists Finitely, the limit being neither 0 nor Infinite, then the Integral Converges if and only if $n > 1$.

EXAMPLES

1. Examine the convergence of

(i) $\int_0^\infty \frac{x\,dx}{(1+x)^3}$ (ii) $\int_1^\infty \frac{dx}{(1+x)\sqrt{x}}$

(iii) $\int_1^\infty \frac{dx}{x^{\frac{1}{3}}(1+x)^{\frac{1}{2}}}$ (iv) $\int_0^\infty \frac{\sin^2 x}{x^2}\,dx.$ *(Dbrugarh 2004)*

Sol. (i) Let $f(x) = x/(1+x)^3$.

We take $\varphi(x) = x/x^3 = 1/x^2$,

so that
$$\lim_{x\to\infty} \frac{f(x)}{\varphi(x)} = \lim_{x\to\infty} \frac{x^3}{(1+x)^3} = 1.$$

Thus the two integrals

$$\int_1^\infty \frac{x\,dx}{(1+x)^3} \text{ and } \int_1^\infty \frac{1}{x^2}\,dx$$

have identical behaviour for convergence at ∞.

By § 9.6.1 the latter integral is convergent. Accordingly, the given integral *viz.*,

$$\int_1^\infty \frac{x\,dx}{(1+x)^3}$$

is also convergent.

(ii) Let $f(x) = 1/(1+x)\sqrt{x}$.

We take $\varphi(x) = 1/x\sqrt{x} = 1/x^{3/2}$

so that

$$\lim_{x\to\infty} \frac{f(x)}{\varphi(x)} = 1.$$

Also $\int_1^\infty \frac{1}{x^{3/2}}\,dx$

is convergent. Thus

$$\int_1^\infty f(x)\,dx \text{ is convergent.}$$

(*iii*) Let $f(x) = 1/x^{1/3}(1+x)^{1/2}$.

We take $\varphi(x) = 1/x^{1/3}\, x^{1/2} = 1/x^{5/6}$.

We have lim $\dfrac{f(x)}{\varphi(x)} = 1$, when $x \to \infty$, and $\int_1^\infty \varphi(x)$ is not convergent.

$\therefore$ $\int_1^\infty f(x)\,dx$ is divergent.

(*iv*) Let $f(x) = \sin^2 x / x^2$.

Here $x^2 f(x) = \sin^2 x$, which does not tend to a limit but is bounded in $[1, \infty[$.

Now, $\dfrac{\sin^2 x}{x^2} \le \dfrac{1}{x^2}$ and $\int_1^\infty \dfrac{1}{x^2}\,dx$ converges.

$\therefore$ $\int_1^\infty \dfrac{\sin^2 x}{x^2}\,dx$ converges. Also $\int_1^\infty \dfrac{\sin^2 x}{x^2}\,dx$ is a proper integral for $\lim\left(\sin^2 x / x^2\right) = 1$ when $x \to 0$, so that, 0, is not a point of infinite discontinuity. Therefore $\int_0^\infty f(x)\,dx$ is convergent.

2. **Gamma function.** *Show that*

$$\int_0^\infty x^{n-1} e^{-x}\,dx \text{ is convergent if, and only if, } n > 0.$$

(*Garhwal 93, 98; M.D.U. 98, 2001; Dibrugarh 2004*)

Sol. Let $f(x) = x^{n-1} e^{-x}$.

Now 0, is a point of infinite discontinuity of the function *f*, if $n < 1$. Thus we have to examine the convergence at ∞ as well as at, 0.

We consider any positive number > 0, say 1, and examine the convergence of the two integrals

$$\int_0^1 f(x)\,dx \text{ and } \int_1^\infty f(x)\,dx,$$

at, 0, and ∞, respectively.

(*i*) *Convergence at 0.* Let $a < 1$. We take

$$\phi(x) = \frac{1}{x^{1-n}}$$

so that

$$\lim_{x \to 0} \frac{f(x)}{\phi(x)} = 1.$$

Also

$$\int_0^1 \frac{1}{x^{1-n}}\,dx$$

converges, if and only if, $(1-n) < 1 \Leftrightarrow 0 < n$.

(*ii*) *Convergence at* ∞.

Now given n we have for sufficiently large values of x.

$$e^x > x^{n+1} \Leftrightarrow x^{n-1}\,e^{-x} < 1/x^2.$$

But $\displaystyle\int_1^\infty \frac{1}{x^2}\,dx$ converges.

Therefore $\displaystyle\int_1^\infty x^{n-1}\,e^{-x}\,dx$ also converges $\forall\, n$.

Thus $\displaystyle\int_0^\infty x^{n-1}\,e^{-x}\,dx$ *converges if, and only if,* $n > 0$.

The integral $\displaystyle\int_0^\infty x^{n-1}\,e^{-x}\,dx$, called **Gamma Function**, is denoted by $\Gamma(n)$. This function is defined only for positive values of n.

3. *Show that*

$$\int_0^\infty \left(\frac{1}{x} - \frac{1}{\sinh x}\right)\frac{dx}{x} \qquad \textit{Garhwal 97; Dibrugarh 2004,}$$

is convergent.

Sol. The point, 0, is not a point of infinite discontinuity of the integrand inasmuch as the integrand $\to \frac{1}{6}$ as $x \to 0$. We have, therefore, to examine convergence at ∞ only. We have

$$\left(\frac{1}{x} - \frac{1}{\sinh x}\right)\frac{1}{x} = \frac{e^x - e^{-x} - 2x}{x^2\left(e^x - e^{-x}\right)}$$

$$< \frac{e^x}{x^2\left(e^x - e^{-x}\right)}, \forall\, x > 0.$$

$$= \frac{e^{2x}}{e^{2x}-1}\cdot\frac{1}{x^2}.$$

Since $e^{2x}/\left(e^{2x}-1\right) \to 1$ as $x \to \infty$, there exists a number k such that

$$e^{2x}/\left(e^{2x}-1\right) < \frac{3}{2}, \forall\, x \geq k.$$

Thus $\forall\, x \geq k$,

$$\left(\frac{1}{x}-\frac{1}{\sinh x}\right)\frac{1}{x}<\frac{3}{2}\cdot\frac{1}{x^2},$$

$$\int_k^{\infty}\left(\frac{1}{x}-\frac{1}{\sinh x}\right)\frac{1}{x}\,dx \text{ is convergent}$$

the given integral is convergent.

4. *Show that the integral*

$$\int_0^{\infty}\left(\frac{1}{1+x}-e^{-x}\right)\frac{dx}{x} \quad \textit{is convergent.}$$

Sol. The integrand $\dfrac{e^x-1-x}{x(1+x)e^x}$ is positive, $\forall\, x>0$.

Also, 0, is not a point of infinite discontinuity for the integrand tends to 0 as $x\to 0$.

For convergence at ∞ we rewrite the integrand as

$$\left(\frac{1}{1+x}-e^{-x}\right)\frac{1}{x}=\frac{e^x-(1+x)}{e^x}\cdot\frac{1}{1+x}\cdot\frac{1}{x}$$

and note that $\left[e^x-(1+x)\right]/e^x\to 1$ as $x\to\infty$. Taking $\phi(x)=1/x$, we may see that the given integral is convergent.

EXERCISES

1. Discuss the convergence of the following :

(i) $\displaystyle\int_0^{\infty}\frac{x^m(1+x^n)}{1+x^p}\,dx$ (ii) $\displaystyle\int_0^{\infty}\frac{x^{2m}}{1+x^{2n}}\,dx,\ (m>0, n>0)$

(Garhwal 94, 99)

(iii) $\displaystyle\int_0^{\infty}\frac{x^{m-1}+x^{n-1}}{1-x}\,dx$ (iv) $\displaystyle\int_1^{\infty}x^{m-1}e^{-nx}\,dx$

(v) $\displaystyle\int_0^{\infty}x^m(\log x)^n\,dx$ (vi) $\displaystyle\int_0^{\alpha}\frac{x^a}{(1+x)^b[1+(\log x)^2]}\,dx.$

2. Show that the improper integral

$$\int_0^{\infty}\log(1+2\,\text{sech}\,x)\,dx,$$

converges.

$\left[\text{Since } \log(1+x)<x\ \forall\, x>0\right]$

$$\log(1+2\,\text{sech}\,x) < 2\,\text{sech}\,x = \frac{22}{e^x+e^{-x}} < \frac{4}{e^x} = 4e^{-x}.\Bigg]$$

3. Show that

$$\int_0^\infty \frac{\cosh bt}{\cosh at}\,dt \qquad a>0, b>0$$

converges if, and only if, $b < a$.

$$\left[\begin{array}{l} b<a \Rightarrow \dfrac{\cosh bt}{\cosh at} = \dfrac{e^{bt}+e^{-bt}}{e^{at}+e^{-at}} < \dfrac{e^{bt}+e^{-bt}}{e^{at}} = 2e^{-(a-b)t} \\ b>a \Rightarrow \dfrac{\cosh bt}{\cosh at} = \dfrac{e^{bt}+e^{-bt}}{e^{at}+e^{-at}} > \dfrac{e^{bt}}{e^{at}+e^{at}} = \dfrac{1}{2}e^{(b-a)t} \end{array}\right]$$

4. Show that

$$\int_0^\infty \frac{\sinh bx}{\sinh ax}\,dx \qquad a>0, b>0,$$

converges if and only if, $a > b$,

$$\left[\begin{array}{l} a>b \Rightarrow \dfrac{\sinh bx}{\sinh ax} = \dfrac{e^{bx}-e^{-bx}}{e^{ax}-e^{-ax}} < \dfrac{b\;e^x}{e^{ax}-1}; \\ a<b \Rightarrow \dfrac{\sinh bx}{\sinh ax} = \dfrac{e^{bx}-e^{-bx}}{e^{ax}-e^{-ax}} < \dfrac{e^{bx}-1}{e^{ax}}. \end{array}\right]$$

5. Discuss the convergence of the improper integral

$$\int_0^\infty \left(\frac{1}{e^x-1}-\frac{1}{x}+\frac{1}{2}\right)\frac{e^{-kx}}{x}\,dx.$$

6. Show that the integrals

$$\int_{-\infty}^\infty e^{-x^2}\,dx \text{ and } \int_{-\infty}^\infty e^{-(x-a/x)^2}\,dx$$

converge.

9.7. CONVERGENCE AT ∞, THE INTEGRAND BEING NOT NECESSARILY POSITIVE

General test for convergence. *The necessary and sufficient condition for the convergence of*

$$\int_a^\infty f(x)\,dx$$

at ∞ is that corresponding to every $\eta > 0$ there exists a number k, such that

$$\left| \int_{t_1}^{t_2} f(x)\, dx \right| < \eta \qquad \forall\, t_1, t_2 \geq k.$$

We write
$$\varphi(t) = \int_a^t f(x)\, dx.$$

The necessary and sufficient condition for lim $\varphi(x)$ to exist finitely is that to every positive number, η, there corresponds a number k, such that, $\forall\, t_1, t_2 > k$

$$|\varphi(t_2) - \varphi(t_1)| < \eta,$$

$$\Leftrightarrow \left| \int_a^{t_2} f(x)\, dx - \int_a^{t_1} f(x)\, dx \right| < \eta \Leftrightarrow \left| \int_{t_1}^{t_2} f(x)\, dx \right| < \eta.$$

9.8. THEOREM

If $\int_a^\infty |f(x)|\, dx$ *converges, then* $\int_a^\infty f(x)\, dx$ *also converges.*

Let η be a positive number. Since $\int_a^\infty |f(x)|\, dx$ converges, there exists a number k such that $\forall\, t_1, t_2 \geq k$

$$\left| \int_{t_1}^{t_2} |f(x)|\, dx \right| < \eta \qquad ...(1)$$

Also,
$$\left| \int_{t_1}^{t_2} f(x)\, dx \right| \leq \left| \int_{t_1}^{t_2} |f(x)|\, dx \right|. \qquad ...(2)$$

From (1) and (2) the result follows.

Def. Absolute convergence. *An improper integral* $\int_a^\infty f(x)\, dx$ *is said to be absolutely convergent, if* $\int_a^\infty |f(x)|\, dx$ *is convergent.*

From the theorem above it follows that *every absolutely convergent improper integral is convergent.*

It will later on be seen that the converse is not necessarily true (See solved Ex. 1, page 313, at the end of this chapter.)

9.8.1. Test for the Absolute Convergence of the Integral of a Product

Let φ *be bounded in* $[a, \infty[$ *and integrable in* $[a, t]$ $\forall\, t \geq a$.

Let $\int_a^\infty f(x)\, dx$ *converge absolutely at* ∞. *Then*

$$\int_a^\infty f(x)\, \varphi(x)\, dx,$$

is absolutely convergent.

Since φ is bounded in $[a, \infty[$ there exists a number A such that

$$|\varphi(x)| \leq A\ \forall\, x \geq a. \qquad ...(1$$

Since the improper integral $\int_a^\infty \left| f(x) \right| dx$ with positive integrand is convergent, there exists a number B, such that

$$\int_a^t \left| f(x) \right| dx \le B \;\forall\; t \ge a. \qquad \text{...(2)}$$

We have from (1) and (2)

$\left| f(x)\,\varphi(x) \right| \le A \left| f(x) \right| \forall\; x \ge a.$

$$\Rightarrow \quad \int_a^t \left| f(x)\,\varphi(x) \right| dx \le A \int_a^t \left| f(x) \right| dx \le AB \;\forall\; t \ge a,$$

$$\Rightarrow \quad \int_a^t \left| f(x)\,\varphi(x) \right| dx \text{ is bounded above } \forall\; t \ge a.$$

$$\Rightarrow \quad \int_a^\infty \left| f(x)\,\varphi(x) \right| dx \text{ is convergent}$$

$$\Rightarrow \quad \int_a^\infty f(x)\,\varphi(x)\, dx \text{ is absolutely convergent.}$$

Ex. Discuss the convergence of the following integrals :

(*i*) $\int_1^\infty \frac{\sin x}{x^2} dx.$ (*ii*) $\int_0^\infty e^{-ax} \cos x \, dx.$

9.9. TESTS FOR CONDITIONAL CONVERGENCE

9.9.1. Abel's Theorem for the Convergence of the Integral of a Product

$\int_a^\infty f(x)\,dx$ *be convergent.*

Then

$\int_a^\infty f(x)\,\varphi(x)\,dx$ *is convergent.* (*M.D.U. 2001*)

The bounded function φ, which is monotonic in $[a, \infty[$, is integrable in $[a, t] \;\forall\; t \ge a.$ Applying the second mean value theorem, we have

$$\int_{t_1}^{t_2} f(x)\,\varphi(x)\,dx = \varphi(t_1) \int_{t_1}^{\xi} f(x)\,dx + \varphi(t_2) \int_{\xi}^{t_2} f(x)\,dx, \qquad \text{...(1)}$$

where $a < t_1 \le \xi \le t_2$.

Let η be a positive number.

Since φ is bounded in $[a, \infty[$, there exists a positive number A such that

$$\left| \varphi(x) \right| \le a \;\forall\; x \ge a.$$

In particular it follows that

$$\left| \varphi(t_1) \right| \le A, \left| \varphi(t_2) \right| \le A. \qquad \text{...(2)}$$

Also, since $\int_a^\infty f(x)\,dx$ is convergent, there exists by § 9.7, a number k such that $\forall\, t_1, t_2 > k$

$$\left| \int_{t_1}^{t_2} f(x)\,dx \right| < \frac{\eta}{2A}.$$

We now suppose that t_1, t_2 are number $\geq k$ so that ξ which lies between t_1 and t_2 is also $\geq k$

$$\therefore \qquad \left| \int_{t_1}^{\xi} f(x)\,dx \right| < \frac{t_1}{2A} \text{ and } \left| \int_{\xi}^{t_2} f(x)\,dx \right| < \frac{\eta}{2A} \qquad ...(3)$$

From (1), (2) and (3), we deduce that there exists a number k such that $\forall\, t_1, t_2 \geq k$

$$\left| \int_{t_1}^{t_2} f(x)\,\varphi(x)\,dx \right| \leq \left| \varphi(t_1) \right| \left| \int_{t_1}^{\xi} f(x)\,dx \right| + \varphi(t_2) \left| \left| \int_{\xi}^{t_2} f(x)\,dx \right| \right.$$

$$< A.\frac{\eta}{2A} + A.\frac{\eta}{2A} = \eta.$$

where η is a positive number assigned arbitrarily.

$$\int_a^\infty f(x)\,\varphi(x)\,dx \text{ converges at } \infty.$$

9.9.2 Dirichilet's Theorem for the Convergence of the Integral of a Product (Kurukshetra 2000, M.D.U. 97)

Let φ be bounded and monotonic in $[a, \infty[$ and let $\varphi(x) \to 0$, when $x \to \infty$. Also let $\int_a^t f(x)\,dx$ be bounded $\forall\, t \geq a$.

Then

$$\int_a^\infty f(x)\,\varphi(x)\,dx$$

is convergent.

The function φ, which is monotonic in $[a, \infty[$, is integrable in $[a, t]\ \forall\, t \geq a$. Applying the second mean value theorem, we have

$$\int_{t_1}^{t_2} f(x)\,\varphi(x)\,dx = \varphi(t_1) \int_{t_1}^{\xi} f(x)\,dx + \varphi(t_2) \int_{\xi}^{t_2} f(x)\,dx, \qquad ...(1$$

where $a < t_1 \leq \xi \leq t_2$.

Since $\int_a^t f(x)\,dx$ is bounded $\forall\, t \geq a$ there exists a number A suc that

$$\left| \int_a^t f(x)\,dx \right| \leq A\ \forall\, t \geq a, \qquad ...(2$$

$$\therefore \quad \left|\int_{t_1}^{\xi} f(x)\,dx\right| = \left|\int_a^{\xi} f(x)\,dx - \int_a^{t_1} f(x)\,dx\right|$$

$$= \left|\int_a^{\xi} f(x)\,dx\right| + \left|\int_a^{t_1} f(x)\,dx\right|$$

$$\leq A + A = 2A.$$

Similarly $$\left|\int_{\xi}^{t_2} f(x)\,dx\right| \leq 2A. \qquad ...(3)$$

Let η be a positive number.

Since $\varphi(x) \to 0$ as $x \to \infty$, there exists a number k such that

$$|\varphi(x)| < \eta / 4A \;\forall\; x \geq k. \qquad ...(4)$$

We now suppose that t_1, t_2 are any two numbers $\leq k$, so that from (4) $\forall\, t_1, t_2 \geq k$

$$|\varphi(t_1)| < \eta / 4A, \; |\varphi(t_2)| < \eta / 4A \qquad ...(5)$$

From (1), (2), (3) and (5), we deduce that there exists a number k such that $\forall\, t_1, t_2 \geq k$

$$\left|\int_{t_1}^{t_2} f(x)\,\varphi(x)\,dx\right| \leq |\varphi(t_1)|\left|\int_{t_1}^{\xi} f(x)\,dx\right| + \varphi(t_2)\left|\left|\int_{\xi}^{t_2} f(x)\,dx\right|\right.$$

$$\leq (\eta / 4A).2A + (\eta / 4A).2A = \eta$$

where η is a positive number arbitrarily assigned.

$\Rightarrow$ $$\int_a^{\infty} f(x)\,\varphi(x)\,dx \text{ converges at } \infty.$$

EXAMPLES

1. *Show that*

$$\int_a^{\infty} \frac{\sin x}{x}\,dx \qquad ...(1)$$

is convergent. *(Ajmer 98; Garhwal 93, 95; Dibrugarh 2005*

Sol. Since the integrand $\to 1$, as $x \to 0$, therefore, 0, is not a point of infinite discontinuity. Now consider the improper integral

$$\int_0^{\infty} \frac{\sin x}{x}\,dx \qquad ...(2)$$

The factor $1/x$ of the integrand is monotonic and $\to 0$ as $x \to \infty$ and

$$\left| \int_1^t \sin x \, dx \right| = \left| -\cos t + \cos 1 \right| \leq \left| \cos t \right| + \left| \cos 1 \right| < 2 \; \forall \, t$$

Hence by § 9.9.2, the integral (2) is convergent. Also since

$$\int_0^1 \frac{\sin x}{x} \, dx \qquad \text{...(2)}$$

is only a proper integral, we see that the given integral (1) is convergent.

2. *Show that* $\int_0^\infty \sin x^2 \, dx$ *is convergent.* (*Garhwal 94; M.D.U. 97*)

Sol. We write

$$\sin x^2 = \frac{1}{2x} 2x \sin x^2$$

and consider the improper integral

$$\int_1^\infty \sin x^2 \, dx, \text{ i.e.}, \int_1^\infty \frac{1}{2x} . 2x \sin x^2 \, dx.$$

Now, $1/2x$ is monotonic and $\to 0$ as $x \to \infty$. Also,

$$\left| \int_1^t 2x \sin x^2 \, dx \right| = \left| -\cos t^2 + \cos 1 \right| < 2.$$

so that

$$\int_1^t 2x \sin x^2 \, dx$$

is bounded $\forall \, t \geq 1$. Hence

$$\int_1^\infty \frac{1}{2x} . 2x \sin x^2 \, dx, \text{ i.e.}, \int_1^\infty \sin x^2 \, dx$$

is convergent.

Since $\int_0^1 \sin x^2 \, dx$ is only a proper integral, we see that the given integral is convergent.

3. *Show that*

$$\int_0^\infty e^{-ax} \frac{\sin x}{x} \, dx, \; a \geq 0,$$ (*Kurukshetra 2000*)

is convergent.

Sol. Here e^{-ax} is monotonic and bounded and $\int_0^\infty (\sin x / x) \, dx$ is convergent. Hence we obtain the result by applying Abel's test.

EXERCISES

1. Test the following for convergence :

(*i*) $\displaystyle\int_a^\infty \frac{\sin kx}{x}\,dx$ (*ii*) $\displaystyle\int_a^\infty \frac{\sin x}{\sqrt{x}}\,dx$

(*iii*) $\displaystyle\int_a^\infty e^{-a^2x^2}\sin 2bx\,\frac{dx}{x}$ (*iv*) $\displaystyle\int_0^\infty \frac{\cos ax\cos bx}{x}\,dx$

(*v*) $\displaystyle\int_0^\infty \frac{\sin x^m}{x^n}\,dx$ (*vi*) $\displaystyle\int_0^\infty \frac{\sin x\,(1-\cos x)}{x^n}\,dx$

(*vii*) $\displaystyle\int_0^\infty \frac{\sin (x+x^2)}{x^n}\,dx$ (*viii*) $\displaystyle\int_0^\infty \frac{x^m\cos ax}{1+x^n}\,dx$

2. Discuss the convergence or divergence of the following :

(*i*) $\displaystyle\int_0^\infty \frac{\cos x}{\sqrt{(x^2+x)}}\,dx$ (*ii*) $\displaystyle\int_0^\infty e^{-a^2x^2}\cos bx\,dx$

(*iii*) $\displaystyle\int_1^\infty \sin x^p\,dx$ (*iv*) $\displaystyle\int_1^\infty \frac{x^p\sin^2 x}{1+x^2}\,dx$

3. Show that the integrals

$$\int_1^\infty \frac{\sin x}{x^p}\,dx \text{ and } \int_1^\infty \cos\left(x^{1+p}\right)dx$$

are convergent if $p > 0$.

EXAMPLES

1. *Show that the improper integral*

$$\int_0^\infty \frac{\sin x}{x}\,dx$$

is not absolutely convergent.

Sol. We have to show that

$$\int_0^\infty \frac{|\sin x|}{x}\,dx$$

is not convergent. Consider the proper integral

$$\int_0^{n\pi} \frac{|\sin x|\,dx}{x}$$

where n is a positive integer. We have

$$\int_0^{n\pi} \frac{|\sin x|}{x}\,dx = \sum_{r=1}^{n} \int_{(r-1)\pi}^{r\pi} \frac{|\sin x|}{x}\,dx.$$

We put $x = (r-1)\pi + y$ so that y varies in $[0, \pi]$. We have

$$\left|\sin\left[(r-1)\pi + y\right]\right| = \left|(-1)^{r-1}\sin y\right| = \sin y.$$

$$\therefore \quad \int_{(r-1)\pi}^{r\pi} \frac{|\sin x|}{x}\,dx = \int_0^{\pi} \frac{\sin y}{(r-1)\pi + y}\,dx,$$

Since, $r\pi$ is the maximum value of $\left[(r-1)\pi + y\right]$ when $y \in [0, \pi]$, we have

$$\int_0^{\pi} \frac{\sin y}{(r-1)\pi + y}\,dy \geq \frac{1}{r\pi}\int_0^{\pi} \sin y\,dy = \frac{2}{r\pi}$$

$$\therefore \quad \int_0^{n\pi} \frac{|\sin x|}{x}\,dx \geq \sum_1^n \frac{2}{r\pi} = \frac{2}{\pi}\sum_1^n \frac{1}{r}.$$

Since $\sum_1^n \frac{1}{r} \to \infty$ as $n \to \infty$, we see that

$$\int_0^{n\pi} \frac{|\sin x|}{x}\,dx \to \infty \text{ as } n \to \infty.$$

Let, now, t be a real number. There exists a positive integer n such that

$$n\pi \leq t < (n+1)\pi.$$

We have

$$\int_0^{t} \frac{|\sin x|}{x}\,dx \geq \int_0^{n\pi} \frac{|\sin x|}{x}\,dx.$$

Let $t \to \infty$ so that n also $\to \infty$. Thus we see that

$$\int_0^{t} \frac{|\sin x|}{x}\,dx \to \infty \Rightarrow \int_0^{\infty} \frac{|\sin x|}{x}\,dx \text{ does not converge.}$$

2. *Show that the improper integral*

$$\int_0^{\infty} \frac{x\,dx}{1 + x^6 \sin^2 x}$$

is convergent.

Sol. The integral is positive for positive values of ξ but the test obtained in § 9.6, page 330 does not enable us to establish the convergence. In order to show that the integral converges, we proceed as follows. Consider the proper integral

$$\int_0^{n\pi} \frac{x\,dx}{1 + x^6 \sin^2 x},$$

and write

$$\int_0^{n\pi} \frac{x\,dx}{1+x^6\sin^2 x} = \sum_{r=1}^{n} \int_{(r-1)\pi}^{r\pi} \frac{x\,dx}{1+x^6\sin^2 x}.$$

Now $\forall\, x \in \left[(r-1)\pi, r\pi\right]$, we have

$$\frac{x}{1+x^6\sin x^2} \leq \frac{r\pi}{1+(r-1)^6\pi^6\sin^2 x}.$$

$$\therefore \quad \int_{(r-1)\pi}^{r\pi} \frac{x\,dx}{1+x^6\sin^2 x} \leq \int_{(r-1)\pi}^{r\pi} \frac{r\pi\,dx}{1+(r-1)^6\pi^6\sin^2 x} = a_r, \text{ say}$$

Putting $x = (r-1)\pi + y$, we see that

$$a_r = \int_0^{\pi} \frac{r\pi\,dy}{1+(r-1)^6\pi^6\sin^2 y} = 2\int_0^{\pi/2} \frac{r\pi\,dy}{1+(r-1)^6\pi^6\sin^2 y}.$$

If, now, A is a positive number, we have

$$\int_0^{\pi/2} \frac{dy}{1+A\sin^2 y} = \int_0^{\pi/2} \frac{\operatorname{cosec}^2 y\,dy}{A+1+\cot^2 y}$$

$$= -\frac{1}{\sqrt{(A+1)}} \left| \tan^{-1} \frac{\cot y}{\sqrt{(A+1)}} \right|_0^{\pi/2} = \frac{\pi}{2}\cdot\frac{1}{\sqrt{(A+1)}}$$

$$\therefore \quad a_r = 2.r\pi.\frac{\pi}{2}.\frac{1}{\sqrt{\left[(r-1)^6\pi^6+1\right]}} = \frac{r\pi^2}{\sqrt{\left[(r-1)^6\pi^6+1\right]}}$$

$$\therefore \quad \int_{(r-1)\pi}^{r\pi} \frac{x}{1+x^6\sin^2 x}\,dx \leq \frac{r\pi^2}{\sqrt{\left[1+(r-1)^6\pi^6\right]}} < \frac{r}{(r-1)^3}\cdot\frac{1}{\pi}\cdot(r \neq 1).$$

Since $\dfrac{r}{(r-1)^3} = \dfrac{1}{(r-1)^2} + \dfrac{1}{(r-1)^3}$,

the two infinite series

$$\sum_{r=2}^{\infty} \frac{1}{(r-1)^2} \text{ and } \sum_{r=2}^{\infty} \frac{1}{(r-1)^3}$$

are both convergent. It follows that

$$\sum_{r=2}^{\infty} \frac{r}{(r-1)^3}\cdot\frac{1}{\pi} \text{ is a convergent series.}$$

$$\therefore \quad \int_0^{n\pi} \frac{x\,dx}{1+x^6 \sin^2 x} \to \text{a finite limit as } n \to \infty.$$

$$\int_0^{\infty} \frac{x\,dx}{1+x^6 \sin^2 x} \text{ converges.}$$

3. *Show that*

$$\int_0^{\infty} \frac{x\,dx}{1+x^4 \sin^2 x} \textit{ is divergent.}$$

Sol. We write

$$\int_0^{n\pi} \frac{x\,dx}{1+x^4 \sin^2 x} = \sum_{r=1}^{n} \int_{(r-1)\pi}^{r\pi} \frac{x\,dx}{1+x^4 \sin^2 x}$$

Now $\forall\, x \in \left[(r-1)\pi, r\pi\right]$, we have

$$\frac{(r-1)\pi}{1+(r\pi)^4 \sin^2 x} \le \frac{x}{1+x^4 \sin^2 x}$$

$$\therefore \quad \int_{(r-1)\pi}^{r\pi} \frac{(r-1)\pi}{1+(r\pi)^4 \sin^2 x}\,dx \le \int_{(r-1)\pi}^{r\pi} \frac{x}{1+x^4 \sin^2 x}\,dx.$$

Now,

$$\int_{(r-1)\pi}^{r\pi} \frac{(r-1)\pi}{1+(r\pi)^4 \sin^2 x}\,dx = (r-1)\pi \int_0^{\pi} \frac{dy}{1+(r\pi)^4 \sin^2 y}.$$

$$= 2\,(r-1)\pi \int_0^{\pi/2} \frac{dy}{1+(r\pi)^4 \sin^2 y}.$$

$$= 2\,(r-1)\pi\, \frac{\pi}{2}\, \frac{1}{\sqrt{\left[1+(r\pi)^4\right]}} = \frac{(r-1)\pi^2}{\sqrt{\left[1+(r\pi)^4\right]}}.$$

The infinite series $\Sigma \dfrac{(r-1)\pi^2}{\sqrt{[1+(r\pi)^4]}}$ diverges, as we may see on comparison with the series $\Sigma\,(1/r)$. Hence the given integral does not converge.

4. Frullani's Integral. *If* φ *is continuous in* $[\,0, \infty\,[$ *and*

$$\lim_{x \to 0} \varphi(x) = \varphi_0 \text{ and } \lim_{x \to 0} \varphi(x) = \varphi_1,$$

then $$\int_0^{\infty} \frac{\varphi(ax) - \varphi(bx)}{x}\,dx = (\varphi_0 - \varphi_1) \log \frac{b}{a}.$$

Sol. We consider the proper integral

$$\int_{\varepsilon}^{t} \frac{\varphi(ax)-\varphi(bx)}{x}\,dx,$$

where ultimately we shall make, ε, tend to 0, and t, to ∞.

We have

$$\int_{\varepsilon}^{t} \frac{\varphi(ax)-\varphi(bx)}{x}\,dx = \int_{\varepsilon}^{t} \frac{\varphi(ax)}{x}\,dx - \int_{\varepsilon}^{t} \frac{\varphi(bx)}{x}\,dx.$$

Putting $ax = u$ and $bx = u$ separately in the two integrals on the right, we obtain

$$\int_{\varepsilon}^{t} \frac{\varphi(ax)-\varphi(bx)}{x}\,dx = \int_{a\varepsilon}^{at} \frac{\varphi(u)}{u}\,du - \int_{b\varepsilon}^{bt} \frac{\varphi(u)}{u}\,du$$

$$= \int_{a\varepsilon}^{b\varepsilon} \frac{\varphi(u)}{u}\,du - \int_{at}^{bt} \frac{\varphi(u)}{u}\,du \qquad \text{...(1)}$$

Applying the first mean value theorem to the two integrals on the right, we see that there exists two numbers ξ, η belonging to the intervals $[a_{\varepsilon}, b\varepsilon], [at, bt]$ respectively such that

$$\int_{\varepsilon}^{t} \frac{\varphi(u)}{u}\,du = \varphi(\xi) \int_{a\varepsilon}^{b\varepsilon} \frac{du}{u} = \varphi(\xi) \log \frac{b}{a}.$$

$$\int_{at}^{bt} \frac{\varphi(u)}{u}\,du = \varphi(\eta) \int_{a\varepsilon}^{b\varepsilon} \frac{du}{u} = \varphi(\eta) \log \frac{b}{a}.$$

$$\therefore \quad \int_{\varepsilon}^{t} \frac{\varphi(ax)-\varphi(bx)}{x}\,dx = [\varphi(\xi)-\varphi(\eta)] \log \frac{b}{a}.$$

Let, now, $\varepsilon \to 0$ and $t \to \infty$. Then, by the given conditions $\varphi(\xi) \to \varphi_0$ and $\varphi(\eta) \to \varphi_1$ when $\xi \to 0$ and $\eta \to \infty$. Thus we see that

$$\int_{0}^{\infty} \frac{\varphi(ax)-\varphi(bx)}{x}\,dx = (\varphi_0 - \varphi_1) \log \frac{b}{a}.$$

Remarks. The following points are worthy of notice :

If $\int_0^1 \frac{\varphi(x)}{x}\,dx$ is convergent at, 0, then, by the general principle of convergence (§ 9.5, page 302)

$$\lim_{\varepsilon \to 0} \int_{a\varepsilon}^{b\varepsilon} \frac{\varphi(x)}{x}\,dx = 0.$$

Also if $\int_1^{\infty} \frac{\varphi(x)}{x} dx$ is convergent at, ∞, then by the general principle of convergence (§ 9.7, page 309)

$$\lim_{t \to \infty} \int_{at}^{bt} \frac{\varphi(x)}{x} dx = 0.$$

Then we obtain from (1), page 318 the following results :

EXERCISES

1. If $\int_0^{\infty} \frac{\varphi(x)}{x} dx$ is convergent at, 0, and $\lim_{n \to \infty} \varphi(x) = \varphi_1$ then

$$\int_0^{\infty} \frac{\varphi(ax) - \varphi(bx)}{x} dx = \varphi_1 \log \frac{a}{b}.$$

2. If $\int_0^{\infty} \frac{\varphi(x)}{x} dx$ is convergent at ∞ and $\lim_{n \to 0} \varphi(x) = \varphi_0$, then

$$\int_0^{\infty} \frac{\varphi(ax) - \varphi(bx)}{x} dx = \varphi_0 \log \frac{b}{a}.$$

3. If $\int_0^{1} \frac{\varphi(x)}{x} dx$ and $\int_1^{\infty} \frac{\varphi(x)}{x} dx$ are both convergent at 0 and ∞ respectively, then

$$\int_1^{\infty} \frac{\varphi(ax) - \varphi(bx)}{x} dx = 0.$$

4. Show that

(*i*) $\int_0^{\infty} \frac{e^{-ax} - e^{-bx}}{x} dx = \log \frac{b}{a}$ $(a > 0,\ b > 0)$

(*ii*) $\int_0^{\infty} \frac{\cos ax - \cos bx}{x} dx = \log \frac{b}{a}.$

5. *Prove that*

$$\lim_{k \to \infty} \int_0^{\infty} \frac{dx}{1 + kx^{10}} = 0.$$

We write

$$\int_0^{\infty} \frac{dx}{1 + kx^{10}} = \int_0^{\varepsilon} \frac{dx}{1 + kx^{10}} + \int_{\varepsilon}^{\infty} \frac{dx}{1 + kx^{10}},$$

where ε is a given positive number. Now, we have

$$\int_0^{\varepsilon} \frac{dx}{1+kx^{10}} < \varepsilon,$$

and
$$\int_0^{\infty} \frac{dx}{1+kx^{10}} < \int_{\varepsilon}^{\infty} \frac{dx}{kx^{10}} = \frac{1}{9k\varepsilon^9} < \varepsilon \text{ if } k > \frac{1}{9\varepsilon^{10}}.$$

$$\therefore \quad 0 < \int_0^{\infty} \frac{dx}{1+kx^{10}} < 2\varepsilon, \text{ if } k > \frac{1}{9\varepsilon^{10}}.$$

Hence the result.

6. *Prove that as* $p \to 0$ *through positive values, then*

$$p\sum_{r=1}^{\infty} \frac{1}{r^{1+p}} \to 1.$$

Sol. Since $(1 + p) > 1$, the infinite series $\Sigma\,(1/r^{1+p})$ is convergent.

Let
$$S_n = \sum_{r=1}^{n} \frac{1}{r^{1+p}}; \; S = \sum_{r=1}^{\infty} \frac{1}{r^{1+p}}. \text{ We have}$$

$$\frac{1}{2^{1+p}} \le \int_1^2 \frac{1}{x^{1+p}}\,dx \le \frac{1}{1^{1+p}}.$$

$$\frac{1}{3^{1+p}} \le \int_2^3 \frac{1}{x^{1+p}}\,dx \le \frac{1}{2^{1+p}};$$

....

....

$$\frac{1}{n^{1+p}} \le \int_{n-1}^{n} \frac{1}{x^{1+p}} \le \frac{1}{(n-1)^{1+p}}$$

$$\therefore \quad (S_n - 1) \le \int_1^n \frac{dx}{x^{1+p}} = \frac{1}{p}\left(1 - \frac{1}{n^p}\right) \le \left(S_n - \frac{1}{n^{1+p}}\right).$$

Let $n \to \infty$.

$$\therefore \quad (S-1) \le 1/p \le S \text{ or } 1 \le pS \le (p+1).$$

$$\lim pS = 1 \text{ as } p \to 0\,(0+0).$$

EXERCISES

1. Show that the following improper integrals converge :

(*i*) $\int_0^1 \log x \log(1+x)\,dx$ (*ii*) $\int_0^1 \frac{\log x}{\sqrt[3]{(1-x)}}\,dx$

(*iii*) $\int_0^{\infty} \frac{x\,dx}{(1+x^2)\sqrt[5]{\sin x}}$ (*iv*) $\int_0^{\infty} e^{-a^2 x} \log \cos^2 x\,dx. \quad (a \ne 0)$

(v) $\displaystyle\int_0^\infty \frac{e^{-ax}}{\sqrt[3]{\sin x}}\,dx, (a>0)$ (vi) $\displaystyle\int_0^1 \frac{x^m + x^{-m}}{x}\log(1+x)\,dx,\ |m|<1$

(vii) $\displaystyle\int_0^\infty x^n e^{-a^2x^2 - b^2/x^2}\,dx.\ (n>-1).$

(viii) $\displaystyle\int_0^\infty e^{-x} x^{n-1} (\log x)^m\,dx.\ (n>0 \text{ and } m\in \mathrm{N}).$

(ix) $\displaystyle\int_0^\infty \frac{x\cosh ax}{\sinh x}\,dx, |a|<1$ (x) $\displaystyle\int_0^\infty \frac{e^x}{e^{4x}\sin^2 x + \cos^2 x}\,dx$

(xi) $\displaystyle\int_0^\infty \frac{\cosh ax \cosh bx}{\cosh x}\,dx.\ \{|a|+|b|\}<1.$

2. Show that

$$\int_0^\infty \frac{dx}{1+x^4\sin^2 x} \text{ converges but } \int_0^\infty \frac{dx}{1+x^2\sin^2 x} \text{ does not.}$$

3. Evaluate

$$\int_0^\infty \frac{\tan^{-1} ax - \tan^{-1} bx}{x}\,dx.$$

4. Show that a, b being positive,

$$\int_0^\infty \frac{\cos ax - \cos bx}{x}\,dx = \log\frac{b}{a},$$

and deduce that

$$\int_0^\infty \frac{\sin ax \sin bx}{x}\,dx = \frac{1}{2}\log\frac{a+b}{a-b}; (a>b>0).$$

5. Show that

$$\int_0^\infty \frac{\sin x}{x}\,dx = \frac{\pi}{2}.$$

(i) $\displaystyle\int_0^\infty \frac{\sin x\,dx}{x} = \lim_{n\to\infty}\int_0^{\frac{1}{2}(2n+1)\pi} \frac{\sin x}{x}\,dx = \lim_{n\to\infty}\int_0^{\pi/2}\frac{\sin(2n+1)x}{x}\,dx.$

(ii) Integrating by parts or otherwise, (See Ex. 8, page 236 and Ex. 9 page 237), show that

$$\lim_{n\to\infty}\int_0^{\frac{1}{2}\pi} \sin(2n+1)x\left(\frac{1}{\sin x} - \frac{1}{x}\right)dx = 0.$$

6. Show that

$$\int_0^{\frac{1}{2}\pi} \frac{\sin(2n+1)x}{\sin x}\,dx = \frac{\pi}{2}.$$

7. Show that

$$\int_0^\infty \frac{\sin ax}{x}\,dx = \frac{\pi}{2}, 0 \text{ or } -\frac{\pi}{2}$$

according as, a, is positive, zero or negative.

8. Show that

$$\int_0^\infty \frac{\sin^2 x}{x^2}\,dx = \frac{\pi}{2}.$$ *(Lucknow 94)*

(Integrate by parts)

9. Show that

$$\int_0^\infty \frac{\cos ax - \cos bx}{x^2}\,dx = \frac{\pi}{2}(b-a). \qquad (a > 0,\ b > 0)$$

(Integrate by parts)

10. Show that

(*i*) $\displaystyle\int_0^\infty \frac{\sin ax \sin x}{x^2}\,dx = \frac{1}{2}\pi a$ when $0 \le a < 1$ and $\frac{1}{2}\pi$ when $a \ge 1$.

(*ii*) $\displaystyle\int_0^\infty \frac{\sin ax \sin^2 x}{x^2}\,dx = \frac{1}{2}\pi a(4-a)$ when $0 \le a \le 2$ and $\frac{1}{2}\pi$ when $a \ge 2$.

11. Find the value of the definite integral

$$\int_0^\pi \frac{\sin x\,dx}{\sqrt{(1-2a\cos x + a^2)}},\ (a > 0).$$

12. Evaluate the integral

$$\int_{-1}^{+1} \frac{\sin x\,dx}{1 - 2x\cos a + x^2}.$$

For what value of, a, is the integral a discontinuous function of.

13. Show that

$$\int_0^1 \frac{dx}{x^2 + 2x\cos a + 1} = \frac{a}{2\sin a}$$

f $-\pi < a < \pi$, except when $a = 0$, when the value of the integral is $\frac{1}{2}$.

14. Prove that

$$\int_0^\infty \log(1 + a^2x^{-2})\,dx = \pi a, \text{ if } a > 0.$$

15. Discuss the convergence of

$$\int_0^{\frac{1}{2}\pi} \cos 2nx \log \sin x\,dx$$

nd evaluate it when n is a positive integer.

[0 is the only point of infinite discontinuity. Now since, $x \to 0$, $\lim \left(\sqrt{x} \log \sin x \cos 2nx\right) = 0$, therefore, for values of x sufficiently near 0,

$$|\cos 2nx \log \sin x| < \varepsilon / \sqrt{x}.$$

For evaluation, proceed integrating by parts.]

16. From the preceding example, deduce that

(i) $\displaystyle\int_0^{\frac{1}{2}\pi} \cos 2nx \log \cos x \, dx = (-1)^{n+1} \frac{\pi}{4n}$.

(ii) $\displaystyle\int_0^{\pi} \cos nx \log 2(1 - \cos x) \, dx = -\frac{\pi}{n}$.

(iii) $\displaystyle\int_0^{\pi} \cos dx \log 2(1 + \cos x) \, dx = (-1)^{n+1} \frac{\pi}{n}$.

17. Prove that if g is bounded and integrable, then

$$\int_a^b g(x) \sin nx \, dx \to 0 \text{ as } a \to \infty.$$

where $[a, b]$ is any finite interval. If further

$$\int_{-\infty}^{+\infty} |g(x)| \, dx$$

is convergent, prove that

$$\int_{-\infty}^{+\infty} g(x) \sin nx \, dx \to 0 \text{ as } n \to \infty.$$

18. Prove that

$$\lim_{p \to (0+0)} \left[\lim_{n \to \infty} \left(\sum_{r=1}^{n} \frac{1}{r^{1+p}} \right) \right] = 1.$$

19. If ϕ is positive and decreasing in $[1, \infty[$, prove that

$$u_n = \sum_{r=1}^{n} f(r) - \int_1^n f(x) \, dx$$

tends to a limit, l, as $n \to \infty$ and that $0 \le l \le f(1)$.

20. Prove that if $s > 1$, then

$$\sum_{r=1}^{n} \left(\frac{s}{r} - \frac{1}{r^3} \right) - s \log n$$

tends to a limit as $n \to \infty$ (s being fixed), and that if this limit is $\varphi\{s\}$ then

$$0 \le [\varphi(s) + (s-1)^{-1}] \le (s-1).$$

OBJECTIVE QUESTIONS

Note. *For each of the following questions, four alternatives are given for the answer. Only one of them is correct. Choose the correct alternative.*

1. Integral $\int_a^b f(x)\,dx$ is said to be improper if :

(*a*) both the limits are finite

(*b*) $f(x)$ is bounded

(*c*) one or both integration limits are infinite

(*d*) none of these

2. If one or both integration limits of the integral $\int_a^b f(x)\,dx$ are infinite then it is an improper integral of

(*a*) first kind (*b*) second kind

(*c*) third kind (*d*) not an improper integral

3. If in $\int_a^b f(x)\,dx$, integration limits are infinite and $f(x)$ is unbounded, then it is an improper integral of :

(*a*) first kind (*b*) second kind

(*c*) third kind (*d*) not an improper integral.

4. $\int_0^\infty \sin x\,dx$ is an improper integral of

(*a*) first kind (*b*) second kind

(*c*) third kind (*d*) none of these

5. $\int_0^\infty \frac{dx}{(1-x)^2}$ is :

(*a*) improper integral of first kind (*b*) improper integral of second kind

(*c*) improper integral of third kind (*d*) a proper integral

6. Integral $\int_0^\infty \frac{4a}{(x^2+4a^2)}\,dx$ is :

(*a*) convergent (*b*) divergent

(*c*) oscillatory (*d*) proper

7. Integral $\int_0^1 \frac{dx}{\sqrt{(1-x)}}$ is :

(*a*) convergent (*b*) divergent

(*c*) oscillatory (*d*) proper

8. $\int_1^\infty \frac{dx}{x}$ is :

(*a*) convergent (*b*) divergent

(*c*) oscillatory (*d*) none of these

9. Integral $\int_a^\infty \frac{dx}{x^n}$, where $a > 0$ is convergent, if

(*a*) $n = 1$ (*b*) $n < 1$

(*c*) $n > 1$ (*d*) never convergent.

10. $\int_\pi^\infty \frac{\sin^2 x}{x^2} dx$ is :

(*a*) convergent (*b*) divergent

(*c*) oscillatory (*d*) proper.

11. Integral $\int_0^\infty \frac{x^{2n}}{1 + x^{2m}} dx$ is convergent if

(*a*) $n < m$ (*b*) $n > m$

(*c*) $n = m$ (*d*) none of these

12. $\int_a^\infty \frac{\sin x}{\sqrt{x}} dx$, where $a > 0$ is :

(*a*) convergent (*b*) divergent

(*c*) oscillatory (*d*) proper.

13. $\int_0^\infty \frac{\sin mx}{a^2 + x^2} dx$:

(*a*) converges but not absolutely (*b*) absolutely convergent

(*c*) divergent (*d*) none of these

14. $\int_b^\infty \frac{x^{3/2}\, dx}{\sqrt{(x^4 - a^4)}}$, where $b > a$ is :

(*a*) convergent (*b*) divergent

(*c*) absolutely convergent (*d*) none of these

15. $\int_a^b \frac{dx}{(x - a)^n}$ is convergent if :

(*a*) $n > 1$ (*b*) $n = 1$

(*c*) $n \leq 1$ (*d*) $n < 1$.

16. Integral $\int_0^1 \frac{\sec x}{x} dx$ is

(*a*) convergent (*b*) absolutely convergent

(*c*) divergent (*d*) proper.

17. $\int_0^{\pi/2} \log \sin x \, dx$ is :

(*a*) convergent (*b*) divergent

(*c*) neither convergent nor divergent (*d*) none of these

18. Integral $\int_0^{\pi/4} \frac{dx}{\sqrt{\tan x}}$:

(*a*) convergent but not absolutely (*b*) absolutely convergent

(*c*) divergent (*d*) proper

19. If $\int_a^{\infty} |f(x)|\, dx$] is convergent then integral $\int_a^{\infty} f(x)\, dx$ is

(*a*) conditionally convergent (*b*) uniformly convergent

(*c*) absolutely convergent (*d*) divergent

20. If, when $x \to \infty$. lim $[f(x)\, x^n]$ exists finitely, the limit being neither 0 nor infinite, then the integral $\int_a^{\infty} \frac{dx}{x^n}; (a > 0)$ converges if

(*a*) $n < 1$ (*b*) $n > 1$

(*c*) $n = 1$ (*d*) $n = 0$.

ANSWERS

1. (*c*)	**2.** (*a*)	**3.** (*c*)	**4.** (*a*)	**5.** (*c*)
6. (*a*)	**7.** (*a*)	**8.** (*b*)	**9.** (*c*)	**10.** (*a*)
11. (*b*)	**12.** (*a*)	**13.** (*b*)	**14.** (*b*)	**15.** (*d*)
16. (*c*)	**17.** (*a*)	**18.** (*a*)	**19.** (*c*)	**20.** (*b*)

10

Fourier Series

10.1. FOURIER SERIES

In this chapter, we shall be concerned with series of the form

$$\frac{1}{2}a_0 + (a_1 \cos x + b_1 \sin x) + \ldots + (a_n \cos nx + b_n \sin nx) + \ldots$$

where the numbers a_n, b_n are constants independent of x.

Every term of a series of this form being periodic with period 2π the sum function must also be necessarily periodic with period 2π. 'A function f with 2π as its period will arise, if it is arbitrarily defined in an interval of length 2π and then periodically extended beyond this interval to the left and to the right so as to satisfy the functional equation $f(x \pm 2\pi) = f(x)$.

Firstly, we proceed to show that on the assumption that the sum function of series of this form is uniformly convergent the constants a_n, b_n can be easily determined. It should, however, be clearly understood that since there is nothing to prove a priori, that this assumption is justifiable, this determination is purely formal.

This determination depends on the following simple facts :

$$\int_{-\pi}^{+\pi} \sin nx\, dx = \int_{-\pi}^{+\pi} \cos nx\, dx; \int_{-\pi}^{+\pi} \sin mx \cos nx\, dx = 0$$

$$\int_{-\pi}^{+\pi} \cos mx \cos nx\, dx = \begin{cases} 0, \text{ if } m \neq n \\ \pi, \text{ if } m = n \end{cases}$$

$$\int_{-\pi}^{+\pi} \sin mx \sin nx\, dx = \begin{cases} 0, \text{ if } m \neq n, \\ \pi, \text{ if } m = n. \end{cases}$$

Integrating term by term from, $-\pi$ to $+\pi$, the equality.

$$f(x) = \frac{1}{2}a_0 + \sum_{n=1}^{\infty} (a_n \cos nx + b_n \sin nx) \qquad \ldots(1)$$

We obtain

$$a_0 = \frac{1}{\pi}\int_{-\pi}^{+\pi} f(x)\, dx.$$

Multiplying the equality (1) by cos nx, and integrating term by term from $-\pi$ to $+\pi$, we obtain

$$a_n = \frac{1}{\pi}\int_{-\pi}^{+\pi} f(x)\cos nx\,dx, \qquad \text{...(2)}$$

which is seen to be true for $n = 0$ also.

Finally, on multiplying with sin nx and integrating term by term from, $-\pi$ to $+\pi$, we obtain

$$b_n = \frac{1}{\pi}\int_{-\pi}^{+\pi} f(x)\sin nx\,dx. \qquad \text{...(3)}$$

Note. The series (1) with the coefficients (2) and (3), will be called the Fourier series corresponding to the function f, in the interval $[-\pi, \pi]$.

Of course, the mere fact that a series corresponding, to a function can be written, does not ensure its convergence, or even if it is convergent, that its sum will be $f(x)$. Complete result in regard to this is given in § 10.4, page 329.

10.2. THEOREM

Let f be bounded and integrable in $[0, a]$ $a > 0$, and let it be monotonic in some interval $[0, \alpha]$ where, α, is any positive number less than a. Then

$$\lim_{n\to\infty}\int_0^a f(x)\frac{\sin nx}{x}\,dx = f(+0)\int_0^\infty \frac{\sin x}{x}\,dx.$$

[Since f is monotonic and bounded in $[0, \alpha]$, therefore $f(+0)$ will exist]

Firstly, we suppose that $f(+0) = 0$. Also, without affecting the result in any way whatsoever, we can suppose that $f(0) = 0$.

Let h be any positive number less than α.

By the second mean value theorem, there exists $h_1 \in [0, h]$ such that

$$\int_0^h f(x)\frac{\sin nx}{x}\,dx = f(0)\int_0^h \frac{\sin nx}{x}\,dx + f(h)\int_{h_1}^h \frac{\sin nx}{x}\,dx,$$

$$= f(h)\int_{h_1}^h \frac{\sin nx}{x}\,dx$$

$$= f(h)\int_{nh_1}^{nh} \frac{\sin y}{y}\,dy = f(h)\int_{nh_1}^{nh} \frac{\sin x}{x}\,dx. \qquad \text{...(1)}$$

Since

$$\int_0^\infty \frac{\sin x}{x}\,dx$$

is convergent, there exists $k > 0$ such that $\forall\, t \geq 0$.

$$\left| \int_0^t \frac{\sin x}{x}\, dx \right| \leq k$$

$$\left| \int_{nh_1}^{nh} \frac{\sin x}{x}\, dx \right| = \left| \int_0^{nh} \frac{\sin x}{x}\, dx - \int_0^{nh_1} \frac{\sin x}{x}\, dx \right| \leq 2k \quad ...(2)$$

Also

$$f(h) \to 0 \text{ as } h \to (0+0).$$

From (1), (2) and (3), we deduce that $\varepsilon > 0$ being given, there exists $\delta > 0$ such that

$$\left| \int_0^a f(x) \frac{\sin nx}{\sin x}\, dx \right| \leq 2k \left| f(h) \right| < \frac{1}{2}\varepsilon \text{ when } 0 < h \leq \delta.$$

Now, we have

$$\int_0^a f(x) \frac{\sin nx}{x}\, dx = \int_0^\delta f(x) \frac{\sin nx}{x}\, dx + \int_\delta^a f(x) \frac{\sin nx}{x}\, dx.$$

*Since, as proved in Ex. 8, page 236, the second integral on the right $\to 0$, as $n \to \infty$, we see that there exists a positive integer m such that $\forall\, n \geq m$,

$$\left| \int_0^a f(x) \frac{\sin nx}{x}\, dx \right| < \frac{1}{2}\varepsilon + \frac{1}{2}\varepsilon = \varepsilon.$$

Thus the theorem is proved in this case.

In the general case, changing $f(x)$ to $[f(x) - f(+0)]$, we see that

$$\lim_{n \to \infty} \int_0^a \left[f(x) - f(+0) \right] \frac{\sin nx}{x}\, dx = 0.$$

Now

$$\int_0^a \frac{\sin nx}{x}\, dx = \int_0^{na} \frac{\sin y}{y}\, dy \to \int_0^\infty \frac{\sin y}{y}\, dy \text{ as } n \to \infty.$$

It follows that

$$\lim_{n \to \infty} \int_0^a f(x) \frac{\sin nx}{\sin x}\, dx = f(+0) \int_0^\infty \frac{\sin x}{x}\, dx.$$

Cor. *If f satisfies the conditions of the above theorem and $0 < \alpha < \pi$, then*

$$\lim \int_0^a f(x) \frac{\sin nx}{\sin x}\, dx = f(+0) \int_0^\infty \frac{\sin x}{x}\, dx.$$

* $f(x)/x$ being bounded and integrable in $[\delta, a]$.

We write

$$f(x)=\frac{\sin nx}{\sin x}=f(x)\frac{x}{\sin x}\cdot\frac{\sin nx}{x}=g(x)\frac{\sin nx}{x},$$

where we assign the values 1 and n to $(x/\sin x)$ and $(\sin nx/x)$ respectively for $x = 0$. Now we know that $(x/\sin x)$ is positive and monotonically increasing in $[0, \frac{1}{2}\pi]$. It follows that if f be monotonically increasing in $[0, \alpha]$, $(\alpha, < 1/2\,\pi)$, then g is also monotonically increasing in $[0, \alpha]$. Therefore

$$\lim_{n\to\infty}\int_0^a f(x)\frac{\sin nx}{\sin x}\,dx=\lim_{n\to\infty}\int g(x)\frac{\sin nx}{x}\,dx$$

$$=g(+0)\int_0^\infty\frac{\sin x}{x}\,dx=f(+0)\int_0^\infty\frac{\sin x}{x}\,dx$$

If f be decreasing, so that $-f$ is increasing, we see that

$$\lim_{n\to\infty}\int_0^a[-f(x)]\frac{\sin nx}{\sin x}\,dx=-f(+0)\int_0^\infty\frac{\sin x}{x}\,dx.$$

$$\lim_{x\to\infty}\int_0^a f(x)\frac{\sin nx}{\sin x}\,dx=f(+0)\int_0^\infty\frac{\sin x}{x}\,dx.$$

10.3. THEOREM

Let f be bounded and integrable in $[-\pi, \pi]$ and monotonic in $[-\alpha, 0[$ and $]0, \alpha]$, (not necessarily in the same sense), where α is some positive number less than π. Then

$$\frac{1}{2}a_0+\sum_{n=1}^{\infty}a_n=\frac{f(+0)+f(-0)}{2}$$

We have

$$\frac{1}{2}a_0+\sum_{n=1}^{m}a_n=\frac{1}{2\pi}\int_{-\pi}^{+\pi}f(x)\,dx+\frac{1}{\pi}\int_{-\pi}^{+\pi}f(x)\sum_{n=1}^{m}\cos n\pi\,dx$$

$$=\frac{1}{2\pi}\int_{-\pi}^{+\pi}f(x)\left[1+2\sum_{n=1}^{m}\cos nx\right]dx$$

$$=\frac{1}{2\pi}\int_{-n}^{+\pi}f(x)\frac{\sin\left(m+\frac{1}{2}\right)x}{\sin\frac{1}{2}x}\,dx$$

$$= \frac{1}{2\pi} \int_0^{\pi} f(-x) \frac{\sin\left(m + \frac{1}{2}\right) x}{\sin \frac{1}{2} x} dx + \frac{1}{2\pi} \int_0^{\pi} f(x) \frac{\sin\left(m + \frac{1}{2}\right) x}{\sin \frac{1}{2} x} dx$$

$$= \frac{1}{2\pi} \cdot 2 \int_0^{\frac{1}{2}\pi} f(-2x) \frac{\sin(2m+1)x}{\sin x} dx$$

$$+ \frac{1}{2\pi} \cdot 2 \int_0^{\frac{1}{2}\pi} f(2x) \frac{\sin(2m+1)x}{\sin x} dx$$

$$= \frac{1}{\pi}\left[f(-0) + f(+0)\right] \int_0^{\infty} \frac{\sin x}{x} dx \text{ as } m \to \infty.$$

Taking $f(x) = 1 \; \forall \; x$, we see that

$$\frac{1}{2} a_0 = \frac{1}{2\pi} \int_{-\pi}^{+\pi} 1.dx = 1 \text{ and } a_n = \frac{1}{\pi} \int_{-\pi}^{+\pi} \cos nx \, dx = 0$$

so that we obtain

$$1 = \frac{1+1}{\pi} \int_0^{\infty} \frac{\sin x}{x} dx \Rightarrow \int_0^{\infty} \frac{\sin x}{x} dx = \frac{\pi}{2}.$$

Hence the theorem.

Note. The value of integral of $(\sin x/x)$ over $[0, \infty[$ has also been already obtained in another way in Ex. 5, p. 321. Still another method will be given later on in Ch. 14.

10.4. MAIN THEOREM

Let f be bounded and integrable in $[-\pi, \pi]$ *and let it be possible to divide* $[-\pi, \pi]$ *into a finite number of open sub-intervals, in each of which f is monotonic. Then*

$$\frac{1}{2} a_0 + \sum_{n=1}^{\infty} (a_n \cos n\xi + b_n \sin n\xi) = \begin{cases} \frac{1}{2}[f(\xi - 0) + f(\xi + 0)] \\ \quad \text{for } -\pi < \xi < \pi \\ \frac{1}{2}[f(\pi - 0) + f(-\pi + 0)], \\ \quad \text{for } \xi = -\pi \text{ and } \xi = \pi. \end{cases}$$

Under the given conditions $f(\xi - 0)$ and $f(\xi + 0)$ necessarily exist.

Lemma. If f is bounded and integrable in every interval and is periodic with 2π as its period, then

$$\int_{-\pi}^{+\pi} f(x) \, dx = \int_{-\pi}^{+\pi} f(a + x) \, dx,$$

a, being any number whatsoever. Putting $a + x = y$, we have

$$\int_{-\pi}^{+\pi} f(a+x)\,dx = \int_{a-\pi}^{a+\pi} f(y)\,dy$$

$$= \int_{a-\pi}^{-\pi} f(y)\,dy + \int_{-\pi}^{+\pi} f(y)\,dy + \int_{+\pi}^{a+\pi} f(y)\,dy.$$

Putting $y = z - 2\pi$, we see that

$$\int_{a-\pi}^{-\pi} f(y)\,dy = \int_{a+\pi}^{\pi} f(z-2\pi)\,dz = -\int_{\pi}^{a+\pi} f(z)\,dz = -\int_{\pi}^{a+\pi} f(y)\,dy$$

Hence the result.

Proof. We have

$$\frac{1}{2}a_0 + \sum_{n=1}^{m}(a_n \cos n\xi + b_n \sin n\xi)$$

$$= \frac{1}{2\pi}\int_{-\pi}^{+\pi} f(x)\,dx + \sum_{n=1}^{m} \frac{1}{\pi}\int_{-\pi}^{+\pi} f(x)\,[(\cos nx \cos n\xi$$

$$+ \sin nx \sin n\xi)]\,dx$$

$$= \frac{1}{2\pi}\int_{-\pi}^{+\pi} f(x)\,[1 + 2\sum_{n=1}^{m} \cos n(x-\xi)]\,dx$$

$$= \frac{1}{2\pi}\int_{-\pi}^{+\pi} f(x+\xi)\,[1 + 2\sum_{n=1}^{m} \cos nx]\,dx \qquad \text{(lemma)}$$

$$= \frac{1}{2\pi}\int_{-\pi}^{+\pi} f(x+\xi)\frac{\sin\left(m+\frac{1}{2}\right)x}{\sin\frac{1}{2}x}\,dx$$

$$= \frac{1}{2\pi}\int_{-\pi}^{0} f(x+\xi)\frac{\sin\left(m+\frac{1}{2}\right)x}{\sin\frac{1}{2}x}\,dx + \frac{1}{2\pi}\int_{0}^{\pi} f(x+\xi)\frac{\sin\left(m+\frac{1}{2}\right)x}{\sin\frac{1}{2}x}\,dx$$

$$= \frac{1}{\pi}\int_{0}^{\frac{1}{2}\pi} f(-2y+\xi)\frac{\sin(2m+1)y}{\sin y}\,dy + \frac{1}{\pi}\int_{0}^{\frac{1}{2}\pi} f(2y+\xi)\frac{\sin(2m+1)y}{\sin y}\,dy$$

$$= \frac{1}{\pi}\left\{\frac{\pi}{2}\cdot f(\xi-0) + \frac{\pi}{2}\cdot f(\xi+0)\right\}$$

$$= \frac{f(\xi+0)+f(\xi-0)}{2}, \text{ as } m \to \infty.$$

Note. The result arrived at above may be re-stated as follows : *If f be bounded and integrable in $[-\pi, \pi]$ and if it be possible to divide $[-\pi, \pi]$ into a finite number of open sub-intervals in each of which f is monotonic, then the Fourier series, corresponding to f converge $\forall\, x$, and if σ denotes the sum function of series, then*

$$\sigma(x) = \frac{1}{2}\left[f(x+0) + f(x-0)\right] \quad \forall\, x \in \,]-\pi, \pi[$$

$$\sigma(x) = \frac{1}{2}\left[f(\pi - 0) + f(-\pi + 0)\right] \text{ for } x = \pm\pi$$

$$\sigma(x + 2\pi) = \sigma(x).$$

The relation $\sigma(x + 2\pi) = \sigma(x)$ enables us to determine the value of the sum function at a point which does not belong to $[-\pi, \pi]$.

Cor. At a point ξ of continuity of f,

$$\frac{1}{2}a_0 + \sum_{n=1}^{\infty}(a_n \cos n\xi + b_n \sin n\xi) = \frac{f(\xi+0) + f(\xi-0)}{2}$$

$$= \frac{f(\xi) + f(\xi)}{2} = f(\xi)$$

Thus we see that if f satisfies the conditions of the theorem of 10.4, in $[-\pi, \pi]$, then the sum of the Fourier series corresponding to f is actually $f(x)$ at all such points x of $[-\pi, \pi]$, where f is continuous; at points x of discontinuity, the sum of the series is

$$\frac{1}{2}\left[f(x+0) + f(x-0)\right].$$

10.5. DIRICHLET'S CONDITION

Following are the assumptions for the expansion in a Fourier's series :

(*i*) The given function $f(x)$ is assumed to be defined and single valued in the given range $(-l, l)$.

(*ii*) $f(x)$ is periodic outside $(-l, l)$ with period $2l$.

(*iii*) We assume that the series $a_0 + \sum_{n=1}^{\infty}(a_n \cos nx + b_n \sin nx)$ is uniformly convergent, so that the term by term integration of the series is possible.

Then the series converges to

(*a*) $f(x)$ if x is a point of continuity

(*b*) $\dfrac{f(x+0) + f(x-0)}{2}$ if x is a point of discontinuity.

Also the series converges to $\dfrac{f(-\pi+0)+f(\pi-0)}{2}$ at $x=\pm\pi$, when these limits exist.

Here $f(x+0)$ is the right hand limit of $f(x)$ at x and represent $\lim\limits_{h\to 0} f(x+h)$. Similarly $f(x-0)$ in the left hand limit of $f(x)$ at x and represents $\lim\limits_{h\to 0} f(x-h)$.

EXAMPLES

1. *Expand in a series of sines and cosines of multiples of x, the function given by*

$$f(x)=x-\pi,\ \textit{when}\ -\pi<x<0;\ f(x)=\pi-x,\ \textit{when}\ 0<x<\pi.$$

What is the sum of the series for $x=\pm\pi$ *and* $x=0$?

(Dibrugarh 2004)

Sol. The given function f satisfies the conditions of the theorem of § 10.4. We have

$$a_n=\frac{1}{\pi}\int_{-\pi}^{+\pi} f(x)\cos nx\,dx$$

$$=\frac{1}{\pi}\int_{-\pi}^{0} f(x)\cos nx\,dx+\frac{1}{\pi}\int_{0}^{\pi} f(x)\cos nx\,dx$$

$$=\frac{1}{\pi}\int_{-\pi}^{0}(x-\pi)\cos nx\,dx+\frac{1}{\pi}\int_{0}^{\pi}(\pi-x)\cos nx\,dx.$$

Integrating by parts, we obtain

$$a_n=\frac{2\left[1-(-1)^n\right]}{x^2\pi},\quad n\neq 0.$$

Also $\quad a_0=-\pi.$

Similarly

$$b_n=\frac{1}{\pi}\int_{-\pi}^{0}(x-\pi)\sin nx\,dx+\frac{1}{\pi}\int_{0}^{\pi}(\pi-x)\sin nx\,dx=\frac{2\left[1-(-1)^n\right]}{n}$$

The coefficients a_n and b_n are zero when n is even.

In $[-\pi,\pi]$ the points $x=0$ and $x=\pm\pi$ are the only points of discontinuity of f. Therefore when x is different from 0 and $\pm\pi$, we have

$$f(x)=-\frac{1}{2}\pi+\frac{4}{\pi}\cdot\frac{1}{1^2}\cos x+\frac{4}{1}\sin x+\frac{4}{\pi}\cdot\frac{1}{3^2}\cos 3x+\frac{4}{3}\sin 3x+\ldots$$

$$=-\frac{1}{2}\pi+\frac{4}{\pi}\left[\frac{\cos x}{1^2}+\frac{\cos 3x}{3^2}+\frac{\cos 5x}{5^2}+\ldots\right]+4\left[\frac{\sin x}{1}+\frac{\sin 3x}{3}+\ldots\right]$$

When $x = 0$ is a point of discontinuity of f, therefore, the sum of the series for $x = 0$

$$= \frac{1}{2}\left[f(+0) + f(-0)\right] = \frac{1}{2}\left[(\pi) + (-\pi)\right] = 0.$$

Thus we obtain a well-known result, *viz.*,

$$\frac{\pi^2}{8} = 1 + \frac{1}{3^2} + \frac{1}{5^2} + \frac{1}{7^2} + \ldots \qquad \ldots(1)$$

For $x = \pm\pi$, the sum of the series

$$= \frac{1}{2}\left[f(\pi - 0) + f(-\pi + 0)\right] = \frac{1}{2}\left[0 + (-2\pi)\right] = -\pi.$$

Putting $x = \pm\pi$, we obtain the same result, *viz.*, (1).

2. *Find a series of sines and cosines of multiples of x, which will represent* $\frac{\pi}{2 \sinh \pi} \cdot e^x$ *in the interval* $-\pi < x < \pi$.

Find also the sum of the series for $x = \pm\pi$. *(Ajmer 98)*

Sol. Suppose $f(x) = a_0 + \sum_{n=1}^{\infty} (a_n \cos nx + b_n \sin nx)$.

Here $f(x) = \frac{\pi}{2 \sinh \pi} e^x$ and interval is $(-\pi, \pi)$

$$a_0 = \frac{1}{2\pi} \int_{-\pi}^{\pi} f(x)\, dx = \frac{1}{2\pi} \int_{-\pi}^{\pi} \frac{\pi}{2 \sinh \pi} e^{ax}\, dx$$

$$= \frac{1}{4 \sinh \pi}\left[e^{\pi} - e^{-\pi}\right] = \frac{1}{2}.$$

$$a_n = \frac{1}{\pi} \int_{-\pi}^{\pi} \frac{\pi}{2 \sinh \pi} e^x \cos nx\, dx$$

$$= \frac{1}{2 \sinh \pi}\left[\frac{e^x}{1 + n^2}(\cos nx + n \sin nx)\right]_{-\pi}^{\pi}$$

$$= \frac{1}{2(1 + n^2) \sinh \pi} \cdot \cos n\pi \left[e^{\pi} - e^{-\pi}\right]$$

$$= \frac{(-1)^n (2 \sinh \pi)}{2(1 + n^2) \sinh y} = \frac{(-1)^n}{1 + n^2}$$

$$b_n = \frac{1}{\pi}\int_{-x}^{\pi} \frac{\pi}{2\sinh\pi} e^x \sin nx\, dx$$

$$= \frac{-n\cos n\pi}{2\left(1+n^2\right)\sinh\pi}\left[e^{\pi} - e^{-\pi}\right] = \frac{-n\cos n\pi\,(2\sin h\pi)}{2\left(1+n^2\right)\sin h\pi}$$

$$= -(-1)^n \cdot \frac{n}{1+n^2}$$

By putting the values of a_0, a_n, b_n, we have

$$f(x) = \frac{\pi}{2\sinh\pi} e^x = \frac{1}{2} + \sum_{n=1}^{\infty} \frac{(-1)^n}{\left(1+n^2\right)} (\cos nx - n\sin nx)$$

At $x = \pm\pi$, the series converges to

$$\frac{1}{2}\left[f(-\pi+0) + f(\pi+0)\right] = \frac{1}{2}\cdot\frac{\pi}{2\sinh\pi}\left[e^{-\pi} + e^{\pi}\right] = \frac{\pi}{2}\coth\pi.$$

3. *Find a series of sines and cosines of multiples of x which will represent* $\left(x+x^2\right)$ *in the interval* $-\pi < x < \pi$. *Hence show that*

$$\frac{\pi^2}{6} = 1 + \frac{1}{2^2} + \frac{1}{3^2} + \dots \qquad \textit{(Ajmer 95)}$$

Sol. $$a_0 = \frac{1}{2\pi}\int_{-\pi}^{\pi} f(x)\,dx = \frac{1}{2\pi}\int_{-\pi}^{\pi}\left(x+x^2\right)dx = \frac{\pi^2}{3}.$$

$$a_n = \frac{1}{\pi}\int_{-\pi}^{\pi}\left(x+x^2\right)\cos nx\,dx = \frac{2}{\pi}\int_0^{\pi} x^2\cos nx\,dx$$

$$= \frac{2}{\pi}\left[2\pi\left(\frac{\cos n\pi}{n^2}\right)\right] = \frac{4}{n^2}(-1)^n$$

$$b_n = \frac{1}{\pi}\int_{-\pi}^{\pi}\left(x+x^2\right)\sin nx\,dx = \frac{2}{\pi}\int_0^{\pi} x\sin nx\,dx$$

$$= \frac{2}{\pi}\left[\pi\left(-\frac{\cos n\pi}{n}\right)+0\right] = -\frac{2}{n}(-1)^n.$$

$$\therefore \quad f(x) = \left(x+x^2\right) = \frac{\pi^2}{3} + \sum_{n=1}^{\infty}\left\{\frac{4}{n^2}(-1)^n\cos nx - \frac{2}{n}(-1)^n\sin nx\right\}$$

Putting $x = \pi$ and $x = -\pi$, successively, we get

$$\pi + \pi^2 = \frac{\pi^2}{3} + 4\left(1 + \frac{1}{2^2} + \frac{1}{3^2} + \dots\right)$$

and $$-\pi + \pi^2 = \frac{\pi^2}{3} + 4\left(1 + \frac{1}{2^2} + \frac{1}{3^2} + \dots\right)$$

Adding these two

$$2\pi^2 = \frac{2\pi^2}{3} + 8\left(1 + \frac{1}{2^2} + \frac{1}{3^2} + \ldots\right)$$

$$\frac{\pi^2}{6} = 1 + \frac{1}{2^2} + \frac{1}{3^2} + \ldots$$

4. *Obtain an expansion in a mixed series of sines and cosines of multiples of x which is zero between* $-\pi$ *and 0 and is equal to* e^x *between zero and* π *and gives its value at three limits.*

Sol. Here,

$$f(x) = \begin{cases} 0, & \text{for } -\pi < x < 0 \\ e^x, & \text{for } 0 < x < \pi. \end{cases}$$

Suppose $f(x) = a_0 + \sum_{n=1}^{\infty} (a_n \cos nx + b_n \sin nx)$

$$a_0 = \frac{1}{2\pi}\int_{-\pi}^{\pi} f(x)\,dx = \frac{1}{2\pi}\left[\int_{-\pi}^{0} f(x)\,dx + \int_0^{\pi} f(x)\,dx\right]$$

$$= \frac{1}{2\pi}\int_0^{\pi} e^x\,dx = \frac{1}{2\pi}\left(e^{\pi} - 1\right)$$

$$a_n = \frac{1}{\pi}\left[\int_{-\pi}^{0} f(x)\cos nx\,dx + \int_0^{\pi} f(x)\cos nx\,dx\right]$$

$$= \frac{1}{\pi}\int_0^{\pi} e^x \cos nx\,dx = \frac{1}{\pi\left(n^2+1\right)}\left[e^{\pi}\cos n\pi - 1\right]$$

$$b_n = \frac{1}{\pi}\left[\int_{-\pi}^{0} f(x)\sin nx\,dx + \int_0^{\pi} f(x)\sin nx\,dx\right]$$

$$= \frac{1}{\pi}\int_0^{\pi} e^x \sin nx\,dx = -\frac{n\left(e^{\pi}\cos n\pi - 1\right)}{\pi\left(1+n^2\right)}$$

$$\therefore \quad f(x) = \frac{\left(e^{\pi}-1\right)}{2\pi} + \frac{1}{n}\sum_{n=1}^{\infty} \frac{\left(e^{\pi}\cos n\pi - 1\right)}{\left(1+n^2\right)}(\cos nx - n\sin nx)$$

This series converges to $\frac{1}{2}\left[f(x+0) + f(x-0)\right]$ at any point where $f(x+0) = \lim_{h\to 0} f(x+h)$ and $f(x-0) = \lim_{h\to 0} f(x-h)$

At $x = 0$,

$$f(0+0) = \lim_{h \to 0} e^{0+h} = 1$$

$$f(0-0) = \lim_{h \to 0} f(0-h) = 0$$

$\therefore$ Sum at $x = 0$ is $\frac{1}{2}(1+0) = \frac{1}{2}$.

At $x = \pm \pi$, the series converges to

$$\frac{1}{2}\left[f(-\pi+0) + f(\pi-0)\right] = \frac{1}{2}\left(0 + e^{\pi}\right) = \frac{1}{2}e^{\pi}.$$

10.6. FOURIER SERIES FOR ODD AND EVEN FUNCTIONS

The Fourier series with sines only. If f be an odd function, *i.e.*, if $f(-x) = -f(x)\ \forall\ x$, then $f(x)$ cos nx gives an odd function and $f(x)$ sin nx an even and hence

$$a_n = \frac{1}{\pi}\int_{-\pi}^{+\pi} f(x) \cos nx\, dx = 0$$

$$b_n = \frac{1}{\pi}\int_{-\pi}^{+\pi} f(x) \sin nx\, dx = \frac{2}{\pi}\int_{0}^{+\pi} f(x) \sin nx\, dx. \qquad ...(2)$$

so that we see that Fourier series corresponding to an odd function, consists of terms with sines only and the coefficients b_n may be computed by (2).

The Fourier series with cosines only. If f is an even function, *i.e.*, if $f(-x) = f(x)\ \forall\ x$ then $f(x)$ cos nx gives an even and $f(x)$ sin nx an odd function and hence

$$a_n = \frac{2}{\pi}\int_{0}^{\pi} f(x) \cos nx\, dx,\ b_n = 0$$

so that we see that the Fourier series corresponding to an even function consists of terms with cosines only.

10.7. HALF RANGE SERIES

From the preceding, we will now deduce the following two results :

10.7.1. If f satisfies the Conditions of the Theorem of § 10.4 in $[0, \pi]$, then the Sum of the sines Series

$$\Sigma\, b_n \sin nx, \textit{ where } b_n = \frac{2}{\pi}\int_0^{\pi} f(x) \sin nx\, dx,$$

is equal to $\frac{1}{2}\left[f(x+0) + f(x-0)\right]$ *at every point x between 0 and* π *and is equal to 0 when* $x = 0$ *and* $x = \pi$.

To see the truth of this result, we define an odd function F in $[-\pi, \pi]$ which is identical with f in $[0, \pi]$. Thus $F(x) = f(x)$ in $[0, \pi]$ and $F(x) = -F(-x) = -f(-x)$ in $[-\pi, 0]$. Clearly F will satisfy the conditions of the theorem of §10.4 in $[-\pi, \pi]$ if does so in $[0, \pi]$. Thus we see that the sum of the series $\Sigma\, b_n \sin nx$,

where
$$b_n = \frac{2}{\pi}\int_0^{\pi} F(x) \sin nx\, dx$$
$$= \frac{2}{\pi}\int_0^{\pi} f(x) \sin nx\, dx.$$

is equal to
$$\frac{1}{2}\left[F(x+0) + F(x-0)\right] = \frac{1}{2}\left[f(x+0) + f(x-0)\right]$$
at every point x between 0 and π.

As $x = 0$, the sum of the series $\frac{1}{2}\left[F(+0) + F(-0)\right] = 0,$ for F is odd Similarly we see that the sum of the series is 0 for $x = \pi$.

10.7.2. If f satisfies the conditions of the theorem of § 10.4 in $[0, \pi]$, then the sum of the series

$$\frac{1}{2}a_0 + \Sigma\, a_n \cos nx, \text{ where } a_n = \frac{2}{\pi}\int_0^{\pi} f(x) \cos nx\, dx,$$

is equal to $\frac{1}{2}\left[f(x+0) + f(x-0)\right]$ *at every point x between 0 and π an[d] is equal to $f(+0)$ for $x = 0$ and $f(\pi - 0)$ for $x = \pi$.*

To prove this result we have to consider an even function F defined i[n] $[-\pi, \pi]$ which is identical with f in $[0, \pi]$.

Note. The sum functions of the half range sine and cosine series a[re] periodic with period 2π.

EXAMPLES

Find (a) the Fourier sine series, and (b) the Fourier cosine seri[es] which represents $f(x) = (\pi - x)$ in $0 < x < \pi$.

To find sine series. We have
$$b_n = \frac{2}{\pi}\int_0^{\pi} (\pi - x) \sin nx\, dx = \frac{2}{n}.$$

Since f is continuous in $0 < x < \pi$, we have
$$\pi - x = 2\Sigma\, \frac{1}{n} \sin nx = 2\left(\frac{\sin x}{1} + \frac{\sin 2x}{2} + \frac{\sin 3x}{3} + \frac{\sin 4x}{4} + \ldots\right.$$

According to § 10.6, the sum of the series must be 0 for $x = 0$ and $x = \pi$ and this fact can be also directly verified. The representation holds for $x = \pi$ but not for $x = 0$.

To find cosine series. We have

$$a_n = \frac{2}{\pi} \int_0^{\pi} (\pi - x) \cos nx \, dx = \frac{2\left[1 - (-1)^n\right]}{\pi n^2}, \quad n \neq 0.$$

Also, $a_0 = \pi$.

Since f is continuous in $0 < x < \pi$, we have

$$\pi - x = \frac{1}{2}\pi + \Sigma \frac{2\left[1 - (-1)^n\right]}{\pi n^2} \cos nx$$

$$= \frac{1}{2}\pi + \frac{4}{\pi}\left(\frac{\cos x}{1^2} + \frac{\cos 3x}{3^2} + \frac{\cos 5x}{5^2} + \dots\right)$$

According to the theorem of § 10.6, the sum of the series must be $f(+0) = \pi$ for $x = 0$ and $f(\pi - 0) = 0$ for $x = \pi$. This gives the result

$$\frac{\pi^2}{8} = 1 + \frac{1}{3^2} + \frac{1}{5^2} + \frac{1}{7^2} + \dots,$$

which was obtained on page 245 also.

Observation. The series in the Ex. on page 245 and the two series obtained above have identical sums for values of x which belong to the open interval $]0, \pi[$ but for values of x belonging to the interval $]-\pi, 0[$; their sums are different. Thus for $-\pi < x < 0$ the sum of the series in the example on page 247, is $(x - \pi)$, whereas the sums of series above are

$$-\left[\pi - (-x)\right] = -(\pi + x) \text{ and } \left[\pi - (-x)\right] = (\pi + x)$$

respectively.

Similar differences in the sums exist for values of x outside the interval $[-\pi, \pi]$.

The sum functions of all the three series are periodic with period 2π.

0.8. OTHER FORMS OF FOURIER SERIES

The particular interval $[-\pi, \pi]$ which we have so far considered had been introduced only as a matter of convenience. We shall now see that is easy to change to any other finite interval.

0.8.1. The Interval $[0, 2\pi]$.

We write $x = y + \pi$ so that y varies in $[-\pi, \pi]$ as x varies in $[0, 2\pi]$. et

$$f(x) = f(y + \pi) = F(y).$$

Let f satisfy the conditions of the theorem of § 10.4 in $[0, 2\pi]$ so that F satisfies the same conditions in $[-\pi, \pi]$. Thus we see that the sum of the series

$$\frac{1}{2}a'_0 + \Sigma\,(a'_n \cos ny + b'_n \sin ny),$$

where $$a'_n = \frac{1}{\pi}\int_{-\pi}^{+\pi} F(y) \cos ny\, ay,\ b'_n = \int_{-\pi}^{+\pi} F(y) \sin ny\, dy.$$

is $\frac{1}{2}[F(y+0)+F(y-0)]$ at any point y between $-\pi$ and π and i $\frac{1}{2}[F(\pi-0)+F(-\pi+0)]$ at $y = \pm\pi$ and is periodic with period 2π.

Changing the variables, we see that

$$a'_n = \frac{1}{\pi}\int_{\pi}^{2\pi} F(x-\pi) \cos n(x-\pi)\, ax = \frac{(-1)^n}{\pi}\int_0^{2\pi} f(x) \cos nx\, d$$

$$b'_n = \frac{1}{\pi}\int_0^{2\pi} F(x-\pi) \sin n(x-\pi)\, dx = \frac{(-1)^n}{\pi}\int_0^{2\pi} f(x) \sin nx\, d$$

Also $\cos ny = (-1)^n \cos nx, \sin ny = (-1)^n \sin nx.$

Finally, we have

$$\frac{1}{2}[F(y+0)+F(y-0)] = \frac{1}{2}[f(y+\pi+0)+f(y+\pi-0)]$$

$$= \frac{1}{2}[f(x+0)+f(x-0)]$$

and $$\frac{1}{2}[F(\pi-0)+F(-\pi+0)] = \frac{1}{2}[f(2\pi-0)+f(+0)].$$

Thus we see that if f satisfies the conditions of the theorem of § 1 in $[0, 2\pi]$, then the sum of the series

$$\frac{1}{2}a_0 + \Sigma\,(a_n \cos nx + b_n \sin nx),$$

where $$a_n = \frac{1}{\pi}\int_0^{2\pi} f(x) \cos nx\, dx;\ b_n = \frac{1}{\pi}\int_0^{2\pi} f(x) \sin nx\, dx$$

is $[\frac{1}{2}f(x+0)+f(x-0)]$ *at every point* x *between* 0 *and* 2π *ar* $\frac{1}{2}[f(2\pi-0)+f(+0)]$ *at* $x = 0$ *and* $x = 2\pi$ *and is periodic with perioa*

10.8.2. The Interval $[-l, l]$

Here l is any real number. By means of the substitution $y = \pi x/$ considering a function F such that

$$f(x) = f(ly/\pi) = F(y),$$

so that y varies in $[-\pi, \pi]$ as x varies in $[-l, l]$, we can show as before *that if f satisfies the conditions of the theorem of § 10.4 in [– l, l], then the sum of the series*

$$\frac{1}{2}a_0 + \Sigma\left[a_n \cos(n\pi x/l) + b_n \sin(n\pi lx/l)\right],$$

where $$a_n = \frac{1}{l}\int_{-l}^{+l} f(x)\cos\frac{n\pi x}{l}\,dx,\quad b_n = \frac{1}{l}\int_{-l}^{+l} f(x)\frac{\sin n\pi x}{l}\,dx$$

is $\frac{1}{2}[f(x+0) + f(x-0)]\ \forall\ x$ between $-l$ and l and is

$$\frac{1}{2}[f(l-0) + f(-l+0)],$$

for x = – l and x = l and is periodic with period 2l.

Note. As in § 10.6, we can have half range sine series at will for a function f given in $[0, l]$.

10.8.3. Any Interval [*a*, *b*]

If f satisfies the conditions of the theorem of § 10.4 in [a, b], then the sum of the series

$$\frac{1}{2}a_0 + \Sigma\left\{a_n \cos\left[2n\pi x/(b-a)\right] + b_n \sin\left[2n\pi x/(b-a)\right]\right\}$$

where

$$a_n = \frac{2}{b-a}\int_a^b f(x)\cos\left(\frac{2n\pi x}{b-a}\right)dx,$$

$$b_n = \frac{2}{b-a}\int_a^b f(x)\sin\left(\frac{2n\pi x}{b-a}\right)dx$$

is $\frac{1}{2}[f(x+0) + f(x-0)]\ \forall\ x$ *between a and b and is* $\frac{1}{2}[f(a+0) +$ *f (b – 0)] and at x = b and is periodic with period (b – a).*

This result follows on writing

$$y^* = \frac{2\pi}{b-a}x - \frac{b+a}{b-a}\pi,$$

so that y varies in $[-\pi, \pi]$ as x varies in $[a, b]$.

---

* This transformation is obtained, if we determine the two constants t and m such that the relation $y = lx + m$ gives $y = -\pi$ when $x = a$ and $y = \pi$ when $x = b$.

Note. It will now be seen that we have obtained four different series which represent $(\pi - x)$ in $0 < x < \pi$ and it will naturally prove of great interest to examine the differences in the sums of these series for values of x other than those which belong to $]0, \pi[$. To do so, the reader would do well to draw the graphs of the four sum functions.

One important difference which must be emphasised is that the sum function of the series obtained above in § 10.8.3, is periodic with period π, whereas the former three sum functions were periodic with period 2π.

EXAMPLES

1. *Show that when* $0 < x < \pi$,

$$\pi - x = \frac{1}{2}x + \frac{\sin 2x}{1} - \frac{\sin 4x}{2} + \frac{\sin 6x}{3} + \ldots$$

Sol. Each term of the series is periodic with period π. Taking $a = 0$ and $b = \pi$ in § 10.8.3, we see that $a_0 = \pi$,

$$a_n = \frac{2}{\pi}\int_0^{\pi} (n - x) \cos 2nx \, dx = 0, \ (n \neq 1).$$

$$b_n = \frac{2}{\pi}\int_0^{\pi} (\pi - x) \sin 2nx \, dx = \frac{1}{n}.$$

Hence the result.

2. *Obtain Fourier's series for the expansion of* $f(x) = x \sin x$ *in the interval* $(-\pi, \pi)$*. Hence deduce*

$$\frac{\pi}{4} = \frac{1}{2} + \frac{1}{1.3} - \frac{1}{3.5} + \ldots$$ (Ajmer 90)

Sol. $f(x) = x \sin x = a_0 + \sum_{n=1}^{\infty} a_n \cos nx$, as $x \sin x$ is an even function of x.

$$a_0 = \frac{1}{\pi}\int_0^{\pi} x \sin x \, dx = \frac{1}{\pi}\cdot\pi = 1$$

and

$$a_n = \frac{2}{\pi}\int_0^{\pi} x \sin x \cos nx \, dx$$

$$= \frac{1}{(n+1)}\cos n\pi + \frac{\cos n\pi}{1-n} = \frac{2\cos n\pi}{(1-n^2)} \text{ if } n > 1$$

If $n = 1$, then

$$a_1 = \frac{2}{\pi}\int_0^{\pi} x \sin x \cos x \, dx = -\frac{1}{2}$$

$$\therefore \quad x \sin x = 1 + 2\left[-\frac{1}{2}\cos x + \sum_{n=2}^{\infty} \frac{\cos n\pi}{1-n^2}\cos nx\right]$$

Putting $x=\dfrac{\pi}{2}$ in both sides, we get

$$\frac{\pi}{2}=1+2\left[0+\frac{1}{1.3}+0-\frac{1}{3.5}+\ldots\right]$$

$$\Rightarrow \qquad \frac{\pi}{4}=\frac{1}{2}+\frac{1}{1.3}-\frac{1}{3.5}+\frac{1}{5.7}-\ldots$$

3. *Obtain Fourier's series whose sum is equal to f (x), where*

$$f(x)=0,\quad -\pi\le x<-\frac{\pi}{2};\quad f\left(-\frac{\pi}{2}\right)=-\frac{\pi}{4}$$

$$f(x)=x,\quad -\frac{\pi}{2}<x<\frac{\pi}{2};\quad f\left(\frac{\pi}{2}\right)=\pi/4,$$

$$f(x)=0,\quad \frac{\pi}{2}<x\le\pi.$$

Sol. Here the interval is $(-\pi,\pi)$.

Let $f(x)=a_0+\sum_{n=1}^{\infty}(a_n\cos nx+b_n\sin nx)$

$$a_0=\frac{1}{2\pi}\int_{-\pi}^{\pi}f(x)\,dx=\frac{1}{2\pi}\left[\int_{-\pi}^{-\pi/2}f(x)\,dx+\int_{-\pi}^{\pi/2}f(x)\,dx+\int_{\pi/2}^{\pi}f(x)\,dx\right]$$

$$=\frac{1}{2}p\left[\int_{-\pi}^{-\pi/2}0.dx+\int_{-\pi/2}^{\pi/2}x\,dx+\int_{\pi/2}^{\pi}0.dx\right]$$

$$=0.$$

$$a_n=\frac{1}{\pi}\left[\int_{-\pi}^{\pi/2}0.\cos nx\,dx+\int_{-\pi/2}^{\pi/2}x\cos nx\,dx+\int_{\pi/2}^{\pi}0.\cos nx\,dx\right]=0$$

$$b_n=\frac{1}{\pi}\int_{-\pi/2}^{\pi/2}x\sin nx\,dx$$

$$=\frac{2}{\pi}\left[-\frac{\pi}{2}\cdot\frac{\cos n\pi/2}{n}+\frac{\sin n\pi/2}{n^2}\right]$$

$$\therefore\quad f(x)=\frac{2}{\pi}\sum_{n=1}^{\infty}\left\{\frac{\sin n\pi/2}{n^2}-\frac{\pi\cos n\pi/2}{2\pi}\right\}\sin nx$$

4. *For all values of x between* $\frac{\pi}{2}$ *and* $-\frac{\pi}{2}$, *prove that*

$$x = \frac{4}{\pi}\left[\sin x - \frac{\sin 3x}{9} + \frac{\sin 5x}{25} + \ldots\right]$$

Sol. Here we shall change the interval $\left[-\frac{\pi}{2}, \frac{\pi}{2}\right]$ to $(0, \pi)$.

Let $x + \frac{\pi}{2} = z$, so that z lies between 0 and π.

Now, we shall expand z as a cosine series.

Let $f(z) = z = a_0 + \sum_{n=1}^{\infty} a_n \cos nz$.

$$a_0 = \frac{1}{\pi}\int_0^{\pi} z\,dz = \frac{\pi}{2}$$

$$a_n = \frac{2}{\pi}\int_0^{\pi} z \cos nz\,dz$$

$$= \frac{2}{\pi}\left[\frac{\cos n\pi - 1}{n^2}\right] = \frac{2}{\pi}\cdot\frac{\left[(-1)^n - 1\right]}{n^2}$$

$$\therefore \quad z = \frac{\pi}{2} - \frac{4}{\pi}\left(\cos z + \frac{1}{3^2}\cos 3z + \frac{1}{5^2}\cos 5z + \ldots\right)$$

Putting $z = \pi/2 + x$,

$$x + \frac{\pi}{2} = \frac{\pi}{2} - \frac{4}{\pi}\left(-\sin x + \frac{1}{3^2}\sin 3x - \frac{1}{5^2}\sin 5x + \ldots\right)$$

$$x = \frac{4}{\pi}\left(\sin x - \frac{1}{3^2}\sin 3x + \frac{1}{5^2}\sin 5x - \ldots\right)$$

5. *Find a series of sines of multiples of x which will represent f (x) in the interval* $(0, \pi)$, *where*

$$f(x) = \frac{\pi}{3}, \quad 0 < x < \frac{\pi}{3};$$

$$f(x) = 0, \quad \frac{\pi}{3} < x < \frac{2\pi}{3};$$

$$f(x) = -\frac{\pi}{3}, \quad \frac{2\pi}{3} < x < \pi.$$

Can you represent this function by a series of cosines of multiples of x as well ? Explain your answer. Draw graphs of these series and find the values of the sine and cosine series where $x = -\pi/3, -2\pi/3, -\pi$.

Sol. *For sine series.*

$$f(x)=\sum_{n=1}^{\infty} b_n \sin nx$$

$$b_n=\frac{2}{\pi}\int_0^{\pi} f(x)\sin nx\,dx$$

$$=\frac{2}{\pi}\left[\int_0^{\pi/3}\frac{\pi}{3}\sin nx\,dx+\int_{\pi/3}^{2\pi/3}0.\sin nx\,dx+\int_{2\pi/3}^{\pi}-\frac{\pi}{3}\sin nx\,dx\right]$$

$$=\frac{2}{3n}\left[-\frac{\cos n\pi}{3}+\cos n\pi-\frac{\cos 2n\pi}{3}\right]$$

$$=\frac{2}{3n}\left[1+\cos n\pi-\frac{\cos n\pi}{3}-\cos\left(n\pi-\frac{n\pi}{3}\right)\right]$$

$$=\frac{2}{3n}\left[(1+\cos n\pi)\left(1-\frac{\cos n\pi}{3}\right)\right]$$

$$=\frac{2}{3n}(1+\cos n\pi).2\sin^2\frac{n\pi}{6}$$

when n is odd, $b_n=0$

when n is even $b_n=\dfrac{8}{3n}\sin^2\dfrac{n\pi}{6}$.

If n is a multiple of 6, then

$$b_n=0.$$

Again if n is not multiple of 6, then $\dfrac{\sin n\pi}{6}\neq 0$

$$\therefore \qquad b_2=\frac{8}{3.2}\sin^2\frac{\pi}{3}=1,\ b_4=\frac{1}{2},\ b_8=\frac{1}{4}\text{ etc.}$$

$$\therefore \qquad f(x)=2\left[\frac{\sin 2x}{2}+\frac{\sin 4x}{4}+\frac{\sin 8x}{8}+\ldots\right]$$

For cosine series.

$$f(x)=a_0+\sum_{n=1}^{\infty} a_n\cos nx$$

$$a_0=\frac{1}{\pi}\left[\int_0^{\pi/3}\frac{\pi}{3}dx+\int_{\pi/3}^{2\pi/3}0.dx+\int_{2\pi/3}^{\pi}-\frac{\pi}{3}dx=0\right]$$

and $$a_n = \frac{2}{\pi}\left[\int_0^{\pi/3} \frac{\pi}{3}\cos nx\,dx + \int_{2\pi/3}^{\pi} -\frac{\pi}{3}\cos nx\,dx\right]$$

$$= \frac{2}{3\pi}\left[\sin\frac{n\pi}{3} + \sin\frac{2n\pi}{3}\right]$$

But $$\frac{\sin 2n\pi}{3} = (-1)^{n-1}\frac{\sin n\pi}{3}$$

$\therefore$ $$a_n = \frac{2}{3n}\left[1+(-1)^{n+1}\right]\frac{\sin n\pi}{3}$$

if n is even, $a_n = 0$

if $n = 3k$, $a_n = 0$

If n is not a multiple of 3 and is odd

$$a_n = \frac{2}{3n}(1+1)\frac{\sin n\pi}{3} = \frac{4}{3n}\frac{\sin n\pi}{3}$$

$\therefore$ $$f(x) = \frac{2}{\sqrt{3}}\left[\cos x - \frac{1}{5}\cos 5x + \frac{1}{7}\cos 7x + \frac{1}{11}\cos 11x + \dots\right]$$

Graph of sine series.

When $\pi < x < 2\pi$, let $x = 2\pi - x'$, *i.e.,* $0 < x' < \pi$.

Let the sine series be denoted by y, then

$$y = -2\left[\frac{1}{2}\sin 2x' + \frac{1}{4}\sin 4x' + \frac{1}{8}\sin 8x' + \dots\right]$$

$$= -f(x').$$

$\therefore$ when $\pi < x < \frac{4\pi}{3}$, $y = \frac{\pi}{3}$,

when $\frac{4\pi}{3} < x < \frac{5\pi}{3}$, $y = 0$ and

when $\frac{5\pi}{3} < x < 2\pi$, $y = -\pi/3$.

Again when $2\pi < x < 3\pi$, let $x = 2\pi + x'$, *i.e.,* $0 < x''$, *i.e.,* $0 < x'' <$ π

and

$$y = 2\left[\frac{1}{2}\sin 2x'' + \frac{1}{4}\sin 4x'' + \dots\right] = f(x'').$$

$\therefore$ when $2\pi < x < \frac{7\pi}{3}$, $y = \frac{\pi}{3}$,

when $\frac{7\pi}{3} < x < \frac{8\pi}{3}$, $y = 0$

and when $\frac{8\pi}{3} < x < 3\pi$, $y = -\pi/3$ etc.

For -ve values of x, when $-\pi < x < 0$, let $x = -z'$,

$\therefore \qquad -\pi < -z' < 0$, *i.e.*, $\pi > z' > 0$,

i.e., $\qquad 0 < z' < \pi$, and $y = -2\left[\frac{1}{2}\sin z' + \frac{1}{4}\sin 2z' + \ldots\right] = -f(z')$

when $\quad -\frac{\pi}{3} < x < 0,\ y = -\pi/3,$

when $\quad -\frac{2\pi}{3} < x < -\frac{\pi}{3},\ y = 0,$

and when $-\pi < x < -\frac{2\pi}{3},\ y = \frac{\pi}{3}$ etc.

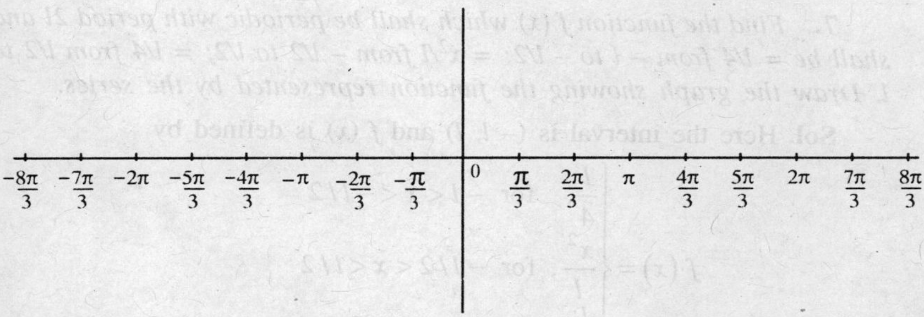

Hence the graph of the sine series is as

Similarly graph of cosine series may be drawn.

6. *If the function defined by* $y = x^2$ *from 0 to* $\frac{\pi}{2}$ *and* $y = 0$ *from* $\frac{\pi}{2}$ *to* π *be represented by a series of sines of multiples of x, show that the coefficient of* $\sin nx$ *is*

$$\left(\frac{4}{\pi n^3} - \frac{\pi}{2n}\right)\cos\frac{1}{2}n\pi + \frac{2}{n^2}\sin\frac{1}{2}n\pi - \frac{4}{\pi n^3}.$$

To what value does the series converge at the point $x = \pi/2$.

Sol. Here the interval is $(0, \pi)$ and let $y = f(x)$ that given

$$f(x) = \begin{cases} x^2, & \text{for } 0 < x < \pi/2 \\ 0, & \text{for } \pi/2 < x < \pi. \end{cases}$$

Let $f(x) \sum_{n=1}^{\infty} b_n \sin nx$

where $\qquad b_n = \frac{2}{\pi}\int_0^{\pi} f(x)\sin nx\, dx$

$$= \frac{2}{\pi}\int_0^{\pi/2} x^2 \sin nx\, dx$$

$$= \frac{2}{\pi}\left[-\frac{\pi^2}{4n}\cos\frac{n\pi}{2} + \frac{\pi}{n^2}\sin\frac{n\pi}{2} + \frac{2}{n^3}\cos\frac{n\pi}{2} - \frac{2}{n^3}\right]$$

$$= \left(\frac{4}{\pi n^3} - \frac{\pi}{2n}\right)\cos\frac{1}{2}n\pi + \frac{2}{n^2}\sin\frac{1}{2}n\pi - \frac{4}{\pi n^3}$$

and this is the coefficient of sin nx.

The series at $x = \pi/2$ converges to

$$\frac{1}{2}\left[f\left(\frac{\pi}{2}+0\right) + f\left(\frac{\pi}{2}-0\right)\right] = \lim_{h\to 0}\frac{1}{2}\left[0 + \left(\frac{\pi}{2}-h\right)^2\right] = \frac{\pi^2}{8}.$$

7. *Find the function f (x) which shall be periodic with period 2l and shall be = l/4 from – l to – l/2; = x^2/l from – l/2 to l/2; = l/4 from l/2 to l. Draw the graph showing the function represented by the series.*

Sol. Here the interval is $(-l, l)$ and $f(x)$ is defined by

$$f(x) = \begin{cases} \dfrac{l}{4}, & \text{for } -l < x < -l/2 \\ \dfrac{x^2}{l}, & \text{for } -l/2 < x < l/2 \\ \dfrac{l}{4}, & \text{for } l/2 < x < l. \end{cases}$$

Obviously, if x be changed to $-x$, $f(x)$ remains unaltered. Therefore we will have only cosine series.

$$a_0 = \frac{1}{l}\int_{-l}^{l} f(x)\,dx$$

$$= \frac{1}{l}\left[\int_{-l}^{-l/2}\frac{l}{4}\,dx + \int_{-l/2}^{l/2}\frac{x^2}{l}\,dx + \int_{l/2}^{l}\frac{l}{4}\,dx\right] = l/3.$$

and
$$a_n = \frac{1}{l}\left[\int_{-l}^{-l/2}\frac{l}{4}\cos\frac{n\pi x}{l}\,dx + \int_{-l/2}^{l/2}\frac{x^2}{l}\cos\frac{n\pi x}{l}\,dx + \int_{l/2}^{l}\frac{l}{4}\cos\frac{n\pi x}{l}\,dx\right]$$

$$= \frac{1}{l}\cdot\frac{2}{l}\left[\frac{l^3}{n^2\pi^2}\cos\frac{n\pi}{2} - \frac{2l^3}{n^3\pi^3}\sin\frac{n\pi}{2}\right]$$

$$= \frac{2l}{n^2\pi^2}\left[\cos\frac{n\pi}{2} - \frac{2}{n\pi}\sin\frac{n\pi}{2}\right]$$

$$\therefore \quad f(x) = \frac{l}{6} + \frac{2l}{\pi^2}\sum_{n=1}^{\infty}\left\{\left(\frac{1}{n^2}\cos\frac{n\pi}{2} - \frac{2}{n^3\pi}\sin\frac{n\pi}{2}\right)\cos\frac{n\pi x}{l}\right\}$$

The graph consists of equal arcs of a parabola $y = x^2/l$ in $(-l/2, l/2)$ and straight lines $y = l/4$ in $(-l, -l/2)$ and $y = l/4$ in $(l/2, l)$.

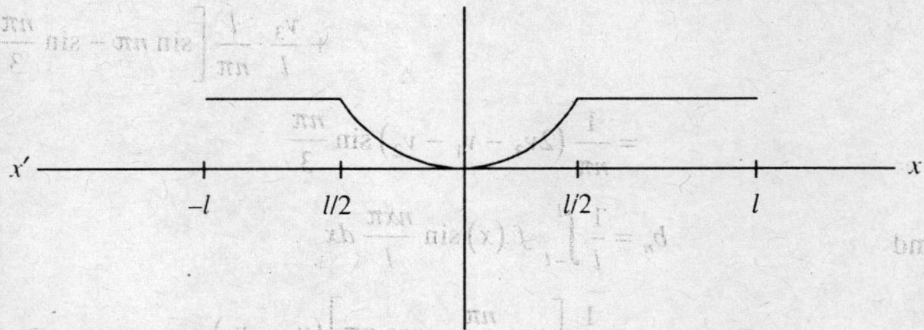

8. *Show that the series*

$$\frac{v_1 + v_2 + v_3}{3} + \frac{l}{\pi} \sum_{n=1}^{\infty} \frac{l}{n} \sin \frac{n\pi}{3} \left\{ 2(v_3 - v_1) \right.$$

$$\left. \sin \frac{2n\pi}{3} \sin \frac{n\pi x}{l} + (2v_3 - v_1 - v_2) \cos \frac{2\pi x}{l} \right\}$$

is equal to v_1 *when* $-l < x < -l/3$,

v_2 *when* $-l/3 < x < l/3$

v_3 *when* $l/3 < x < l$

where v_1, v_2, v_3 *are constants.*

Sol. The interval is $(-l, l)$.

Let
$$f(x) = \frac{a_0}{2} + \sum_{n=1}^{\infty} \left(a_n \cos \frac{n\pi x}{l} + b_n \sin \frac{n\pi x}{l} \right)$$

$$a_0 = \frac{1}{l} \int_{-l}^{l} f(x)\, dx$$

$$= \frac{1}{l} \int_{-l}^{-l/3} v_1\, dx + \frac{1}{l} \int_{-l/3}^{l/3} v_2\, dx + \frac{1}{l} \int_{l/3}^{l} v_3\, dx$$

$$= \frac{2}{3}(v_1 + v_2 + v_3)$$

$$a_n = \frac{1}{l} \left[\int_{-l}^{-l/3} v_1 \cos \frac{nx\pi}{l}\, dx + \int_{-l/3}^{l/3} v_2 \cos \frac{nx\pi}{l}\, dx \right.$$

$$\left. + \int_{l/3}^{l} v_3 \cos \frac{nx\pi}{l}\, dx \right]$$

$$= \frac{v_1}{l} \cdot \frac{l}{n\pi} \left[\sin n\pi - \sin \frac{n\pi}{3} \right] + \frac{v_2}{l} \cdot \frac{l}{n\pi} \cdot 2 \sin \frac{n\pi}{3}$$

$$+ \frac{v_3}{l} \cdot \frac{l}{n\pi} \left[\sin n\pi - \sin \frac{n\pi}{3} \right]$$

$$= \frac{1}{n\pi} (2v_3 - v_1 - v_2) \sin \frac{n\pi}{3}$$

and $$b_n = \frac{1}{l} \int_{-l}^{l} f(x) \sin \frac{nx\pi}{l} dx$$

$$= \frac{1}{n\pi} \left[\cos \frac{n\pi}{3} - \cos n\pi \right] (v_3 - v_1)$$

$$= \frac{2}{n\pi} (v_3 - v_1) \sin \frac{2n\pi}{3} \sin \frac{n\pi}{3}.$$

Hence the result.

EXERCISES

1. If $f(x) = -\frac{1}{4}\pi$ when $-\pi < x < 0$, $f(x) = \frac{1}{4}\pi$ when $0 < x < \pi$, $f(-\pi) = f(0) = f(x) = 0$ and $f(x + 2\pi) = f(x)$ for all x, show that $\forall\ x$

$$f(x) = \sin x + \frac{1}{3} \sin 3x + \frac{1}{5} \sin 5x + \ldots$$

Deduce that

$$\frac{1}{4}\pi = 1 - \frac{1}{3} + \frac{1}{5} - \frac{1}{7} + \ldots$$

2. If $f(x) = cx$ when $-\pi < x < \pi$, $f(-\pi) = f(\pi) = 0$, and $f(x + 2\pi) = f(x) \forall x$, show that

$$f(x) = 2c \left[\sin x - \frac{1}{2} \sin 2x + \frac{1}{3} \sin 3x - \frac{1}{4} \sin 4x + \ldots \right]$$

for all values of x. Draw the graph of $f(x)$.

3. Find the Fourier series of the following function

$$f(x) = \begin{cases} -\pi & \text{for } -\pi < x < 0 \\ x & \text{for } 0 < x < \pi \end{cases}$$

Hence deduce that

$$\frac{\pi^2}{8} = \frac{1}{1^2} + \frac{1}{3^2} + \frac{1}{5^2} + \ldots$$ *(Ajmer 9*

4. Obtain Fourier series which will be equal to $-\pi - x$ when $-\pi \le x < -\frac{1}{2}\pi$, equal to x when $-\frac{1}{2}\pi \le x < \frac{1}{2}\pi$ and equal to $n - x$ when $\frac{1}{2}\pi \le x \le \pi$. Explain graphically how to obtain the sum of the series for any value of x.

5. $f(x) = \cos x$ for $0 < x < \pi$ and $f(x) = -\cos x$ for $-\pi < x < 0$; show that the Fourier series which converges to $f(x)$ is

$$\frac{\pi}{4}\left(\frac{2}{1.3}\sin 2x + \frac{4}{3.5}\sin 4x + \frac{6}{5.7}\sin 6x + \ldots\right).$$

Draw the graph of the sum function of the series for $-2\pi \le x \le 2\pi$.

(Ajmer 97)

6. Find the Fourier series which represents $|\sin x|$ in $-\pi \le x \le \pi$.

7. (*a*) Show that for $-\pi < x < \pi$,

$$e^x = \frac{e^n - e}{\pi}\left[\frac{1}{2} + \sum_{n=1}^{\infty} \frac{(-1)^n}{n^2+1}(\cos nx - n \sin nx)\right]$$

What is the sum of the series for $x = \pm\pi$.

(*b*) Show that for $0 < x < \pi$,

$$e^x = \begin{cases} \dfrac{2}{\pi}\displaystyle\sum_{n=1}^{\infty}\left(1-(-1)^n e\pi\right)\frac{n \sin nx}{n^2+1} \\ \dfrac{e^{\pi}-1}{\mu} - \dfrac{2}{\pi}\displaystyle\sum_{n=1}^{\infty}\left(1-(-1)^n\right) e\pi \frac{\cos nx}{n^2+1}. \end{cases}$$

What are the sums of these series for $-\pi \le x \le 0$.

8. Show that when $-\pi \le x \le \pi$,

$$\cos kx = \frac{\sin k\pi}{\pi}\left(\frac{1}{k} - \frac{2k\cos x}{k^2 - 1^2} + \frac{2k \cos 2x}{k^2 - 2^2} \ldots\right)$$

k being non-integral. Deduce that

$$\pi \cot k\pi = \frac{1}{k} + 2k\,\Sigma \frac{1}{k^2 - n^3},$$

and

$$\frac{\pi}{\sin k\pi} = \Sigma(-1)^n\left(\frac{1}{n+k} + \frac{1}{n+1-k}\right).$$

9. Obtain a sine series which will be equal to x^2 for $0 \le x < \pi$. Draw a graph of the sum function of the series for $-2\pi \le x \le 2\pi$.

10. Show that in $[0, \alpha]$,

$$f(x)=\frac{\pi^2}{16}-2\sum_{n=1}^{\infty}\frac{\cos(4n-2)x}{(4n-2)^2},$$

where $f(x)=\frac{1}{4}\pi x$ when $0 \le x \le, \frac{1}{2}\pi$ and $f(x)=\frac{1}{4}\pi(\pi-x)$, when $\frac{1}{2}\pi < x \le \pi$.

11. Prove that when $\theta < 0 < 2\pi$,

$$\frac{1}{2}(n-\theta)=\sum_{n=1}^{\infty}\frac{\sin n\theta}{n}.$$

What is the sum of the series for $\theta = 0$ and $\theta = 2\pi$?

12. Prove that when $-1 < x < 1$,

$$x+x^2=\frac{1}{3}+\frac{2}{\pi}\sum_{n=1}^{\infty}(-1)^n\left[\frac{2\cos n\pi x}{n^2\pi}-\frac{\sin n\pi x}{n}\right].$$

13. Find (*i*) the series of sines, (*ii*) the series of cosines, which will represent $f(x)$ in $]\,0, l\,[$, where

$$f(x)=kx, \text{ when } 0 \le x \le \frac{1}{2}l,\ f(x)=k\left(l-x \text{ when } \frac{1}{2}l < x \le l.\right)$$

14. If ϕ be a periodic function of period 4 such that $\phi(x)=0$ when $0 < x < 1$ and $\phi(x)=x-1$ when $1 \le x < 2$, and $\phi(2)=0$, express $\phi(x)$ as a series of sines of multiples of x. Draw the graph of $\phi(x)$.

15. Show that for $0 \le x \le 1$,

$$x-\frac{1}{2}=-\frac{1}{\pi}\left\{\sum_{n=1}^{\infty}\frac{\sin 2n\pi x}{n}\right\},\ x^2-x+\frac{1}{6}=\frac{1}{x^2}\left\{\sum_{n=1}^{\infty}\frac{\cos 2n\pi x}{n^2}\right\}.$$

16. If f be a periodic function of period $\frac{1}{2}\pi$ such that $f(x) = \sin x$ for $0 \le x \le \frac{1}{4}\pi$ and $f(x) = \cos x$ for $\frac{1}{4}\pi \le x \le \frac{1}{2}\pi$, express $f(x)$ as a Fourier series.

17. If $f(x) = 1$ when $0 < x < 1$, $f(x) = 2$ when $1 < x < 3$ and $f(x)=\frac{3}{4}$ when $x = 0$, 1 and 3 and $f(x+3)=f(x)\ \forall\, x$; show that

$$f(x)=\frac{5}{9}-\frac{1}{\pi}\sum_{n=1}^{\infty}\frac{2\sin\frac{1}{3}n\pi}{n}\cos\frac{n\pi(2x-1)}{}\ \forall\, x.$$

18. If $f(x)=\frac{1}{4}a-x$ when $0 \le x \le \frac{1}{2}a$ and $f(x)=x-\frac{3}{4}a$ where $\frac{1}{2}a \le x \le a$, show that $\forall\, x$ in $0 \le x \le a$,

$$f(x)=\frac{2a}{\pi^2}\sum_{n=1}^{\infty}\frac{1}{(2n-1)^2}\cos\frac{(4n-2)\pi x}{a}.$$

19. Show that if a function f is bounded in $\left[0, \frac{1}{2}\pi\right]$ and is such that it is possible to divide $\left[0, \frac{1}{2}\pi\right]$ into a finite number of open sub-intervals in each of which f is monotonic, then $f(x)$ can be expressed in the form

$$b_1 \sin x + b_3 \sin 3x + b_5 \sin 5x + \ldots$$

for $0 < x < \frac{1}{2}\pi$ and obtain the coefficients b_n and the sum of the series outside $\left]0, \frac{1}{2}\pi\right[$.

If $f(x) = \sin x$ when $0 \le x \le \frac{1}{4}\pi$ and $f(x)$ cos x when $\frac{1}{4}\pi \le x \le \frac{1}{2}\pi$, show that

$$f(x) = \frac{\sin x}{2} + \frac{2}{\pi} + \left(\frac{\sin 2x}{1.2} - \frac{\sin 5x}{2.3} + \frac{\sin 11x}{5.6} - \frac{\sin 13x}{6.7} + \ldots\right)$$

20. Expand in a Fourier series of the form $a_n \sin n\pi x$ a function f given by

$$f(x) = \sinh \pi x, \text{ for } 0 \le x < \frac{1}{2}, \; f(0) = 0 \text{ for } \frac{1}{2} < x \le 1.$$

Deduce that

$$\sum_{n=1}^{\infty} \frac{1}{n^2 + (n+1)^2} = \frac{\pi}{2} \tanh \frac{\pi}{2}.$$

21. Show that the graph of the equation

$$y^2 = \frac{a^2}{3} + \frac{16a^2}{\pi^2} \sum_{n=1}^{\infty} \frac{(-1)^{n-1}}{(2n-1)^3} \cos \frac{(2n-1)\pi x}{2a} + \frac{8a^2}{\pi^2} \sum_{n=1}^{\infty} \frac{(-1)^{n-1}}{(2n)^2} \cos \frac{n\pi x}{a}$$

is a series of circles of radius a connected by straight lines of length $2a$, the origin being the centre of one of the circles.

[Expand $f(x)$ in [$-2a$, $2a$] where $f(x) = 0$ where $-2a \le x \le -a$, $f(x) = a^2 - x^2$ when $-a \le x \le a$; $f(x) = 0$ when $a \le x \le 2a$.]

OBJECTIVE QUESTIONS

Note. *For each of the following questions, four alternatives are given for the answer. Only one of them is correct. Choose the correct answer.*

1. When a periodic function is symmetric about the time-axis, its steady level A_0 will be

(*a*) finite (*b*) constant

(*c*) zero (*d*) indeterminate

2. A function can be expanded by Fourier series, if it is

(*a*) finite and continuous only

(*b*) single valued only

(*c*) finite, single-valued and continuous

(*d*) none of these

3. If f be an odd function, then Fourier series contains the terms :

(*a*) with sines only (*b*) with cosines only

(*c*) with sines and cosines both (*d*) constant only

4. The sum functions of the half range sine and cosine Fourier series are

(*a*) non-periodic (*b*) periodic with period π

(*c*) periodic with period 2π (*d*) periodic with period $\pi/2$

ANSWERS

1. (*c*) **2.** (*c*) **3.** (*a*) **4.** (*c*)

11

Euclidean Space. Open and Closed Sets. Compact Sets

11.1. INTRODUCTION

In chapters 4, 5 and 6 we have been concerned with the continuity, derivability and integrability properties of functions whose domains and ranges are sub-sets of the set **R** of real numbers. These properties are basically dependent upon what may be called the Algebraic and Topological structures of **R**; the algebraic structure referring to the addition and multiplication compositions satisfying certain properties of commutativity, associativity, distributivity etc., and the topological structure to that aspect of **R** which renders the notion of a 'Neighbourhood' of a point meaningful. The importance of the notion of 'Neighbourhood' of a point **R** may be seen from the fact that this notion is basic in our formulation of all the different forms of limit with which we have been so far concerned.

The rest of the book is to be devoted to the study of real valued functions of several real variables. Thus while the range of such a function is a sub-set of **R**, the domain will be of a different type. It is thus natural to first learn to describe such domains and equip the same with topological structure.

11.2. THE SET R^n

The set of all ordered n-tuples of real numbers is denoted by the symbol $\mathbf{R}^n$. Thus the n-tuples

$$(x_1, \ldots, x_n)$$

where $x_1, \ldots, x_n$ are real numbers are the members of $\mathbf{R}^n$. Each of the numbers $x_1, \ldots, x_n$ is called a Coordinate or Component of the n-tuple.

We shall denote the elements of $\mathbf{R}^n$ by lower case black letters **x, y, z,** etc., so that each of **x, y, z,** etc. stands for an ordered n-tuple of real numbers.

Algebraic structure. We now define two compositions in the set $\mathbf{R}^n$ *viz.,* addition and scalar multiplication.

If $$\mathbf{x} = (x_1, ..., x_n), \quad \mathbf{y} = (y_1, ..., y_n)$$
we define
$$\mathbf{x} + \mathbf{y} = (x_1 + y_1, ..., x_n + y_n),$$
and
$$cX = (cx_1, ..., cx_n)$$
for any real number c.

Also we write
$$\mathbf{0} = (0, ..., 0),$$
$$-\mathbf{x} = (-x_1, ..., -x_n)$$

Ex. Show that $\forall\, \mathbf{x}, \mathbf{y}, \mathbf{x} \in \mathbf{R}^n, \quad a, b \in \mathbf{R}$,
$$\mathbf{x} + \mathbf{y} = \mathbf{y} + \mathbf{x}, \qquad \mathbf{x} + (\mathbf{y} + \mathbf{z}) = (\mathbf{x} + \mathbf{y}) + \mathbf{z}$$
$$\mathbf{x} + \mathbf{0} = \mathbf{x}, \ \mathbf{x} + (-\mathbf{x}) = \mathbf{0}.$$
$$(a + b)\,\mathbf{x} = a\mathbf{x} + b\mathbf{x}, \ a\,(\mathbf{x} + \mathbf{y}) = a\mathbf{x} + a\mathbf{y}$$
$$ab\,(\mathbf{x}) = a\,(b\mathbf{x}).$$
$$1\mathbf{x} = \mathbf{x}.$$

Note. A point in $\mathbf{R}^2$ is often denoted as (x, y) in place of (x_1, x_2) and that in $\mathbf{R}^3$ as (x, y, z) in place of (x_1, x_2, x_3).

11.2.1. Distance in $\mathbf{R}^n$

Before introducing the notion of neighbourhood of points of $\mathbf{R}^n$, we shall define distance in $\mathbf{R}^n$.

If $$\mathbf{x} = (x_1, ..., x_n), \quad \mathbf{y} = (y_1, ..., y_n),$$
we write
$$d\,(\mathbf{x}, \mathbf{y}) = \sqrt{\left[\sum_{i=1}^{n} (x_i - y_i)^2\right]}$$
and describe $d\,(\mathbf{x}, \mathbf{y})$ as the distance between the points $\mathbf{x}, \mathbf{y}$.

Norm. If $\mathbf{x} = (x_1, ..., x_i, ..., x_n) \in \mathbf{R}^n$, we write
$$\|x\| = \sqrt{\left(\sum_{i=1}^{n} x_i^2\right)}$$
so that $\|x\|$ is a non-negative real number. This number $\|x\|$ which denotes the distance between $\mathbf{x}$ and $\mathbf{o}$ is called the **Norm** of $\mathbf{x}$. It will be seen that the norm plays the same part as the absolute value of a real number.

Also $$d\,(\mathbf{x}, \mathbf{y}) = \|\mathbf{x} - \mathbf{y}\|.$$

As we proceed, it will be seen that *precise* definition of distance between two points is not of much significance and it is only a few salient properties of the distance which are needed. These properties are given below.

The Euclidean space $\mathbf{R}^n$. The set $\mathbf{R}^n$ equipped with the algebraic structure and the distance relation defined as above will be called the Euclidean space of dimension *n*.

We shall show that $d(\mathbf{x}, \mathbf{y})$ as defined above satisfies the following properties :

(*i*) $d(\mathbf{x}, \mathbf{y}) \geq 0 \ \forall\ \mathbf{x}, \mathbf{y} \in \mathbf{R}^n$ and $d(\mathbf{x}, \mathbf{y}) = 0 \Leftrightarrow \mathbf{x} = \mathbf{y}$,

(*ii*) $d(\mathbf{x}, \mathbf{y}) = d(\mathbf{y}, \mathbf{x}) \ \forall\ \mathbf{x}, \mathbf{y} \in \mathbf{R}^n$,

(*iii*) $d(\mathbf{x}, \mathbf{y}) \leq d(\mathbf{x}, \mathbf{z}) + d(\mathbf{z}, \mathbf{y}) \ \forall\ \mathbf{x}, \mathbf{y}, \mathbf{z} \in \mathbf{R}^n$.

The properties (*i*) and (*ii*) are immediate consequences of the definition of $d(\mathbf{x}, \mathbf{y})$. We shall now prove in § 11.2.2 that $d(\mathbf{x}, \mathbf{y})$ also satisfies the property (*iii*) which is known as *Triangle Inequality.*

11.2.2. Triangle Inequality

To prove

$$d(\mathbf{x}, \mathbf{y}) \leq d(\mathbf{x}, \mathbf{z}) + d(\mathbf{z}, \mathbf{y})$$

$\forall\ \mathbf{x}, \mathbf{y}, \mathbf{z} \in \mathbf{R}^n$.

Let $\mathbf{x} = (x_1, \ldots, x_i, \ldots, x_n)$, $\mathbf{y} = (y_1, \ldots, y_i, \ldots, y_n)$.

Firstly we prove what is called Cauchy-Schwartz inequality, *viz.*,

$$\left| \sum_{i=1}^{n} x_i y_i \right| \leq \sqrt{\left[\sum_{i=1}^{n} x_i^2\right]} \sqrt{\left[\sum_{i=1}^{n} y_i^2\right]}.$$

Now $\forall\ x \in \mathbf{R}$, we have

$$\sum_{i=1}^{n} (x_i . x + y_i)^2 \geq 0$$

$$\left(\sum_{i=1}^{n} x_i^2\right) x^2 + 2\left(\sum_{i=1}^{n} x_i y_i\right) x + \sum_{i=1}^{n} y_i^2 \geq 0$$

$$Ax^2 + 2Bx + C \geq 0$$

where we have written

$$A = \sum_{i=1}^{n} x_i^2, \quad B = \sum_{i=1}^{n} x_i y_i, \quad C = \sum_{i=1}^{n} y_i^2.$$

Surely $A \geq 0$. We suppose that $A \neq 0$, for $A = 0 \Rightarrow x = 0$.

Now $\qquad \mathbf{A}x^2 + 2\mathbf{B}x + \mathbf{C} = \left[(\mathbf{A}x + \mathbf{B})^2 + (\mathbf{AC} - \mathbf{B}^2) / \mathbf{A} \ \forall\ x\right]$

Thus $Ax^2 + 2Bx + C \geq 0 \Rightarrow A > 0$ and $AC - B^2 \geq 0.$

Also $AC - B^2 \geq 0 \Rightarrow AC \geq B^2 \Rightarrow (AC) \geq |B|$

so that the Cauchy Schwartz inequality is established.

Now $$\|\mathbf{x} + \mathbf{y}\| = \sum_{i=1}^{n} (x_i + y_i)^2$$

$$= \sum_{i=1}^{n} x_i^2 + \sum_{i=1}^{n} y_i^2 + 2\sum_{i=1}^{n} x_i y_i$$

$$\leq \sum_{i=1}^{n} x_i^2 + \sum_{i=1}^{n} y_i^2 + 2\sqrt{\left[\sum_{i=1}^{n} x_i^2\right]}\sqrt{\left[\sum_{i=1}^{n} y_i^2\right]}$$

$$= \left\{\sqrt{[\Sigma x_i^2]} + \sqrt{[\Sigma y_i^2]}\right\}^2 = \{\|\mathbf{x}\| + \|\mathbf{y}\|\}^2$$

$$\Rightarrow \qquad \|\mathbf{x} + \mathbf{y}\| \leq \|\mathbf{x}\| + \|\mathbf{y}\|.$$

Replacing **x** and **y** by **x** – **z** and **z** – **y** respectively, we obtain

$$\|\mathbf{x} - \mathbf{y}\| \leq \|\mathbf{x} - \mathbf{z}\| + \|\mathbf{z} - \mathbf{y}\|$$

$$\Leftrightarrow \qquad \mathbf{d}(\mathbf{x}, \mathbf{y}) \leq \mathbf{d}(\mathbf{x}, \mathbf{z}) + \mathbf{d}(\mathbf{y}, \mathbf{z}).$$

11.3. NEIGHBOURHOODS OF A POINT

Spheres in $\mathbf{R}^n$. If $\mathbf{a} \in \mathbf{R}^n$ and r is any positive real number, we write

$$S(\mathbf{a}, r) = \{\mathbf{x} : \|\mathbf{x} - \mathbf{a}\| < r\}$$

and call the set an *open sphere* with centre **a** and radius r.

The set

$$\{\mathbf{x} : \|\mathbf{x} - \mathbf{a}\| \leq r\}$$

is called a *closed sphere* with centre **a** and radius r and is denoted by

$$\bar{S}(\mathbf{a}, r).$$

Spherical Neighbourhoods of a point. Any open sphere with **a** as its centre is called a spherical neighbourhood of the point **a.** Thus S (**a**, **r**) is a spherical neighbourhood of the point $\mathbf{a} \ \forall \ r > 0$.

Rectangles. If

$$\mathbf{a} = (a_1, ..., a_i, a_n), \text{ and } \mathbf{b} = (b_1, ..., b_i, ..., b_n)$$

be two points such that

$$a_1 < b_1, ..., a_n < b_n$$

then the set of points

$$\mathbf{x} = (x_1, ..., x_i, ..., x_n)$$

where

$$a_1 < x_1 < b_1, ..., a_n < x_n < b_n$$

is called an open rectangle in $\mathbf{R}^n$ denoted by

$$]\,\mathbf{a}, \mathbf{b}\,[\text{ or }]\,a_1, b_1; \;...;\; a_n, b_n\,[$$

Again the set of points **x** such that

$$a_1 \le x_1 \le b_1, ..., a_n \le x \le b_n$$

is called a *closed rectangle* denoted by

$$[\mathbf{a}, \mathbf{b}] \text{ or } [a_1, b_1; ..., a_n, b_n].$$

Rectangular neighbourhoods. Any open rectangle containing a point is also called a rectangular neighbourhood of the point.

We shall now show that *every spherical neighbourhood of a point contains a rectangular neighbourhood and vice-versa.*

Let $S(\mathbf{a}, r)$ be any spherical neighbourhood of **a**, If we write

$$\mathbf{r} = \left(\frac{1}{2}r, ..., \frac{1}{2}r\right),$$

then it may be seen that the open rectangle

$$\left]\mathbf{a} - \frac{1}{n}\mathbf{r}, \mathbf{a} + \frac{1}{n}\mathbf{r}\right[$$

is contained in the open sphere

$$S(\mathbf{a}, r).$$

In fact, if **x** be any point of this open rectangle, we have

$$\|\mathbf{x} - \mathbf{a}\|^2 < n \cdot \frac{1}{n^2} r^2 = \frac{r^2}{n} < r^2$$

$\Rightarrow \quad \mathbf{x} \in S(\mathbf{a}, r)$. Thus

$$\left]\mathbf{a} - \frac{1}{n}r, \mathbf{a} + \frac{1}{n}r\right[\subset S(\mathbf{a}, r).$$

It follows that every given spherical neighbourhood of a point contains rectangular neighbourhoods of the point.

Now consider any rectangular neighbourhood] **b, c** [of a point **a**. We shall show that it contains a spherical neighbourhood.

Let $\quad r = \min.\{a_i - b_i, c_i - a_i\},$

$1 \le i \le n$

where $\quad \mathbf{a} = (a_1, ..., a_n), \mathbf{b} = (b_1, ..., b_n), \mathbf{c} = (c_1, ..., c_n).$

We have $r \le a_i - b_i,\ r \le c_i - a_i;\ 1 \le i \le n$

so that $b_i \le a_i - r,\ c_i \ge a_i + r.$ ...(*i*)

We shall show that S (**a**, r) is contained in [b, c [

i.e., $S(\mathbf{a}, r) \subset]\,\mathbf{b}, \mathbf{c}\,[.$

If **x** be any point of S (**a**, r), we have

$$\|\mathbf{x} - \mathbf{a}\|^2 < r^2$$

$$|x_i - a_i| < r,\ 1 \le i \le n$$

$$a_i - r < x_i < a_i - r, |, 1 \le i < n \qquad \text{...(ii)}$$

From (*i*) and (*ii*), we obtain

$$b_i < x_i < c_i, 1 \le i \le n$$

$$\Rightarrow \quad \mathbf{x} \in]\,\mathbf{b}, \mathbf{c}\,[\;\Rightarrow\; S(\mathbf{a}, r) \subset [\,\mathbf{b}, \mathbf{c}]$$

Note. We shall in the following use the word 'Neighbourhood' instead of Rectangular or Spherical neighbourhood. In fact because of the result : *Every spherical neighbourhood of a point contains a rectangular neighbourhood and vice-versa,* its suffices only to consider one type of neighbourhoods and no generality is to be gained by considering both the types. In fact, we shall often be dealing with spherical neighbourhoods inasmuch as such neighbourhoods have a very natural relationship with the distance defined in $\mathbf{R}^n$.

11.4. LIMIT POINTS OF A SET

Let E be any sub-set of $\mathbf{R}^n$. *A point* **a** $\in$ $\mathbf{R}^n$ *is called a limit point of E, if every neighbourhood of the point* **a** *contains an infinite number of points of the set E.*

Illustrations

1. In $\mathbf{R}^2$, the point (0, 0) is a limit point of the set

$$\left\{\left(\frac{1}{m}, \frac{1}{n}\right);\ m \in \mathbf{N}, n \in \mathbf{N}\right\}.$$

2. Every point of $\mathbf{R}^2$ is a limit point of the set

$$\mathbf{Q}^2 = \{(x, y) : x \in \mathbf{Q}, y \in \mathbf{Q}\}.$$

Ex. 1. Show that a point **a** is a limit point of $E \subset \mathbf{R}^n$ if and only if every neighbourhood of **a** contains a point of **E** other than **a**.

Ex. 2. What do you mean by saying that **a** is not a limit point of **E.**

Ex. 3. If **A** $\in$ **B,** show that every limit point of **A** is as well a limit point of **B.**

Closed sets. *A non-void set is said to be closed if every limit point thereof belongs to the set.*

A void set is also called a closed set.

It may be shown that closed spheres and closed rectangles are closed sets.

Open sets. *A non-void set is said to be open if some neighbourhood of every point of the set is contained in the set.*

A void set is also called open.

Open spheres and open rectangles are open sets.

Theorem. *A set is closed if and only if its complement is open.*

Let A be a closed set and let B its complement. Let $\mathbf{x} \in B$. Of course $\mathbf{x} \notin A$.

Since A is closed, the point **x**, being not a number of **A**, is not a limit point of A. Thus there exists a neighbourhood of **x** containing no point of A, and as such, contained in the complement of A. It follows that the set B contains a neighbourhood of each of its points and is as such an open set.

Now suppose that A is an open set. Let **x** be any limit point of the complement B of A. If **x** does not belong to B, it belongs to the open set A and as such there exists a neighbourhood of **x** contained in A and therefore containing no point of B. This implies that **x** is not a limit point of B so that we arrive at a contradiction. Thus the complement B of an open set A is closed.

Theorem. *The union of any family of open sets is open.*

The proof which is very simple is left to the reader.

Theorem. *The intersection of any finite family of open sets is open.*

Let

$$G_1 \ldots\ldots\ldots\ldots G_k \qquad \ldots(i)$$

be any k open sets.

The intersection, if void, is open.

Suppose that the intersection which we denote by G is not void. Let

$$\mathbf{a} \in G.$$

Since **a** belongs to each of the open sets (1), there exists open spheres

$$S(\mathbf{a}, \mathbf{r}_1), \ldots, S(\mathbf{a}, \mathbf{r}_k)$$

such that

$$S(\mathbf{a}, r_1) \subset G_1, \ldots, S(\mathbf{a}, r_k) \subset G_k$$

Let $\mathbf{r} = \min \{r_1, \ldots, r_k\}$.

Clearly then $S(\mathbf{a}, \mathbf{r}) \subset G$.

Thus G is an open set.

Theorem. *The intersection of any family of closed sets is closed.*

Theorem. *The union of any finite family of closed sets is closed.*

The proofs, being simple and straightforward, are left to the reader.

Bounded sets. Def. A set is *said to be bounded, if it is contained in some sphere S* (**0**, *k*).

Here, **0** is the *n*-tuple (0, 0, ..., 0).

This means that a set E is bounded if there exists $k > 0$ such that

$$\|\mathbf{x}\| < k \ \forall \ \mathbf{x} \in E.$$

Ex. For a given set E, there exists a number $k > 0$ such that

$$\|\mathbf{x} - \mathbf{y}\| < k \ \forall \ \mathbf{x}, \mathbf{y} \in E.$$

Show that E is a bounded set.

11.5. BOLZANO-WEIERSTRASS THEOREM

Every infinite and bounded set has at least one limit point.

While the idea constituting the basis of the proof for subsets of $\mathbf{R}^n$ is the same as that for subsets of $\mathbf{R}^2$, we shall give the proof for the subsets of $\mathbf{R}^2$. The proof for the subsets of $\mathbf{R}^n$ which is only a little more cumbersome in details may easily be supplied by the reader once he has followed the proof for subsets of $\mathbf{R}^2$.

Let A be any infinite and bounded subset of $\mathbf{R}^2$. Being bounded, there exists $k > 0$ such that

$$A \subset S(\mathbf{0}, k),$$

We have, as may be easily seen,

$$S(\mathbf{0}, k) \subset [-k, k; \ -k, k]$$

We write $\quad [-k, k; \ -k, k] = [a, b; \ c, d].$

Now $\quad [a, b; \ c, d]$

is the union of the four sub-rectangles

$$\left[a, \frac{1}{2}(a+b) : c, \frac{1}{2}(c+d)\right], \left[a, \frac{1}{2}(a+b); \frac{1}{2}(c+d), d\right]$$

$$\left[\frac{1}{2}(a+b), b; c, \frac{1}{2}(c+d)\right], \left[\frac{1}{2}(a+b), b; \frac{1}{2}(c+d), d\right]$$

one at least of which must contain an infinite number of points of A. Let $a_1, b_1; c_1, d_1]$ denote any one of the rectangles containing an infinite number of points of A. We thus have

$$a \le a_1 < b_1 \le b, \ c \le c_1 < d_1 \le d;$$

$$b_1 - a_1 = \frac{1}{2}(b-a); \ d_1 - c_1 = \frac{1}{2}(d-c),$$

$[a_1, b_1; c_1, d_1]$ contains an infinite number of points of A.

Again, we express $[a_1, b_1; c_1, d_1]$ as a union of four rectangles and denote by $[a_2, b_2; c_2, d_2]$ any one of those which contain an infinite number of points of A.

Thus proceeding, we obtain a sequence

$$\{[a_n, b_n; c_n, d_n]\}$$

of rectangles.

Now the sequence $\{a_n\}$ of real numbers is monotonically increasing and bounded above and the sequence $\{b_n\}$ of real numbers is monotonically decreasing and bounded below. These sequences are thus convergent. Let

$$\xi_1 = \lim a_n, \ \xi_2 = \lim b_n.$$

In fact we have

$$\xi_1 = \sup.\{a_n\}, \ \xi_2 = \inf.\{b_n\}.$$

Now

$$b_n - a_n = \frac{1}{2^n}(b-a)$$

$$0 = \lim(b_n - a_n) = \xi_2 - \xi_1 \Rightarrow \xi_1 = \xi_2 = \xi, \text{ say}$$

Similarly we may arrive at a number η such that

$$\sup.\{c_n\} = \lim\{c_n\} = \eta = \lim\{d_n\} = \inf.\{d_n\}$$

Now $\quad a_n \le \xi \le b_n;\ c_n \le \eta \le d_n \ \forall\ n \Rightarrow (\xi, \eta) \in [a_n, b_n; c_n, d_n]\ \forall\ n.$

We shall show that (ξ, η) is a limit point of A.

Consider any neighbourhood

$$S[(\xi, \eta), r]$$

of (ξ, η). This spherical neighbourhood contains the rectangular neighbourhood

$$\left]\xi - \frac{1}{2}r,\ \xi + \frac{1}{2}r;\ \eta - \frac{1}{2}r,\ \eta + \frac{1}{2}r\right[$$

Also there exists m such that

$$[a_m, b_m; c_m, d_m] \subset \left]\xi - \frac{1}{2}r, \xi + \frac{1}{2}r;\ \eta - \frac{1}{2}r, \eta + \frac{1}{2}r\right]$$

As $[a_m, b_m; c_m, d_m]$ contains an infinite number of points of A, we see that $S[(\xi, \eta), r]$ contains an infinite number of points of A. Thus we prove that (ξ, η) is a limit point of A.

Compact sets. Def. *A set is said to be compact if it is both bounded and closed.*

Every finite set is compact. Also every closed sphere and closed rectangle is compact.

The set $\mathbf{R}^n$ is not compact.

Ex. Give some examples of compact sets and of sets which are not compact.

11.6. THEOREM

A set is compact if and only if every infinite sub-set thereof has a limit point in the set.

Let A be a compact set and let B be any infinite sub-set of A.

Being a sub-set of bounded set, B is itself bounded. Now, B being an infinite and bounded set, has, by Bolzano-Weierstrass theorem, a limit point. This limit point of B will also be a limit point of A.

Also, A being closed, the limit point will belong to A. We have thus shown that every infinite sub-set of a compact set has a limit point in the set.

Conversely, suppose that A is a set such that every infinite sub-set of A has a limit point in A. We shall show that A is a closed and bounded set.

Let $\mathbf{a}$ be a limit point of A. We shall show that $\mathbf{a} \in A$. To do this we shall construct an infinite sub-set of A having $\mathbf{a}$ set as its *only* limit point.

Consider the infinite sub-set of A defined as follows :

$$\left\{\mathbf{x}_k : k \in \mathbf{N}, \mathbf{x}_k \in S\left(\mathbf{a}, 1/k\right), \mathbf{x}_{k+1} \neq \mathbf{x}_k\right\}.$$

It may be seen that $\mathbf{a}$ is a limit point of this set and as a matter of fact the only limit point of A. Thus $\mathbf{a} \in A$ and A is closed.

Now suppose that A is not bounded. Consider an infinite sub-set of A defined as follows :

$$\{\mathbf{y}_k : k \in \mathbf{N}, \| \mathbf{y}_k \| > k\}.$$

This infinite sub-set of A has, as may be easily seen, no limit point and as such we contradict the given property of A. Thus A must be bounded. Hence the result.

11.7. HEINE-BOREL PROPERTY

Def. *A set is said to have the Heine-Borel property if every open cover of the set admits of a finite sub-cover.*

Theorem. *A set is compact if and only if it has Heine-Borel property.*

This theorem will be proved in two steps :

I. A set with Heine-Borel property is compact.

II. A compact set has the Heine-Borel property.

We shall firstly prove that *a set A with Heine-Borel property is compact.*

We shall show that the set A is bounded. Consider the family of open spheres

$$\{S(\mathbf{x}, 1) : \mathbf{x} \in A\}.$$

This family constitutes an open cover of A, so that there exists a finite sub-cover consisting of, say, the members

$$S(\mathbf{x}_1, 1), S(\mathbf{x}_2, 1) \ldots, S(\mathbf{x}_m, 1).$$

Let $\qquad k = \max \| \mathbf{x}_p - \mathbf{x}_q \|, 1 \le \mathbf{p} \le \mathbf{m}, 1 \le \mathbf{q} \le \mathbf{m}.$

Consider now any two point $\mathbf{x}$, $\mathbf{y}$ of A.

Let $\qquad \mathbf{x} \in S(\mathbf{x}_p, 1), \ \mathbf{y} \in S(\mathbf{x}_q, 1).$

We have
$$\| \mathbf{x} - \mathbf{y} \| \le \| \mathbf{x} - \mathbf{x}_p \| + \| \mathbf{x}_p - \mathbf{x}_q \| + \| \mathbf{x}_q - \mathbf{y} \|.$$
$$< 1 + k + 1 = k + 2$$

If we suppose that y is some fixed point of A, we have

$$\| \mathbf{x} \| < \| \mathbf{x} - \mathbf{y} \| + \| \mathbf{y} \| < k + 2 + \| \mathbf{y} \|.$$

This shows that the set A is bounded.

Now to show that A is closed, we shall show that the complement of A is open. Let B denote the complement of A.

Let $\mathbf{b}$ be any point of B.

If $\mathbf{x} \in A$, there exist spherical neighbourhoods of $\mathbf{x}$ and $\mathbf{b}$ which are disjoint. Let these be

$$S(\mathbf{x}, r_\mathbf{x}), \ S(\mathbf{b}, r'_\mathbf{b}).$$

The spheres $\qquad S(\mathbf{x}, r_\mathbf{x}), \ \mathbf{x} \in A$

constitute an open cover of A, admitting of a finite sub-cover.

Let the spheres constituting this finite sub-cover of A be

$$S(\mathbf{x}_1, r_{\mathbf{x}_1}), \ldots, S(\mathbf{x}_m, r_{\mathbf{x}_m}).$$

Consider now the corresponding spheres

$$S(\mathbf{b}, r'_{\mathbf{x}_1}), \ldots, S(\mathbf{b}, r'_{\mathbf{x}_m})$$

around $\mathbf{b}$. Let $\rho_\mathbf{b}$ be the smallest of the radii

$$r'_{\mathbf{x}_1}, \ldots r'_{\mathbf{x}_m}$$

of these spheres. Then

$$S(\mathbf{b}, \rho_\mathbf{b})$$

is a sphere around **b** containing no point of A and as such contained in its complement B. Thus the set B is such that it contains a neighbourhood of each of its points and as such B is open so that **A** is closed.

Note 1. Having proved that every set with the Heine-Borel property is bounded and closed and, therefore, compact, it only remains to show that every compact set has the Heine-Borel property. Thus we have to show that every open cover of a closed and bounded set admits of a finite sub-cover. This result which is known as the 'Heine-Borel' theorem will be proved in a number of steps. The proof to be given for compact sets in $\mathbf{R}^n$ could be adopted almost word for word for compact sets in **R**.

Note 2. Rational Spheres. A sphere S (**a**, r) will be called rational if the co-ordinates of its centre **a** are all rational numbers and its radius **r** is also a rational number. The notion of rational spheres will be utilised in the proof of the Heine-Borel Theorem. Firstly we show that

the set of all rational spheres is countable.

We shall only give a brief outline of the proof of this result.

To start with, we may easily see that *every subset of a countable set is at the most countable.*

If A and B are two countable sets, then the set $A \times B = \{(x, y); x \in A, x \in B\}$ is also countable. We write

$$A\{a_n : n \in \mathbf{N}\},\ B = \{b_n : n \in \mathbf{N}\}.$$

We notice that any natural number can be expressed as the sum of two natural numbers in an infinite number of ways.

We now associate members of **N** to the members $(a_n, b_m) \in A \times B$ on the basis of the sum $n + m$ of the suffixes.

Thus we associate to the members 1, 2, 3, ... of **N** the members

$$(a_1, b_1), (a_1, b_2), (a_2, b_1), (a_1, b_3), (a_2, b_2), (a_3, b_1) \ldots$$

of $A \times B$. This shows that $A \times B$ is countable.

Now consider the set of positive rational numbers p/q; where p, q are in their lowest terms.

To each such number we associate an ordered pair (p, q) of members of **N** and see that the set of positive rational numbers is countable.

Suppose that the countable set of positive rational numbers is described as

$$\{r_n : n \in \mathbf{N}\}$$

We now associate the members of **N** to the members of **Q** as follows

$$0, r_1, -r_1, r_2, -r_2, \ldots, r_n, -r_n \ldots$$

and see that **Q** is countable.

It is now an immediate consequence of the preceding that the set $S(\mathbf{a}, r)$ of rational spheres is countable.

11.8. HEINE-BOREL THEOREM

This will be proved in a number of steps.

I. *There exists a rational sphere containing a point contained in any spherical neighbourhood of the point.*

Let $\mathbf{a}$ be any given point and let $S(\mathbf{a}, r)$ be any spherical neighbourhood of the point.

We have to show that there exists a rational sphere containing $\mathbf{a}$ and contained in $S(\mathbf{a}, r)$.

Let $\qquad \mathbf{a} = (a_1, \ldots, a_n)$

and let $p_1, \ldots, p_n$ be rational numbers such that

$$a_i - \frac{1}{2n} r < p_i < a_i + \frac{1}{2n} r, 1 \le i \le n.$$

We write

$$\mathbf{p} = (p_1, \ldots, p_n).$$

We have

$$\| \mathbf{a} - \mathbf{p} \|^2 < \frac{nr^2}{4n^2} = \frac{r^2}{4n} \le \frac{r^2}{4}$$

$$\| \mathbf{a} - \mathbf{p} \|^2 < \frac{1}{2} r.$$

Let ρ be a rational number such that

$$\| \mathbf{a} - \mathbf{p} \|^2 < \rho < \frac{1}{2} r.$$

Now consider the rational sphere

$$S(\mathbf{p}, \rho).$$

We have $\qquad \| \mathbf{a} - \mathbf{p} \|^2 \le \rho \Rightarrow \mathbf{a} \in S(\mathbf{p}, \rho).$

Let $\mathbf{x}$ be any point of $S(\mathbf{p}, \rho)$. We have

$$\| \mathbf{x} - \mathbf{a} \| \le \| \mathbf{x} - \mathbf{p} \| + \| \mathbf{p} - \mathbf{a} \|$$

$$< \rho \| \mathbf{p} - \mathbf{a} \| < \frac{1}{2} r + \frac{1}{2} r = r$$

$$\| \mathbf{x} - \mathbf{a} \| < r \Rightarrow \mathbf{x} \in S(\mathbf{a}, r)$$

Thus $\qquad S(\mathbf{p}, \rho) \subset S(\mathbf{a}, r).$

Hence the result.

II. *Every open cover of a set admits of a countable sub-cover.*

Let **i** be any open cover of a set A so that every point of **A** belongs to some member of **i**.

Let β denote the countable family of rational spheres. Let **a** be any point of **A**.

Now there exists a member, say *F* of **i** containing **a** and a sphere *S* (**a**, *r*) such that

$$\mathbf{a} \in S(\mathbf{a}, r) \subset F.$$

By I above there exists a rational sphere *S* (**p**, ρ) such that

$$a \in S(p, \rho) \subset S(a, r) \subset F.$$

Thus we associate with the set *A* a sub-set of the countable set β consisting of rational spheres and with this latter sub-set of β a sub-set of **i**. A sub-set of a countable set being countable, we arrive at a countable sub-set of **i** covering *A*.

III. *Let* $\{G_n\}$ *be a sequence of non-empty closed sets such that* (*i*) $G_{n+1} \subset G_n \ \forall n$, (*ii*) G_1 *is bounded.*

Then the intersection

$$\cap G_n, n \in \mathbf{N},$$

is not-empty, i.e., there exists at least one point which belongs to $G_n \ \forall n$.

We select $\mathbf{x}_n \in G_n$

so that we obtain a bounded set

$$A = \{\mathbf{x}_n : n \in \mathbf{N}\}.$$

Now if this set *A* is finite, at least one member thereof would belong to G_n for an infinite number of values of *n*. Again, because of (*i*), this point would belong to $G_n \ \forall n$.

Now suppose that *A* is an infinite set. Being bounded and infinite, it has at least one limit point, say **a**.

Again the limit point **a** of *A*, would also be a limit point of the complement of each of the finite sub-sets of *A*. Also each G_r containing, because of (*i*) the complement of a finite subset of *A* we see that **a** is a limit point of $G_n \ \forall n$. Finally G_n being closed $\forall n$, we see that

$$a \in G_n \ \forall n.$$

IV. Finally we prove that

Every open cover of a compact set admits of a finite sub-cover.

Let **i** be any given open cover of a compact set *A*. There exists by II a countable sub-cover. Let

Let ι be any given open cover of a compact set A. There exists by II a countable sub-cover. Let

$$H_1, H_2, H_3, ..., H_n, ...; n \in \mathbf{N}$$

denote the members of this countable sub-cover.

We define a family of open sub-sets as follows :

$$K_1 = H_1$$

$$K_n = \cup H_i . 1 \le t \le n$$

Being the union of open sets, K_n is open $\forall n$.

We denote by G_n the intersection with A of the complement of K_n so that G_n is the set of the points of which do not belong to K_n and which is closed $\forall n$.

Also $$K_{n+1} \supset K_n \Rightarrow G_{n+1} \subset G_n.$$

Being compact, A is bounded so that K_1, being a sub-set of A, is also bounded.

We shall show that G_n is empty for some n.

Let, if possible G_n, be non-empty $\forall n$. By III above, there exists a point which belongs to $G_n \forall n$. This implies that there exists a point of A which does not belong to any K_n and therefore to any H_n and as such we arrive at a contradiction.

Thus there exists m such that G_m is empty. This means that K_m covers A, *i.e.*, the set of members

$$H_1, H_2, ..., H_m$$

of ι which are finite in number is a cover of A.

Hence the result.

OBJECTIVE QUESTIONS

Note. *For each of the following questions, four alternatives are given for the answer. Only one of them is correct. Choose the correct alternative.*

1. If $x, y, z \in R^n$, then

(*a*) $d(x, y) + d(y, z) + d(x, z) = 0$ (*b*) $d(x, y) \le d(x, z) + d(z, y)$

(*c*) $d(x, y) \ge d(x, z) + d(z, y)$. (*d*) $d(x, y) = d(x, z) + d(z, y)$.

2. If every limit point of a set belongs to the set, then the set is

(*a*) closed (*b*) open

(*c*) derived (*d*) none of these

3. The number of limit points in every infinite and bounded set is

(*a*) infinite (*b*) one

(*c*) atleast one (*d*) any

4. If a set is both bounded and closed, it is called :

(*a*) countable set (*b*) compact set

(*c*) cover (*d*) weirstrass' set.

5. If every open cover of a set admits of a finite sub-cover, it is said to :

(*a*) have the Heine-Borel property

(*b*) follow Bolzano-Weirstrass theorem

(*c*) be a open sphere

(*d*) none of these.

ANSWERS

1. (*b*) **2.** (*a*) **3.** (*c*) **4.** (*b*) **5.** (*a*)

12

Real Valued Functions of Several Real Variables

Limit. Continuity

12.1. REAL VALUED FUNCTIONS

In this chapter, we shall be concerned with the notions of *Limit* and *Continuity* of real valued functions of several real variables, *i.e.,* with functions having sub-sets of $\mathbf{R}^n$ as their domains; n being any given natural number.

If $D \subset \mathbf{R}^n$ be the domain of a function f, we shall denote a point of D as $\mathbf{x}$ and the f value at $\mathbf{x}$ by $f(\mathbf{x})$. Thus we have

$$f(\mathbf{x}) = f(x_1, x_2, ..., x_n).$$

12.2. LIMIT

Let f be a real valued function defined in a domain D. Let $\mathbf{a}$ be any limit point of D. We say that

$$\lim_{\mathbf{x} \to a} f(\mathbf{x})$$

exists and is equal to l, if to any given $\varepsilon > 0$, there corresponds $\delta > 0$ such that

$$|f(\mathbf{x}) - l| < \varepsilon,$$

for every point $\mathbf{x} \neq \mathbf{a}$ of D satisfying the condition

$$\|\mathbf{x} - \mathbf{a}\| < \delta.$$

In this case, we write

$$\lim_{\mathbf{x} \to \mathbf{a}} \mathbf{f}(\mathbf{x}) = \mathbf{l}, \qquad ...(1)$$

If $\mathbf{x} = (x_1, x_2, ..., x_n)$, $\mathbf{a} = (a_1, a_2, ..., a_n)$, we may re-write (1) as

$$\lim_{(x_1, x_2, ..., x_n) \to (a_1, a_2, ..., a_n)} f(x_1, x_2, ..., x_n) = l.$$

Note. The notation of limit introduced above should be carefully distinguished from what are usually called 'Repeated limits.'

Let us consider a real valued function of two variables defined in a certain neighbourhood of a point $(a, b) \in \mathbf{R}^2$ excluding possibly the point (a, b) itself.

Then while the notions

$$\lim_{y \to b} \lim_{x \to a} f(x, y), \qquad \lim_{x \to a} \lim_{y \to b} f(x, y) \qquad \text{...(1)}$$

should be distinguished from each other, they should also be further distinguished from

$$\lim_{(x, y) \to (a, b)} f(x, y). \qquad \text{...(2)}$$

For the sake of distinction, while the limit in (2) may be referred to as the '*Simultaneous Limit*' those in (1) will be referred to as '*Repeated Limits*'.

In any given case, the two repeated limits, even if they exist, may not be equal.

Consider a function f defined by

$$f(x, y) = \frac{x^2 - y^2}{x^2 + y^2}; \ (x, y) \neq (0, 0).$$

We have, as may be seen,

$$\lim_{x \to 0} f(x, y) = -1 \ \forall \ y$$

$$\Rightarrow \qquad \lim_{y \to 0} \lim_{x \to 0} f(x, y) = -1.$$

Again,

$$\lim_{y \to 0} f(x, y) = 1 \ \forall \ x$$

$$\Rightarrow \qquad \lim_{x \to 0} \lim_{y \to 0} f(x, y) = 1$$

Thus while the two repeated limits exist in this case, they are unequal. It may also be seen that in this case

$$\lim_{(x, y) \to (0, 0)} f(x, y)$$

does not exist.

Note. Non-existence of limit. In general, to determine whether a simultaneous limit exists or not is a difficult matter but a simple consideration which we now describe, often enables us to decide about the non-existence of a limit.

It may be easily seen that if

$$\lim_{(x, y) \to (a, b)} f(x, y) = l,$$

and if ϕ is any function of a single real variable such that

$$\lim_{x \to a} \phi(x) = b,$$

then
$$\lim_{x \to a} f[x, \phi(x)] = l.$$

Thus if we can determine two functions ϕ_1 and ϕ_2 such that

$$\lim_{x \to a} f(x, \phi_1, (x)) \neq \lim_{x \to a} f(x, \phi_2(x)),$$

then we shall conclude that the simultaneous limit

$$\lim_{(x, y) \to (a, b)} f(x, y)$$

does not exist.

In fact in the case of the function, f given by

$$f(x, y) = (x^2 - y^2) / (x^2 + y^2), \quad \lim_{(x, y) \to (0, 0)} f(x, y)$$

does not exist, since taking $\phi(x) = mx$, we see that

$$\lim_{x \to 0} f(x, mx) = \frac{1 - m^2}{1 + m^2}$$

which depends on m.

EXAMPLES

1. *Show that the following limit does not exist :*

$$\lim \frac{2xy}{x^2 + y^2}$$

when $(x, y) \to (0, 0)$.

Sol. First we put $y = m_1 x$ and let $(x, y) \to (0, 0)$ along this path, hence

$$\lim_{x \to 0} f(x, m_1, x) = \frac{2m_1}{1 + m_1^2}$$

Again, by approaching origin along $y = m_2 x$,

$$\lim_{x \to 0} f(x, m_2 x) = \frac{2m_2}{1 + m_2^2}$$

Since limit depends on m_1, m_2, *i.e.*, the path chosen, hence limit does not exist.

2. *Show that*

$$\lim_{(x, y) \to (0, 0)} xy\left(\frac{x^2 - y^2}{x^2 + y^2}\right) = 0 \qquad \textit{(Utkal 2003)}$$

Sol. Changing to polar coordinates

$$\left| xy \frac{x^2 - y^2}{x^2 + y^2} \right| = \left| r^2 \sin\theta \cos\theta \sin 2\theta \right|$$

$$= \left| \frac{r^2}{4} \sin 4\theta \right| \le \frac{r^2}{4} \qquad \because \sin 4\theta \le 1$$

$$\le \frac{x^2+y^2}{4} \le \frac{x^2}{4} + \frac{y^2}{4}$$

$$< \varepsilon, \text{ if we choose } \frac{x^2}{4} < \frac{\varepsilon}{2} \text{ and } \frac{y^2}{4} < \frac{\varepsilon}{2}$$

or $$x < \sqrt{(2\varepsilon)} = \delta;\ y < \sqrt{(2\varepsilon)} = \delta$$

Thus for each $\varepsilon > 0$, there exists $\delta > 0$, such that

$$\left| xy \frac{x^2-y^2}{x^2+y^2} - 0 \right| < \varepsilon, \text{ where } |x| < \delta,\ |y| < \delta$$

$$\lim_{(x,y)\to(0,0)} xy \frac{x^2-y^2}{x^2+y^2} = 0.$$

3. *Show that the double limit as* $(x, y) \to (0, 0)$ *for the function*

$$f(x,) = \frac{x-y}{x+y},\ x \ne 0, y \ne 0$$

can not exist.

Sol. By repeated or iterated method,

$$\lim_{x\to 0} \lim_{y\to 0} f(x, y) = \lim_{x\to 0} \left\{ \lim_{y\to 0} \frac{x-y}{x+y} \right\}$$

$$= \lim_{x\to 0} \left\{ \frac{x}{x} \right\} = 1$$

Also, $$\lim_{x\to 0} \lim_{y\to 0} f(x, y) = \lim_{y\to 0} \left\{ \lim_{x\to 0} \frac{x-y}{x+y} \right\}$$

$$= \lim_{y\to 0} \left(\frac{-y}{y} \right) = -1.$$

Since repeated limits are unequal, double limit does not exist.

4. *Evaluate the limit for the function*

$$f(x, y) = \frac{x^2-y^2}{x^2+y^2},\ \textit{where } (x, y) \to (0, 0)$$

Sol. $$\lim_{x\to 0} \lim_{y\to 0} f(x, y) = \lim_{x\to 0} \left\{ \lim_{y\to 0} \frac{x^2-y^2}{x^2+y^2} \right\}$$

$$= \lim_{x\to 0} \left\{ \frac{x^2}{x^2} \right\} = 1$$

Also, $$\lim_{y\to 0}\lim_{x\to 0} f(x, y) = \lim_{y\to 0}\left\{\lim_{x\to 0}\frac{x^2 - y^2}{x^2 + y^2}\right\}$$

$$= \lim_{y\to 0}\left\{\frac{-y^2}{y^2}\right\} = -1.$$

double limit does not exist.

EXERCISES

1. Show that the following limit does not exist

$$\lim \frac{xy^3}{x^2 + y^6}$$

as $(x, y) \to (0, 0)$. (*Indore 98*)

2. Evaluate the following limits or show that the limits do not exist :

(*i*) $\lim \dfrac{xy^2}{x^2 + y^4}$ (*ii*) $\lim\,(x + y)\dfrac{y + (x + y)^2}{y - (x + y)^2}$

(*iii*) $\lim \dfrac{xy}{\sqrt{(x^2 + y^2)}}$ (*iv*) $\lim\left(y\sin\dfrac{1}{x} + x\sin\dfrac{1}{y}\right)$,

when, in each case, $(x, y) \to (0, 0)$.

3. Show that for the function f of two real variable defined by

$$f(x, y) = \frac{x^2 y^2}{x^2 y^2 + (x - y)^2}$$

the two repeated limits exist and are equal but the simultaneous limit does not exist when $(x, y) \to (0, 0)$. (*Indore 2000*)

12.3. ALGEBRA OF LIMITS

Let f and ϕ be two real valued functions with domain $D \subset \mathbf{R}^n$. Then as in § 4.1.1, p. 97 we can define the functions

$$f \pm \phi,\ f\,\phi,\ f \div \phi$$

with domain D.

Let $\mathbf{a}$ be any limit point of D and let

$$\lim_{\mathbf{x}\to\mathbf{a}} f(\mathbf{x}) \text{ and } \lim_{\mathbf{x}\to\mathbf{a}} \phi(\mathbf{x})$$

exist. Then exactly, as in § 4.2.1 it can be shown that

(*i*) $\lim\,(f \pm \phi)(\mathbf{x}) = \lim f(\mathbf{x}) \pm \lim \phi(\mathbf{x})$,

(*ii*) $\lim\,(f\,\phi)(\mathbf{x}) = \lim f(\mathbf{x}) \lim \phi(\mathbf{x})$,

(*iii*) $\lim\,(f \div \phi)(\mathbf{x}) = \lim f(\mathbf{x}) \div \lim \phi(\mathbf{x})$ if $\lim \phi(\mathbf{x}) \neq 0$

where $\mathbf{x} \to \mathbf{a}$ is to be understood in each case.

12.4. CONTINUITY

(Indore 97, 98, 99, Sagar 99, Bilaspur 99)

Let f be a real valued function with domain D and let $a \in D$.

We say that f is continuous at **a**, if to each $\varepsilon > 0$ there corresponds $\delta > 0$ such that

$$|f(\mathbf{x}) - f(\mathbf{a})| < \varepsilon$$

when $$\|\mathbf{x} - \mathbf{a}\| < \delta \text{ and } \mathbf{x} \in D.$$

It will be easily seen that if **a** is not a limit point of D so that **a** is what may be called an Isolated point of D, then the function f is continuous at **a**.

In case **a** is a limit point of D, then f is continuous at **a** if and only if

$$\lim_{\mathbf{x} \to \mathbf{a}} f(\mathbf{x})$$

exists and equals $f(\mathbf{a})$.

It may be shown that if f and ϕ are two functions continuous at **a**, then

$$f \pm \phi, \ f\phi$$

are also continuous at **a**. Also if $\phi(\mathbf{a}) \neq 0$, then $f \div \phi$ is continuous at **a**.

(Lucknow 94, 95, 96, 97)

12.5. CONTINUITY IN A DOMAIN

If a function is continuous at each point of its domain, we say that the function is continuous.

Note. Let f be a real valued function defined in a domain $D \subset \mathbf{R}^n$ and continuous at a point

$$a = (a_1, \ldots, a_n) \in D$$

Consider

$$f_1(x) = f(x, a_2, \ldots, a_n)$$

so that f_1 is a real valued function of a single real variable defined in a domain consisting of points x such that

$$(x, a_2, \ldots, a_n) \in D$$

Then f_1 is continuous at a_1.

Let $\varepsilon > 0$ be given. As f is continuous at **a** there exists $\delta > 0$ such that

$$|f(\mathbf{x}) - f(\mathbf{a})| < \varepsilon \text{ when } \|\mathbf{x} - \mathbf{a}\| < \delta$$

This implies that

$$|f(x) - f_1(a_1)| < \varepsilon \text{ when } |x - \mathbf{a_1}| < \delta$$

so that f_1 is continuous at a_1.

We may similarly consider other functions

$$f_2, f_3, \ldots, f_n.$$

These results are, rather loosely, expressed by saying that *a continuous function of n real variables is continuous in each variable separately.*

The converse is not true, as may be seen on considering

$$f(x, y) = \begin{cases} x^2 + y^2 & \text{when } xy \neq 0 \\ 0 & \text{when } xy = 0 \end{cases}$$

and the point (0, 0)

EXAMPLES

1. *Show that the function*

$$f(x, y) = x^2 + 2y, \ (x, y) \neq (1, 2)$$
$$f(x, y) = 0, \ (x, y) = (1, 2)$$

has a removable discontinuity at (1, 2).

Sol. We have $\lim\limits_{x \to 1} \lim\limits_{y \to 2} f(x, y) = \lim\limits_{x \to 1} \left\{ \lim\limits_{y \to 2} x^2 + 2xy \right\}$

$$= \lim_{x \to 1} \left\{ x^2 + 4x \right\} = 5$$

$$\lim_{y \to 2} \lim_{x \to 1} f(x, y) = \lim_{y \to 2} \left\{ \lim_{x \to 1} \left(x^2 + 2xy \right) \right\}$$

$$\lim_{y \to 2} \{1 + 2y\} = 5.$$

$\therefore$ $$\lim_{(x, y) \to (1, 2)} f(x, y) = 5.$$

But $$f(1, 2) \neq 5.$$

Hence point (1, 2) is a point of removable discontinuity.

2. *Show that the function*

$$f(x, y) = \frac{2xy}{\sqrt{\left(x^2 + y^2\right)}}; \ (x, y) \neq (0, 0)$$
$$f(x, y) = 0, \ (x, y) = (0, 0)$$

is continuous at origin.

Sol. Let $x = r \cos \theta$, $y = r \sin \theta$, we have

$$\left| f(x, y) - f(0, 0) \right| = \left| f(r \cos \theta, \ r \sin \theta) - f(0, 0) \right|$$
$$= \left| 2r \sin \theta \cos \theta \right|$$
$$< \varepsilon \text{ if } |r| < \varepsilon / 2.$$

Hence by definition, function is continuous at origin.

3. *Examine the continuity of*

$$f(x, y) = \frac{xy^2}{x^2 + y^2}, \ (x, y) \neq (0, 0)$$
$$= 0, \ (x, y) = (0, 0). \quad \textit{(Indore 98; Bilaspur 97; Sagar 96)}$$

Sol. Let point (x, y) approach the origin (0, 0) along $y = mx$.

$$\lim_{(x,y)\to(0,0)} f(x, y) = \lim_{x\to 0} \lim_{y\to mx} f(x, y)$$
$$= \lim_{x\to 0}\left(\frac{m^2 x}{1 + m^2}\right) = 0.$$

This is independent of m, hence there exists a limit. Also functional value is equal to the limit, hence the function is continuous at (0, 0).

4. *Examine the continuity of the function* $f(x, y) = \sqrt{|xy|}$, *at the origin.*

Sol. Let $x = r\cos\theta, \ y = r\sin\theta$, then

$$|f(x, y) - f(0, 0)| = |f(r\cos\theta, r\sin\theta) - f(0, 0)|$$
$$= \left| r\sqrt{(\cos\theta\sin\theta)} - 0 \right|$$
$$= \left| r\sqrt{\sin\theta}\sqrt{\cos\theta} \right|$$
$$< \varepsilon \text{ if } |r| < \varepsilon.$$

Hence the function is continuous.

EXERCISES

1. If f is continuous at **a** and $f(\mathbf{a}) \neq 0$, show that there exists a neighbourhood of **a** for every point **x** of which $f(\mathbf{x})$ has the sign of $f(\mathbf{a})$.

2. Discuss the continuity and discontinuity of the function of two real variables as follows :

$$f(x, y) = 2xy^2 / (x^3 + y^3) \text{ when } (x, y) \neq (0, 0) \text{ and } f(0, 0) = 0.$$

(Lucknow 94, 96)

3. Show that the function f is continuous at (0, 0) where

$$f(x, y) = e^{-|x-y|/(x^2 - 2xy + y^2)}, \text{ when } (x, y) \neq (x, x) \text{ and } f(x, x) = 0$$

4. Show that the function given by

$$f(x, y) = xy\frac{x^2 - y^2}{x^2 + y^2} \text{ if } (x, y) \neq (0, 0), \ f(0, 0) = 0$$

is continuous at (0, 0).

5. Can $f(0, 0)$ be defined so that

$$f(x, y) = \frac{x^2 y}{x^4 + y^2}, \ (x, y) \neq (0, 0)$$

and f is continuous at (0, 0) ? *(Lucknow 97)*

6. Show that the functions f and φ defined as below are discontinuous at (0, 0) :

(*i*) $f(x, y) = x^4 y^4 / (x^2 + y^4)^3$ when $(x, y) \neq (0, 0)$ and $f(0, 0) = 0$.

(*ii*) $\varphi(x, y) = x^2 / (x^2 + y^2 - x)$ when $(x, y) \neq (0, 0)$ and $\varphi(0, 0) = 0$.

$$\left[\textbf{Hint. } \lim (x, y) = \frac{1}{8}, \lim \varphi(x, y) = 1 \text{ if } (x, y) \to (0, 0) \text{ along } y^2 = x\right].$$

7. The functions f and g of one variable each are continuous in the intervals $[a, b]$ and $[c, d]$ respectively.

Show that a function h of two variables defined by

$$h(x, y) = \max. \{f(x), g(y)\},$$

is continuous in the rectangle $[a, b; c, d]$.

8. Show that a polynomial in n variables is continuous in $\mathbf{R}^n$.

12.5. FUNCTIONS CONTINUOUS ON COMPACT SETS

12.6.1. Theorem

The range of a function continuous on a compact set is compact.

Thus we have to prove that the range of a continuous function on a bounded and closed set is bounded and closed.

Let f be a real valued function continuous on a compact set $D \subset \mathbf{R}^n$.

Let E denote the range of the function.

If E is finite, we have finished in as much as a finite set is compact.

Now suppose that E is an infinite set. Let B denote any infinite subset of E.

To each $\mathbf{y} \in \mathbf{B}$ there corresponds at least one $\mathbf{x} \in \mathbf{D}$ such that $f(\mathbf{x}) = y$. We select one such $\mathbf{x}$ and name the subset of D having these as members as A. Thus to each $y \in B$ there corresponds one and only one $\mathbf{x} \in A$ such that $f(\mathbf{x}) = y$.

Since A is an infinite sub-set of the compact set D, the set A has a limit point, say $\mathbf{a}$ which $\in D$. Let

$$b = f(\mathbf{a}) \Rightarrow b \in \mathbf{E}.$$

We shall show that b is a limit point of the set E. Consider a neighbourhood

$$]b - \varepsilon, b + \varepsilon[$$

of b. Then there exists $\delta > 0$

such that

$$\|\mathbf{x} - \mathbf{a}\| < \delta \Rightarrow f(\mathbf{x}) \in]b - \varepsilon, b + \varepsilon[.$$

Since $\|\mathbf{x} - \mathbf{a}\| < \delta$ contains an infinite number of points of A, $]b - \varepsilon, b + \varepsilon[$ contains an infinite number of points of the set B. Thus **b** is a limit point of **B**.

Thus every infinite sub-set **B** of the range **E** having a limit point in **E**, we see that the range E is compact.

Cor. 1. *A real valued function continuous on a compact set is bounded and attains its bounds.*

[A function with a bounded range is said to be bounded].

Cor. 2. *A real valued function continuous on a closed rectangle*

$$[\mathbf{a}, \mathbf{b}]$$

is bounded and attains its bounds.

12.6.2. Uniform continuity

Let f be a real valued function defined in any domain D. Then f is said to be uniformly continuous in D if to each $\varepsilon > 0$ there corresponds $\delta > 0$ such that

$$|f(\mathbf{x}) - f(\mathbf{y})| < \varepsilon, \text{ when } \|\mathbf{x} - \mathbf{y}\| < \delta.$$

It may be seen at once that a function which is uniformly continuous in its domain is also continuous in the domain. The converse is not true in general. In case, however, the domain D is compact, continuity also implies uniform continuity. *(Lucknow 96)*

Theorem. *A function continuous on a compact domain is uniformly continuous.*

Let the domain D of a continuous function f be compact and let E be the range of f,

Let $\varepsilon > 0$ be given.

Let $\mathbf{x} \in D$.

Now there exists $\delta(\mathbf{x}) > 0$ such that

$$\|\mathbf{x} - \mathbf{t}\| < \delta(\mathbf{x}) \Rightarrow |f(\mathbf{x}) - f(\mathbf{t})| < \frac{1}{2}\varepsilon.$$

Now the spheres

$$S\left[\mathbf{x}, \frac{1}{2}\delta(\mathbf{x})\right], \ \mathbf{x} \in D$$

constitute an open cover of the compact set D. Thus there exists a finit[e] sub-cover. Let this finite sub-cover consist of the spheres.

$$S\left[\mathbf{x}_1, \frac{1}{2}\delta(\mathbf{x}_1)\right], ..., S\left[\mathbf{x}_p, \frac{1}{2}\delta(\mathbf{x}_p)\right].$$

We write

$$\delta = \frac{1}{2}\min\left[\delta(\mathbf{x}_1), ..., \delta(\mathbf{x}_p)\right].$$

Now suppose that **x, y** are two points of D such that

$$\|\mathbf{x} - \mathbf{y}\| < \delta.$$

Suppose that

$$\mathbf{x} \in S\left[\mathbf{x}_i, \frac{1}{2}\delta(\mathbf{x}_i)\right].$$

We have $\quad \|\mathbf{y} - \mathbf{x}_i\| \le \|\mathbf{y} - \mathbf{x}\| + \|\mathbf{x} - \mathbf{x}_i\| < \delta + \frac{1}{2}\delta(\mathbf{x}) < \delta(\mathbf{x}_i).$

Thus the points **x** and **y** belong to the same sphere *viz.*,

$$S\left[\mathbf{x}_i, \delta(\mathbf{x}_i)\right].$$

$$|f(\mathbf{x}) - f(\mathbf{y})| \le |f(\mathbf{x}) - f(\mathbf{x}_i)| + |f(\mathbf{x}_i) - f(\mathbf{y})|$$

$$< \frac{1}{2}\varepsilon + \frac{1}{2}\varepsilon = \varepsilon.$$

Thus we have shown that given $\varepsilon > 0$, there exists $\delta > 0$ such that

$$\|\mathbf{x} - \mathbf{y}\| < \delta \Rightarrow |f(\mathbf{x}) - f(\mathbf{y})| < \varepsilon.$$

12.7. INTERMEDIATE VALUE THEOREM

Let f be a real valued continuous function with domain $D \subset \mathbf{R}^n$. *Let the domain D be such that*

$$\mathbf{x} \in D,\ \mathbf{y} \in D \Rightarrow t\,\mathbf{x} + (1-t)\,\mathbf{y} \in D\ \forall\ \mathbf{t} \in [0, 1].$$

Then the function f assumes every value between f (**x**) *and f* (**y**).

Consider a real valued function ϕ with domain [0, 1], defined by

$$\phi(t) = f\left(t\,\mathbf{x} + (1-t)\,\mathbf{y}\right);$$

x, y being given fixed members of D.

Firstly we shall show that ϕ is a continuous functions.

Let $c \in [0, 1]$ and $\varepsilon > 0$ be given. Since f is continuous, there exists $\delta_o > 0$ such that

$$\left|f\left(t\mathbf{x} + (1-t)\,\mathbf{y}\right) - f\left(c\mathbf{x} + (1-c)\,\mathbf{y}\right)\right| < \varepsilon$$

when $\quad \left\|\left(t\mathbf{x} + (1-t)\,\mathbf{y}\right) - \left(c\mathbf{x} + (1-c)\,\mathbf{y}\right)\right\| < \delta_o.$

Now $$\left\| \left(t\mathbf{x} + (1-t)\mathbf{y}\right) - \left(c\mathbf{x} + (1-c)\mathbf{y}\right) \right\| = \left\| (t-c)(\mathbf{x}-\mathbf{y}) \right\|$$
$$\leq |t-c| \left(\|\mathbf{x}\| + \|\mathbf{y}\| \right) < \delta_o.$$

if $$|t-c| < \delta_o / \left(\|\mathbf{x}\| + \|\mathbf{y}\| \right) = \delta, \text{ say.}$$

Thus we see that there exists $\delta > 0$ such that

$$|t-c| < \delta \Rightarrow \left| f\left(t\mathbf{x} + (1-t)\mathbf{y}\right) - f\left(c\mathbf{x} + (1-c)\mathbf{y}\right) \right| < \varepsilon.$$

so that ϕ is continuous at c and as such in $[0, 1]$.

Now $$\phi(0) = f(\mathbf{y}),\ \phi(1) = f(\mathbf{x}).$$

Thus the continuous function ϕ of one variable assumes every values between $f(\mathbf{x})$ and $f(\mathbf{y})$.

Hence the result.

Note. Convex Sets. A set $D \subset \mathbf{R}^n$ is said to be convex, if it is such that

$$\mathbf{x} \in D,\ \mathbf{y} \in D \Rightarrow t\mathbf{x} + (1-t)\mathbf{y} \in D\ \forall\ t \in [0, 1].$$

The set of points

$$\left\{ t\mathbf{x} + (1-t)\mathbf{y} : \mathbf{t} \in [0, 1] \right\}$$

is called the line segment joining the points **x, y**. The motivation for this name may be seen when we consider $\mathbf{R}^2$ and $\mathbf{R}^3$.

It may be easily seen that Spheres and Rectangles are convex sets.

Note. It will thus be seen that the Intermediate value theorem holds for continuous functions with convex domains.

Ex. 1. Show that the following sub-sets of $\mathbf{R}^2$ are convex :

(*i*) $\left\{(x, y);\ x \geq 0,\ y \geq 0,\ x + y \leq 1\right\}$

(*ii*) $\left\{(x, y) : x^2/a^2 + y^2/b^2 \leq 1\right\}$.

(*iii*) $\left\{(x, y) : y^2 \leq 4ax, x \leq a\right\}$.

Ex. 2. Show that

$$\left\{(x, y, z) : x^2/a^2 + y^2/b^2 + z^2/c^2 \leq 1\right\}.$$

is a convex sub-set of $\mathbf{R}^3$.

Ex. 3. Give examples of sets in $\mathbf{R}^2$ which are not convex.

12.8. Having so far considered functions with domains in $\mathbf{R}^n$ and ranges in R, we now consider functions with domains in $\mathbf{R}^n$ and ranges in $\mathbf{R}^m$.

Let **f** be a function with domain $D \subset \mathbf{R}^n$ and range $E \subset \mathbf{R}^m$. Then

$$\mathbf{x} \in D \Rightarrow f(\mathbf{x}) = \mathbf{y} \in E \subset \mathbf{R}^m.$$

We write

$$\mathbf{y} = (y_1, y_2, \dots, y_m).$$

Thus we see that to each $\mathbf{x} \in D$, there corresponds an ordered set of m real numbers $y_1, y_2, \dots, y_m$ and as such are defined m real valued functions, say

$$f_1, f_2, \dots, f_m$$

of n variables each with a common domain D such that

$$y_1 = f_1(\mathbf{x}), \dots, y_m = f_m(\mathbf{x}).$$

We write $\quad \mathbf{f} = (f_1, \dots, f_m)$.

so that the function $\mathbf{f}$ can be thought of as an m-tuple of real valued functions.

Thus for example, if we have

$$x_1 = f_1(x_1, x_2), \quad y_2 = f_2(x_1, x_2), \quad y_3 = f_3(x_1, x_2)$$

we may re-write these relations in a condensed form as

$$\mathbf{y} = (y_1, y_2, y_3) = \mathbf{f}(\mathbf{x}) \text{ where } \mathbf{x} = (x_1, x_2).$$

Here $\mathbf{x} \in R^2$ and $\mathbf{y} \in R^3$ and $\mathbf{f} = (f_1, f_2, f_3)$.

12.8.1. Limits of Functions

Let $\mathbf{f}$ be a function with domain $D \subset \mathbf{R}^n$ and range $E \subset \mathbf{R}^m$.

We suppose that $\mathbf{a}$ is a limit point of D. We say that

$$\lim_{\mathbf{x} \to \mathbf{a}} \mathbf{f}(\mathbf{x}) : \mathbf{x} \in D \subset \mathbf{R}^n, \ \mathbf{f}(\mathbf{x}) \in \mathbf{E} \subset \mathbf{R}^m,$$

exists, if there exists $\mathbf{b} \in \mathbf{R}^m$ such that to each $\varepsilon > 0$ there corresponds $\delta > 0$ such that

$$\| \mathbf{f}(\mathbf{x}) - \mathbf{b} \| < \varepsilon,$$

when $\mathbf{x}$ is a point D, different from $\mathbf{a}$, such that

$$\| \mathbf{x} - \mathbf{a} \| < \delta.$$

Also we write

$$\lim_{\mathbf{x} \to \mathbf{a}} \mathbf{f}(\mathbf{x}) = \mathbf{b}.$$

Theorem. *Let*

$$\lim_{\mathbf{x} \to \mathbf{a}} \mathbf{f}(\mathbf{x}) = \mathbf{b}.$$

and let $\quad \mathbf{b} = (b_1, \dots, b_m), \ \mathbf{f} = (f_1, \dots, f_m).$

Then $\quad \displaystyle\lim_{\mathbf{x} \to \mathbf{a}} f_i(\mathbf{x}) = b_i, \ 1 \le i \le m$

and conversely.

Let $\varepsilon > 0$ be given. Then there exists $\delta > 0$ such that

$$\| \mathbf{f}(\mathbf{x}) - \mathbf{b} \| < \varepsilon$$

where $\mathbf{x}$ is a point of D different from $\mathbf{a}$ such that

$$\| \mathbf{x} - \mathbf{a} \| < \delta.$$

Now $\| f(\mathbf{x}) - \mathbf{b} \| = \sqrt{\left[\sum_{i=1}^{m} | f_i(\mathbf{x}) - b_i|^2\right]} < \varepsilon \Rightarrow \left[f_i(\mathbf{x}) - b_i\right] < \varepsilon,$

We thus see that to a given $\varepsilon > 0$, there corresponds $\delta > 0$ such that

$$f_i(\mathbf{x}) - b_i \,| < \varepsilon$$

where $\mathbf{x}$ is a point of D different from a such that

$$\| \mathbf{x} - \mathbf{a} \| < \delta$$

Thus we see that

$$\lim_{\mathbf{x} \to \mathbf{a}} f_i(\mathbf{x}) = b_i.$$

The proof of the converse which is also simple is left to the reader.

Ex. Give the definition of the equality

$$\lim_{\mathbf{x} \to \mathbf{a}} \mathbf{f}(x) = \mathbf{b}$$

in terms of neighbourhood of the points **a** and **b**.

12.8.2. Continuity of Function

Let **f** be a function with domain $D \subset \mathbf{R}^n$ and range $E \subset \mathbf{R}^m$. Let $\mathbf{a} \in D$.

We say that **f** is continuous at **a**, if to each $\varepsilon > 0$ there corresponds $\delta > 0$ such that

$$\| \mathbf{f}(\mathbf{x}) - \mathbf{f}(\mathbf{a}) \| < \varepsilon$$

when $\mathbf{x}$ is a point of D such that

$$\| \mathbf{x} - \mathbf{a} \| < \delta.$$

It will be seen that **f** will be continuous at **a** if **a** is an isolated point of D, *i.e.*, is not a limiting point of D. Also in case **a** is a limit point of D, **f** will be continuous at **a** and only if

$$\lim_{\mathbf{x} \to \mathbf{a}} \mathbf{f}(x) \text{ exists and equals } \mathbf{f}(\mathbf{a}).$$

Also it may be seen that if

$$\mathbf{f} = (f_1, \ldots, f_m)$$

then the continuity of the function **f** at **a** is equivalent to the simultaneous continuity of each of the real valued functions

$$f_1, \ldots, f_m$$

at **a**.

OBJECTIVE QUESTIONS

Note. *For each of the following questions, four alternatives are given for the answer. Only one of them is correct. Choose the correct alternative.*

1. $\lim_{(x,y)\to(a,b)} f(x,y)$ is said to be

   (*a*) simultaneous limit (*b*) repeated limit
   (*c*) left-right limit (*d*) double limit

2. A function $f(x, y)$ is said to be continuous if
   (*a*) It is continuous at isolated point only
   (*b*) It is continuous at each point of its domain.
   (*c*) It is continuous at some deleted neighbourhood of domain
   (*d*) none of these

3. In $\mathbf{R}^n$, a polynomial in n variables is :
   (*a*) continuous
   (*b*) discontinuous
   (*c*) continuous only at same isolated points
   (*d*) none of these.

4. The range of a function continuous on a compact set is
   (*a*) cover (*b*) sub-cover
   (*c*) compact (*d*) none of these

5. A function continuous on a compact domain is
   (*a*) discontinuous (*b*) uniformly continuous
   (*c*) absolutely continuous (*d*) continuous but not uniformly

ANSWERS

1. (*a*) **2.** (*b*) **3.** (*a*) **4.** (*c*) **5.** (*b*)

13

Partial Derivatives

13.1. PARTIAL DERIVATIVES

Let f be a real valued function defined in a domain $D \subset \mathbf{R}^n$ such that D contains a neighbourhood of a point a. We have

$$f(x) = f(x_1, ..., x_n)$$

Then

$$\lim_{t \to 0} \frac{f(\mathbf{a} + t\mathbf{u}_1) - f(\mathbf{a})}{t}$$

if it exists, is called the *first order partial derivative* or simply the *partial derivative* of the function f with respect to x_1 at $\mathbf{a}$ and denoted by

$$D_1 f(\mathbf{a}) \text{ or } f_{x_1}(\mathbf{a}) \text{ or } \frac{\partial f}{\partial x_1}(\mathbf{a}).$$

Here $\mathbf{u}_1$ denotes the point (1, 0, 0..., 0) with first coordinate 1 and the others all zero.

It may be seen that

$$\frac{f(\mathbf{a} + t\mathbf{u}_1) - f(\mathbf{a})}{t} = \frac{f(a_1 + t, a_2, ..., a_n) - f(a_1, ..., a_n)}{t}$$

so that $D_1 f(\mathbf{a})$, if it exists, may also be thought of as

$$\lim_{t \to 0} \frac{f(a_1 + t, a_2, ..., a_n) - f(a_1, ..., a_n)}{t}$$

We may similarly define the other first order partial derivatives of the

$$D_i f(\mathbf{a}), \ 2 \le i \le n$$

function f at the point $\mathbf{a}$.

Note. It may be noticed that the existence of each of the first order partial derivatives of a function at a point may *not* imply the continuity of the function at the point. Consider, for example the function f with domain $\mathbf{R}^2$ defined by

$$f(x, y) = \begin{cases} 0 & \text{if } xy = 0 \\ 1 & \text{if } xy \neq 0. \end{cases}$$

While the function f is not continuous at (0, 0), each of the two first order partial derivatives exist there at and is, in fact, 0.

There is nothing surprising in this situation inasmuch as while the continuity at a point takes account of the function values at all points in its neighbourhood, the partial derivatives take account of the function values in only a proper sub-set of this neighbourhood.

EXAMPLES

1. *Let ϕ be a function of two variables defined as follows :*

$$\phi(x, y) = \left(x^3 + y^3\right) / (x - y) \text{ when } x \neq y,$$

$$\phi(x, y) = 0 \text{ when } x = y.$$

Show that ϕ is discontinuous at the origin but that the first order partial derivatives exist at that point. (*Lucknow 94*)

Sol. Taking $y = x - mx^3$, we see that

$$\underset{x \to 0}{\phi} (x, y) = 2/m$$

so that the limit is different for different values of m. Thus

$$\lim_{(x, y) \to (0, 0)} \phi(x, y)$$

does not exist and as such the function ϕ is not continuous at (0, 0).

Also we have

$$\phi_x(0, 0) = \lim_{t \to 0} \frac{\phi(0 + t, 0) - \phi(0, 0)}{t} = 0$$

and $\phi_y(0, 0) = 0.$

2. *If*

$$f(x, y) = xy\left(x^2 - y^2\right) / \left(x^2 + y^2\right) \text{ when } (x, y) \neq (0, 0) \text{ and } f(0, 0) = 0,$$

show that

$$f_x(x, 0) = 0 \ \forall\ x \text{ and } f_y(0, y) = 0 \ \forall\ y.$$

13.1.1. Partial Derivatives as Functions

Let f be a real valued function defined in an open set $D \subset \mathbf{R}^n$ so that D contains a neighbourhood of each of its points.

If the function f possesses first order partial derivatives at each point of its domain D, we say that f is partially derivable in D. Thus if f is partially derivable in D, there are defined, each with domain D, n real valued functions

$$D_1 f, \ldots, D_n f,$$

such that the value of any of these functions at a point of D is the value of the corresponding first order partial derivative of f at the point.

13.1.2. Functions with Continuous Partial Derivatives

If a real valued function f defined in an open set $D \subset \mathbf{R}^n$ possesses first order partial derivatives at each point of D and the functions

$$D_1 f, \ldots, D_n f,$$

with domain D are all continuous in D, we say that f is *continuously derivable* in D.

13.2. DIRECTIONAL DERIVATIVES

Let f be a real valued function defined in a domain $D \subset \mathbf{R}^n$ and let $\mathbf{a}$ be a point of D such that D contains a neighbourhood of $\mathbf{a}$.

Let $\mathbf{u}$ be such that $\| \mathbf{u} \| = 1$. Thus if

$$\mathbf{u} = (\alpha_1, \ldots, \alpha_n)$$

we have

$$\| \mathbf{u} \| = \sqrt{\left[\sum_{i=0}^{n} \alpha_i^2 \right]} = 1.$$

Now consider

$$\frac{f(\mathbf{a} + t\mathbf{u}) - f(\mathbf{a})}{t}.$$

As D contains a neighbourhood of $\mathbf{a}$, it also contains $\mathbf{a} + t\mathbf{u}$ for sufficiently small values of t. Thus if $S(\mathbf{a}, r) \subset D$, we have

$$\mathbf{a} + t\mathbf{u} \in D \text{ when } |t| < r.$$

Then

$$\lim_{t \to 0} \frac{f(\mathbf{a} + t\mathbf{u}) - f(\mathbf{a})}{t},$$

if it exists, is called the *Directional derivative* of the function f at the point $\mathbf{a}$ in the direction of $\mathbf{u}$ and is written as

$$D_{\mathbf{u}} f(\mathbf{a}) \text{ or } f_{\mathbf{u}}(\mathbf{a}).$$

First order partial derivatives are also directional derivatives.

It may be seen that $D_{\mathbf{u}} f(\mathbf{a})$, if it exists, can also be written as

$$\lim_{t \to 0} \frac{f(a_1 + t\alpha_1, \ldots, a_n + t\alpha_n) - f(a_1, \ldots, a_n)}{t}$$

where

$$\mathbf{a} = (a_1, \ldots, a_n), \ \mathbf{u} = (\alpha_1, \ldots, \alpha_n), \| \mathbf{u} \| = 1.$$

Note. It should be seen that even the existence of all the directional derivatives at a point may not imply the continuity of the function at the point.

Consider the function f defined in $\mathbf{R}^2$ such that

$$f(x, y) = xy/\left(x^2 + y^2\right) \text{ when } (x, y) \neq (0, 0) \text{ and } f(0, 0) = 0.$$

Consider the point $(\cos\theta, \sin\theta)$ with unit norm.

We have

$$\frac{f(\mathbf{0} + t\mathbf{u}) - f(\mathbf{0})}{t} = \cos\theta \sin\theta, \; t \neq 0$$

so that $f_{\mathbf{u}}(\mathbf{0})$ exists. Thus we see that the function admits of every directional derivative at the origin (0, 0). But as may be seen, the function is not continuous there at.

13.2.1. Existence of Directional Derivatives

Let f be a real valued function with an open set $D \subset \mathbf{R}^n$ as its domain. Then the function admits of directional derivatives at every point where it admits of continuous first order partial derivatives.

Let f admit of continuous partial derivatives at a point $\mathbf{a} \in D$.

Now there exists a neighbourhood of $\mathbf{a}$ contained in D. Also, because of the continuity of partial derivatives at $\mathbf{a}$, there exists a neighbourhood of $\mathbf{a}$ at every point of which the function admits of partial derivatives.

Let this neighbourhood be $S(\mathbf{a}, r)$. Let

$$\mathbf{u} = \sum_{i=1}^{n} \alpha_i \mathbf{u}_i$$

where u_i is the point each of whose coordinate is 0 except the ith which is 1.

Consider

$$\frac{f(\mathbf{a} + t\mathbf{u}) - f(\mathbf{a})}{t}, \; |t| < r.$$

We write $\mathbf{v}_i = \alpha_1 \mathbf{u}_1 + \ldots + \alpha_i \mathbf{u}_i, 1 \leq i \leq n.$

where $\mathbf{v}_0 = 0, \mathbf{v}_n - \mathbf{u}.$

As may be easily seen,

$$\frac{f(\mathbf{a} + t\mathbf{u}) - f(\mathbf{a})}{t} = \sum_{i=1}^{n} \frac{f(\mathbf{a} + t\mathbf{v}_i) - f\left(\mathbf{a} + t\mathbf{v}_{i-1}\right)}{i}$$

Now

$$f(\mathbf{a} + t\mathbf{v}_i) - f\left(\mathbf{a} + t\mathbf{v}_{i-1}\right) = f\left(\mathbf{a} + t\mathbf{v}_{i-1} + t\alpha_i \mathbf{u}_i\right) - f\left(\mathbf{a} + t\mathbf{v}_{i-1}\right),$$

so that by Lagrange's mean value theorem, there exists $\theta_i, 0 < \theta_i < 1,$ such that

$$f\left(\mathbf{a} + t\mathbf{v}_{i-1} + t\alpha_i \mathbf{u}_i\right) - f\left(\mathbf{a} + t\mathbf{v}_{i-1}\right) = t\alpha_i D_i f\left(\mathbf{a} + t\mathbf{v}_{i-1} + \theta_i t\alpha_i \mathbf{u}_i\right).$$

Thus

$$\frac{f(\mathbf{a}+t\mathbf{u})-f(\mathbf{a})}{t}=\sum_{i=1}^{n}\alpha_i D_i f\left(\mathbf{a}+t\mathbf{v}_{i-1}+\theta_i t\alpha_i \mathbf{u}_i\right)$$

The partial derivatives being continuous at **a**, we obtain, as $t \to 0$.

$$\lim_{t\to 0}\frac{f(\mathbf{a}+t\mathbf{u})-f(\mathbf{a})}{t}=\sum_{i=1}^{n}\alpha_i D_i f(\mathbf{a}).$$

Thus we have shown that if the function f possesses continuous first order partial derivatives at **a**, it admits of every directional derivative and that the directional derivative along the direction of **u** where

$$\mathbf{u}=\sum_{i=1}^{n}\alpha_i \mathbf{u}_i$$

is given by

$$D_u f(\mathbf{a})=\sum_{i=1}^{n}\alpha_i D_i f(\mathbf{a}).$$

Note. The use of Lagrange's mean value theorem would be seen more clearly, if we write

$$f\left(\mathbf{a}+t\mathbf{v}_{i-1}+t\alpha_i\mathbf{u}_i\right)-f\left(\mathbf{a}+t\mathbf{v}_{i-1}\right)$$
$$=f\left(a_1+t\alpha_1, a_2+t\alpha_2, \ldots, a_{i-1}+t\alpha_{i-1} a_i+t\alpha_i, a_{i+1}, \ldots, a_n\right)$$
$$=f\left(a_1+t\alpha_1, a_2+t\alpha_2, \ldots, a_{i-1}+t\alpha_{i-1}, a_i, a_{i+1}, \ldots, a_n\right)$$

so that we have x_i varying in $\left[a_i, a_i+t\alpha_i\right]$ and $x_1, \ldots, x_{i-1}, x_{i+1}, x_n$ all fixed.

Note. The reader would be advised to re-write the proof for two and three variables. This will enable him to see the point in the proof without being burdened with details.

13.3. COMPOSITE FUNCTIONS (LINEAR CASE)

We suppose that f is a real valued function with continuous first order partial derivatives in an open domain $D \subset \mathbf{R}^n$. Also we write

$$\mathbf{x}=\mathbf{a}+t\mathbf{u}, \qquad \ldots(1)$$

where **a** and **u** are fixed points of D and $t \in [c, d]$.

We suppose that

$$\mathbf{a}+t\mathbf{u} \in D \;\forall\; t \in [c, d].$$

Thus we define a real valued function say, g, of a single real variable $t \in [c, d]$ such that

$$g(t)=f(\mathbf{a}+t\mathbf{u}).$$

It will now be shown that g is derivable in $[c, d]$ and that $\forall\, t \in [c, d]$

$$g'(t) = \sum_{i=1}^{n} \alpha_i D_i\, f(\mathbf{x});\ \ \mathbf{x} = \mathbf{a} + t\mathbf{u}$$

where

$$\mathbf{u} = \sum_{i=1}^{n} \alpha_i \mathbf{u}_i.$$

We proceed as we did in § 13.2.1.

Let

$$\mathbf{v}_i = \alpha_1 \mathbf{u}_1 + \ldots + \alpha_i \mathbf{u}_i.$$

We have

$$g(t+h) - g(t) = f(a + t\mathbf{u} + h\mathbf{u}) - f(a + t\mathbf{u})$$

$$= f(\mathbf{x} + h\mathbf{u}) - f(\mathbf{x});\ \ \mathbf{x} = \mathbf{a} + t\mathbf{u}$$

$$= \sum_{i=1}^{n} \left[f(\mathbf{x} + h\mathbf{v}_i) - f\left(\mathbf{x} + h\mathbf{v}_{i-1}\right)\right]$$

$$= \sum_{i=1}^{n} \left[f\left(\mathbf{x} + h\mathbf{v}_{i-1} + h\alpha_i \mathbf{u}_i\right) - f\left(\mathbf{x} + h\mathbf{v}_{i-1}\right)\right]$$

$$= \sum_{i=1}^{n} \mathbf{h}\, \alpha_i D_i\, f\left(\mathbf{x} + h\mathbf{v}_{i-1} + \theta_i \mathbf{h} \alpha_i \mathbf{u}_i\right), (0 < \theta < i).$$

Thus we obtain

$$g'(t) = \lim \frac{g(t + \mathbf{h}) - g(\mathbf{t})}{\mathbf{h}} = \sum_{i=1}^{n} \alpha_i D_i\, f(\mathbf{x}).$$

13.3.1. Mean Value Theorem

Let f be a real valued function with an open convex domain $D \subset \mathbf{R}^n$ and possess continuous first order partial derivative in D. (Sagar 2003)

Thus if $\mathbf{a} \in D, \mathbf{b} \in D$ *are any two points, there exists a number* $\theta, 0 < \theta < 1,$ such that

$$f(\mathbf{b}) - f(\mathbf{a}) = \sum_{i=1}^{n} (b_i - a_i)\, D_i\, f\left[\mathbf{a} + \theta(\mathbf{b} - \mathbf{a})\right]$$

where $\mathbf{a} = (a_1, \ldots, a_n),\ \mathbf{b} = (b_1, \ldots, b_n).$

We write

$$\mathbf{u} = \frac{\mathbf{b} - \mathbf{a}}{\| \mathbf{b} - \mathbf{a} \|} = \sum_{i=1}^{n} \frac{b_i - a_i}{\mathbf{r}}\, \mathbf{u}_i$$

where $r = \| \mathbf{b} - \mathbf{a} \|$

and $$\mathbf{u}_i = (0, 0, ..., 1, ..., 0)$$

so that the ith coordinate of $\mathbf{u}_i$ is 1 and others are all zero.

We write

$$g(t) = f(\mathbf{a} + t\mathbf{u}),\ t \in [0, r]$$

$$\Rightarrow \quad f(\mathbf{b}) - f(\mathbf{a}) = g(r) - g(0).$$

By Lagrange's mean value theorem there exists θ, $0 < \theta < 1$ such that

$$g(r) - g(0) = rg'(\theta r).$$

Also

$$g'(t) = \sum_{i=1}^{n} \frac{b_i - a_i}{r} D_i f(\mathbf{a} + t\mathbf{u}).$$

Thus we have

$$f(\mathbf{b}) - f(\mathbf{a}) = rg'(\theta r)$$

$$= \sum_{i=1}^{n} (b_i - a_i) D_i f[\mathbf{a} + \theta(\mathbf{b} - \mathbf{a})] \quad ...(1)$$

Taking $\mathbf{b} = \mathbf{a} + \mathbf{h}$, we may re-state the equality (1) as follows :

$$f(\mathbf{a} + \mathbf{h}) = f(\mathbf{a}) + \sum_{i=1}^{n} h_i D_i f(\mathbf{a} + \theta\mathbf{h})$$

where

$$\mathbf{h} = (h_1, ..., h_n).$$

Note. While for the validity of the mean value theorem for real value functions of a single real variable, the mere existence of the derivative of the functions was considered sufficient, it is not so in relation to functions of severa variables. It will be seen, however, that in the formulation of the mean valu theorem for several variables, we have assumed not only the existence of th derivatives but also their continuity.

We give below an example to illustrate that mean value theorem ma not be valid when the derivatives are not continuous.

If $f(x, y) = \sqrt{|xy|}$ in $\mathbf{R}^2$, *show that*

$$f_x(x, y) = \begin{cases} \frac{1}{2} \cdot \frac{\sqrt{|y|}}{\sqrt{|x|}} & \text{if } x > 0 \\ -\frac{1}{2} \cdot \frac{\sqrt{|y|}}{\sqrt{|x|}} & \text{if } x < 0 \end{cases}$$

$$f_y(x, y) = \begin{cases} \dfrac{1}{2} \cdot \dfrac{\sqrt{|x|}}{\sqrt{|y|}} & if\ y > 0 \\ -\dfrac{1}{2} \cdot \dfrac{\sqrt{|x|}}{\sqrt{|y|}} & if\ y < 0 \end{cases}$$

and prove that the mean value theorem is not valid about the point (x, x) in any domain enclosing the origin (0, 0).

We have

$$f_x(x, y) = \lim_{h \to 0} \frac{f(x+h, y) - f(x, y)}{h}$$

$$= \lim_{h \to 0} \frac{\sqrt{|x+h|}\sqrt{|y|} - \sqrt{|x|}\sqrt{|y|}}{h}$$

$$= \lim_{h \to 0} \frac{|x+h||y| - |x||y|}{h\left[\sqrt{|x+h|}\sqrt{|y|} + \sqrt{|x|}\sqrt{|y|}\right]}$$

Suppose that, x is positive. We can then take, h, so small that $x + h$ is also positive. Thus, for $x > 0$.

$$f_x(x, y) = \lim_{h \to 0} \frac{(x+h)|y| - x|y|}{h\left[\sqrt{|x+h|}\sqrt{|y|} - \sqrt{|x|}\sqrt{|y|}\right]}$$

$$= \lim_{h \to 0} \frac{|y|}{\sqrt{|x+h|}\sqrt{|y|} + \sqrt{|x|}\sqrt{|y|}}$$

$$= \frac{1}{2} \cdot \frac{\sqrt{|y|}}{\sqrt{|x|}}.$$

Similarly other cases may be disposed of.

Now consider a neighbourhood of (0, 0) and let (x, x), $(x + p, y + p)$ be two points of this neighbourhood such that x and $x + p$ have opposite signs. Assuming the validity of the mean value theorem, we see that there exists $\theta, 0 < \theta < 1$ such that

$$f(x+p, x+p) = f(x, x) + p\left[f_x(x+\theta p, x+\theta p) + f_y(x+\theta p, x+\theta p)\right]$$

or $$|x+p| = \begin{cases} |x| + p & \text{if } x + \theta p > 0 \\ |x| - p & \text{if } x + \theta p < 0 \\ |x|, & \text{if } x + \theta p = 0. \end{cases} \qquad ...(1)$$

Thus we have either

$$|x+p| = x+p, \quad |x| = -x,$$

or $$|x+p| = -(x+p), \quad |x| = x.$$

It may now be seen that none of the inequalities (1) can be true. It will be seen that the partial derivatives of f are not continuous at (0, 0).

13.4. DERIVABILITY OF COMPOSITES

Let f be a real valued function with continuous first order partial derivatives in an open convex domain $D \subset \mathbf{R}^n$. Also let $\mathbf{g}$ be vector valued function of a single real variable with domain $[c, d]$ and with its range contained in the domain D of the function f.

We write

$$\mathbf{x} = \mathbf{g}(t),$$

so that $$t \in [c, d] \Rightarrow \mathbf{x} \in D,$$

and we have a real valued function say h of a single real variable with domain $[c, d]$ defined as

$$h(t) = f(\mathbf{g}(t)).$$

We write

$$\mathbf{g}(t) = \mathbf{x} = [g_1(t), ..., g_n(t)]$$

and suppose that the real valued functions

$$g_1, ..., g_n$$

of a real variable with domain $[c, d]$ admit of continuous derivatives in $[c, d]$.

We shall show that the function h admits of continuous derivatives in $[c, d]$ and that $\forall\, t \in [c, d]$,

$$h'(t) = \sum_{i=1}^{n} D_i f(\mathbf{x})\, g_i'(t).$$

We write

$$h(t+p) - h(t) = f(\mathbf{g}(t+p)) - f(\mathbf{g}(t)).$$

Let $$\mathbf{x} = \mathbf{g}(t), \mathbf{g} + \mathbf{q} = \mathbf{g}(t+p)$$

$\therefore$ $$h(t+p) - h(t) = f(\mathbf{x}+\mathbf{q}) - f(\mathbf{x})$$

$$= \sum_{i=1}^{n} q_i D_i f(\mathbf{x} + \theta\mathbf{q}), \; 0 < \theta < 1$$

[By the mean value theorem

where $$\mathbf{q} = (q_1, ..., q_n).$$

We have $$\mathbf{q} = \mathbf{g}(t+p) - \mathbf{g}(t).$$

If $$\mathbf{x} = \mathbf{g}(t) = (x_1, ..., x_n),$$

we write

$$q_j = g_i(t+p) - g_i(t) = pg'(t+\theta_i p),\ 0<\theta_i<1.$$

$$\therefore \quad \frac{h(t+p)-h(t)}{p} = \sum_{i=1}^{n} g_i{}'(t+\theta_i p)\, D_i f(\mathbf{x}+\theta\mathbf{q}).$$

Thus in the limit, derivatives involved being all continuous, we obtain

$$h(t) = \sum_{i=1}^{n} g_i{}'(t)\, D_i f(\mathbf{x}) = \sum_{i=1}^{n} D_i f(\mathbf{x})\, g_i{}'(t).$$

Cor. Let f be a real valued function with continuous first order partial derivatives in an open convex domain $D \subset \mathbf{R}^n$. Also let g be a vector valued function with its domain $E \subset \mathbf{R}^m$ and with its range contained in D. Let $\mathbf{t} \in E \subset \mathbf{R}^m$ so that

$$\mathbf{g}(\mathbf{t}) \in D \subset \mathbf{R}^n.$$

We write

$$h(\mathbf{t}) = f(\mathbf{g}(\mathbf{t})).$$

Let

$$\mathbf{t} = (t_1, t_2, ..., t_j, ..., t_m),$$

$$\mathbf{g} = (g_1, ..., g_i, ..., g_n).$$

Applying the result of the preceding result and taking t_j in place of t, we obtain

$$D_j h(t) = \sum_{i=1}^{n} D_i f(\mathbf{x})\, D_j g_i(\mathbf{t})\, 1 \le j \le m.$$

$$= \sum_{i=1}^{n} D_j q_i(\mathbf{t})\, D_i f(\mathbf{x}).$$

In the preceding it has been assumed that each of $g_1, ..., g_n$ is a real valued function with continuous partial derivatives in the common domain E.

13.5. DIFFERENTIABILITY AT A POINT

Let f be a real valued function defined in a domain $D \subset \mathbf{R}^n$ and let $\mathbf{a}$ be an interior point of D. Thus there exists a sphere $S(\mathbf{a}, r)$ such that

$$\mathbf{a} \in S(\mathbf{a}, r) \subset D$$

Let

$$\mathbf{a}+\mathbf{h} \in S(\mathbf{a}, r).$$

We say that the function f is *differentiable* at $\mathbf{a}$, if there exists a elation of the form

$$f(\mathbf{a}+\mathbf{h})-f(\mathbf{a})=\sum_{i=1}^{n}\alpha_i h_i+\|\mathbf{h}\|\phi(\mathbf{h}), \quad \text{...(1)}$$

where

(*i*) $\alpha_1, \alpha_2, ..., \alpha_n$ are numbers independent of $\mathbf{h}$,

(*ii*) $\mathbf{h} = (h_1, h_2, ..., h_n)$,

(*iii*) ϕ is a real valued function such that $\lim_{\mathbf{h}\to\mathbf{0}} \phi(\mathbf{h}) = 0$.

We shall now obtain two results which may be thought of as outlining two necessary conditions for differentiability at a point.

Theorem. *A function which is differentiable at a point admits of partial derivatives at the point.* (*Patna 2003*)

Let f be differentiable at $\mathbf{a}$ so that we have a relation of the form (1) above.

We take

$$\mathbf{h} = t\mathbf{u}_1$$

where $\mathbf{u}_1$, has its usual meaning, *i.e.*, $\mathbf{u}_1$ is the point whose first coordinate is 1, and each of the other coordinates is 0, and obtain,

$$f(\mathbf{a}+t\mathbf{u}_1)-f(\mathbf{a})=t\alpha_1+|t|\phi(t\mathbf{u}_1)$$

i.e.,
$$\frac{f(\mathbf{a}+t\mathbf{u}_1)-f(\mathbf{a})}{t}=\alpha_1+\frac{|t|}{t}\phi(t\mathbf{u}_1)$$

Letting $t\to 0$, we obtain

$$D_1 f(\mathbf{a})=\alpha_1.$$

We may similarly prove that in general

$$D_i f(\mathbf{a})=\alpha_i, \quad 1\le i\le n.$$

Thus we have not only proved that f admits of partial derivatives at but have also obtained an interpretation for the constants α_i, $1\le i\le$ inasmuch as

$$\alpha_i = D_i f(\mathbf{a}), \quad 1\le i\le n.$$

Theorem. *A function which is differentiable at a point is al continuous at the point.* (*Lucknow 97, Utkal 200*

Because of differentiability at $\mathbf{a}$, we have a relation of the form

$$f(\mathbf{a}+\mathbf{h})-f(\mathbf{a})=\sum_{i=1}^{n}h_i D(\mathbf{a})+\|\mathbf{h}\|\phi(\mathbf{h}),$$

where
$$\lim_{\mathbf{h}\to\mathbf{0}}\phi(\mathbf{h})=0.$$

Now
$$\left|\mathbf{f}(\mathbf{a}+\mathbf{h})-\mathbf{f}(\mathbf{a})\right| \le \sum_{i=1}^{n} |\mathbf{h}_i| \left| D_i\, \mathbf{f}(\mathbf{a}) \right| + \|\mathbf{h}\| \left| f(\mathbf{h}) \right|$$
$$\le \|\mathbf{h}\| \sum_{i=1}^{n} \left| \mathbf{D_i\, f(a)} \right| + \|\mathbf{h}\| \left| \phi(\mathbf{h}) \right|,$$
for
$$|\mathbf{h_i}| \le \|\mathbf{h}\|,\ 1 \le \mathbf{i} \le \mathbf{n}.$$

It now follows that
$$\lim_{\mathbf{h}\to\mathbf{0}} \left|\mathbf{f}(\mathbf{a}+\mathbf{h})-\mathbf{f}(\mathbf{a})\right| = \mathbf{0} \quad i.e., \quad \lim_{\mathbf{h}\to\mathbf{0}} \mathbf{f}(\mathbf{a}+\mathbf{h}) = \mathbf{f}(\mathbf{a}).$$

Thus the function f is continuous at **a**.

13.5.1. Sufficient Condition for Differentiability

Theorem. *A function f will be differentiable at* **a**, *if it admits of continuous partial derivatives at* **a**. (*Lucknow 89*)

Since f admits of continuous partial derivatives at **a**, there exists a neighbourhood, say S (**a, r**) at every point of which the function admits of partial derivatives. Let **a + h** be any point of this neighbourhood.

We have
$$\mathbf{h} = \sum_{i=1}^{n} h_i \mathbf{u}_i.$$

Also we write
$$\mathbf{v}_i = h_1\mathbf{u}_1 + \ldots + h_i\mathbf{u}_i \text{ and } \mathbf{v}_0 = \mathbf{0}.$$

We have
$$f(\mathbf{a}+\mathbf{h}) - f(\mathbf{a}) = \sum_{i=1}^{n} \left[f(\mathbf{a}+\mathbf{v}_i) - f(\mathbf{a}+\mathbf{v}_{i-1}) \right]$$
$$= \sum_{i=1}^{n} \left[f(\mathbf{a}+\mathbf{v}_{i-1}+h_i\mathbf{u}_i) - f(\mathbf{a}+\mathbf{v}_{i-1}) \right]$$
$$= \sum_{i=1}^{n} h_i D_i\, f(\mathbf{a}+\mathbf{v}_{i-1}+\theta_i h_i \mathbf{u}_i),\ 0<\theta_i<1$$
$$= \sum_{i=1}^{n} h_i D_i\, f(\mathbf{a})$$
$$+ \|\mathbf{h}\| \sum_{i=1}^{n} \frac{h_i}{\|\mathbf{h}\|} \Big[D_i\, f(\mathbf{a}+\mathbf{v}_{i-1}+\theta_i h_i \mathbf{u}_i)$$
$$- D_i f(\mathbf{a}) \Big]$$

If we write

$$\phi(\mathbf{h}) = \sum_{i=1}^{n} \frac{h_i}{\|\mathbf{h}\|}\left[D_i f\left(\mathbf{a} + \mathbf{v}_{i-1} + \theta_i h_i \mathbf{u}_i\right) - D_i f(\mathbf{a})\right],$$

we see that because of the continuity of each of the partial derivative at **a**,

$$\lim_{\mathbf{h} \to 0} \phi(\mathbf{h}) = 0.$$

We may also see that $|h_i|/\|\mathbf{h}\| < 1, 1 \le i \le n$. Thus we show that the function f is differentiable at **a**.

Note. While in the above, we have established the differentiability of a function at a point on the assumption of the function admitting of continuous partial derivatives at the point, actually the condition can be slightly improved inasmuch as we may only insist on the continuity of all the partial derivatives except any one and the mere existence of the latter.

We see that the first term on the right in the equality

$$f(\mathbf{a}+\mathbf{h}) - f(\mathbf{a}) = \sum_{i=1}^{n} \left[f(\mathbf{a}+\mathbf{v}_i) - f\left(\mathbf{a}+\mathbf{v}_{i-1}\right)\right]$$

obtained on putting $i = 1$ is

$$f(\mathbf{a}+h_1\mathbf{u}_1) - f(\mathbf{a}).$$

Again

$$\lim_{h \to 0} \frac{f(\mathbf{a}+h_1\mathbf{u}_1) - f(\mathbf{a})}{h_1} = D_1 f(\mathbf{a}),$$

so that

$$f(\mathbf{a}+h_1\mathbf{u}_1) - f(\mathbf{a}) - h_1 D_1 f(\mathbf{a})$$

can be made as small as we like by making $|h_1|$ sufficiently small merely on the basis of the existence of the partial derivative $D_1 f(\mathbf{a})$.

Changing the notation we may see the validity of the statement made in the beginning of this note.

It will be a good exercise for the reader to prove the following :

Theorem. *Let f be defined in a domain* $D \subset \mathbf{R}^2$. *Let (a, b) be an interior point of D and let*

(i) f_x *exist at (a, b)*

(ii) f_y *be continuous at (a, b)*

Then f is differentiable at (a, b).

We say that (a, b) is an interior point of D, if a neighbourhood of (a, b) is contained in D.

Note. It should be seen that the conditions as stated in the theorem or as improved in the note are *only sufficient* for differentiability. We give below a counter-example in support of this assertion.

Let

$$f(x, y) = x^2 \sin(1/x) + y^2 \sin(1/y) \text{ when } xy \neq 0,$$
$$f(x, 0) = x^2 \sin(1/x) \text{ when } x \neq 0,$$
$$f(0, y) = y^2 \sin(1/y) \text{ when } y \neq 0,$$
$$f(0, 0) = 0.$$

It may be easily seen that

$$f_x(x, y) = 2x \sin(1/x) - \cos(1/x) \text{ when } x \neq 0 \text{ and } f_x(0, y) = 0$$
$$f_y(x, y) = 2y \sin(1/y) - \cos(1/y) \text{ when } y \neq 0 \text{ and } f_y(x, 0) = 0$$

so that neither f_x nor f_y is continuous at (0, 0). Again, we have

$$f(h, k) - f(0, 0) = h^2 \sin(1/h) + k^2 \sin(1/k)$$
$$= 0h + 0k + \sqrt{(h^2 + k^2)}\, \phi(h, k)$$

where

$$\phi(h, k) = h \frac{h}{\sqrt{(h^2 + k^2)}} \sin\frac{1}{h} + k \frac{k}{\sqrt{(h^2 + k^2)}} \sin\frac{1}{k}$$

so that

$$\lim_{(h, k) \to (0, 0)} \phi(h, k) = 0.$$

Thus the function f is differentiable at (0, 0) in spite of f_x and f_y merely existing and not being continuous at (0, 0).

13.5.2. Differential of a Differentiable Function

Let f be differentiable at a point **a** so that we have a relation of the form

$$f(\mathbf{a} + \mathbf{h}) - f(\mathbf{a}) = \sum_{i=1}^{n} h_i D_i f(\mathbf{a}) + \| \mathbf{h} \| \phi(\mathbf{h}), \qquad \text{...(1)}$$

where $\lim_{\mathbf{h} \to 0} \phi(\mathbf{h}) = 0.$

Then the part

$$\sum_{i=1}^{n} h_i D_i f(\mathbf{a})$$

of the difference $f(\mathbf{a} + \mathbf{h}) - f(\mathbf{a})$ is called the **Differential** *of f* at the point **a** and is written as $df(\mathbf{a})$. It should be seen, however, that the differential at a point **a**, besides depending on the point, also depends on a second variable **h** so that it would be more pertinent to denote the differential at **a** as $df(\mathbf{a}, \mathbf{h})$ in place of $df(\mathbf{a})$. In general, however, **h** is supposed to be understood.

In the event of differentiability at **a**, there exists a neighbourhood say $S(\mathbf{a}, r)$ such that (1) holds when $\mathbf{a}+\mathbf{h} \in S(\mathbf{a}, r)$, *i.e.*, for all **h** such that $\|\mathbf{h}\| < r$.

It should be noticed that the differential $df(\mathbf{a}, \mathbf{h})$ is *linear* in **h**, *i.e.*, it satisfies the conditions

$$df(\mathbf{a}, \mathbf{h}_1 + \mathbf{h}_2) = df(\mathbf{a}, \mathbf{h}_1) + df(\mathbf{a}, \mathbf{h}_2)$$

$$df(\mathbf{a}, k\mathbf{h}) = k\,df(\mathbf{a}, \mathbf{h}).$$

It is of course understood that $\mathbf{h}_1, \mathbf{h}_2, \mathbf{h}_1 + \mathbf{h}_2, k\mathbf{h}$ are suitably restricted.

It is usual to write $df(\mathbf{a})$ as

$$\sum_{i=1}^{n} D_i f(\mathbf{a})\, dx_i$$

where we have taken $\mathbf{h} = (dx_1, \ldots, dx_n)$.

Ex. Given that f is differentiable at **a**, show that there exists a neighbourhood $S(\mathbf{a}, r)$ of **a** and a positive constant k such that

$$|f(\mathbf{a}+\mathbf{h}) - f(\mathbf{a}) - df(\mathbf{a}, \mathbf{h})| < k\|\mathbf{h}\|.$$

$\forall\, \mathbf{h}$ such that $\|\mathbf{h}\| < r$.

EXAMPLES

1. *Given*

$$f(x, y) = \frac{xy}{x^2 + y^2} \text{ when } (x, y) \neq (0, 0)$$

$$f(0, 0) = 0.$$

Show that both the partial derivatives exist at (0, 0) but the function not continuous at (0, 0). (Lucknow 95; Bhopal 98; Jiwaji 95

Sol. By definition

$$f_x(0, 0) = \lim_{h \to 0} \frac{f(0+h, 0) - f(0, 0)}{h}$$

$$= \lim_{h \to 0} \frac{0}{h} = 0$$

$$f_y(0, 0) = \lim_{k \to 0} \frac{f(0+k, 0) - f(0, 0)}{k}$$

$$= \lim_{k \to 0} \frac{0}{k} = 0$$

Thus both the partial derivatives exist at (0, 0).

Approaching (0, 0) along $y = mx$, we get

$$\lim_{x \to 0} f(x, y) = \lim_{x \to 0} \frac{x.mx}{x^2 + m^2 x^2} = \frac{m}{1+m^2}.$$

Since the value of limit depends on m and is different for different path, therefore does not exist. Thus the function $f(x, y)$ is not continuous at (0, 0).

2. *Given*

$$f(x, y) = xy / \sqrt{(x^2 + y^2)} \ \text{when}\ (x, y) \neq (0, 0)$$

$$f(0, 0) = 0,$$

show that f is continuous, possesses partial derivatives but is not differentiable at (0, 0). *(Patna 2003, Utkal 2003)*

Sol. We have

$$\left| \frac{xy}{\sqrt{(x^2 + y^2)}} - 0 \right| = \frac{|xy|}{\sqrt{(x^2 + y^2)}}.$$

Also

$$x^2 + y^2 - |xy| = \left(|x| - \frac{1}{2}|y| \right)^2 + \frac{3}{4}|y^2| \geq 0$$

$$\Rightarrow \quad x^2 + y^2 - |xy| \geq 0$$

$$\Rightarrow \quad \frac{|xy|}{\sqrt{(x^2 + y^2)}} \leq \sqrt{(x^2 + y^2)} \ \forall (x, y) \neq (0, 0).$$

Thus

$$\left| \frac{xy}{\sqrt{(x^2 + y^2)}} - 0 \right| < \varepsilon \ \text{if}\ \sqrt{(x^2 + y^2)} < \varepsilon,$$

nd accordingly f is continuous at (0, 0).

Also it is easy to see that $f_x(0, 0) = 0 = f_y(0, 0)$ so that if f were ifferentiable at (0, 0), we should have, by definition,

$$\frac{hk}{\sqrt{(h^2 + k^2)}} = 0h + 0k + \sqrt{(h^2 + k^2)}\ \phi(h, k) \qquad \text{...(1)}$$

here $\phi(h, k) \to 0$ as $(h, k) \to (0, 0)$.

We have

$$\phi(h, k) = \frac{hk}{h^2 + k^2}$$

where $\lim_{(h,k)\to(0,0)} \phi(h, k) = 0.$

Thus we arrive at a contradiction inasmuch as

$$\lim_{(h,k)\to(0,0)} \frac{hk}{h^2 + k^2}$$

does not exist.

3. *Show that the function*

$$f(x, y) = \sqrt{|xy|}$$

is not differentiable at the origin, but both the partial derivatives exist at (0, 0).

At (0, 0), we have

$$f_x(0,0) = \lim_{h\to 0} \frac{f(h,0) - f(0,0)}{h} = 0$$

$$f_y(0,0) = \lim_{k\to 0} \frac{f(0,k) - f(0,0)}{k} = 0.$$

Now to prove differentiability consider

$$df(x, y) = f(h, k) - f(0,0) = Ah + Bk + h\phi + ky$$

where ϕ and ψ are functions of h, k and tend to zero as (h, k) approaching (0, 0).

$$\therefore \quad df(x, y) = h\phi + k\phi \text{ as } A = f_x(0,0) = 0,\ B = f_y(0,0) = 0$$

On putting $h = \rho \cos\theta,\ k = \rho \sin\theta$ and dividing by ρ, we get

$$\sqrt{\cos\theta \sin\theta} = \phi \cos\theta + \psi \sin\theta$$

For arbitrary $\theta,\ \rho \to 0$ indicates that $(h, k) \to (0,0)$. Now on taking limit as $\rho \to 0$, we get

$$\cos\theta . \sin\theta = 0$$

which is not true for all arbitrary values of θ.

Therefore, $f(x, y)$ is not differentiable at (0, 0), although both the partial derivatives exist at (0, 0)

Ex. 4. Show that f, where $f(x, y) = |x| + |y|$ is continuous but no differentiable at (0, 0).

Ex. 5. Discuss the differentiability of the following functions at (0, 0) :

$$f(x, y) = xy^2 \left(x^2 + y^2\right) \text{ when } (x, y) \neq (0,0)\ f(0, 0) = 0.$$

Examine for continuity also. *(Lucknow 81)*

Ex. 6. If $f(x, y) = x \sin 1/x + y \sin 1/y$ where $xy \neq 0$, $f(x, 0) = x \sin 1/x$ when $x \neq 0$, $f(0, y) = y \sin 1/y$ when $y \neq 0$; $f(0, 0) = 0$. Show that f is continuous but not differentiable at (0, 0).

Ex. 7. Investigate the differentiability of the function f where $f(x, y) = (xy)^{1/3}$ at all points $(x, y) \in \mathbf{R}^2$.

Ex. 8. Find the value of k for which the function f where

$$f(x, y) = \{|x + y| + (x + y)\}^k$$

is everywhere differentiable.

13.6. PARTIAL DERIVATIVES OF HIGHER ORDERS

We suppose that f admits $D_i f$ in a neighbourhood of a point **a**. Then

$$D_j D_i f(\mathbf{a}),$$

if it exists, is called a second order partial derivative of f at **a** and is denoted by

$$D_{ji} f(\mathbf{a}) \quad \text{or} \quad f_{ji}(\mathbf{a}).$$

Thus a real valued function of n variables may have as many as n^2 second order partial derivatives at a point.

We may similarly define partial derivatives of third and higher orders.

It should be seen that

$$D_{ji} f(\mathbf{a}) \text{ and } D_{ij} f(\mathbf{a}),\ j \neq i$$

may not be equal.

Note. It is easy to see *a priori* why $f_{yx}(a, b)$ may be different from $f_{xy}(a, b)$.

We have

$$f_{xy}(a, b) = \lim_{h \to 0} \frac{f_y(a+h, b) - f_y(a, b)}{h}$$

$$f_y(a+h, b) = \lim_{k \to 0} \frac{f(a+h, b+k) - f(a+h, b)}{k}$$

$$f_y(a, b) = \lim_{k \to 0} \frac{f(a, b+k) - f(a, b)}{k}$$

$$\therefore \quad f_{xy}(a, b) = \lim_{h \to 0} \lim_{k \to 0} \frac{f(a+h, b+k) - f(a+h, b) - f(a, b+k) + f(a, b)}{hk}$$

$$= \lim_{h \to 0} \lim_{k \to 0} \frac{\phi(h, k)}{hk}, \text{ say.}$$

It may similarly be shown that

$$f_{yx}(a, b) = \lim_{k \to 0} \lim_{h \to 0} \frac{\phi(h, k)}{hk}.$$

Thus we see that f_{yx} (a, b) and f_{xy} (a, b) appear as repeated limits of the same expression and as such equality of the two derivatives may not be taken for granted.

EXAMPLES

1. *If* $f(x, y) = xy(x^2 - y^2)/(x^2 + y^2)$ *when* $(x, y) \neq (0, 0)$ and $f(0, 0) = 0$,

show that

$$f_{xy}(0,0) \neq f_{yz}(0,0). \qquad \textit{(Lucknow 91, 96)}$$

Sol. We have

$$f_{xy}(0,0) = \lim_{h \to 0} \frac{f_y(0+h,0) - f_y(0,0)}{h}.$$

But $$f_y(0,0) = \lim_{k \to 0} \frac{f(0,0+k) - f(0,0)}{k} = \lim_{k \to 0} \frac{0}{k} = 0,$$

and $$f_y(h,0) = \lim_{k \to 0} \frac{f(h,0+k) - f(h,0)}{k} = \lim_{k \to 0} \frac{hk(h^2 - k^2)}{k(h^2 + k^2)}$$

$$\therefore \quad f_{xy}(0,0) = \lim_{h \to 0} \frac{h - 0}{h} = 1.$$

Again

$$f_{yx}(0,0) = \lim_{k \to 0} \frac{f_x(0,0+k) - f_x(0,0)}{k}$$

But $$f_x(0,0) = \lim_{h \to 0} \frac{f(0+h,0) - f(0,0)}{h} = 0$$

and $$f_x(0,k) = \lim_{h \to 0} \frac{f(0+h,k) - f(0,k)h}{h}$$

$$= \lim_{h \to 0} \frac{hk(h^2 - k^2)}{h(h^2 + k^2)} = -k.$$

$$\therefore \quad f_{xy}(0,0) = \lim_{k \to 0} \frac{-k - 0}{k} = -1.$$

Thus $f_{xy}(0,0) \neq f_{yx}(0,0)$.

2. *If* $f(x, y) = (x^2 + y^2)\tan^{-1}\left(\frac{y}{x}\right)$ when $x \neq 0 = \frac{\pi}{2} y^2$, *when* $x = ($

Show that $f_{xy}(0,0) \neq f_{yx}(0,0)$. (*Lucknow 90, 9*

Sol. $$f_y(0,0)=\lim_{k\to 0}\frac{f(0,0+k)-f(0,0)}{k}$$

$$=\lim_{k\to 0}\frac{k^2\left(\dfrac{\pi}{2}\right)-0}{k}=0$$

$$f_y(h,0)=\lim_{k\to 0}\frac{f(h,0+k)-f(h,0)}{k}$$

$$=\lim_{k\to 0}\left(h^2+k^2\right)\tan^{-1}\left(\frac{k}{h}\right)-0$$

$$=h\lim_{k\to 0}\frac{\tan^{-1}(k/h)}{k/h}=h\times 1=h$$

Again,

$$f_x(0,0)=\lim_{h\to 0}\frac{f(0+h,0)-f(0,0)}{h}=\lim_{h\to 0}\left(\frac{0-0}{h}\right)=0$$

$$f_x(0,k)=\lim_{h\to 0}\frac{f(0+h,k)-f(0,k)}{h}$$

$$=\lim_{h\to 0}\frac{\left(h^2+k^2\right)\tan^{-1}(k/h)-\dfrac{\pi k^2}{2}}{h}$$

$$=\lim_{h\to 0}h.\tan^{-1}(k/h)+k^2\lim_{h\to 0}\left[\frac{\tan^{-1}(k/h)-\dfrac{1}{2}\pi}{h}\right]$$

$$=\lim_{h\to 0}h\tan^{-1}(k/h)+k\lim_{h\to 0}\left[\frac{\cot^{-1}(k/h)}{h}\right]$$

$$=0.\left(\frac{\pi}{2}\right)+k^2\lim_{h\to 0}\left[\frac{-k/h^2}{1+(k/h)^2}\right]=-k$$

by L′ Hospital rule. ...(1)

Again, we know

$$f_{xy}(0,0)=\frac{\partial}{\partial x}f_y(0,0)=\lim_{h\to 0}\frac{f_y(0+h,0)-f_y(0,0)}{h} \quad ...(2)$$

$$f_{yx}(0,0)=\frac{\partial}{\partial y}f_x(0,0)=\lim_{h\to 0}\frac{f_x(0,0+k)-f_x(0,0)}{k}$$

Substituting the values of the four partial derivatives from above into (1) and (2), we get

$$f_{xy}(0,0)=\lim_{h\to 0}\frac{h-0}{h}=1;$$

$$f_{yx}(0,0)=\lim_{k\to 0}\left(\frac{-k-0}{k}\right)=-1.$$

Therefore, $f_{xy}(0,0)\neq f_{yx}(0,0)$

Ex. 3. Given that

$$f(x,y)=xy \text{ if } |y|\leq |x| \text{ and } f(x,y)=-xy \text{ if } |y|>|x|;$$

show that

$$f_{xy}(0,0)\neq f_{yx}(0,0).$$

Ex. 4. Examine for change in the order of derivation at the origin if

(*i*) $f(x,y)=|x^2-y^2|$,

(*ii*) $f(x,y)=\begin{cases}x^2\tan^{-1}(y/x)-y^2\tan^{-1}(x/y); & x\neq 0, y\neq 0.\\ 0 & \text{elsewhere}\end{cases}$

Ex. 5. Examine the equality of $f_{xy}(0, 0)$ and $f_{yx}(0, 0)$ if

$$f(x,y)=\sqrt{(x^2+y^2)}\sin 2\phi$$

where $f(0,0)=0$ and $\phi=\tan^{-1}(y/x)$.

13.6.1. Sufficient Condition for the Validity of Reversal in the Order of Derivation

(*Sagar 99; Jiwaji 98*)

We now prove two theorems which lay down sufficient conditions fo[r] the equality of f_{xy} and f_{yx} at a point (a, b).

Schwarz's theorem. *If (a, b) be a point of the domain $D\subset \mathbf{R}^2$ o[f a] real valued function f such that*

(*i*) *f_x exists in a certain neighbourhood of (a, b);*

(*ii*) *f_{xy} is continuous at (a, b),*

then $f_{yx}(a, b)$ exists and is equal to $f_{xy}(a, b)$.(Lucknow 90, 92, 94, 95, 9[7]

The given conditions imply that there exists a certain neighbourhoo[d] of (a, b) at every point (x, y) of which $f_x(x, y)$, $f_y(x, y)$ and $f_{xy}(x, y)$ exis[t]. Let $(a + h, b + k)$ be any point of this neighbourhood. We write

$$\varphi(h,k)=f(a+h,b+k)-f(a+h,b)-f(a,b+k)+f(a,b),$$

$$g(y)=f(a+h,y)-f(a,y),$$

so that

$$\varphi(h,k)=g(b+k)-g(b). \quad \ldots$$

Since f_y exists in a neighbourhood of (a, b) the function g of one variable is derivable in $[b, b + k]$, and, therefore, by applying the mean value theorem to the expression on the right of (1), we have

$$\varphi(h, k) = kg'(b + \theta k) \qquad (0 < \theta < 1)$$
$$= k\left[f_y(a + h, b + \theta k) - f_y(a, b + \theta k)\right]. \qquad \ldots(2)$$

Again since f_{xy} exists in a neighbourhood of (a, b), we have, applying the mean value theorem to the right of (2),

$$\varphi(h, k) = hkf_{xy}(a + \theta' h, b + \theta k), \qquad (0 < \theta' < 1)$$

$$\Rightarrow \quad \frac{1}{k}\left[\frac{f(a+h, b+k) - f(a, b+k)}{h} - \frac{f(a+h, b) - f(a, b)}{h}\right]$$
$$= f_{xy}(a + \theta' h, b + \theta k).$$

Since f_{xy} exists in a neighbourhood of (a, b), this gives when $h \to 0$,

$$\frac{f_x(a, b+k) - f_x(a, b)}{k} = \lim_{h \to 0} f_{xy}(a + \theta' h, b + \theta k).$$

Let, now, $k \to 0$. Since f_{xy} is continuous at (a, b), we obtain

$$f_{yx}(a, b) = \lim_{h \to 0} \lim_{h \to 0} f_{xy}(a + \theta' h, b + \theta k) = f_{xy}(a, b).$$

Cor. 1. *If f_{xy} and f_{yx} are both continuous at (a, b), then*

$$f_{xy}(a, b) = f_{yx}(a, b).$$

Young's theorem. *If (a, b) be a point of the domain contained in* $\mathbf{R}^2$ *of a function f such that f_x and f_y are both differentiable at (a, b), then*

$$f_{xy}(a, b) = f_{yx}(a, b). \qquad \text{(Lucknow 96)}$$

The differentiability of f_x and f_y at (a, b) implies that they exist in a certain neighbourhood of (a, b) and that

$$f_{xx}, f_{yx}, f_{xy}, f_{yy}$$

exist at (a, b). Let $(a + h, b + h)$ be a point of this neighbourhood.

We write

$$\begin{cases} \varphi(h, h) = f(a + h, b + h) - f(a + h, b) - f(a, b + h) + f(a, b), \\ g(y) = f(a + h, y) - f(a, y), \end{cases}$$

$$\Rightarrow \quad \varphi(h, h) = g(b + h) - g(b). \qquad \ldots(1)$$

Since f_y exists in a neighbourhood of (a, b), we, by applying the mean value theorem to the expression on the right of (1), obtain

$$\varphi(h, h) = hg'(b + \theta h) \qquad (0 < \theta < 1)$$

$$= h\left[f_y(a + h, b + \theta h) - f_y(b, b + \theta h)\right]. \qquad ...(2)$$

Since f_y is differentiable at (a, b) we have, by definition,

$$f_y(a + h, b + \theta h) - f_y(a, b) = hf_{xy}(a, b) + \theta hf_{yy}(a, b) + |h|\sqrt{(1 + \theta^2)}\,\phi_1(h, h) \qquad ...(3)$$

and $\quad f_y(a, b + \theta h) - f_y(a, b) = \theta hf_{yy}(a, b) - \theta\,|h|\,\phi_2(h, h) \qquad ...(4)$

where $\qquad \phi_1, \phi_2 \to 0$ as $h \to 0$.

From (2), (3) and (4) we obtain

$$\varphi(h, h)\,h^2 = f_{xy}(a, b) + \sqrt{(1 + \theta^2)}\;\phi_1(h, h) - \theta\phi_2(h, h) \qquad ...(5)$$

In a similar manner considering

$$F(x) = f(x, b + h) - f(x, b),$$

we can show that

$$\varphi(h, h)h^2 = f_{yx}(a, b) + \sqrt{(1 + \theta'^2)}\;\psi_1(h, h) - \theta'\psi_2(h, h) \qquad ...(6)$$

where

$$\psi_1\,\psi_2 \to 0 \text{ as } h \to 0.$$

Equating the right hand sides of (5) and (6) and making $h \to 0$, we obtain

$$f_{xy}(a, b) = f_{yx}(a, b)$$

EXAMPLES

1. *Show that the function*

$$f(x, y) = \frac{xy\left(x^2 - y^2\right)}{\left(x^2 + y^2\right)}, \; (x, y) \neq (0, 0)$$

$$= 0, \text{ when } (x, y) = (0, 0) \qquad (\textit{Ravishankar 2001})$$

does not satisfy conditions of Schwarz's theorem.

Sol. $\quad f_x(x, y) = \dfrac{\left(x^2 + y^2\right)y\left(3x^2 - y^2\right) - 2x^2y\left(x^2 - y^2\right)}{\left(x^2 - y^2\right)^2}$

$$= \frac{y\left(x^4 + 4x^2y^2 - y^4\right)}{\left(x^2 - y^2\right)^2}$$

$$\text{Again } f_{yx}(x, y) = \frac{x^6 + 9x^4y^2 - 9x^2y^4 - y^6}{(x^2 + y^2)^3} \quad \text{(on simplification)}$$

Now, on taking limit along $y = mx$, it may be shown that

$$\lim_{(x, y)\to(0, 0)} f_{yx}(x, y) = -1 = f_{yx}(0, 0)$$

Thus $f_{yx}(x, y)$ is not continuous at (0, 0).

Similarly, we can show $f_{xy}(x, y)$ is also not continuous at (0, 0). Thus the conditions of Schwarz's theorem are not verified.

Ex. 2. If

$$f(x, y) = x^2y^2/(x^2 + y^2) \text{ for } x^2 + y^2 \neq 0$$

and is equal to zero otherwise, verify that f_{xy}, and f_{yx} exist in a neighbourhood of (0, 0) but are not continuous at (0, 0) and yet are equal at (0, 0).

Ex. 3. $f(x, y) = (x^2 + y^2) \log (x^2 + y^2)$ for $(x, y) \neq (0, 0)$ and $f(0, 0) = 0$; show that f_{xy} and f_{yx} are not continuous at (0, 0) but $f_{xy}(0, 0) = f_{yx}(0, 0)$.

Ex. 4. Evaluate f_{xy} at a point (x, y) for the function

$$f(x, y) = x^2 + y^2 + xy.$$

Verify that the function $f(x, y)$ satisfies the requirements of Schwarz's theorem and hence evaluate f_{yx}. *(Lucknow 95)*

13.6.2. Generalised Reversal Theorem

If f possesses continuous partial derivatives of the nth order at a point (a, b),

then

$$f_{p_n p_{n-1} \cdots p_1 p_2}(a, b) = f_{q_n q_{n-1} \cdots q_1 q_2}(a, b),$$

where each p and q is either x or y and the number of x's among p's is the same as the number of x's among q's and similarly about the y's.

The proof which is simple, may be obtained by the principle of mathematical induction from the cor. of § 13.6.1, page 402.

For example, the theorem shows that when $n = 3$,

$$f_{xxy} = f_{xyx} = f_{yxx}.$$

13.7. SECOND AND HIGHER ORDER DERIVATIVES OF COMPOSITES. LINEAR CASE

We suppose that f is a real valued function with an open domain $D \subset \mathbf{R}^n$. We further suppose that f admits of second order continuous partial derivatives at each point of D. Because of the continuity of second order partial derivatives, we see that

$$D_{ji}(a) = D_{ij}(a),\ a \in D,\ i \neq j.$$

Again let

$$\mathbf{x} = \mathbf{a} + t\mathbf{u};\ t \in [c, d]$$

and let

$$\mathbf{x} = \mathbf{a} + t\mathbf{u} \in D \ \forall\ t \in [c, d].$$

We have seen that under the given conditions

$$g(t) = f(\mathbf{a} + t\mathbf{u})$$

is derivable $\forall\ t \in [c, d]$ and that

$$g'(t) = \sum_{i=1}^{n} \alpha_i D_i f(\mathbf{x})$$

where

$$\mathbf{u} = \sum_{i=1}^{n} \alpha_i\ \mathbf{u}_i.$$

Again, because of the partial derivatives $D_i f$ admitting of continuous first order partial derivative *viz.*, the second order continuous partial derivatives of f, we see that $g''(t)$ exists and that

$$g''(t) = \sum_{i=1}^{n} \alpha_i \sum_{i=1}^{n} \alpha_i\ D_{ji} f(\mathbf{x})$$

$$= \sum_{j=1}^{n} \sum_{i=1}^{n} \alpha_i\, \alpha_j\ D_{ji} f(\mathbf{x}),\ D_{ji} f(\mathbf{x}) = D_{ij} f(\mathbf{x}).$$

Again supposing f to admit of continuous third order partial derivatives, we see that

$$g'''(t) = \sum_{k=1}^{n} \sum_{j=1}^{n} \sum_{i=1}^{n} \alpha_i\, \alpha_j\, \alpha_k\ D_{kji} f(\mathbf{x}).$$

Proceeding in this manner and assuming f to admit of continuous partial derivatives of order m, we may show that $g'''(t)$ exists.

Note. We have seen that if a real valued function with domain $D_n \subset \mathbf{R}^n$ be differentiable at x, then the differential

$$\sum_{i=1}^{n} D_i f(\mathbf{x})\, t_i$$

is essentially dependent on two n-dimensional variables, *viz.*, $\mathbf{x}$ and $\mathbf{t}$ where

$$\mathbf{x} = (x_1, ..., x_n), \ \mathbf{t} = (t_1, ..., t_n)$$

and as such the differential should really be denoted as $df(\mathbf{x}, \mathbf{t})$. It is this notation which is essential in some cases as will be so in our statement of Taylor's theorem in the following section.

Also we agree to write

$$d^2 f(\mathbf{x}, \mathbf{t}) = \sum_{j=1}^{n} \sum_{i=1}^{n} t_i t_j D_{ij} f(\mathbf{x})$$

$$d^3 f(x, t) = \sum_{k=1}^{n} \sum_{j=1}^{n} \sum_{i=1}^{n} t_i t_j t_k D_{ijk} f(\mathbf{x})$$

and so on.

13.8. TAYLOR'S THEOREM IN R^n *(Utkal 2003)*

Let f be a real valued function with an open convex domain $D \subset \mathbf{R}^n$. Further let f admit of continuous partial derivatives of every order upto and including m.

Let $\mathbf{a} \in D, \mathbf{b} \in D$.

Then there exists θ, $0 < \theta < 1$ such that

$$f(\mathbf{b}) = f(\mathbf{a}) + d f(\mathbf{a}, \mathbf{u}) + \frac{1}{2!} d^2 f(\mathbf{a}, \mathbf{u}) + ...$$

$$+ ... \frac{1}{(m-1)!} d^{m-1} f(\mathbf{a}, \mathbf{u}) + \frac{1}{m!} d^m f[a + \theta(\mathbf{b} - \mathbf{a})]$$

where $\mathbf{u} = \mathbf{b} - \mathbf{a}$.

The domain D being convex, we have

$$\mathbf{a} \in D, \mathbf{b} \in D \Rightarrow t\mathbf{b} + (1-t)\mathbf{a} \in D, \ 0 \le t \le 1$$

$$\Rightarrow \mathbf{a} + t(\mathbf{b} - \mathbf{a}) \in D, \ 0 \le t \le 1.$$

We write

$$x = \mathbf{a} + t(\mathbf{b} - \mathbf{a}) \text{ and } g(t) = f[\mathbf{a} + t(\mathbf{b} - \mathbf{a})].$$

From the preceding, we know that Taylor's theorem with remainder after $m - 1$ terms is applicable to the function g in the interval [0, 1]. Thus, we see that there exist θ, $0 < \theta < 1$ such that

$$g(t) = g(0) + tg'(0) + \frac{1}{2!} t^2 g''(0) + ...$$

$$+ \frac{1}{(m-1)!} t^{m-1} g^{m-1}(0) + \frac{1}{m!} t^m g^m(\theta t).$$

Putting $t = 1$, we obtain

$$g(1) = g(0) + g'(0) + \frac{1}{2!} g''(0) + \dots$$
$$+ \frac{1}{(m-1)!} g^{m-1}(0) + \frac{1}{m!} g^m(\theta).$$

Restating this equality in terms of the function f, we obtain

$$f(\mathbf{b}) = f(\mathbf{a}) + d\mathbf{f}(\mathbf{a}, \mathbf{u}) + \frac{1}{2!} \mathbf{d}^2 \mathbf{f}(\mathbf{a}, \mathbf{u}) + \dots$$
$$+ \frac{1}{(m-1)!} d^{m-1} f(\mathbf{a}, \mathbf{u}) + \frac{1}{m!} d^m f[\mathbf{a} + \theta\mathbf{u}].$$

Ex. Give full statements of the Taylor's theorem for (*i*) $n = 2, m = 2$. (*ii*) $n = 3, m = 2$, (*iii*) $n = 3, m = 3$.

13.9. EXTREME VALUES OF A FUNCTION

We suppose that f is a real valued function defined in a certain neighbourhood of a point $\mathbf{a}$.

We say that f has a maximum value at $\mathbf{a}$ or that $f(\mathbf{a})$ is a maximum value of f, if there exists a neighbourhood $S(\mathbf{a}, r)$ of $\mathbf{a}$ such that

$$\mathbf{x} \neq \mathbf{a},\ \mathbf{x} \in S(\mathbf{a}, r) \Rightarrow f(\mathbf{x}) < f(\mathbf{a}),$$

i.e., if $f(\mathbf{a})$ is greater than $f(\mathbf{x})$ for all points $\mathbf{x}$ other than $\mathbf{a}$ in a certain neighbourhood of $\mathbf{a}$.

We may similarly give a meaning to the statement :

$f(\mathbf{a})$ is a minimum value of f.

A function f is said to have an extreme value at $\mathbf{a}$, if $f(\mathbf{a})$ is either maximum or a minimum value of the function.

A necessary condition for an extreme value.

If f has an extreme value at $\mathbf{a}$ *and the first order partial derivatives* $D_i f(\mathbf{a})$, $1 \leq i \leq n$, all exist, then

$$D_i f(\mathbf{a}) = 0, 1 \leq i \leq n.$$

Let f be maximum at $\mathbf{a}$ so that there exists a neighbourhood $\mathbf{S}(\mathbf{a}, r)$ of $\mathbf{a}$ such that

$$\mathbf{x} \neq \mathbf{a},\ \mathbf{x} \in S(\mathbf{a}, r) \Rightarrow f(\mathbf{x}) < f(\mathbf{a}).$$

Let $\quad \mathbf{a} = (a_1, \dots, a_n),$

We take $\quad \mathbf{x} = (x_1, a_2, \dots, a_n)$ where $x_1 \in]a_1 - r, a_1 + r[$.

Thus we see that the function f_1 of a single real variable such that

$$f_1(x_1) = f(x_1, a_2, \dots, a_n)$$

has a maximum value at a_1, and as such

$$f_1'(a) = D_1 f(\mathbf{a}) = 0.$$

We may similarly show that, in general,

$$D_1 f(\mathbf{a}) = 0, \ 1 \leq i \leq n.$$

The case of a minimum value at **a** may be similarly treated.

Sufficient conditions for an extreme value of a real valued function of two variables. We shall firstly examine the case of a real valued function of *two* variables.

We suppose that f is a real valued function of two variables. Let $D \subset \mathbf{R}^2$ be the domain of the function and let (a, b) be an interior point of D such that f admits of second order continuous partial derivatives in this neighbourhood. We suppose that

$$f_x(a, b) = 0 = f_y(a, b).$$

We write

$$A = f_x{}^2(a, b), \ B = f_{xy}(a, b), \ C = f_y{}^2(a, b).$$

If $(a + h, b + k)$ be any point of the neighbourhood of (a, b) referred to above, we have, by Taylor's theorem, $(m = 2)$,

$$f(a+h, b+k) = f(a, b) + \left[hf_x(a, b) + kf_y(a, b)\right]$$
$$+ \frac{1}{2}\left[h^2 f_{x^2}(\alpha, \beta) + 2hkf_{xy}(\alpha, \beta) + k^2 f_{y^2}(\alpha, \beta)\right]$$

where

$$\alpha = a + \theta h, \ \beta = b + \theta k; \ 0 < \theta < 1.$$

Thus, we have

$$f(a+h, b+k) - f(a, b) = \frac{1}{2}\left[h^2 f_{x^2}(\alpha, \beta) + 2hkf_{xy}(\alpha, \beta) + k^2 f_{y^2}(\alpha, \beta)\right].$$

We define a function D' of two variables as follows :

$$D(x, y) = f_{x^2}(x, y) f_{y^2}(x, y) - \left[f_{xy}(x, y)\right]^2$$

We have $\quad D(a, b) = AC - B^2.$

Surely D is continuous in the neighbourhood of (a, b) referred to above. We write

$$p = f_{x^2}(\alpha, \beta), \ q = f_{xy}(\alpha, \beta), \ r = f_{y^2}(\alpha, \beta).$$

I. *Let* $A > 0$, $AC - B^2 > 0$. Because of continuity, there exists a neighbourhood of (a, b) contained in the neighbourhood referred to above such that for every point (x, y) of the same

$$f_{x^2}(x, y) > 0, \ D(x, y) > 0.$$

We suppose that $(a + h, b + k)$ is a point in this neighbourhood and as such (α, β) also belongs to it so that

$$p > 0 \text{ and } pr - q^2 > 0.$$

We have

$$f(a+h, b+k) - f(a, b) = \frac{1}{2}\left[ph^2 + 2qhk + rk^2\right]$$

$$= \frac{1}{2p}\left[(ph + qk)^2 + \left(pr - q^2\right)k^2\right] > 0.$$

We thus see that there exists a neighbourhood of (a, b) for every point $(a + h, b + k)$ of which we have

$$f(a+h, b+k) > f(a, b)$$

so that f is a *minimum* at (a, b).

II. *Let* $A < 0$, $AC - B^2 > 0$. As in the preceding case, we may show that f is a maximum at (a, b).

It should be seen that $AC - B^2 > 0 \Rightarrow A \neq 0,\ C \neq 0$.

III. *Let* $AC - B^2 < 0$. We suppose that A, C are not both zero. Let $A \neq 0$.

Let $$A > 0,\ AC - B^2 < 0.$$

Now there exists a neighbourhood of (a, b) for every point (x, y) of which

$$f_{x^2}(x, y) > 0,\ D(x, y) < 0.$$

We suppose that $(a + h, b + k)$ belongs to this neighbourhood and as such (α, β) is also a point of the same. We have

$$p > 0,\ pr - q^2 < 0.$$

We have

$$f(a+h, b+k) - f(a, b) = \frac{1}{2p}\left[(ph + qk)^2 + \left(pr - q^2\right)k^2\right]$$

Now

$$k = 0, h \neq 0 \Rightarrow f(a+h, b+k) > f(a, b),$$

$$h \neq 0, h = -ak / p \Rightarrow f(a+h, b+k) < f(a, b),$$

so that in every neighbourhood of (a, b) there are points function values at which are greater than the function value at (a, b) and points function values at which are smaller than the function value at (a, b).

Thus $f(a, b)$ is neither a maximum nor a minimum value of the function f in this case.

We may similarly show that if $A < 0$ and $AC - B^2 < 0$, $f(a, b)$ is not an extreme value.

We have still to consider the case :

$$AC - B^2 < 0, A = 0, C = 0, B \neq 0.$$

Now

$$f(a+h, b+k) - f(a, b) = ph^2 + 2qhk + rk^2.$$

Taking $k = h$, we have

$$\lim_{h \to 0} \frac{f(a+h, b+h) - f(a, b)}{h^2} = A + 2B + C = 2B.$$

Taking $k = -h$, we have

$$\lim_{h \to 0} \frac{f(a+h, b-h) - f(a, b)}{h^2} = -2B$$

so that the two limits are of opposite signs.

Thus we see that in every neighbourhood of (a, b), the function assumes values greater than $f(a, b)$ and smaller than $f(a, b)$.

Thus f does not have an extreme value at (a, b) in this case.

Note. In case $AC - B^2 = 0$, we have to consider higher degree terms in the Taylor's expansion.

Case of a real valued function of *n* variables. Let f be a real valued function with domain $D \subset \mathbf{R}^n$ and let $\mathbf{a}$ be an interior point of D. We suppose that f admits of first and second order continuous partial derivatives in a certain neighbourhood of $\mathbf{a}$. Also suppose that

$$D_i f(\mathbf{a}) = 0, \ 1 \leq i < n.$$

Let $\mathbf{a} + \mathbf{h}$ be a point of the neighbourhood referred to above.

We have by Taylor's theorem $(m = 2)$:

$$f(\mathbf{a} + h) = f(\mathbf{a}) + \sum_{i=1}^{n} h_i D_i f(\mathbf{a}) + \frac{1}{2} \sum_{j=1}^{n} \sum_{i=1}^{n} h_i h_j D_{ij} f(\mathbf{a} + \theta \mathbf{h})$$

where

$$\mathbf{h} = (h_1, ..., h_n), \ 0 < \theta < 1.$$

We write

$$\mathbf{a} + \theta h = \mathbf{c}.$$

$$\therefore \quad f(\mathbf{a} + \mathbf{h}) - f(\mathbf{a}) = \frac{1}{2} \sum_{j=1}^{n} \sum_{i=1}^{n} h_i h_j D_{ij} f(\mathbf{c}).$$

We write

$$D_{ji} f(\mathbf{a}) = d_{ij},$$

As in the case of two variables, the investigation now depends on the nature of the quadratic form

$$\sum_{j=1}^{n} \sum_{i=1}^{n} d_{ij}, h_i h_j$$

It may be known from the theory of quadratic forms that this form is positive definite, *i.e.*, assumes only positive values except when $\mathbf{h} = \mathbf{0}$ if and only if the determinants

$$d_{11} \begin{vmatrix} d_{11} & d_{12} \\ d_{21} & d_{22} \end{vmatrix}, \begin{vmatrix} d_{11} & d_{12} & d_{13} \\ d_{21} & d_{22} & d_{23} \\ d_{31} & d_{32} & d_{33} \end{vmatrix}, \text{ etc.}$$

are all positive.

As in the case of two variables it may be shown that f (**a**) is a minimum value of f in this case.

Again it is known from the theory of quadratic forms that the form is negative definite, *i.e.*, assumes only negative values except when $\mathbf{h} = \mathbf{0}$ if and only if the determinant

$$d_{11} \begin{vmatrix} d_{11} & d_{12} \\ d_{21} & d_{22} \end{vmatrix}, \begin{vmatrix} d_{11} & d_{12} & d_{13} \\ d_{21} & d_{22} & d_{23} \\ d_{31} & d_{32} & d_{33} \end{vmatrix}, \text{ etc.}$$

are alternatively negative and positive.

The function f will assume a maximum value at **a** in this case.

EXAMPLES

1. *Find all the maxima and minima of the function given by*

$$f(x, y) = x^3 + y^3 - 63(x + y) + 12xy.$$

Sol. We have

$$f_x(x, y) = 3x^2 - 63 + 12y, \; f_y(x, y) = 3y^2 - 63 + 12x;$$
$$f_{xx}(x, y) = f_{xy}(x, y) = 12, \; f_{yy}(x, y) = 6y.$$

The equations

$$f_x(x, y) = 3x^2 - 63 + 12y = 0, \; f_y(x, y) = 3y^2 - 63 + 12x = 0,$$

$$\Leftrightarrow \begin{cases} f_x(x, y) = 3x^2 - 63 + 12y = 0 \\ f_x(x, y) - f_y(x, y) = 3(x^2 - y^2) + 12(y - x) = 0 \end{cases}$$

$$\Leftrightarrow \begin{cases} x^2 - 21 + 4y = 0 \\ (y - x)(4 - y - x) = 0 \end{cases}$$

$$\Leftrightarrow \begin{cases} x^2 - 21 + 4x = 0 \\ y - x = 0 \end{cases} \text{ and } \begin{cases} x^2 - 21 + 4y = 0 \\ 4 - y - x = 0 \end{cases}$$

These give $(-7,-7),\ (3,3),\ (5,-1),\ (-1,5)$
as 4 pairs of solutions. Thus the function has four stationary points.

(A point where the first order partial derivatives all vanish is called a *Stationary* point.)

At (– 7, – 7),

$$f_{xx} = 42 < 0 \text{ and } f_{xy}\, f_{yy} - \left(f_{xy}\right)^2 = 1620 > 0$$

so that the function is maximum at (– 7, – 7).

At (3, 3),

$$f_{xx} = 18 > 0 \text{ and } f_{xx}\, f_{yy} - \left(f_{xy}\right)^2 = 180 > 0,$$

so that the function is minimum at (3, 3).

At each of the other points (5, – 1) and (– 1, 5),

$$f_{xx} f_{yy} - \left(f_{xy}\right)^2 = -324 < 0$$

so that the function is neither a maximum nor a minimum.

2. *Prove that the function f where*

$$f(x, y) = x^2 - 2xy + y^2 + x^3 - y^3 + x^5$$

has neither a maximum nor a minimum at the origin.

Sol. It may be easily seen that at (0, 0)

$$f_x = 0,\ f_y = 0$$

$$f_{xx} = 2,\ f_{xy} = -2,\ f_{yy} = 0 \text{ so that } f_{xx}\, f_{yy} - \left(f_{xy}\right)^2 = 0.$$

Thus we need to further examine the question. We have

$$f(0, 0) = 0.$$

Since

$$f(x, y) = (x - y)^2 + (x - y)\left(x^2 + xy + y^2\right) + x^5,$$

we see that for points along the line $y - x = 0$, we have

$$f(x, y) = x^5$$

and accordingly in every neighbourhood of (0, 0) there are points wherefor $f(x, y)$ is positive and there are points wherefor $f(x, y)$ is negative.

Hence $f(0, 0)$ is neither a maximum nor a minimum value.

3. *Discuss maximum or minimum of*

$$f(x, y) = 2 \sin \frac{x-y}{2} \cos \frac{x-y}{2} + \cos(x + y)$$

for maxima or minima

$$f_x = \cos x - \sin(x + y) = 0 \Rightarrow \sin(x + y) = \cos x \quad ...(i)$$

and $$f_y = \cos y - \sin(x + y) = 0 \Rightarrow \sin(x + y) = \cos y \quad ...(ii)$$

Sol. From (*i*) and (*ii*)

$$x = y$$

$$\sin 2x - \cos x = 0, \text{ i.e.,}$$

$$\cos x\,(2 \sin x - 1) = 0$$

and from (*i*) $\quad \cos x = 0, \; x = \frac{1}{2}\pi, \frac{3}{2}\pi, \ldots, \frac{1}{2}(2n+1)\pi$

and when $\quad \sin x = \frac{1}{2}, \; x = n\pi + (-1)^n \frac{1}{6}\pi.$

Now to discuss maxima or minima, we have

$$f_{xx} = -\sin x - \cos(x+y), \quad f_{xy} = -\cos(x+y)$$

$$f_{yy} = -\sin y - \cos(x+y)$$

so that when $\quad x = y = \frac{1}{2}\pi, \; f_{xx} = 0, f_{yy} = 0$

i.e., $f(x, y)$ is neither maxima nor minima at $x = y = \frac{1}{2}\pi.$

When $\quad x = y = \frac{3}{2}\pi, \; f_{xx} = 2, \; f_{yy} = 2, \; f_{xy} = 1$

so that $f_{xx} f_{yy} - (f_{xy})^2 = 3 = +\text{ ve}$ and f_{xx} is positive, so there is a minimum at $x = y = \frac{3}{2}\pi.$

When $x = y = n\pi + (-1)^n \frac{1}{6}\pi, \; f_{xx} = -1, f_{yy} = -1, f_{xy} = -\frac{1}{2}$, so that

$$f_{xx} f_{yy} - (f_{xy})^2 = +\text{ ve and } f_{xx} = -\text{ ve.}$$

Hence there is a maximum at $x = y = n\pi + (-1)^n \frac{1}{6}\pi.$

4. *Find a point within a triangle such that the sum of the squares of its distances from the three vertices is a minimum.*

Sol. Let (x_1, y_1), (x_2, y_2), (x_3, y_3) be the vertices of the triangle and (x, y) be a point inside the triangle.

If $f(x, y)$ denotes the sum of the squares of the distances of (x, y) from three vertices, then

$$f(x, y) = \left[(x - x_1)^2 + (y - y_1)^2\right] + \left[(x - x_2)^2 + (y - y_2)^2\right] + \left[(x - x_3)^2 + (y - y_3)^2\right]$$

For maxima and minima

$$f_x = 2(x - x_1) + 2(x - x_2) + 2(x - x_3) = 0$$

$$\Rightarrow \qquad x = \frac{x_1 + x_2 + x_3}{3}$$

Similarly, $f_y = 2(y - y_1) + 2(y - y_2) + 2(y - y_3) = 0$

$$\Rightarrow \quad y = \frac{y_1 + y_2 + y_3}{3}$$

Thus, the point is $\left(\dfrac{x_1 + x_2 + x_3}{3}, \dfrac{y_1 + y_2 + y_3}{3}\right)$

Again

$$f_{xx} = 6, \ f_{xy} = 0, \ f_{yy} = 6$$

$$\therefore \qquad f_{xx} f_{yy} - \left(f_{xy}\right)^2 = +\text{ ve and } f_{xx} \text{ is } +\text{ ve}$$

Hence $f(x, y)$ is minimum.

5. *Show that the maxima and minima of the fraction*

$$u(x, y) = \frac{ax^2 + by^2 + 2hxy + 2gx + 2fy + c}{a'x^2 + b'y^2 + 2h'xy + 2g'x + 2f'y + c'}$$

are given by the roots of the equation

$$\begin{vmatrix} a - a'u & h - h'u & g - g'u \\ h - h'u & b - b'u & f - f'u \\ g - g'u & f - f'u & c - c'u \end{vmatrix} = 0$$

Sol. Differentiating u partially w.r.t. x and y and putting u_x and u_y equal to zero for maxi. and mini., we have

$$u(a'x + h'y + g') = ax + hy + g \qquad \text{...}(i)$$

and

$$u(h'x + b'y + f') = hx + by + f \qquad \text{...}(ii)$$

Multiplying (*i*) by x, (*ii*) by y and subtracting their sum from given eqn.

$$u(g'x + f'y + c') = gx + fy + c \qquad \text{...}(iii)$$

Equation (*i*), (*ii*) and (*iii*) can be written as

$$(a - a'u)\,x + (h - uh')\,y + g - g'u = 0$$

$$(h - h'u)\,x + (b - b'u)\,y + f - f'u = 0$$

and

$$(g - g'u)\,x + (f - f'u)\,y + c - c'u = 0$$

Eliminating x and y from these equations, the maximum and minimum values of u are given by

$$\begin{vmatrix} a - a'u & a - h'u & g - g'u \\ h - h'u & b - b'u & f - f'u \\ g - g'u & f - f'u & c - c'u \end{vmatrix} = 0.$$

6. *Examine*

$f(x, y, z) = 2xyz - 4zx - 2yz + x^2 + y^2 + z^2 - 2x + 4y + 4z$ *for extreme values.*

Sol. We have

$$f_x(x, y, z) = 2yz - 4z + 2x - 2,$$
$$f_y(x, y, z) = 2zx - 2z + 2y + 4,$$
$$f_z(x, y, z) = 2xy - 4x - 2y + 2z + 4.$$

The stationary points (x, y, z) are thus given by the system of equations

$$\left.\begin{array}{l} yz - 2z + y - 1 \quad = 0 \\ zx - z + y - 2 \quad = 0 \\ xy - 2x - y + z + 2 = 0 \end{array}\right\}$$

Adding the last two equations, we see that system is equivalent to

$$\begin{cases} yz - 2z + x - 1 = 0, \\ zx - z + y - 2 = 0 \\ zx + xy - 2x = 0 \Rightarrow x(z + y - 2) = 0. \end{cases}$$

Thus stationary points are given by the two systems of equations

$$\left.\begin{array}{r} yz - 2z + x - 1 = 0 \\ zx - z + y - 2 = 0 \\ x = 0 \end{array}\right\}. \qquad \left.\begin{array}{r} yz - 2z + x - 1 = 0 \\ zx - z + y - 2 = 0 \\ z + y - 2 = 0 \end{array}\right\}$$

These give $(0, 3, 1), (0, 1, -1), (1, 2, 0), (2, 1, 1), (2, 3, -1)$ as the stationary points of the function.

Again, we have at any point (x, y, z)

$$f_{xx} = 2, f_{yy} = 2, f_{zz} = 2; f_{xy} = 2z, \ f_{yz} = 2x - 2, \ f_{zx} = 2y - 4.$$

For (0, 3, 1) the matrix of the quadratic form is

$$\begin{bmatrix} 2 & 2 & 2 \\ 2 & 2 & -2 \\ 2 & -2 & 2 \end{bmatrix}.$$

Its principal minors

$$2, \begin{vmatrix} 2 & -2 \\ -2 & 2 \end{vmatrix} \begin{bmatrix} 2 & 1 & 2 \\ 2 & 2 & -2 \\ 2 & -2 & 2 \end{bmatrix}.$$

are 2, 0, and – 32 respectively.

Thus the function is neither a maximum nor a minimum at (0, 3, 1).

It may similarly be shown that the function is neither a maximum nor a minimum at the stationary points (0, 1, – 1), (2, 1, 1) and (2, 3, – 1).

At (1, 2, 0) we have

$$d^2f = 2h^2 + 2k^2 + 2l^2$$

which is clearly positive definite. Thus the function is a minimum at this point.

7. *Find the maximum value of*

$$f(x, y, z) = (ax + by + cz)^{-\alpha^2x^2 - \beta^2y^2 - \gamma^2z^2}.$$

Sol. We have

$$\log f(x, y, z) = \log(ax + by + cz) - \left(\alpha^2x^2 + \beta^2y^2 + \gamma^2z^2\right)$$

For max. or min. of f,

$$\frac{1}{f} f_x(x, y, z) = \frac{a}{ax + by + cz} - 2\alpha^2 x = 0$$

$$\Rightarrow \qquad x(ax + by + cz) = \frac{a}{2\alpha^2}$$

Similarly $y(ax + by + cz) = \dfrac{b}{2\beta^2}$

$$z(ax + by + cz) = \frac{c}{2\gamma^2}.$$

Multiplying these by a, b, c respectively and adding,

$$(ax + by + cz)^2 = \frac{1}{2}\left(\frac{a^2}{\alpha^2} + \frac{b^2}{\beta^2} + \frac{c^2}{\gamma^2}\right) = \lambda^2 \text{ say}$$

$$\Rightarrow \qquad ax + by + cz = \lambda$$

Then $\qquad x\lambda = \dfrac{a}{2\alpha^2} \Rightarrow x = \dfrac{a}{2\lambda\alpha^2},\ y = \dfrac{b}{2\lambda\beta^2},\ z = \dfrac{c}{2\lambda\gamma^2}.$

Again $\dfrac{1}{f} f_{xx}(x, y, z) - \dfrac{1}{f^2}\{f_x(x, y, z)\}^2$

$$= -\frac{a^2}{(ax + by + cz)^2} - \alpha^2$$

or $\qquad f_{xx}(x, y, z) = -f\left[\dfrac{a^2}{(ax + by + cz)^2} + 2\alpha^2\right]$

as $f_x(x, y, z) = 0$ for maxi. or mini.

Since $f_{xx}(x, y, z)$ is – ve, the value of $f(x, y, z)$ is maximum.

Putting the values of x, y, z, in $f(x, y, z)$

$$f_{\max} = \left(\frac{a^2}{2\lambda\alpha^2} + \frac{b^2}{2\lambda\beta^2} + \frac{c^2}{2\lambda\gamma^2}\right) e^{-\frac{1}{4\lambda^2}\left(\frac{a^2}{\alpha^2} + \frac{b^2}{\beta^2} + \frac{c^2}{\gamma^2}\right)}$$

$$= \sqrt{\left\{\frac{1}{2}\left(\frac{a^2}{\alpha^2} + \frac{b^2}{\beta^2} + \frac{c^2}{\gamma^2}\right)\right\}}\ e^{-1/2} \quad \text{as}$$

$$\lambda^2 = \frac{1}{2}\left(\frac{a^2}{\alpha^2} + \frac{b^2}{\beta^2} + \frac{c^2}{\gamma^2}\right)$$

$$= \sqrt{\left\{\frac{1}{e}\left(\frac{a^2}{\alpha^2} + \frac{b^2}{\beta^2} + \frac{c^2}{\gamma^2}\right)\right\}}$$

EXERCISES

1. Examine the following functions for extreme values :

(i) $y^2 + 4xy + 3x^2 + x^3$.

(ii) $3x^4y^2 - 5x^2y^2 + 3x^4 + 2y^3 - 6x^2 - 3y^2 + 1$.

(iii) $y^2 + x^2y + ax^4$. *(iv)* $x^2 + xy + y^2 + ax + by$.

(v) $x^3y^3 (12 - 3x - 4y)$. *(vi)* $xy(a - x - y)$.

(vii) $x^3 - y^3 + 3x^2 + 3y^2 - 9x$. *(viii)* $\left(x^2 + y^2\right)^2 - a^2\left(x^2 - y^2\right)$.

(ix) $\left(x^2 + y^2 - 4\right)^2 - x^2$. *(x)* $\left(x^2 + y^2 - 1\right)(xy + 4)$.

(xi) $3x^4y^2 - 6x^2y^2 + 3x^4 + 2y^2 - 6x^2 - 3y^2 + 1$.

(xii) $\left(x^2 + y^2\right)e^{6x + 2x^2}$. *(xiii)* $(x - y)^3\left(x^2 + y^2 - 2\right)$

2. Show that

$$f(x, y) = (y - x)^4 + (x - 2)^4$$

has a minimum at (2, 2).

3. Show that

$$f(x, y) = y^2 + x^2y + x^4,$$

has a minimum at (0, 0) where $f_{xx}, f_{yy} - \left(f_{xy}\right)^2 = 0$.

4. Show that

$$f(x, y) = 2x^4 - 3x^2y + y^2, \qquad \textit{(Utkal 2003)}$$

has neither a maximum nor a minimum at (0, 0) where $f_{xx}, f_{yy} - \left(f_{xy}\right)^2 = 0$.

5. Determine for each of the following whether there is a maximum or a minimum at the origin :

(i) $x^2 + 3xy + y^2 + x^3 + y^3$.

(ii) $x^2 + 2xy + y^2 + x^2 + x^2y - xy^3 - y^3 + x^4$.

(iii) $x^2 + 4xy + 4y^2 + x^3 + 2x^2y + x^4$.

6. Show that there are five sets of real values of x and y for which the polynomial

$$x^4 + y^4 - 6(x^2 + y^2) + 8xy$$

is stationary and determine the nature of each of the stationary values.

7. If r is the distance between two points P and Q situated respectively on the curves

$$y^2 = 4ax \text{ and } (x - 3a)^2 + y^2 = 12a^2,$$

find the positions of P and Q where r is a maximum or a minimum and interpret the results geometrically.

8. Examine for extreme values

(*i*) $2xyz + x^2 + y^2 + z^2$. (*ii*) $2x^2 + 3y^2 + 4z^2 - 3xy + 8z$.

9. Show that the function f where

$$f(x, y, z) = 3\log\left(x^2 + y^2 + z^2\right) - 2x^3 - 2y^3 - 2z^3$$

has only one extreme value, *viz.*, $\log(3/e^2)$.

10. Show that

$$u = (x + y + z)^3 - 3(x + y + z) - 24xyz + a^3$$

has min. at (1, 1, 1) and max. at (– 1, – 1, – 1).

OBJECTIVE QUESTIONS

Note. *For each of the following questions, four alternatives are given for the answer. Only one of them is correct. Choose the correct alternative.*

1. If a real valued function f defined is an open set Δ C $\mathbf{R}^n$ possesses first order partial derivatives at each point of Δ and the functions $D_1 f, ..., D_n f$ are all continuous, then f is

(*a*) derivable in Δ (*b*) partially derivable in Δ

(*c*) continuously derivable in Δ (*d*) not derivable in Δ

2. A function aduits of directional derivatives at every point where it aduits

(*a*) continuous first order partial derivatives

(*b*) zero first order partial derivatives

(*c*) bounded first order partial derivatives

(*d*) no partial derivative

3. If $f(a, b)$ be a point of the domain contained in $\mathbf{R}^2$ of a function f such that f_x and f_y are both differentiable at (a, b), then

(*a*) $f_{xy}(a, b) \neq f_{yx}(a, b)$ (*b*) $f_{xy}(a, b) = f_{yx}(a, b)$

(*c*) f_{xy} and f_{yx} do not exist at (a, b) (*d*) none of these.

4. $\lim\limits_{h \to 0} \dfrac{f(x+h, y) - f(x, y)}{h}$, if exists is called the partial derivative of f with respect to :

(*a*) x at (a, b) (*b*) x at (x, y)

(*c*) y at (a, b) (*d*) y at (x, y)

5. If $f(x, y) = c$, then $\dfrac{dy}{dx}$ is :

(*a*) $\dfrac{\partial f}{\partial x}$ (*b*) $\dfrac{\partial f}{\partial y}$

(*c*) $-\dfrac{\partial f/\partial x}{\partial f/\partial y}$ (*d*) $\dfrac{\partial f/\partial x}{\partial f/\partial y}$

6. The necessary conditions for a function $f(x, y)$ to have a maximum or minimum value at a point (a, b) are that

(*a*) $\dfrac{\partial f}{\partial x} = \dfrac{\partial f}{\partial y}$ at (a,b) (*b*) $\dfrac{\partial f}{\partial x} \neq \dfrac{\partial f}{\partial y}$ at (a,b)

(*c*) $\dfrac{\partial f}{\partial x} = 0 = \dfrac{\partial f}{\partial y}$ at (a,b) (*d*) $\dfrac{\partial^2 f}{\partial x^2} = 0 = \dfrac{\partial^2 f}{\partial y^2}$ at (a,b)

ANSWERS

1. (*c*) **2.** (*a*) **3.** (*b*) **4.** (*b*) **5.** (*c*)

14

Invertible Functions Implicit Functions

14.1 INVERTIBLE FUNCTIONS

Let $\mathbf{f}$ be a vector valued function with both its domain D and range E as subsets of $\mathbf{R}^n$.

We write

$$\mathbf{y} = \mathbf{f}(\mathbf{x}), \mathbf{x} \in D, \mathbf{y} \in E. \qquad ...(1)$$

If we write

$$\mathbf{x} = (x_1, ..., x_n), \ \mathbf{y} = (y_1, ..., y_n), \ \mathbf{f} = (f_1, ..., f_n),$$

the single equation (1) may be seen as equivalent to a system of n equations

$$y_1 = f_1(x_1, ..., x_n),$$
$$y_2 = f_2(x_1, ..., x_n),$$
$$\dots\dots\dots\dots\dots\dots$$
$$\dots\dots\dots\dots\dots\dots$$
$$y_n = f_n(x_1, ..., x_n).$$

The function $\mathbf{f}$ may thus be thought of as a transformation which transforms the set D to the set E.

Now to each point of D, there corresponds a point of E. It is possible, however, that in any given case two different points of D correspond to the same point of E. Thus we distinguish between (*i*) One-one transformations, (*ii*) Many-one transformation.

A transformation $\mathbf{f}$ from D to E is one-one if while to each point of D there corresponds a point of E, there is one and only one point of D which corresponds to a given point of E.

The transformation is many-one if there exist atleast two different points of D which correspond to the same point of E.

A one-one function **f** with domain D and range E is called **invertible** inasmuch as there is a natural way of defining a function g with domain E and range **D** such that

$$\mathbf{y}=\mathbf{f}(\mathbf{x}) \Leftrightarrow \mathbf{x}=\mathbf{g}(\mathbf{y}).$$

The function **g** is called the *Inverse* of the function **f.**

Of course not every function admits of the inverse function.

In § 4.4, p. 124, we considered a criteria for a real valued function of a single real variable to be invertible.

14.2. LOCALLY INVERTIBLE TRANSFORMATIONS

In applications we often come across what may be called *locally invertible* transformations which we shall now define.

Consider a function

$$\mathbf{y}=\mathbf{f}(\mathbf{x}),$$

with its domain D and the range E as subsets of $\mathbf{R}^n$.

Let $\mathbf{a} \in D$, so that $\mathbf{f}(\mathbf{a}) \in E$.

We write

$$\mathbf{f}(\mathbf{a})=\mathbf{b}, \ \mathbf{a} \in D, \mathbf{b} \in E.$$

We say that the transformation or the function **f** is locally invertible at **a** if there exists a neighbourhood S of **a** and a neighbourhood T of **b** such that

(*i*) $\mathbf{x} \in S \Rightarrow f(\mathbf{x}) \in T,$

(*ii*) $t=\mathbf{f}(S)=\{f(\mathbf{x}) : \mathbf{x} \in S\},$

(*iii*) $\mathbf{x}_1 \in S, \ \mathbf{x}_2 \in S, \mathbf{x}_1 \neq \mathbf{x}_2 \ \Rightarrow \ \mathbf{f}(\mathbf{x}_1) \neq \mathbf{f}(\mathbf{x}_2).$

This means that while to each point of neighbourhood S of the point **a**, there corresponds a point of the neighbourhood T of **b**, there exists one and only one point of S which corresponds to a given point of T.

It may thus be noticed that if the function **f** is locally invertible at **a**, there may exist a point of E belonging to T which corresponds also to a point not belonging to S.

Thus if the function **f** is locally invertible at **a**, there exist neighbourhoods S and T of **a** and **b** = **f (a)** respectively such that if we restrict **f** to S, it is invertible.

There exist transformations which are not invertible as such but which are locally invertible at some points. Transformations which are invertible are known as globally invertible with a view to distinguish them from locally invertible transformations.

Illustrations. Consider the transformation defined by

$$u = x^2 - y^2,\ v = xy.$$

This transformation is not globally invertible inasmuch as two points (x, y) and $(-x, -y)$ correspond to the same point (u, v).

This transformation is, however, locally invertible at each point other than $(0, 0)$.

However small a neighbourhood of $(x, y) = (0, 0)$ we may take some two points thereof must necessarily correspond to the same point (u, v). If, however, we take any point.

$$(h, k) \neq (0, 0),$$

We can mark a neighbourhood of the same such that no two points there of correspond to the same point (u, v).

2. The transformation defined by

$$u = e^x \cos y.\ v = e^x \sin y$$

is not globally invertible even though it is locally invertible at each point.

Note. In section 14.5, P. 425 we shall find sufficient conditions for a given transformation to be locally invertible at any given point.

14.3. JACOBIAN OF A TRANSFORMATION

Consider a vector valued function **f** with its domain D and range E as subsets of $\mathbf{R}^n$.

We suppose that the domain D of **f** is an open set. We write

$$\mathbf{y} = \mathbf{f}(\mathbf{x});$$

Let **a** belong to the domain of **f**. Let

$$\mathbf{f} = (f_1, ..., f_n)$$

so that $f_1, ..., f_n$ are n real valued functions with domain D.

We suppose that each of the n real valued functions $f_1, ..., f_n$ possesses each of the n first order partial derivatives at **a**.

In this case we say that the function **f** possesses first order partial derivatives at **a**.

Consider the determinant

$$\begin{vmatrix} D_1 f_1(a) & D_2 f_1(a) & \ldots\ldots & D_n f_1(a) \\ \ldots & \ldots & \ldots & \ldots \\ \ldots & \ldots & \ldots & \ldots \\ D_1 f_n(a) & D_2 f_n(a) & \ldots\ldots & D_n f_n(a) \end{vmatrix}$$

of order n such that the element in the ith row and jth column of this determinant is

$$D_i f_j(\mathbf{a}).$$

This determinant is called the **Jacobian** of the function **f** at the point **a** and denoted by

$$\mathbf{J_f}\,(\mathbf{a}).$$

If $$\mathbf{a} = (a_1, ..., a_n),$$

the Jacobian is also denoted by

$$\partial\,(f_1, ..., f_n) \,/\, \partial\,(a_1, ..., a_n),$$

The nature of the Jacobian **J** is very intimately related to nature of the transformation brought out by **f**.

14.3.1. Resultants of Transformations

Consider two functions **f** and **g** with domains and ranges as subsets of $\mathbf{R}^n$. We further suppose that the range of **f** is contained in the domain of **g**. Then the composite

$$\mathbf{g \ o \ f}$$

has a meaning and is such that

$$(\mathbf{g \ o \ f})\,(\mathbf{x}) = \mathbf{g}\left[\mathbf{f}\,(\mathbf{x})\right].$$

The composite **g o f** is also called the resultant of **g** and **f**.

We write

$$\mathbf{g \ o \ f = h}.$$

Let **a** belong to the domain of **f**. Let

$$\mathbf{f}\,(\mathbf{a}) = \mathbf{b},\ \ \mathbf{g}\,(\mathbf{b}) = \mathbf{c}.$$

We suppose that **f** and **g** possess continuous first order partial derivatives in some neighbourhoods of **a** and **b** respectively. Then as seen in § 13.4, P. 389, the function **h** also possesses continuous first order partial derivatives in a neighbourhood of **a**. We shall show that

$$\mathbf{J_h}\,(\mathbf{a}) = \mathbf{J_f}\,(\mathbf{a})\,\mathbf{J_g}\,(\mathbf{b}),$$

so that the Jacobian of the composite **h** is the product of Jacobians of **f** and **g**. Let

$$\mathbf{f} = (f_1, ..., f_n),\ \ \mathbf{g} = (g_1, ..., g_n),\ \ \mathbf{h} = (h_1, ..., h_n).$$

We have

$$D_i h_k\,(\mathbf{a}) = D_1 g_k\,(\mathbf{b})\,D_i f_1\,(\mathbf{a}) + D_2 g_k\,(\mathbf{b})\,D_i f_2\,(\mathbf{a}) + D_3 g_k\,(\mathbf{b})\,D_i f_3\,(\mathbf{a}) + ...$$

$$= \sum_{j=1}^{n} D_i g_k\,(\mathbf{b})\,D_i f_j\,(\mathbf{a})$$

so that the $(i,\ k)$th element of

$$\mathbf{J_h}\,(\mathbf{a})$$

is the sum of the products of the corresponding elements in the j-the row of

$$\mathbf{J_g}\,(\mathbf{b})$$

and the j-th column of

$$\mathbf{J_f}\,(\mathbf{a}).$$

Hence the result.

Note. For fixing ideas the reader may consider the cases $n = 2$ and $n = 3$.

14.4. LINEAR FUNCTIONS : LINEAR TRANSFORMATIONS

Let $\mathbf{f}$ be a function with domain $\mathbf{R}^n$ and range in $\mathbf{R}^n$. Then we say that $\mathbf{f}$ is a linear function, if

$$\mathbf{f}\,(\mathbf{x}+\mathbf{y})=\mathbf{f}\,(\mathbf{x})+\mathbf{f}\,(\mathbf{y})\ \forall\ \mathbf{x},\mathbf{y}\in\mathbf{R}^n$$

$$\mathbf{f}\,(\lambda\,\mathbf{x})=\lambda\,\mathbf{f}\,(\mathbf{x})\ \forall\ \lambda\in\mathbf{R},\ \forall\ \mathbf{x}\in\mathbf{R}^n.$$

We shall now proceed to investigate the general form of linear functions. Let

$$\mathbf{x}\in\mathbf{R}^n.$$

We have a relation

$$\mathbf{x}=\sum_{i=1}^{n} x_i\,\mathbf{u}_i.$$

Then, because of the linearity of $\mathbf{f}$, we have

$$\mathbf{f}\,(\mathbf{x})=\mathbf{f}\sum_{i=1}^{n}(x_i\,\mathbf{u}_i)=\sum_{i=1}^{n} x_i\,\mathbf{f}\,(\mathbf{u}_i).$$

Now $\mathbf{f}\,(\mathbf{u}_i)\in\mathbf{R}^n$. We write

$$\mathbf{f}\,(\mathbf{u}_i)=\sum_{j=1}^{n} c_{ij}\,\mathbf{u}_j.$$

Thus we have

$$\mathbf{f}\,(\mathbf{x})=\sum_{i=1}^{n}\sum_{j=1}^{n} x_i\,c_{ij}\,\mathbf{u}_j.$$

Since $\mathbf{f}\,(\mathbf{x})\in\mathbf{R}^n$, we write

$$\mathbf{f}\,(x)=\sum_{j=1}^{n} y_i\,u_j$$

and obtain

$$\sum_{j=1}^{n} y_j\,\mathbf{u}_j=\sum_{i=1}^{n}\sum_{j=1}^{n} x_i\,c_{ij}\,\mathbf{u}_j.$$

This gives

$$y_j = \sum_{i=1}^{n} x_i\, c_{ij},\ 1 \le j \le n.$$

Thus we see that if f is a linear function from $\mathbf{R}^n$ into $\mathbf{R}^n$ and $\mathbf{f}(\mathbf{x}) = \mathbf{y}$, where

$$\mathbf{x} = (x_1, ..., x_n),\ y = (y_1, ..., y_n)$$

we have a relation

$$y_j = \sum_{i=1}^{n} x_i\, c_{ij},\ 1 \le j \le n$$

so that writing in details

$$y_1 = c_{11}x_1 + c_{21}x_2 + c_{31}x_3 + ... + c_{n1}x_n$$
$$y_2 = c_{12}x_1 + c_{22}x_2 + c_{32}x_3 + ... + c_{n2}x_n$$
$$\dots\dots\dots\dots\dots\dots\dots\dots$$
$$\dots\dots\dots\dots\dots\dots\dots\dots$$
$$\dots\dots\dots\dots\dots\dots\dots\dots$$
$$y_n = c_{1n}x_1 + c_{2n}x_2 + c_{3n}x_3 + ... + c_{nn}x_n.$$

We may thus see that each linear function **f** from $\mathbf{R}^n$ into $\mathbf{R}^n$ determines $n \times n$ matrix

$$\mathbf{c} = [c_{ij}],\ 1 \le i \le n,\ 1 \le j \le n$$

such that **y** = **xc**

where **x, y** are the row matrices

$$[x_1, ..., x_n], [y_1, ..., y_n]$$

respectively.

We may also easily show that each $n \times n$ matrix determines a linear function from $\mathbf{R}^n$ into $\mathbf{R}^n$. This matrix **c** may be referred to as the matrix of the linear transformation **f**.

Note. For the fixation of ideas, the reader may consider the cases $n = 2$ and $n = 1$.

14.4.1. The Jacobian of a Linear Transformation

Let a linear transformation **f** be determined by the system of n equations

$$y_j = \sum_{i=1}^{n} x_i\, c_{ij} = f_j(\mathbf{x}).$$

We have

$$D_i\, f_i(\mathbf{x}) = c_{ij}$$

so that we see that the Jacobian

$$\mathbf{J_f}(\mathbf{x})$$

of the linear function **f** is the determinant of its matrix *i.e.*,

$$\mathbf{J_f}(\mathbf{x}) = \det \mathbf{c}$$

which is as such a constant.

14.4.2. Inverse function theorem for linear functions

Let **f** be a linear function on $\mathbf{R}^n$ into $\mathbf{R}^n$. We then have a relation of the form

$$\mathbf{y} = \mathbf{x}\,\mathbf{c},$$

where **c** is a $n \times n$ matrix,

$$\mathbf{x} = (x_1, \ldots, x_n),\ y = (y_1, \ldots, y_n)$$

and **f** (**x**) = **y**.

From the theory of matrices we may know that if **c** is a non-singular matrix, *i.e.*, $\det \mathbf{c} \neq 0$, the matrix admits of inverse $\mathbf{c}^{-1}$ so that (1) can be re-written as

$$\mathbf{x} = \mathbf{y}\,\mathbf{c}^{-1}.$$

This means that to each $\mathbf{y} \in \mathbf{R}^n$ there corresponds **x** such that **f** (**x**) = **y**, *i.e.*, the range of the function **f** is actually $\mathbf{R}^n$.

Also it shows that the function is such that if $\mathbf{y} \in \mathbf{R}^n$, there exists one and only one $\mathbf{x} \in \mathbf{R}^n$ such that **f** (**x**) = **y**.

Thus we see that if **f** is a linear function on $\mathbf{R}^n$ such that the matrix **c** corresponding to **f** is non-singular, then **f** is a one-one onto function and is as such invertible.

In particular the means that if $\det \mathbf{c} \neq 0$ *i.e.*, Jacobian of the transformation is non-zero, and **y** = 0 then we must necessarily have **x** = **0**.

14.5. INVERSE FUNCTION THEOREM. SUFFICIENT CRITERION FOR LOCALLY INVERTIBLE TRANSFORMATIONS

Theorem. *Let* **f** *be a vector valued function with its domain D and range E as sub-sets of* $\mathbf{R}^n$.

Let $\mathbf{a} \in D$, $f(\mathbf{a}) = \mathbf{b} \in E$. *Let* **f** *admit of continuous first order partial derivatives in a neighbourhood of* **a** *and let the Jacobian*

$$\mathbf{J_f}(\mathbf{a}) \neq 0.$$

Then the function **f** *is locally invertible at the point* **a**.

Also the local inverse **g** *of* **f** *admits of continuous first order partial derivatives in a neighbourhood of* **b**.

We have thus to show that under the given conditions on **f**, there exists neighbourhoods U and V of **a** and **b** respectively such that

(*i*) $\mathbf{f}(U)=\{\mathbf{f}(\mathbf{x}):\mathbf{x}\in U\}=V$.

(*ii*) No two different points of U correspond to the same point of V.

(*iii*) **f (a) = b.**

Also if we define **g** as the inverse of **f** with domain V and range U so that

$$\mathbf{y}=\mathbf{f}(\mathbf{x}) \Leftrightarrow \mathbf{x}=\mathbf{g}(\mathbf{y});\ \mathbf{x}\in U,\ \mathbf{y}\in V$$

the function **g** admits of continuous first order partial derivatives in V.

Consider the determinant

$$\begin{vmatrix} D_1f_1(\mathbf{x}_1) & D_2f_1(\mathbf{x}_1),\ldots, D_nf_1(\mathbf{x}_1) \\ D_1f_2(\mathbf{x}_2) & D_2f_2(\mathbf{x}_2),\ldots, D_nf_2(\mathbf{x}_2) \\ \cdots\ \cdots & \cdots\ \cdots\qquad \cdots\ \cdots \\ \cdots\ \cdots & \cdots\ \cdots\qquad \cdots\ \cdots \\ D_1f_n(\mathbf{x}_n) & D_2f_n(\mathbf{x}_n),\ldots, D_nf_n(\mathbf{x}_n) \end{vmatrix}$$

where $\mathbf{x}_1$, $\mathbf{x}_2$, ... $\mathbf{x}_n$ are points in a neighbourhood of **a**. We look upon this determinant as a function of n^2 variables *viz.*, the n^2 coordinates of the n-points $\mathbf{x}_1$, ..., $\mathbf{x}_n$, such that its value when $\mathbf{x}_1$, ..., $\mathbf{x}_n$ all take the same value **a** is non-zero. Also the determinant being a polynomial in its elements is a continuous function.

Thus there exists a neighbourhood of the point **a** such that if

$$\mathbf{p}_1, \ldots, \mathbf{p}_n$$

are any n arbitrary points of this neighbourhood, we have

$$\left| D_jf_i(\mathbf{p}_i)\right| \neq 0.$$

Again there exists $r > 0$ such that the closed sphere

$$S\ (\mathbf{a},\ \mathbf{r})$$

is contained in the neighbourhood referred to above.

I. Let **x**, **y** be two different points of the sphere S (**a**, **r**). We shall show that

$$\mathbf{x}\neq\mathbf{y} \Rightarrow \mathbf{f}(\mathbf{x})\neq\mathbf{f}(\mathbf{y})$$

Let if, possible,

$$\mathbf{f}(\mathbf{x})=\mathbf{f}(\mathbf{y})$$

$$\Leftrightarrow \qquad f_i(\mathbf{x}) = f_i(\mathbf{y}), \; 1 \le i \le n$$

By the mean value theorem, there exists $\theta_i, 0 < \theta_j < 1$ such that

$$0 = f_i(\mathbf{x}) - f_i(\mathbf{y}) = \sum_{j=1}^{n} (y_i - x_j) D_j f_i [\mathbf{x} + \theta_i (\mathbf{y} - \mathbf{x})].$$

We write

$$\mathbf{x} + \theta_i (\mathbf{y} - \mathbf{x}) = \mathbf{p}_i$$

so that we obtain

$$\sum_{j=1}^{n} (y_j - x_j) D_j f_i (\mathbf{p}_i) = 0.$$

We thus obtain a system of n-linear equations

$$(y_1 - x_1) D_1 f_1 (\mathbf{p}_1) + (y_2 - x_2) D_2 f_1 (\mathbf{p}_1) + \ldots + (y_n - x_n) D_n f_1 (\mathbf{p}_1) = 0$$

$$\ldots \quad \ldots \quad \ldots \quad \ldots \quad \ldots \quad \ldots \quad \ldots$$

$$\ldots \quad \ldots \quad \ldots \quad \ldots \quad \ldots \quad \ldots \quad \ldots$$

$$\ldots \quad \ldots \quad \ldots \quad \ldots \quad \ldots \quad \ldots \quad \ldots$$

$$(y_1 - x_1) D_1 f_n (\mathbf{p}_n) + (y_2 - x_2) D_2 f_n (\mathbf{p}_n) + \ldots + (y_n - x_n) D_n f_n (\mathbf{p}_n) = 0$$

whose determinant

$$\left| D_j f_i (\mathbf{p}_i) \right| \ne 0, 1 \le i \le n, \quad 1 \le j \le n.$$

This implies that

$$y_1 - x_1 = 0, \ldots, y_n - x_n = 0$$

$$\mathbf{x} = \mathbf{y},$$

so that we arrive at a contradiction.

Thus we prove that

$$\mathbf{x} \ne \mathbf{y} \Rightarrow \mathbf{f}(\mathbf{x}) \ne \mathbf{f}(\mathbf{y}).$$

We denote by $\mathbf{f}(S)$ the set

$$\{\mathbf{f}(\mathbf{x}) : \mathbf{x} \in \bar{S}\}$$

and see that $\mathbf{f}$ is a one-one function from $\bar{S}$ onto $\mathbf{f}(S)$.

II. We shall now *show that* $\mathbf{f}(\bar{S})$ *contains a neighbourhood of the point* $\mathbf{b} = \mathbf{f}(\mathbf{a})$.

Let T denote the boundary of the sphere $\bar{S}$, *i.e.*, the set

$$\{\mathbf{x} : \|\mathbf{x} - \mathbf{a}\| = r\}$$

Now consider

$$\|\mathbf{f}(\mathbf{x}) - \mathbf{b}\|; \; \mathbf{x} \in T.$$

Here we have a real valued function with values

$$\|\mathbf{f}(\mathbf{x}) - \mathbf{f}(\mathbf{a})\| = \|\mathbf{f}(\mathbf{x}) - \mathbf{b}\|$$

with domain T. As the function **f** is one-one with domain T and **a** is not a member of T, we see that

$$\|\mathbf{f}(\mathbf{x}) - \mathbf{b}\|$$

is positive $\forall\, \mathbf{x} \in T$. As T is a compact set and the real valued function, in question, is continuous $\forall\, \mathbf{x} \in T$, we see that the infimum of the function is necessarily positive. Let this infimum be denoted by $2k$, so that we have

$$2k \leq \|\mathbf{f}(\mathbf{x}) - \mathbf{b}\|; \forall\, \mathbf{x} \in T.$$

We now define a set V as follows :

$$V = \{y : \|y - b\| < k\}$$

and show that V is contained in $\mathbf{f}(\bar{S})$.

Let $\mathbf{y} \in V$ so that we have to show that there exists $\mathbf{x} \in \bar{S}$ such that y = **f** (**x**). We have

$$2k \leq \|\mathbf{f}(\mathbf{x}) - \mathbf{b}\| \leq \|\mathbf{f}(\mathbf{x}) - \mathbf{y}\| + \|\mathbf{y} - \mathbf{b}\|$$

$$\leq \|\mathbf{f}(\mathbf{x}) - \mathbf{y}\| + k$$

so that

$$\|\mathbf{f}(\mathbf{x}) - \mathbf{y}\| > k\ \forall\, \mathbf{y} \in V,\ \forall\, \mathbf{x} \in T \qquad \text{...(I)}$$

Also $$\|\mathbf{f}(\mathbf{a}) - \mathbf{y}\| = \|\mathbf{b} - \mathbf{y}\| < \mathbf{k}. \qquad \text{...(II)}$$

Keeping y fixed, we consider the real valued function with value

$$\|\mathbf{f}(\mathbf{x}) - \mathbf{y}\|^2,\ \mathbf{x} \in \bar{S}.$$

From (I) and (II), we see that the infimum of this function is assumed at an interior point of $\bar{S}$ and as such this infimum is a minimum value.

We have

$$\sum_{i=1}^{n} [f_i(\mathbf{x}) - y_i]^2 = \|\mathbf{f}(\mathbf{x}) - \mathbf{y}\|$$

Thus by § 13.9, p. 405, we see that

$$\sum_{i=1}^{n} [f_i(\mathbf{x}) - y_i] D_j f_i(\mathbf{x}) = 0, 1 \leq j \leq n.$$

Re-writing these equations in details, we obtain

$$\left[f_1(\mathbf{x})-y_1\right]D_1 f_1(\mathbf{x})+\ldots+\left[f_n(\mathbf{x})-(y_n)\right]D_1 f_n(\mathbf{x})=0$$

$$\ldots \quad \ldots \quad \ldots \quad \ldots$$

$$\ldots \quad \ldots \quad \ldots \quad \ldots$$

$$\left[f_1(\mathbf{x})-y_1\right]D_n f_n(\mathbf{x})+\ldots+\left[f_n(\mathbf{x})-(y_n)\right]D_n f_n(\mathbf{x})=0.$$

The determinant of this set of linear equations being non-zero we see that

$$f_1(\mathbf{x})-\mathbf{y}_1=0,\ f_2(\mathbf{x})-y_2=0,\ldots,\ f_n(\mathbf{x})-y_n=0$$

which imply

$$f(\mathbf{x})=\mathbf{y}.$$

We thus see that V is contained in $\mathbf{f}(S)$.

III. Let U denote the set of points $\mathbf{x}$ such that

$$\mathbf{f}(\mathbf{x})\in V.$$

This set U is open. This is a consequence of the fact that $\mathbf{f}$ is continuous and V is an open set.

We thus see that there exist open sets U and V containing $\mathbf{a}$ and $\mathbf{f}(\mathbf{a})=\mathbf{b}$ respectively such that $\mathbf{f}$ is a one-one function from U onto V *i.e.*,

$$\mathbf{x}\in U \Rightarrow \mathbf{f}(\mathbf{x})\in V.$$

and to $y\in V$ there corresponds one and only one

$$\mathbf{x}\in U \text{ such that } \mathbf{f}(\mathbf{x})=\mathbf{y}.$$

Thus the function $\mathbf{f}$ with domain U and range V is inversible. Let $\mathbf{g}$ denote the inverse of the function $\mathbf{f}$ so that V is its domain and U its range. We thus have

$$\mathbf{y}=\mathbf{f}(\mathbf{x})\Leftrightarrow \mathbf{x}=\mathbf{g}(\mathbf{y}),\ \mathbf{x}\in U,\mathbf{y}\in V.$$

IV. We have lastly to show that the function $\mathbf{g}$ itself admits of continuous first order partial derivatives. Let

$$\mathbf{y}=\mathbf{f}(\mathbf{x})\Leftrightarrow \mathbf{x}=\mathbf{g}(\mathbf{y}).$$

For sufficiently small values of λ,

$$\mathbf{y}+\lambda\mathbf{u}_k\in V.$$

Let

$$\mathbf{x}'=\mathbf{g}(\mathbf{y}+\lambda\mathbf{u}_k)\Leftrightarrow \mathbf{y}+\lambda\mathbf{u}_k=\mathbf{x}'.$$

$$\therefore \qquad \mathbf{f}(\mathbf{x}')-\mathbf{f}(\mathbf{x})=\lambda\mathbf{u}_k,$$

$$\Leftrightarrow \qquad f_i(\mathbf{x}')\,f_i(\mathbf{x})=\lambda\delta_{ik},\ 1\le i\le n.$$

By the mean value theorem, we obtain

$$f_i(\mathbf{x}') - f_i(\mathbf{x}) = \sum_{j=1}^{n} \left(x'_j - x_j\right) D_j f_i \left[x + \theta_i (\mathbf{x}' - \mathbf{x})\right]$$

so that we obtain

$$\sum_{j=1}^{n} \left(x'_j - x_i\right) D_j f_i \left[\mathbf{x}' + \theta_i (\mathbf{x}' - \mathbf{x})\right] = \lambda\, \delta_{ik}.$$

$$\Leftrightarrow \qquad \sum_{j=1}^{n} \frac{\left[g_j(\mathbf{y} + \lambda \mathbf{u}_k) - g_j(\mathbf{y})\right]}{\lambda} D_j f_i \left[x + \theta_i (\mathbf{x}' - \mathbf{x})\right] = \delta_{ik}$$

As the determinant of this system of linear equations is non-zero, we see that

$$\frac{g_j(\mathbf{y} + \lambda \mathbf{u}_k) - g_j(\mathbf{y})}{\lambda}$$

is determined $1 \le j \le n$. Imagining these equations solved and letting **x'** tend to **x** which is equivalent to λ tending to 0, we obtain

$$\sum_{j=1}^{n} D_k g_j(\mathbf{y})\, D_j f_i(\mathbf{x}) = \delta_{ik}$$

Thus $D_k g_j(\mathbf{y})$ exists $\forall\, 1 \le k \le n, 1 \le j \le n$ and $\forall\, y \in V$, so that **g** admits of partial derivatives in V.

Again because the partial derivatives

$$D_k g_j(\mathbf{y})$$

are expressibe as linear combinations of

$$D_j f_i(x),$$

we see that the partial derivatives $D_k g_j(\mathbf{y})$ are continuous in V.

Hence the theorem.

Note. While the statement as also the proof as given above holds for $\mathbf{R}^n$ for arbitrary n the reader is advised, for the fixation of ideas, to visualise the cases $n = 2$, $n = 3$ and keep the same at the back of his mind. This will help the reader see the motive underlying different steps in the proof.

14.6. IMPLICITLY DEFINED FUNCTIONS

Let f be a real valued function of two variables so that its domain D is a sub-set of $\mathbf{R}^2$. We are now interested in the sub-set E of D defined as follows :

$$E = \{(x, y) : (x, y) \in D \text{ and } f(x, y) = 0\}.$$

Now there may or may not exist a real valued function ϕ of single variable with some domain $A \subset \mathbf{R}$ such that

$$E = \{(x, \phi(x)) : x \in A\}.$$

If there exists a function such as ϕ defined above, we may say that $y=\phi(x)$ is a solution of $f(x, y) = 0$.

Again if ϕ exists, one is naturally interested in the continuity and derivability properties of the function ϕ. Of course, as one may expect, the continuity and derivability properties of ϕ will be related to the corresponding properties of the given function f.

Now the problem as described above is of a *Global character*. In general this is not capable of a solution; nor is it of much usefulness and interest. There is, however, a corresponding problem of *Local nature* which is of great interest in applications.

This problem is, however, capable of generalisation in two ways. In the first instance, f may be a real valued function of $n + 1$ variables instead of just 2 and we may be interested in a solution of

$$f(x_1, ..., x_n, y)=0$$

in the form

$$y=\phi(x_1, ..., x_n).$$

Again we may have a number of functional equations, say

$$\begin{aligned} f_1(x_1, ..., x_n; y_1, y_2, ..., y_m)&=0 \\ f_2(x_1, ..., x_n; y_1, y_2, ..., y_m)&=0 \\ \dots \quad \dots \quad \dots \quad \dots \quad & \dots \\ \dots \quad \dots \quad \dots \quad \dots \quad & \dots \\ f_n(x_1, ..., x_n; y_1, y_2, ..., y_m)&=0 \end{aligned}$$

and seek solutions of the form

$$\begin{aligned} y_1 &= \phi_1(x_1, ..., x_n). \\ \dots \quad & \dots \quad \dots \\ \dots \quad & \dots \quad \dots \\ y_m &= \phi_m(x_1, ..., x_n). \end{aligned}$$

The general problem will be precisely stated and proved in the following section. The solution will be a deduction from the Inverse function theorem and proved in § 13.5.

We first introduce some new notation.

If

$$\mathbf{x}=(x_1, ..., x_n), \mathbf{y}=(y_1, ..., y_m),$$

we shall write the pair

$$(\mathbf{x}, \mathbf{y})$$

to denote a point of $\mathbf{R}^{n+m}$. Whenever we come across the symbol $(\mathbf{x}, \mathbf{y})$ it will mean that the first member $\mathbf{x} \in \mathbf{R}^n$ and the second member $\mathbf{y} \in \mathbf{R}^m$.

Implicit function theorem. *Let* **f** *be a function with open domain* $D \subset \mathbf{R}^{n+m}$ *and range* $\subset \mathbf{R}^n$. *Let* $(\mathbf{a}, \mathbf{b}) \in D$, *so that* $\mathbf{a} \in \mathbf{R}^n$ *and* $\mathbf{b} \in \mathbf{R}^m$ *and let* **f** *possess continuous first order partial derivatives in a neighbourhood of* (**a**, **b**) *and let* **f** (**a**, **b**) = *0 and*

$$\text{det. } D_i f_j (\mathbf{a}, \mathbf{b}) \neq 0, 1 \leq \mathbf{i} \leq \mathbf{n}, 1 \leq \mathbf{j} \leq \mathbf{n}$$

where

$$\mathbf{f} = (f_1, ..., f_n).$$

Then there exists a unique function h with domain in $\mathbf{R}^m$ *and range in* $\mathbf{R}^n$ *such that there exists a neighbourhood B of* **b** *in which* **h** *is defined and admits of continuous first order partial derivatives and*

(*i*) **h** (**b**) = **a**

(*ii*) **f** (**h** (**y**), **y**) = **0**, **y** ∈ B.

We consider the transformation defined as follows :

$$z_1 = f_1 (x_1, ..., x_n; y_1, ..., y_m),$$
$$z_2 = f_2 (x_1, ..., x_n; y_1, ..., y_m),$$
$$... \quad ... \quad ... \quad ...$$
$$... \quad ... \quad ... \quad ...$$
$$z_n = f_n (x_1 ..., x_n; y_1 ..., y_m).$$
$$z_{n+1} = y_1,$$
$$z_{n+2} = y_2,$$
$$.....................$$
$$.....................$$
$$z_{n+m} = y_m.$$

We denote this transformation whose domain and range are sub-sets of $\mathbf{R}^{n+m}$ by **F**. Now

$$\mathbf{F} (\mathbf{a}, \mathbf{b}) = (\mathbf{0}, \mathbf{b})$$

and $$J_{\mathbf{f}} (\mathbf{a}, \mathbf{b}) = \text{det. } D_i f_j (\mathbf{a}, \mathbf{b}) \neq 0.$$

Also **F** possess continuous first order partial derivatives in a neighbourhood of (**a**, **b**).

Thus **F** is locally invertible at (**a**, **b**). Let **G** denote the inverse of **F** and let *U* and *V* be the neighbourhoods of (**a**, **b**) and (**0**, **b**) respectively such that **G** and **F** are inverses of each other in these neighbourhoods. We write

$$(\mathbf{z}', \mathbf{z}'') = \mathbf{F} (\mathbf{x}, \mathbf{y}) \quad \Leftrightarrow \quad (\mathbf{x}, \mathbf{y}) = \mathbf{G} (\mathbf{z}', \mathbf{z}'').$$

Now $$(\mathbf{x}, \mathbf{y}) = \mathbf{G} (\mathbf{z}', \mathbf{z}'')$$

can be re-written as

$$\mathbf{x} = \mathbf{g}(\mathbf{z}', \mathbf{z}''),\ \mathbf{y} = \mathbf{z}''.$$

We now write

$$\mathbf{x} = \mathbf{h}(\mathbf{y}) = \mathbf{g}(\mathbf{0}, \mathbf{y});\ \mathbf{y} \in B$$

where B is a neighbourhood of **b** such that $(\mathbf{0}, \mathbf{y}) \in V$ if $\mathbf{y} \in B$.

Now

$$\mathbf{f}\left[\mathbf{g}(\mathbf{z}', \mathbf{z}''), \mathbf{y}\right] = \mathbf{z}'$$

$$\Rightarrow \qquad \mathbf{f}\left[\mathbf{h}(\mathbf{y}), \mathbf{y}\right] = \mathbf{0}.$$

Also since

$$\mathbf{g}\ (\mathbf{0}, \mathbf{b}) = \mathbf{a},$$

we have

$$\mathbf{h}\ (\mathbf{b}) = \mathbf{a}.$$

To prove the uniqueness of **h**, we may see that the points (**x**, **y**) of U are in one-one correspondence with the points **f** [(**x**, **y**), **y**] of V.

Thus if

$$f(\mathbf{x}', \mathbf{y}) = \mathbf{f}(\mathbf{x}, \mathbf{y})$$

for some

$$(\mathbf{x}, \mathbf{y}) \in U \text{ and } (\mathbf{x}', \mathbf{y}) \in U,$$

we must have $\mathbf{x}' = \mathbf{x}$.

Note. For the benefit of those of the readers, who may not be interested in the above general discussion, we give below the cause of a real valued function of two variables.

14.7. IMPLICIT FUNCTION DETERMINED BY A SINGLE FUNCTIONAL EQUATION. CASE OF TWO VARIABLES

Let f be a function of two variables and let (a, b) be a point of its domain such that

(*i*) $f(a, b) = 0$

(*ii*) *the function possesses continuous partial derivatives* f_x *and* f_y *in a certain neighbourhood of (a, b) and*

(*iii*) $f_y(a, b) \neq 0$;

then there exists a rectangle $[a - h, a + h; b - k, b + k]$ *about (a, b) such that* $\forall\ x \in [a - h, a + h]$, *the equation* $f(x, y) = 0$ *determines one and only one solution* $y = \varphi(x)$, *lying in the interval* $[b - k, b + k]$, *with the following properties :*

1. $b = \varphi(a)$,

2. $f\{x, \varphi(x)] = 0 \;\forall\; x \in [a - h, a + h]$.

3. ... *derivable and both* φ *as well as* φ' *are continuous in*

$$[a - h, a + h].$$

(The reader is advised to follow the following line of argument by means of the diagram attached.)

Without any loss of generality, we suppose that $f_p\ (a, b) > 0$; for, otherwise, we should only have to replace $f(x, y)$ by $-f(x, y)$ and this change would leave the equation $f(x, y) = 0$ unaltered.

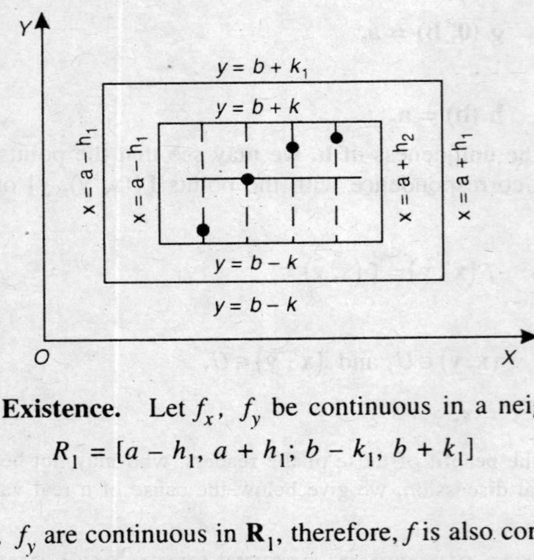

Unique Existence. Let f_x, f_y be continuous in a neighbourhood.

$$R_1 = [a - h_1, a + h_1; b - k_1, b + k_1]$$

of (a, b).

Since f_x, f_y are continuous in $\mathbf{R}_1$, therefore, f is also continuous in R_1.

Since f_y is continuous at (a, b) and $f_y\ (a, b) > 0$, there exists a rectangle

$$R_2 = [a - h_2, a + h_2; b - k, b + k], \quad h_2 < h_1, k < k_1$$

such that for every point (x, y) of this rectangle R_2, $f_y\ (x, y) > 0$.

Now, since

$$f_y\,(x, y) > 0 \;\forall\, (x, y) \in R_2,$$

therefore,

$$\forall\; x \in [a - h_2, a + h_2],$$

the function f of y strictly increases as y increases from $b - k$ to $b + k$. In particular, since $f(a, b) = 0$, we have

$$f(a, b - k) < 0, f(a, b + k) > 0.$$

In view of this and the fact that f is continuous, there exists an interval $[a - h, a + h]$, $(h < h_2)$ such that for every x of this interval, we have

$$f(x, b - k) < 0, f(x, b + k) > 0.$$

Now for every fixed value of x in $[a - h, a + h]$, the continuous function of y *strictly* increases from a negative to a positive value as y increases from $b - k$ to $b + k$ and therefore there exists *one and only one* value of y for which the function f vanishes.

Hence for each value of x in $[a - h, a + h]$ there is a uniquely determined value $y = \phi(x)$ for which $f(x, y) = 0$; this y is such that the properties (1) and (2) are true.

This completes the proof of the *existence* and the *uniqueness* of the implicit function φ.

Continuity. We now prove that φ is continuous in $[a - h, a + h]$. Let x_0 be any point of this interval and let $y_0 = \varphi(x_0)$. Let ε be any given positive number.

Let $\quad R' = [x_0 - \delta_1, x_0 + \delta_1; y_0 - \varepsilon, y_0 + \varepsilon]$

be a rectangle entirely lying within the rectangle

$$R = [a - h, a + h; b - k, b + k]$$

found above. For this rectangle, we can carry out exactly the same process as before in order to obtain a solution of $f(x, y) = 0$. Since the solution was uniquely determined in R which encloses R', we see that $y = \varphi(x)$ is a solution in R' also. Thus there exists an interval

$$]x_0 - \delta, x_0 + \delta[, (\delta \leq \delta_1)$$

such that for every value of x in this interval $\varphi(x)$ lies between $y_0 - \varepsilon$ and $y_0 + \varepsilon$, *i.e.*,

$$|y - y_0| = |\varphi(x) - \varphi(x_0)| < \varepsilon.$$

$$|x - x_0| \leq \delta.$$

Hence φ is *continuous* at x_0 and therefore, in $[a - h, a + h]$.

Derivability. Let x be a point of the interval $[a - h, a + h]$ and let $x + p$ be another point of the same interval.

Let $\quad y = \varphi(x), y + q = \varphi(x + p)$.

$$\Rightarrow \quad f(x, y) = 0, f(x + p, y + q) = 0.$$

$$\begin{aligned}\Rightarrow \quad 0 &= f(x + p, y + q) - f(x, y)\\ &= f(x + p, y + q) - f(x + p, y) + f(x + p, y) - f(x, y).\\ &= qf_y(x + p, y + \theta_1 q) + pf_x(x + \theta_2 p, y).\end{aligned}$$

(*By mean value theorem*)

Since $f_y \neq 0$ in R and $(x + p, y + \theta_1 q)$ is a point of R, we have

$$\frac{q}{p} = -\frac{f_x(x+\theta_2 p, y)}{f_y(x+p, y+\theta_1 q)}.$$

Since φ is continuous, $q \to 0$ as $p \to 0$. Therefore f_x and f_y being continuous, we obtain from above, when $(p, q) \to (0, 0)$.

$$\varphi'(x) = -\frac{f_x(x, y)}{f_y(x, y)}.$$

Thus φ is derivable and $\varphi'(x) = -f_x(x, y)/f_y(x, y)$. Also this formula for $\varphi'(x)$ shows that it is continuous.

Note. 1. The function $y = \varphi(x)$ is said to be the unique solution of $f(x, y) = 0$ *near* (a, b) or the unique implicit function determined by $f(x, y) = 0$ *near* (a, b).

Note. 2. It is very important to bear in mind the fact that the theorem just proved is essentially of a *Local Character*. It states that if f is a function of two variable satisfying certain assumptions in respect of continuity and derivability in a neighbourhood of a point (a, b) and if $(a, b) = 0$, then there exists a neighbourhood

$$[a - h, a + h; b - k, b + k] \text{ of } (a, b)$$

such that for each x belonging to $[a - h, a + h]$ there exists a unique y belonging to $[b - k, b + k]$ such that $f(x, y) = 0$ *i.e.*, $f(x, y) = 0$ defines a function $y = \varphi(x)$ in $[a - h, a + h]$; y lying in $[b - k, b + k]$ and this function φ of x is derivable.

Ex. Consider $f(x, y) = x^2 + y^2 - 1$ and a point $(0, 1)$ so that $f(0, 1) = 0$, $f_y(0, 1) \neq 0$. How do you reconcile the conclusion of the theorem with the fact that $x^2 + y^2 - 1 = 0$ defines *two* functions, *viz.*

$$y = +\sqrt{(1 - x^2)},\ y = -\sqrt{(1 - x^2)}.$$

EXERCISES

1. If f is a continuous function of each variable x and y separately in a certain neighbourhood of (a, b), $f(a, b) = 0$, f_y is continuous at (a, b) and $f_y(a, b) \neq 0$, then the equation $f(x, y) = 0$ determines a unique continuous implicit solution $y = \varphi(x)$ near (a, b).

2. If f is a continuous function of each variable x and y separately in a neighbourhood of (a, b), $f(a, b) = 0$ and $f_y(a, b)$ exists $\neq 0$ then the equation $f(x, y) = 0$ determines at *least one* continuous solution $y = \varphi(x)$ near (a, b).

3. f is a continuous function of x and y in the neighbourhood of (a, b) such that $f(a, b) = 0$; and further f is, for all x in the neighbourhood of (a, b), strictly increasing function of y. Prove that there exists a unique solution $y = \varphi(x)$, which when substituted in the equation $f(x, y) = 0$, satisfies it identically for all values of x in the neighbourhood of a, and that φ is continuous for all values of x in the neighbourhood of a.

If $f(x, y) = y^4 - y^2 + x^2$, discuss the existence of the function $y = \varphi(x)$ in the neighbourhood of $x = 0$, $y = 0$.

4. Show that the following equations determine unique solutions near the points indicated; find also the first derivatives of the solution :

(*i*) $x^3 + y^3 - 3xy + y = 0$, $(0, 0)$

(*ii*) $xy \sin x + \cos y = 0$, $\left(0, \frac{1}{2}\pi\right)$.

(*iii*) $y^3 \cos x + y^2 \sin^2 x = 7$, $\left(\frac{1}{3}\pi, 2\right)$.

(*iv*) $2xy - \log xy = 2$, $(1, 1)$.

5. Examine the following equations for the existence of unique solutions near the points indicated and verify by direct calculations :

(*i*) $y^2 - yx^2 - 2x^5 = 0$ near $(0, 0)$ and $(1, -1)$

(*ii*) $y^4 + x^2y^2 + 2x^5 = 0$ near $(1, -1)$.

(*iii*) $y^2 + 2x^2y + x^5 = 0$ near $(1, -1)$.

(*iv*) $y^2 + yx^3 + x^2 = 0$ near $(0, 0)$

6. Show that the least positive root of $xy = \tan y$ is a continuous function of x throughout the interval $[1, \infty[$ and increases steadily from 0 to $\frac{1}{2}\pi$ as y increases from 1 towards ∞.

OBJECTIVE QUESTIONS

Note. *For each of the following questions, four alternatives are given for the answer. Only one of them is correct. Choose the correct alternative.*

1. If f be a vector valued function and if $a \in \Delta$ (domain of f) and $f(a) = b \in E$ (range of f) then the function f is locally invertible at a if the Jacobian

(*a*) $J_f(a) = 0$ (*b*) $J_f(a) \neq 0$

(*c*) $J_f(a)$ may or may not be zero (*d*) $J_f(a)$ is necessarily zero.

2. The transformation defined by $u = e^x \cos y$, $v = e^x \sin y$ is

(*a*) globally invertible (*b*) locally invertible at $(x, y) = (0, 0)$

(*c*) locally invertible at each point (*d*) not invertible

3. Transformation defined by $u = x^2 - y^2$, $v = xy$ is

(*a*) globally invertible (*b*) locally invertible at each point

(*c*) locally invertible at $(0, 0)$ only

(*d*) locally invertible at each point other than $(0, 0)$

ANSWERS

1. (*b*) **2.** (*c*) **3.** (*d*)

15

Integrals as Functions of a Parameter

15.1. DEFINITE INTEGRAL AS FUNCTION OF A PARAMETER

Let f be a continuous function of two variables with rectangle

$$[a, b; c, d] \subset \mathbf{R}^2,$$

as its domain.

For a *fixed* value of $y \in [c, d]$, the function of x with domain $[a, b]$ is continuous and therefore the integral

$$\int_a^b f(x, y)\, dx$$

exists and defines a function of a single variable y, say ϕ, with domain $[c, d]$.

It is now proposed to investigate the nature of the function ϕ in relation to continuity and derivability.

15.1.1. Theorem

The function ϕ is continuous in $[c, d]$.

Let, y, $y + k$ be any two points $[c, d]$. We have

$$\phi(y+k) - \phi(y) = \int_a^b \left[f(x, y+k) - f(x, y)\right] dx.$$

Let ε be any positive number.

Since f is continuous in $[a, b; c, d]$, it is uniformly continuous and as such there exists a positive number δ such that

$$\left| f(x_2, y_2) \right| - f(x_1, y_1) < \varepsilon / (b - a),$$

when $|x_2 - x_1| \le \delta, | y_2 - y_1| \le \delta$.

In particular, we see that when $|k| \le \delta$ and x is any number $\in [a, b]$, we have

$$| f(x, y + k) - f(x, y) | < \varepsilon / (b - a).$$

Thus

$$|\phi(y+k)-\phi(y)| = \left|\int_a^b [f(x,y+k)-f(x,y)]\,dx\right|$$

$$\le \int_a^b |f(x,y+k)-f(x,y)|\,dx$$

$$\le [\varepsilon/(b-a)][(b-a)] = \varepsilon, \text{ when } |k| \le \delta.$$

Hence ϕ is a continuous function in $[c, d]$.

Ex. If f is continuous in $[a, b; c, d]$ and F is a function of x which is bounded and integral in $[a, b]$, then $\int_a^b f(x, y)\, F(x)\, dx$ is a continuous function of y in $[c, d]$.

15.1.2. Theorem

If, in addition to f being continuous, f_y also exists and is continuous in $[a, b\ ;\ c, d]$, then ϕ is derivable and

$$\phi'(y) = \int_a^b f_y(x,y)\,dx \ \forall\ y \in [c,d]$$

$$\Rightarrow \quad \frac{d}{dy}\left\{\int_a^b f(x,y)\,dx\right\} = \int_a^b f_y(x,y)\,dx,$$

so that the inversion of the operations of differentiation and integration is valid. *(Lucknow 94, 95, 96)*

We have

$$\phi(y+k)-\phi(y) = \int_a^b [f(x,y+k)-f(x,y)]\,dx.$$

By the Lagrange's mean value theorem, we have

$$f(x,y+k)-f(x,y) = kf_y(x,y+\theta k)$$

$$= k[f_y(x,y+\theta k)-f_y(x,y)+f_y(x,y)].$$

Thus

$$\frac{\phi(y+k)-\phi(y)}{k} - \int_a^b f_y(x,y)\,dx = \int_a^b [f_y(x,y+\theta k)-f_y(x,y)]\,dx.$$

Let ε be any positive number. Because of the continuity and consequent uniform continuity of f_y in $[a, b; c, d]$, there exists a positive number δ, such that when $|k| \le \delta$ and x has any value, then

$$|f_y(x, y+k) - f_y(x, y)| < \varepsilon / (b-a).$$

Thus there exists a positive number δ such that

$$\left|\frac{\phi(y+k)-\phi(y)}{k} - \int_a^b f_y(x,y)\,dx\right| < \varepsilon \text{ when } |k| \le \delta.$$

Hence the result.

Ex. If f and f_y be continuous in $[a, b; c, d]$ and F be bounded and integrable in $[a, b]$, then

$\int_a^b f(x, y) F(x)\, dx$ is derivable in $[c, d]$ with derivative $\int_a^b f_y(x, y) F(x)\, dx$.

Note. In the above investigation, the limits a, b of integration were considered constants independent of y. The case where they are themselves function values is considered below.

15.1.3. Theorem

Let f, f_y be continuous in $[a, b; c, d]$; and let g_1, g_2 be two functions derivable in $[c, d]$ such that $\forall\ y \in [c, d]$ the points

$$[g_1(y), y] \text{ and } [g_2(y), y]$$

belong to the rectangle $[a, b; c, d]$. Then the function ϕ such that

$$\varphi(y) = \int_{g_1(y)}^{g_2(y)} f(x, y)\, dx$$

is derivable in $[c, d]\ \forall\ y \in [c, d]$ and

$$\varphi'(y) = \left\{\int_{g_1(y)}^{g_2(y)} f_y(x, y)\, dx\right\} - g_1'(y) f[g_1(y), y] + g_2'(y) f[g_2(y), y].$$

We have

$$\varphi(y+k) - \varphi(y) = \int_{g_1(y)}^{g_2(y)} [f(x, y+k) - f(x, y)]\, dx$$

$$- \int_{g_1(y)}^{g_1(y+k)} f(x, y+k)\, dx + \int_{g_2(y)}^{g_2(y+k)} f(x, y+k)\, dx.$$

Applying the result of Cor. 2 to § 6.7.8, page 155 to each of the two integrals and dividing by k, we get

$$\frac{\varphi(y+k) - \varphi(y)}{k} = \int_{g_1(y)}^{g_2(y)} f_y(x, y+\theta k)\, dx - \frac{g_1(y+k) - g_1(y)}{k} f(\xi, y+k)$$

$$+ \frac{g_2(y+k) - g_2(y)}{k} f(\eta, y+k),$$

where ξ, η lie between $g_1(y), g_1(y+k)$, and $g_2(y), g_2(y+k)$ respectively. As in the previous section, it can be shown that, when $k \to 0$, then

$$\int_{g_1(y)}^{g_2(y)} f_y(x, y+\theta k)\, dx \to \int_{g_1(y)}^{g_2(y)} f_y(x, y)\, dx.$$

Thus, on taking limits when $k \to 0$, we deduce that

$$\varphi'(y) = \int_{g_1(y)}^{g_2(y)} f_y(x, y)\, dx - g_1'(y) f[g_1(y), y] + g_2'(y) f[g_2(y), y].$$

15.1.4. Inversion of the Order of Integration

If f be continuous in [a, b; c, d], then

$$\int_c^d \left\{\int_a^b f(x, y)\, dx\right\} dy = \int_a^b \left\{\int_c^d f(x, y)\, dy\right\} dx,$$

i.e., the two repeated integrals are equal.

Because of the continuity of f, the integrals of $f(x, y)$ over $[a, b]$ with respect to x and over $[c, d]$ with respect to y exist and are continuous functions of y and x respectively and as such both the repeated integrals exist.

We consider two functions of ϕ and ψ defined as follows :

$$\varphi(t) = \int_c^t \left\{\int_a^b f(x, y)\, dx\right\} dy;\ \psi(t) = \int_a^b \left\{\int_c^t f(x, y)\, dx\right\} dx,$$

so that $\varphi(c) = \psi(c) = 0.$

Now $\forall\, t \in [c, d],$

$$\varphi'(t) = \int_a^b f(x, t)\, dx. \qquad \text{(§ 6.8.2, page 228)}$$

Also $\forall\, t \in [c, d].$

$$\psi'(t) = \int_a^b \left\{\frac{\partial}{\partial t}\int_c^t f(x, y)\, dy\right\} dx, \qquad \text{(§15.1.2, page 327)}$$

$$= \int_a^b f(x, t)\, dx \qquad \text{(§ 6.8.2, page 158)}$$

Since $\forall\, t \in [c, d], \varphi'(t) = \psi'(t)$, therefore φ and ψ differ by a constant. Also, since $\varphi(c) = \psi(c)$, we see that $\forall\, t \in [c, d], \varphi(t) = \psi(t)$. Putting $t = d$, we obtain the required equality.

Ex. If f is continuous in $[a, b\ ;\ c, d]$ and

$$F(x, y) = \int_c^y \left\{\int_a^x f(x, y)\, dx\right\} dy,$$

then show that

$$F_{xy} = F_{yx} = f\ \forall\, (x, y) \in [a, b; c, d].$$

Deduce that

$$\int_a^b \left\{\int_c^d f(x, y)\, dy\right\} dx = F(a, c) + F(b, d) - F(a, d) - F(b, c)$$

$$= \int_c^d \left\{\int_a^b f(x, y)\, dx\right\} dy.$$

Note. The theorems above enable us to evaluate certain types of definite integrals without the knowledge of the corresponding primitives. A few examples of such evaluation are given below.

EXAMPLE

If $|a| < 1$, *show that*

$$\int_0^{\pi} \frac{\log(1 + a\cos x)}{\cos x}\, dx = \pi \sin^{-1} a.$$

We write

$$f(a, x) = \frac{\log(1 + a\cos x)}{\cos x}.$$

It may be seen that the integrand has only a removable discontinuity at $\pi/2$ inasmuch as

$$\lim_{x \to \pi/2} \frac{\log(1 + a\cos x)}{\cos x} = a.$$

Also

$$f_a(a, x) = \frac{1}{1 + a\cos x}$$

We see that f and f_a are both continuous functions of two variables in the domain $\{(a, x) : 0 \le x \le \pi, |a| < 1\}$, if we assign to f, the value a for $x = \pi/2$.

We write

$$\phi(a) = \int_0^{\pi} f(a, x)\, dx. \quad \text{..(1)}$$

$$\Rightarrow \quad \phi'(a) = \int_0^{\pi} f_a(a, x) = \int_0^{\pi} \frac{1}{1 + a\cos x}\, dx.$$

Putting $\tan(x/2) = t$, we may see that

$$\int_0^{\pi} \frac{dx}{1 + a\cos x} = \frac{\pi}{\sqrt{(1 - a^2)}}$$

$$\Rightarrow \quad \phi'(a) = \frac{\pi}{\sqrt{(1 - a^2)}}.$$

From this we obtain

$$\phi(a) = \pi \sin^{-1} a + c, \quad \text{...(2)}$$

where, c, is an arbitrary constant.

From (1), we have

$$\phi(0) = 0.$$

Putting $a = 0$ in (2), we obtain $c = 0$.

Hence $\phi(a) = \pi \sin^{-1} a.$

EXERCISES

1. Starting from a suitable integral, show that

$$\int_0^x \frac{dx}{\left(x^2+a^2\right)^2} = \frac{1}{2a^3}\tan^{-1}\frac{x}{a} + \frac{x}{2a^2\left(x^2+a^2\right)}.$$

2. Find the value of

$$\int_0^\pi \frac{dx}{a+b\cos x}. \qquad a > 0,\ |b| < a$$

and deduce that

$$\int_0^p \frac{dx}{(a+b\cos x)^2} = \frac{pa}{\left(a^2-b^2\right)^{3/2}} \text{ and } \int_0^\pi \frac{\cos x\,dx}{(a+b\cos x)^2}.$$

3. Show that

$$\int_0^{\frac{1}{2}\pi} \log\left(1-x^2\cos^2\theta\right)d\theta = \pi\log\left[1+\sqrt{\left(1-x^2\right)} - \pi\log 2,\right]$$

if $x^2 \le 1$.

4. If $|a| \le 1$, show that

$$\int_0^\pi \log(1+a\cos x)\,dx = \pi\log\left[\frac{1}{2}+\frac{1}{2}\sqrt{\left(1-a^2\right)}\right].$$

5. Show that

$$\int_0^a \frac{\log(1+ax)}{1+x^2}\,dx = \frac{1}{2}\log\left(1+a^2\right)\tan^{-1}a.$$

Hence or otherwise show that

$$\int_0^1 \frac{\log(1+x)}{1+x^2}\,dx = \frac{\pi\log 2}{8}.$$

6. Show that

$$\int_0^{\pi/2} \log\left(a^2\cos^2\theta + b^2\sin^2\theta\right)d\theta = \pi\log\left[\frac{1}{2}(a+b)\right],$$

$(a > 0,\ b > 0)$

7. Prove that

$$\int_0^{\frac{1}{2}\pi} \frac{\log(1+\cos\alpha\cos x)}{\cos x}\,dx = \frac{\pi^2-4\alpha^2}{8}. \qquad \textit{(Lucknow 96)}$$

8. Show that, if $a > b$,

$$\int_0^\pi \log\left(\frac{a+b\sin\theta}{a-b\sin\theta}\right)\operatorname{cosec}\theta\,d\theta = \pi\sin^{-1}\frac{b}{a}.$$

9. Prove that

$$\int_{\frac{1}{2}\pi-\alpha}^{\pi} \sin\theta \cos^{-1}(\cos\alpha \operatorname{cosec}\theta)\, dx = \frac{\pi}{2}(1-\cos\alpha)$$

10. Show that

$$\int_0^{\frac{1}{2}\pi} \frac{\log\left(1+y\sin^2 x\right)}{\sin^2 x}\, dx = \pi\left[\sqrt{(1+y)}-1\right].$$

11. Verify that

$$y = \frac{1}{k}\int_0^x f(t)\sin k(x-t)\, dt,$$

satisfies the differential equation

$$\frac{d^2y}{dx^2} + k^2 y = f(x),$$

where k is a constant.

12. Show that the theorem of § 15.1, p. 436 does not hold for $y = 0$, when $a = 0$, $b = 1$ and $f(x, y) = xy^3/(x + y^2)^3$.

13. If

$$f(x, y) = y^2 / \left(x^2 + y^2\right) \text{ and } g(y) = \int_0^1 f(x, y)\, dx,$$

show that the right and left hand derivatives, $g'(+0)$ and $g'(-0)$, of g at $y = 0$ differ from each other and from

$$\int_0^1 f_y(x, 0)\, dx.$$

14. Show that

$$\int_0^1 \frac{x^y - 1}{\log x}\, dx = \log(1+y), \int_0^1 \frac{x-1-\log x}{(\log x)^2}\, dx = 2\log 2 - 1.$$

15.2. UNIFORM CONVERGENCE OF IMPROPER INTEGRALS

The operations of differentiation and integration may not be invertible when the integrals, in question, are improper so that it should be carefully understood that while proving various theorems in the preceding § 15.1 we have constantly regarded the integral

$$\int_a^b f(x, y)\, dx,$$

as proper. The situation here is analogous to that for infinite series in which case derivative (integral) of the sum may not be equal to the sum of the derivative (integrals). It has been seen in § 7.4, that with the help of the notion of uniform convergence of a series, it is possible to lay down sufficient conditions for the invertibility of the operation of derivatio (integration) and the summation of a series. It will now be seen that the corresponding notion of uniform convergence of improper integrals will do the same job for improper integrals.

15.2.1. Definitions. Uniform Convergence of Improper Integrals

Range Infinite. *The convergent improper integral*

$$\varphi(y) = \int_a^\infty f(x, y)\, dx, \qquad \text{...(1)}$$

is said to converge uniformly in the interval, [c, d], if corresponding to any positive number, η, there exists a positive number t which does not depend on y, such that $\forall\, t \geq t_1$.

$$\left|\int_a^t f(x, y)\, dx - \varphi(y)\right| = \left|\int_t^\infty f(x, y)\, dx\right| < \eta.$$

15.2.2. The Integrand Unbounded

Let $f(x, y)$ tend to ∞ as $x \to a$ for some or all $y \in [c, d]$. *The convergent improper integral*

$$\varphi(y) = \int_a^b f(x, y)\, dx \qquad \text{...(2)}$$

is said to converge uniformly in the interval [c, d], if corresponding to any positive number, η, there exists a positive number δ, independent of y, such that

$$\left|\int_{a+\varepsilon}^b f(x, y)\, dx - \varphi(y)\right| < \eta,$$

for $0 < \varepsilon \leq \delta$.

15.3. THE TESTS FOR UNIFORM CONVERGENCE

The following straightforward analogues of the tests for the uniform convergence of series may be proved in a similar manner.

15.3.1. General Test

The necessary and sufficient condition for the uniform convergence of the improper integral (1) is that corresponding to any positive number η, there exists a positive number t_0 **independent of y** *such that.*

$$\left|\int_{t_1}^{t_2} f(x, y)\, dx\right| < \eta, \text{ where } t_1, t_2 \geq t_0.$$

15.3.2. Weierstrass's '*M*' Test

If there exists a function M with domain [a, ∞ [which is positive $\forall$ x and which is such that

$$\int_a^\infty M(x)\, dx,$$

converges and $\forall\ y, \in [c, d]$ and $\forall\ x \in [a, \infty[$,

$$|\, f(x, y)\,| \le M(x),$$

then the integral (1) *of § 15.21 is uniformly convergent.*

Note. Similar theorem holds for the uniform convergence of

$$\int_a^b f(x, y)\, dx.$$

EXAMPLES

1. Since $|\, e^{-x^2} \cos yx\,| \le e^{-x^2}\ \forall\ y$, and

$$\int_0^\infty e^{-x^2}\, dx,$$

is convergent, therefore

$$\int_0^\infty e^{-x^2} \cos yx\, dx,$$

is uniformly convergent in the interval $]-\infty, \infty[$.

2. Since $\left|\cos yx / \sqrt{\left(1-x^2\right)}\right| \le 1/\left(1-x^2\right), \forall\ y \in \mathbf{R}$, and

$$\int_{-1}^{+1} \frac{1}{\sqrt{\left(1-x^2\right)}}\, dx$$

is convergent, therefore

$$\int_{-1}^{+1} \frac{\cos yx}{\sqrt{\left(1-x^2\right)}}\, dx,$$

is uniformly convergent in $]-\infty, \infty[$.

3. The uniformity of convergence of

$$\int_1^\infty \frac{x \sin xy}{1+x^2}\, dx,$$

cannot be established by M-test.

For $x \ge 1$, $x/(1+x^2)$ is monotonically decreasing and tends to 0 a[s] $x \to \infty$ so that employing the Bonnett's form of second mean valu[e] theorem, we have

$$\left|\int_{t_1}^{t_2} \frac{x \sin xy}{1+x^2}\, dx\right| = \left|\frac{t_1}{1+t_1^{\,2}} \int_{t_1}^{\xi} \sin xy\, dx\right| \le \frac{2}{k\left(1+t_1^{\,2}\right)},$$

if $y \ge k$, where k is any fixed positive number.

Thus we see that the integral is uniformly convergent in $[k, \infty[, k > 0$ (§ 15.3.1).

EXERCISES

1. Establish the uniform convergence of the following integrals :

(i) $\int_0^{\infty} \frac{y\,dx}{x^2 + y^2}, \ (0 < c \le y \le d)$ (ii) $\int_0^{\infty} e^{-xy} \sin x\,dx, (y \ge k > 0)$.

(ii) $\int_0^{\infty} e^{-xy} \frac{\sin x}{x}\,dx, (y \ge 0)$ (iv) $\int_0^{\infty} e^{\left(-x^2 - y^2/x^2\right)}\,dx,$

(v) $\int_0^{\infty} e^{-xy}\,dx, (y \ge k > 0)$.

15.4. THEOREM

Uniformly convergent improper integral of a continuous function is itself a continuous function.

Consider the improper integral (1) of § 15.2.1, page 443. We choose number t such that

$$\left|\int_t^{\infty} f(x, y)\,dx\right| = \left|\int_a^t f(x, y)\,dx - \varphi(y)\right| < \frac{\varepsilon}{3}.$$

We have

$$|\varphi(y+k) - \varphi(y)| \le \left|\int_a^t [f(x, y+k) - f(x, y)]\,dx\right| + \frac{2\varepsilon}{3}.$$

The range $[a, t]$ being finite and f being continuous, we can choose a positive number δ, as in § 15.1.1, page 436, so that the finite integral on the right is less than $\varepsilon/3$ when $|k| \le \delta$. thus φ is continuous.

15.4.1. *If f be a continuous function when $c \le y \le d$ and $x \ge a$ and the integral*

$$\varphi(y) = \int_a^{\infty} f(x, y)\,dx,$$

is uniformly convergent, then φ can be integrated under the integral sign, i.e.,

$$\int_c^d \left\{\int_a^{\infty} f(x, y)\,dx\right\} dy = \int_c^d \varphi(y)\,dy = \int_a^{\infty} \left\{\int_c^d f(x, y)\,dy\right\} dx$$

We choose a number t such that

$$\left|\int_t^{\infty} f(x, y)\,dx\right| < \varepsilon / (d - c) \text{ when } t \ge i_1$$

$$\Rightarrow \qquad \int_c^d \left\{\left|\int_t^{\infty} f(x, y)\,dx\right|\right\} dy < \varepsilon.$$

We have

$$\int_c^d \varphi(y)\,dy = \int_c^d \left\{\int_a^t f(x, y)\,dx\right\} dy$$

$$+ \int_c^d \left\{\int_t^\infty f(x, y)\,dx\right\} dy.$$

In the first integral on the right, the order can be interchanged. We, therefore, have

$$\left| \int_c^d \varphi(y)\,dy - \int_a^t \left\{\int_c^d f(x, y)\,dx\right\} dy \right| \le \varepsilon \text{ when } t \ge t_1.$$

Hence the result.

15.4.2. Theorem

If in addition to the conditions of the preceding theorem, the integral

$$\int_a^\infty f_y(x, y)\,dx,$$

also converges uniformly in $[c, d]$, *then*

$$\varphi'(y) = \int_a^\infty f_y(x, y)\,dx.$$

Let $$\psi(y) = \int_a^\infty f_y(x, y)\,dx.$$

As proved above the order of integration can be changed so that we have

$$\int_c^y \phi(y)\,dy = \int_c^y dy \int_a^\infty f_y(x, y)\,dx$$

$$= \int_a^\infty dx \int_c^y f_y(x, y)\,dy$$

$$= \int_a^\infty \left[f(x, y) - f(x, c)\right] dx = \varphi(y) - \varphi(c).$$

Differentiating, we get

$$\psi(y) = \varphi'(y),$$

as was to be proved.

Note. Results similar to those above hold also when the range of integration is finite but the integrand has a point of finite discontinuity.

EXAMPLES

1. *Evaluate*

$$f(\alpha, \beta) = \int_a^\infty e^{-\alpha x} \frac{\sin \beta x}{x}\,dx, \text{ where } \alpha \ge 0$$

and deduce that

$$\int_0^\infty \frac{\sin \beta x}{x}\,dx = \begin{cases} +\frac{1}{2}\pi, & \text{if } \beta > 0, \\ 0, & \text{if } \beta = 0, \\ -\frac{1}{2}\pi, & \text{if } \beta < 0 \end{cases}$$

Sol. When $x > 0$, we have

$$\left| e^{-\alpha x} \sin(\beta x)/x \right| \le e^{-\alpha x}/x,$$

and the integral of $(e^{-\alpha x}/x)$ is convergent at ∞ if $\alpha > 0$. Thus we see that the integral is uniformly convergent with respect to the parameter β varying in $]-\infty, \infty[$; α, being any fixed positive number.

Again, the derivative of the integrand with respect to β is

$$e^{-\alpha x} \cos \beta x.$$

For a fixed $\alpha > 0$, we have

$$\left| e^{-\alpha x} \cos \beta x \right| \le e^{-\alpha x},$$

which is independent of β.

Since, when $\alpha > 0$, the integral of $e^{-\alpha x}$ converges over the range $[0, \infty[$, therefore, for fixed positive α, the integral

$$\int_0^\infty e^{-\alpha x} \cos \beta x\, dx,$$

is uniformly convergent w.r. to β in $]-\infty, \infty[$.

Thus the differentiation under the integral sign is valid and we have

$$f_\beta(\alpha, \beta) = \int_0^\infty e^{-\alpha x} \cos \beta x\, dx$$

$$= \left| e^{-\alpha x} \frac{\beta \sin \beta x - \alpha \cos \beta x}{\alpha^2 + \beta^2} \right|_0^\infty = \frac{\alpha}{(\alpha^2 + \beta^2)}.$$

$\therefore\ f(\alpha, \beta) = \tan^{-1}(\beta/\alpha) + c$, when c is a constant.

To determine c, we have, putting $\beta = 0$ in the given integral

$$0 = f(\alpha, 0) = 0 + c, \Rightarrow c = 0.$$

It will now be shown that $f(\alpha, \beta)$ is a continuous function of α also (for $\alpha > 0$) and for a fixed β. We have

$$\left| \int_{t_1}^{t} e^{-\alpha x} \frac{\sin \beta x}{x}\, dx \right| = \left| e^{-\alpha t} \int_{t_1}^{\xi} \frac{\sin \beta x}{x}\, dx \right| \le \left| \int_{t_1}^{\xi} \frac{\sin \beta x}{x}\, dx \right|.$$

Now since $\displaystyle\int_0^\infty \frac{\sin \beta x}{x}\, dx,$

is known to be convergent we deduce that the integral is uniformly convergent with α as parameter; α being ≥ 0.

$$\therefore \quad f(0, \beta) = \lim_{\alpha \to 0} f(\alpha, \beta) = \frac{1}{2}\pi, \text{ when } \alpha \to (0+0); \quad (\beta > 0)$$

$$\Leftrightarrow \quad \int_0^\infty \frac{\sin \beta x}{x} dx = \frac{1}{2}\pi. \quad (\beta > 0)$$

The other results, now, at once follow.

2. *Evaluate*

$$f(\alpha) = \int_0^\infty e^{-\alpha^2} \cos \alpha x \, dx.$$

Sol. Since

$$\left| xe^{-\alpha^2} \sin \alpha x \right| \le xe^{-\alpha^2},$$

we see that the differentiation under the integral sign is justified.

$$\therefore \quad f'(\alpha) = -\int_0^\infty xe^{-\alpha^2} \sin \alpha x \, dx.$$

Integrating by parts, we have

$$f'(\alpha) = \left| \frac{1}{2} e^{-\alpha^2} \sin \alpha x \right|_0^\infty - \frac{\alpha}{2} \int_0^\infty e^{-\alpha^2} \cos \alpha x \, dx = -\frac{\alpha}{2} f(\alpha).$$

$$\therefore \quad \log f(\alpha) = -\frac{1}{4}\alpha^2 + c_1 \Rightarrow f(\alpha) = ce^{-\frac{1}{4}\alpha^2}$$

Now, $f(0) = \int_0^\infty e^{-\alpha^2} dx = \frac{\sqrt{\pi}}{2}$ (See Chapter 16

$$\therefore \quad c = f(0) = \sqrt{\pi / 2}.$$

so that $f(\alpha) = \frac{1}{2}\sqrt{\pi}\, e^{-\frac{1}{4}\alpha^2}$

3. *Evaluate*

$$f(y) = \int_0^\infty \frac{\cos xy}{1+x^2} dx. \quad ...(i$$

Sol. It is easy to see that differentiation and integration under integra sign are valid. We, therefore, have

$$f'(y) = -\int_0^\infty \frac{x \sin xy}{1+x^2} dx = -\int_0^\infty \frac{\sin xy}{x} dx + \int_0^\infty \frac{\sin xy}{x(1+x^2)} dx$$

Again

$$\int_0^y f(y)\, dy = \int_0^y dy \int_0^\infty \frac{\cos xy}{1+x^2} dx = \int_0^\infty dx \int_0^\infty \frac{\cos xy}{1+x^2} dy$$

$$= \int_0^\infty \frac{\sin xy}{x(1+x^2)} dx.$$

$$\therefore \qquad f'(y) = -\frac{\pi}{2} + \int_0^y f(y)\,dx. \; (y > 0) \qquad \text{...(ii)}$$

Since f is continuous, we have on differentiation,

$$f''(y) = f(y)$$

$\therefore f(y) = Ae^{y} + Be^{-y}$, where A, B are constants.

Since $f(0)$ is easily seen to be $\pi/2$, we have, on making $y \to 0$,

$$\frac{1}{2}\pi = A + B.$$

Also, since from (*ii*), $f'(0) = -\pi/2$, we see that

$$-\frac{1}{2}\pi = A - B.$$

Thus $A = 0$, and $B = \dfrac{1}{2}\pi$ so that we have, when $y > 0$.

$$\int_0^\infty \frac{\cos xy}{1+x^2}\,dx = \frac{1}{2}\pi e^{-y} \text{ and } \int_0^\infty \frac{\sin xy}{x\left(1+x^2\right)}\,dx = \frac{1}{2}\pi\left(1-e^{-y}\right).$$

The necessary modification can easily be made when y is negative.

4. *Show that* $I = \displaystyle\int_0^\infty \frac{\tan^{-1} ax}{x\left(1+x^2\right)}\,dx = \frac{\pi}{2}\log(1+a)$, *if* $a \geq 0$, *and find the value of the integral if* $a < 0$.

Sol. We have

$$\frac{dI}{da} = \int_0^\infty \frac{1}{\left(1+x^2\right)\left(1+a^2x^2\right)}\,dx$$

$$= \frac{1}{\left(1-a^2\right)}\int_0^\infty \left[\frac{1}{\left(1+x^2\right)} - \frac{a^2}{1+a^2x^2}\right]dx$$

$$= \frac{1}{\left(1-a^2\right)}\left[\tan^{-1} x - a\tan^{-1}(ax)\right]_0^\infty$$

$$= \frac{1}{\left(1-a^2\right)}\left[\frac{\pi}{2} - a\cdot\frac{\pi}{2}\right] = \frac{\pi}{2(1+a)} \qquad \text{...(i)}$$

Here, since a is a positive, we have

$$\lim_{x\to\infty}\left(\tan^{-1} ax\right) = \frac{1}{2}\pi.$$

It can be seen easily that (*i*) is valid for $a = 0$ also.

Integrating (*i*) with respect to a, we obtain

$$I = \frac{1}{2}\pi \log(1+a) + c$$

Initially, when $a = 0$, $I = 0$. Substituting this, we get $c = 0$. Hence

$$I = \frac{1}{2}\pi \log(1+a).$$

When a is negative, we have

$$\frac{dI}{da} = \frac{1}{1-a^2}\left[\tan^{-1} x - a\tan^{-1}(ax)\right]_0^\infty$$

$$= \frac{1}{(1-a^2)}\left[\frac{\pi}{2} - a\left(-\frac{\pi}{2}\right)\right] = \frac{\pi}{2(1-a)}$$

because, when a is negative, we have

$$\lim_{x\to\infty} \tan^{-1}(ax) = -\frac{\pi}{2}.$$

Now, integrating

$$I = -\frac{1}{2}\pi \log(1-a) + c'.$$

When $a = 0$, $I = 0$, from this $c' = 0$. Hence

$$I = -\frac{1}{2}\pi \log(1-a).$$

5. *Show that* $I = \int_0^\infty e^{-x^2}\, dx = \frac{\sqrt{\pi}}{2}.$ (*Lucknow 9C*

Sol. Writing αx for x in the given integral, we have

$$I = \int_0^\infty e^{-\alpha^2 x^2}\, d(\alpha x)$$

$$\therefore \qquad Ie^{-\alpha^2} = \int_0^\infty e^{-\alpha^2(1+x^2)} . \alpha\, dx.$$

Integrating both sides with respect to α between the limit 0 to ∞, w have

$$I \cdot \int_0^\infty e^{-\alpha^2} d\alpha = \int_0^\infty \left\{\int_0^\infty e^{-\alpha^2(1+x^2)}\, \alpha\, d\alpha\right\} dx$$

i.e.,

$$I^2 = \int_0^\infty \left\{\frac{e^{-\alpha^2(1+x^2)}}{-2(1+x^2)}\right\}_{\alpha=0}^\infty dx$$

$$= \frac{1}{2}\int_0^\infty \frac{dx}{1+x^2} = \frac{1}{2}\left[\tan^{-1} x\right]_0^\infty = \frac{\pi}{4}$$

Hence $$I = \int_0^\infty e^{-x^2}\, dx = \frac{\sqrt{\pi}}{2}.$$

6. *Show that*

$$I=\int_0^\infty \frac{\tan^{-1}(\alpha x)\tan^{-1}(\beta x)}{x^2}\,dx$$

$$=\frac{\pi}{2}\log\left\{\frac{(\alpha+\beta)^{\alpha+\beta}}{\alpha^\alpha\beta^\beta}\right\}$$

Sol. We have

$$\frac{dI}{d\alpha}=\int_0^\infty \frac{\tan^{-1}(\beta x)}{x\left(1+\alpha^2x^2\right)}\,dx$$

and
$$\frac{d^2I}{d\beta\,d\alpha}=\int_0^\infty \frac{dx}{\left(1+\alpha^2x^2\right)\left(1+\beta^2x^2\right)}$$

$$=\frac{\alpha^2}{\alpha^2-\beta^2}\int_0^\infty \frac{dx}{1+\alpha^2x^2}+\frac{\beta^2}{\beta^2-\alpha^2}\int_0^\infty \frac{dx}{1+\beta^2x^2}$$

$$=\frac{1}{\left(\alpha^2-\beta^2\right)}\left[\alpha\tan^{-1}(\alpha x)-\beta\tan^{-1}(\beta x)\right]_{x=0}^{\infty}$$

$$=\frac{1}{\left(\alpha^2-\beta^2\right)}\left[\alpha.\frac{\pi}{2}-\beta.\frac{\pi}{2}\right]=\frac{\pi}{2(\alpha+\beta)}$$

assuming that α and β are both positive.

Integrating with respect to β, we get

$$\frac{dI}{d\alpha}=\frac{\pi}{2}\log(\alpha+\beta)+c_1 \qquad ...(iii)$$

Initially, when $\beta=0$, $\frac{dI}{d\alpha}=0$,

then
$$c_1=-\frac{\pi}{2}\log\alpha.$$

Hence,
$$\frac{dI}{d\alpha}=\frac{\pi}{2}\log(\alpha+\beta)-\frac{\pi}{2}\log\alpha.$$

Integrating once again, with respect to α, we get

$$I=\frac{\pi}{2}\left[\left\{\alpha\log(\alpha+\beta)-\int_0^\infty\frac{d\alpha}{\alpha+\beta}\right\}-\alpha\log\alpha+\alpha\right]$$

$$=\frac{1}{2}\pi\left[\alpha\log(\alpha+\beta)+\log(\alpha+\beta)-\alpha\log\alpha\right]+c_2$$

$$=\frac{\pi}{2}\left[\log(\alpha+\beta)^{\alpha+\beta}-\log\alpha^\alpha\right]+c^2$$

Again, when $\alpha = 0$, $I = 0$. Hence

$$0 = \frac{\pi}{2} \log \beta^{\beta} - 0 + c_2 \text{ or } c_2 = -\frac{\pi}{2} \log \beta^{\beta}$$

Since $\lim\limits_{\alpha \to 0} \alpha \log \alpha = \lim\limits_{\alpha \to 0} \dfrac{\log \alpha}{1/\alpha} = \lim\limits_{\alpha \to 0} \dfrac{1/\alpha}{1/\alpha^2} = 0.$

Hence, $I = \dfrac{\pi}{2} \log (\alpha + \beta)^{\alpha + \beta} - \dfrac{\pi}{2} \log \alpha^{\alpha} \beta^{\beta}$

$$= \frac{\pi}{2} \log \left\{ \frac{(\alpha + \beta)^{\alpha + \beta}}{\alpha^{\alpha} \beta^{\beta}} \right\}.$$

EXERCISES

1. Show, by differentiating under integral sign, that

$$\int_0^{\infty} e^{-x^2 - a^2/x^2} dx = \frac{1}{2} \sqrt{\pi e^{-2|a|}}$$

2. Establish the right to integrate under integral sign

$$\int_0^{\infty} e^{-xy} \cos mx \, dx,$$

and deduce that

$$\int_0^{\infty} \frac{e^{-ax} - e^{-bx}}{x} \cos mx \, dx = \frac{1}{2} \log \frac{b^2 + m^2}{a^2 + m^2} \qquad a, b > 0$$

3. By considering the identity

$$\frac{1}{y} = \int_0^{\infty} e^{-xy} dx, \qquad y > 0$$

show that

$$\int_0^{\infty} \frac{e^{-ax} - e^{-bx}}{x} dx = \log \frac{b}{a} \qquad (a > 0, b > 0)$$

4. Show that

$$\left[\int_0^{\infty} \frac{\operatorname{sech} ax - \operatorname{sech} bx}{x} dx = \log \frac{b}{a} \right. \qquad (a > 0; b > 0)$$

Change the order of integration in

$$\left. \int_a^b dx \int_0^{\infty} \operatorname{sech} xy \tanh xy \, dx. \right]$$

5. Under certain conditions to be stated, show that

$$\int_0^{\infty} \frac{f(ax) - f(bx)}{x} dx = f(0) \log \frac{b}{a}.$$

6. Show that

$$\int_0^\infty e^{-ax^2}\,dx = \frac{1}{2}\sqrt{\left(\frac{\pi}{a}\right)}, \qquad (a > 0).$$

Deduce that

$$\int_0^\infty e^{-ax^2} x^{2n}\,dx = \frac{1.3.5.\dots(2n-1)}{2^{n+1}}\,\frac{\sqrt{\pi}}{a^{n+1/2}}.$$

7. $\displaystyle\int_0^{\pi/2} \frac{dx}{1+\alpha\cos x} = \frac{\cos^{-1}\alpha}{\sqrt{1-\alpha^2}}, \qquad 0 \le \alpha < 1,$

show that for $0 \le a < 1$, $0 \le b < 1$.

$$\int_0^{\pi/2} \sec x \log\left(\frac{1+a\cos x}{1+b\cos x}\right) dx = \frac{1}{2}\left[\left(\cos^{-1} a\right)^2 - \left(\cos^{-1} b\right)^2\right],$$

and deduce that

$$\int_0^{\pi/2} \sec x \log\left(1+\frac{1}{2}\cos x\right) dx = \frac{5\pi^2}{72}.$$

OBJECTIVE QUESTIONS

Note. *For each of the following questions, four alternatives are given for the answer. Only one of them is correct. Choose the correct alternative.*

1. If f is continuous in $[a, b; c, d]$ and f is a function of x which is bounded and integrable in $[a, b]$, then $\int_a^b f(x, y) F(x)\,dx$ is :

(*a*) continuous function of y is $[c, d]$

(*b*) discontinuous function of y is $[c, d]$

(*c*) continuous function of y is $[a, b]$

(*d*) constant

2. Uniformly convergent improper integral of continuous function :

(*a*) has a finite number of discontinuities

(*b*) has infinite discontinuities

(*c*) is itself a continuous function

(*d*) none of these

3. In definite integrals the order of integration can be changed if :

(*a*) limits are independent of the variables

(*b*) limits depend on the variables

(*c*) integrand is constant

(*d*) none of these

4. Value of $\displaystyle\int_0^\infty \frac{e^{-ax}\sin mx}{x}\,dx$ is (*a*) $-\tan^{-1}\dfrac{m}{a}$ (*b*) $-\tan^{-1}\dfrac{a}{m}$

(*c*) $\tan^{-1}\dfrac{m}{a}$ (*d*) $\tan^{-1}\dfrac{a}{m}$

5. Value of integral $\int_0^\infty e^{-x^2}\,dx$ is :

(*a*) $\dfrac{\pi}{2}$ (*b*) $\dfrac{\sqrt{\pi}}{2}$

(*c*) $\sqrt{\dfrac{\pi}{2}}$ (*d*) $\dfrac{\pi}{4}$

6. If $\int_0^\infty e^{-ax}\,dx = \dfrac{1}{a}$, then $\int_0^\infty e^{-ax}\,x^n\,dx =$

(*a*) $\dfrac{n!}{a^n}$ (*b*) $\dfrac{n!}{a^{n+1}}$

(*c*) $\dfrac{(n+1)!}{a^n}$ (*d*) $\dfrac{(n-1)!}{a^n}$

ANSWERS

1. (*a*) **2.** (*c*) **3.** (*a*) **4.** (*c*) **5.** (*b*) **6.** (*b*

16

Integration in $\mathbf{R}^2$ Line Integrals, Double Integrals

16.1. INTRODUCTION

This chapter will deal with two types of integrals in $\mathbf{R}^2$, *viz.*,

(*i*) Line integrals, (*ii*) Double integrals.

The chapter 18 will deal with three types of integrals in $\mathbf{R}^3$ *viz.*, (*i*) Line integrals (*ii*) Surface integrals, (*iii*) Volume integrals.

16.2. THE CONCEPT OF A PLANE CURVE

Let ϕ, ψ be two real valued functions with domain $[\alpha, \beta]$. Then the set

$$\{(x, y);\ x = \phi(t),\ y = \psi(t);\ t \in [\alpha, \beta]\}$$

of point in $\mathbf{R}^2$ is called a curve in $\mathbf{R}^2$.

We say that

$$x = \phi(t),\ y = \psi(t)$$

is this curve. The curve is said to be closed, if the points corresponding to the end point α, β of the interval coincide, *i.e.*, if

$$\phi(\alpha) = \phi(\beta);\ \psi(\alpha) = \psi(\beta).$$

Let f be a real valued function with domain $[a, b]$, then the set of points

$$\{(x, y) : y = f(x);\ x \in [a, b]\}$$

is also a curve, as we may see on setting

$$x = t,\ y = f(t).$$

In respect of integration, it is appropriate to think of a curve as *oriented* so that we say that a curve $x = \phi(t)$, $y = \psi(t)$ is oriented one way or the other according as t varies from α to β or from β to α. Thus in a way we think of the set of points constituting a curve as *ordered*, the order being induced by the order of the set of numbers t. If the curve as oriented in one way is denoted C, then the curve oriented in the second way is denoted by $-C$.

16.2.1. Line Integral

Let

$$x = \phi(t),\ y = \psi(t)$$

be a curve, C, where ϕ and ψ are functions defined in an interval $[\alpha, \beta]$.

Let f be a function defined in a domain containing the curve C. *To define the line integral*

$$\int_C f(x, y)\, dx.$$

Let

$$\{\alpha = t_0 < t_1 < t_2 < \dots < t_{r-1} < t_r < \dots < t_{n-1} < t_n = \beta\}$$

be any division D of $[\alpha, \beta]$. Let

$$x_r = \phi(t_r),\ y_r = \psi(t_r).$$

and let ξ_r be any member of the interval $[t_{r-1}, t_r]$.

We form the sum

$$S = \Sigma f[\phi(\xi_r), \psi(\xi_r)]\,(x_r - x_{r-1}). \qquad \dots(1)$$

If, as the norm of D tends to zero, this sum, S, tends to a finite limit which is independent of the choice of the points ξ_r, we denote the limit by the symbol

$$\int_C f(x, y)\, dx$$

and call it the *line integral* of the function f along the curve C.

This means that there exists a number I such that to each $\varepsilon > 0$ there corresponds $\delta > 0$ such that for every division D of norm $< \delta$ and for every choice of the points ξ_1, ξ_2 etc. in the subintervals of the division D,

$$\left|\Sigma f\left[\phi(\xi_r), \psi(\xi_r)\right]\left(x_r - x_{r-1}\right) - I\right| < \varepsilon.$$

Then we have

$$I = \int_C f(x, y)\, dx.$$

16.2.2. A Sufficient Condition for the Existence of the Integral

Assuming that f, φ and ψ are continuous and φ possesses a continuous derivative φ', we now show that this limit exists.

There exists a point $\eta_r \in [t_{r-1}, t_r]$ such that

$$(x_r - x_{r-1}) = \varphi(t_r) - \varphi(t_{r-1}) = (t_r - t_{r-1})\, \phi'(\eta_r)\, \delta_r.$$

We have

$$S = \Sigma f\left[\varphi(\xi_r), \phi(\xi_r)\right]\left(x_r - x_{r-1}\right)$$

$$= \Sigma f\left[\varphi(\xi_r), \psi(\xi_r)\right]\varphi'(\xi_r)\,\delta_r + \Sigma f\left[\varphi(\xi_r), \psi(\xi_r)\right]\left[\varphi'(\eta_r) - \varphi'(\xi_r)\right]\delta_r$$

$$= S_1 + S_2.$$

Since ϕ' and $f[\varphi, \psi]$ are continuous, it follows that the sum $S_1 \rightarrow$ a finite limit, *viz.*,

$$\int_\alpha^\beta f\left[\varphi(t), \psi(t)\right]\varphi'(t)\, dt,$$

as the norm of D tends to zero.

Since $f[\varphi, \psi]$ is continuous, it is bounded. Let k be a positive number such that

$$\left|f\left[\varphi(t), \psi(t)\right]\right| \leq k \;\forall\; t \in [\alpha, \beta].$$

Since, φ, is continuous it is integrable and accordingly there exists a positive number δ such that for every division D of $[\alpha, \beta]$ of norm less than or equal to δ, the oscillatory sum of φ' is less than the arbitrarily small positive number ε/k.

$$\therefore \quad |S_2| \leq k\, \Sigma \left|\varphi'(\eta_r) - \varphi'(\xi_r)\right|\delta_r$$

$$\leq k\, \Sigma (M_r - m_r)\,\delta_r < k(\varepsilon/k) = \varepsilon,$$

where M_r, m_r are the bounds of φ' in $[t_{r-1}, t_r]$.

Thus $S_2 \rightarrow 0$ as the norm of D tends to zero.

Hence we see that the line integral, in question, exists and further we have the equality

$$\int_C f(x, y)\, dx = \int_\alpha^\beta f\left[\varphi(t), \psi(t)\right]\varphi'(t)\, dt,$$

where we have an ordinary on the right.

Note. We may similarly define and examine the existence of the line integral

$$\int_C g(x, y)\, dy,$$

and the line integral

$$\int_C \left[f(x, y)\, dx + g(x, y)\, dy\right].$$

Note. It is easy to see that the line integral

$$\int_C g(x, y)\, dy$$

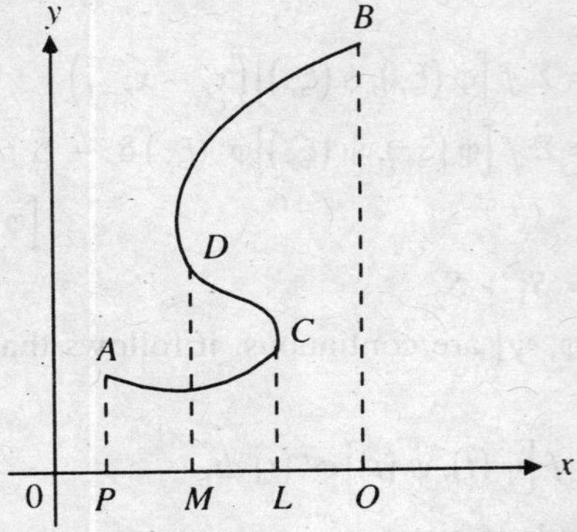

along the curve C, $y = \varphi(x)$, $a \le x \le b$, is equal to the ordinary integral

$$\int_a^b g[x, \varphi(x)]\, \varphi'(x)\, dx.$$

Note. It may be seen that

$$\int_{-C} f\, dx + g\, dy = -\left[\int_C f\, dx + g\, dy\right].$$

Note. The ordinary integral is a special case of a line integral, where we take an interval of the x or y axis as the path of integration.

16.2.3. *It can be easily proved that*

(i) line integrals are additive for arcs, i.e.,

$$\int_{AB} f(x, y)\, dx = \int_{AC} f(x, y)\, dx + \int_{CD} f(x, y)\, dx + \int_{DB} f(x, y)\, dx,$$

and *(ii)*
$$\int_{AC} f(x, y)\, dx = \int_{CA} f(x, y)\, dx.$$

EXAMPLES

1. *Evaluate* $\int_C (3xy\, dx - y^2\, dy)$, *where C is the curve in the xy-plane* y = $2x^2$ *from (0, 0) to (1, 2).*

Sol. In the given line integral, put $y = 2x^2 \Rightarrow dy = 4x\, dx$ and x varies from 0 to 1.

$$\therefore \quad \int_C (3xy\, dx - y^2\, dy) = \int_0^1 3x(2x^2)\, dx - \int_0^1 (2x^2)^2 .4x\, dx$$

$$= \left[\frac{6}{4} x^4 - \frac{16}{6} x^6 \right]_0^1$$

$$= -7/6.$$

2. *Evaluate* $\int_C (xy\,dx + yz\,dy + zx\,dz)$, *where C is* $x = t$, $y = t^2$, $z = t^3$, *t varying from – 1 to + 1.*

$dx = dt$, $dy = 2t\,dt$, $dz = 3t^2 dt$

$$\therefore \quad \int_C (xy\,dx + yz\,dy + zx\,dz) = \int_{-1}^{1} \left(t^3 + 5t^6 \right) dt = \frac{10}{7}.$$

3. *Evaluate* $\int_C \left(2x^2 + y^2\right) dx + (3y - 4x)\,dy$ *around the triangle ABC whose vertices are* $A \equiv (0, 0), B \equiv (2, 0), C \equiv (2, 1)$.

Sol. Along AB, $y = 0$, $\therefore$ $dy = 0$ and x varies from 0 to 2.

$$\therefore \quad \int_{AB} \left(2x^2 + y^2\right) dx + (3y - 4x)\,dy = \int_0^2 2x^2\,dx = \frac{16}{3}$$

Equation of BC is $x = 2$, $\therefore$ $dx = 0$

$$\therefore \text{ Line integral along } BC = \int_0^1 (3y - 8)\,dy$$

$$= -\frac{13}{2}.$$

Equation of CA = $x = 2y \Rightarrow dx = 2dy$

$\therefore$ Line integral along CA

$$= \int_1^0 \left(18y^2 - 5y\right) dy$$

$$= -\frac{7}{2}$$

$$\therefore \quad \int_c \left(2x^2 + y^2\right) dx + (3y - 4x)\,dx$$

$$= \frac{16}{3} - \frac{13}{2} - \frac{7}{2} = -\frac{14}{3}.$$

EXERCISES

1. Evaluate

$$\int_C \left(x^2 + y^2\right) dx \text{ and } \int_C \left(x^2 + y^2\right) dy$$

where C is the arc of the parabola $y^2 = 4ax$ between (0, 0) and $(a, 2a)$.

2. Show that

$$\int_C \left[(x - y)^3\,dx + (x - y)^3\,dy \right] = 3\pi a^4,$$

taken along the circle $x^2 + y^2 = a^2$ in the counter clockwise sense.

3. Evaluate

$$\int_C \frac{dx}{x+y},$$

where C is the curve $x = at^2$, $y = 2at$. $0 \le t \le 2$.

4. Show that the value of the integral

$$\int \left(xy^2 dy - x^2 y dx\right),$$

taken in the counter-clockwise sense along the Cardioide $r = a\ (1 + \cos \theta)$ is $35a^4\pi / 16$.

5. Find the value of

$$\int \left(x + y^2\right) dx + \left(x^2 - y\right) dy,$$

taken in the clockwise sense along the closed curve C formed by $y^2 = x$ and $y = x$ between (0, 0) and (1, 1).

6. Find the value of

$$\int \left(x^2 y\, dx + y^2 x\, dy\right)$$

taken in the clockwise sense along the hexagon whose vertices are

(*i*) $(\pm\, 3a,\, 0)$; $\left(\pm\, 2a, \pm \sqrt{3a}\right)$, (*ii*) $(2, \pm\, 3a)$, $\left(\pm \sqrt{3a}, \pm\, 2a\right)$

7. Show that

$$\int_C \frac{y\, dx - x\, dy}{x^2 + y^2} = -\, 2\pi$$

round the circle C :

$$x^2 + y^2 = 1.$$

or any simple closed curve containing the origin in its interior.

16.3. THE AREA OF A SUBSET OF R²

The area of a rectangle $[a, b; c, d]$. As suggested by intuitive considerations, *we define* the area of a rectangle $[a, b; c, d]$ to be the number $(b - a)\ (d - c)$.

The area of any bounded set E. Since E is bounded, there exists a rectangle R which completely encloses E. We divide R into sub-rectangles.

Let

(*i*) *s (D) denote the sum of the areas of the sub-rectangles which consist* **entirely** *of points of E, and*

(*ii*) *S (D) denote the sum of the areas of the sub-rectangles which have at least one point in common with E.* Clearly

$$s\ (D) \le S\ (D).$$

To each mode of division D of R there correspond a pair of sums $S(D)$ and $s(D)$. Clearly the sets of these sums are both bounded.

Inner area and outer area. The area.

The supremum of the set of sums $s(D)$ will be called the *Inner area* of E and the infimum of the set of sums $S(D)$ will be called the the Outer area of E.

The region E will be said to possess an area, if its inner and outer areas are equal and the common value will be called the area of E.

In § 16.8.3, p. 471 will be obtained a sufficient condition for a set to possess an area.

16.3.1. Integrability of a Bounded Function over a Rectangle

Let f be a bounded function with domain $[a, b; c, d]$.

Let

$$D_1 = \{a = x_0 < x_1 < x_2 < \ldots < x_{r-1} < x_r \ldots < x_n = b\}$$

and
$$D_2 = \{c = y_0 < y_1 < y_2 < \ldots < y_{s-1} < y_s \ldots < y_n = d\},$$

be any divisions of the intervals $[a, b]$ and $[c, d]$ respectively. These divisions of the two intervals give rise to a division of the rectangle R into mn sub-rectangles $[x_{r-1}, x_r; y_{s-1}, y_s]$ where r, s take up all positive integral values from 1 to n and 1 to m respectively. We will denote the rectangle $[x_{r-1}, x_r; y_{s-1}, y_s]$ as well as its area

$$(x_r - x_{r-1})(y_s - y_{s-1})$$

by the same symbol, w_{rs}. Let M_{rs}, m_{rs} denote the bounds of the function f in w_{rs}. Consider the two sums

$$S(D) = \sum_{s=1}^{s=m} \sum_{r=1}^{r=n} M_{rs}\, w_{rs},\; s(D) = \sum_{s=1}^{s=m} \sum_{r=1}^{r=n} m_{rs}\, w_{rs}.$$

It is easy to see that for every division D of R into sub-rectangles, we have

$$m(b-a)(d-c) \le s(D) \le S(D) \le M(b-a)(d-c),$$

where M, m are the bounds of f in $[a, b; c, d]$.

Thus the two sets of the sums $S(D)$ and $s(D)$ are bounded.

Upper and lower integrals. Def. *The infimum of the set of the upper sums $S(D)$ is called the upper integral and the supremum of that of lower sums $s(D)$ is called the lower integral of f over R and are denoted by the symbols*

$$U = \overline{\int\int_R} f(x, y)\, dx\, dy,\; L = \underline{\int\int_R} f(x, y)\, dx\, dy$$

respectively.

In case these two upper and lower integrals are equal, then f is said to be integrable and the common value, which is denoted by the symbol

$$I = \int \int_R f(x, y)\, dx\, dy,$$

is said to be the double integral of f over R.

Note. Norm of a division of a rectangle. By the norm of a division of a rectangle into sub-rectangles $\left[x_{r-1}, x_r\,;\, y_{s-1}, x_s\right]$ is meant the greatest member of the set of numbers $\left(x_r - x_{r-1}\right)$ and $\left(y_s - y_{s-1}\right)$, $r = 1, 2..., n$; $s = 1, 2..., m$.

16.3.2. Darboux's Theorem

To every pre-assigned positive number ε, there corresponds a positive number δ, such that for every division D whose norm is ≤ δ.

$$S(D) < U + \varepsilon;\; s(D) > L - \varepsilon.$$

The proof is exactly similar to the corresponding proof for functions of a single variable.

Cor. *The upper integral ≥ the lower integral.*

16.4. CONDITIONS FOR INTEGRABILITY

16.4.1. First Form

The necessary and sufficient condition for the integrability of a bounded function f over a rectangle R is that to every positive number ε, there corresponds a positive number δ, such that for every division D of R whose norm ≤ δ, the oscillatory sum S (D) – s (D) is less than ε.

The condition is necessary. The bounded function f being integrable,

$$U = L = I.$$

If ε be any positive number, then by the Darboux's theorem there exists a positive number δ such that for every division D of norm ≤ δ,

$$S(D) < U + \frac{1}{2}\varepsilon = I + \frac{1}{2}\varepsilon,\; s(D) > L - \frac{1}{2}\varepsilon = I - \frac{1}{2}\varepsilon,$$

$$\Rightarrow \qquad I - \frac{1}{2}\varepsilon < s(D) \le S(D) < I + \frac{1}{2}\varepsilon,$$

$$\Rightarrow \qquad S(D) - s(D) < \varepsilon.$$

The condition is sufficient. There exists a division such that if S, s be the corresponding upper and lower sums, then

$$S - s = (S - U) + (U - L) + (L - s) < \varepsilon.$$

Since each of the three numbers

$$(S - U), (U - L), (L - s)$$

is non negative, it follows that

$$0 \le (U - L) < \varepsilon.$$

As ε is an arbitrary positive number, we deduce that

$$U - L = 0 \quad \Rightarrow \quad U = L$$

Therefore the function is integrable.

16.4.2. Second Form

The necessary and sufficient condition for the integrability of a bounded function f is that to every preassigned positive number ε, there corresponds a division for which the oscillatory sum is less than ε.

The proof which is quite similar to that of the first form is left to the reader.

16.5. PARTICULAR CLASSES OF BOUNDED INTEGRABLE FUNCTIONS

16.5.1. Every Continuous Function is Integrable

Let f be continuous in a rectangle $R = [a, b; c, d]$. Let ε be any positive number. There exists, because of uniform continuity, a division such that the oscillation of f in every sub-rectangle of the division is

$$< \varepsilon / (b - a)(d - c).$$

For such a division, the oscillatory sum

$$S - s = \Sigma\Sigma\left(M_{rs} - m_{rs}\right) w_{rs} \le \left[\varepsilon / (b - a)(d - c)\right].\Sigma\Sigma\, w_{rs} = \varepsilon.$$

Hence f is integrable in the rectangle R.

16.5.2. *If a function f is bounded in R = [a, b : c, d] and is such that its points of discontinuity can be enclosed in a finite number of rectangles the sum of whose areas is less than a given positive number, then f is integrable in R.*

Let ε be any positive number. We enclose the points of discontinuity in rectangles the sum of whose areas is $< \varepsilon/2\,(M - m)$. The part of the oscillatory sum $(S - s)$ arising from these rectangles or from the sub-rectangles into which they may be further divided is $< \varepsilon/2$.

On producing the sides of the rectangles which enclose the points of discontinuity, we obtain a division of R into subrectangles. These sub-rectangles are of two types :

(*i*) those which include points of discontinuity, and

(*ii*) those which do not include any point of discontinuity.

The sub-rectangles of the latter types can be further sub-divided such that the part of $(S - s)$ arising from them is $< \varepsilon/2$.

Thus we have a division of R such that the corresponding oscillatory sum is less than the given positive number ε.

Hence the result. § (16.4.2)

Cor. *If a function f is bounded in $R = [a, b; c, d]$ and its points of discontinuity lie on a finite number of curves of the form $y = \varphi(x)$, $x = \psi(y)$ etc., where φ, ψ etc., are continuous, then f is integrable in R.*

Let, p, be the number of the curves in question. Let ε be any positive number.

Since φ is continuous in $[a, b]$, there exists a division

$$D = \{a = x_0 < x_1 < x_2 < x_3 \ldots < x_{r-1} < x_r \ldots < x_n = b\} \text{ of } [a, b]$$

such that the oscillation of φ in every sub-interval is

$$< \varepsilon/p\,(b - a).$$

The points of $y = j(x)$ which correspond to the values of x of the sub-interval $[x_{r-1}, x_r]$ of the division D clearly belong to the rectangle $[x_{r-1}, x_r; m_r, M_r]$ of area $(x_r - x_{r-1})(M_r - m_r)$; M_r, m_r being the bounds of φ in $[x_{r-1}, x_r]$. Thus all the points of $y = \varphi(x)$ have been enclosed in a system of rectangles of total area.

$$\Sigma (x_r - x_{r-1})(M_r - m_r)$$

which is clearly less than ε/p.

Applying the same process to the other curves, we see that they can all be enclosed in a finite number of rectangles of total area less than any given positive number ε.

Hence the result.

16.6. THE CALCULATION OF A DOUBLE INTEGRAL

Equivalence of a double with repeated integrals. Theorem. *If the double integral*

$$\iint_R f(x, y)\,dx\,dy$$

exists where R is the rectangle [a, b; c, d] and if also

$$\int_a^b f(x, y)\,dx$$

exist $\forall\, y \in [c, d]$, *then the repeated integral*

$$\int_c^d \left\{\int_a^b f(x, y)\,dx\right\} dy$$

exists and is equal to the double integral.

[**Observation**. The proof depends upon a simple consideration, *viz.*, that if D be any division of $[a, b]$ and K_r, k_r be upper and lower bounds of a function φ in any sub-interval δ_n, of the division, then

$$\int_a^{\bar{b}} \varphi(x)\,dx \le S(D) \le \Sigma\, k_r\, \delta_r \quad\text{and}\quad \int_{\bar{a}}^{b} \varphi(x)\,dx \ge s(D) \ge \Sigma\, k_r\, S_r \Bigg]$$

Let U and L denote the upper and lower integrals of f over R. Let ε be any positive number. There exists a division of R into sub-rectangles

$$\left[x_{r-1}, x_r; y_{s-1}, y_s\right]$$

such that

$$\Sigma\Sigma\, M_{rs}\left(x_r - x_{r-1}\right)\left(y_s - y_{s-1}\right) < U + \varepsilon \qquad \text{...(1)}$$

Since for every fixed value of y in $\left[y_{s-1}, y_s\right] M_{rs}$, is bound of f in $\left[x_{r-1}, x_r\right]$, therefore

$$\int_a^{\bar{b}} f(x, y)\,dx \le \sum_{r=1}^{n} M_{rs}\left(x_r - x_{r-1}\right) \text{ when } y_{s-1} \le y \le y_s \qquad \text{...(2)}$$

Since, from (2), $\displaystyle\sum_{r=1}^{n} M_{rs}\left(x_r - x_{r-1}\right)$

is an upper bound of the function φ, where

$$\varphi(y) = \int_a^{\bar{b}} f(x, y)\,dx,\ y \in \left[y_{s-1}, y_s\right],$$

we have by an application of the same reasoning,

$$\int_c^{\bar{d}} \left\{\int_a^{\bar{b}} f(x, y)\,dx\right\} dy \le \sum_{s=1}^{m}\sum_{r=1}^{n} M_{rs}\left(x_r - x_{r-1}\right)\left(y_s - y_{s-1}\right) < U + \varepsilon. \qquad \text{...(3)}$$

As ε is an arbitrary positive number, and, by hypothesis,

$$\int_a^{\bar{b}} f(x, y)\,dx = \int_{\bar{a}}^{b} f(x, y)\,dx = \int_a^{b} f(x, y)\,dx,$$

therefore we have

$$\int_c^{\bar{d}} \left\{\int_a^{b} f(x, y)\,dx\right\} dy = \int_c^{\bar{d}} \left\{\int_a^{\bar{b}} f(x, y)\,dx\right\} dy$$

$$\le U = \overline{\iint_R} f(x, y)\,dx\,dy. \qquad \text{...(4)}$$

We can similarly prove that

$$\int_{\underline{c}}^{d} \left\{\int_a^{b} f(x, y)\,dx\right\} dy = \int_{\underline{c}}^{d} \left\{\int_{\underline{a}}^{b} f(x, y)\,dx\right\} dy$$

$$\ge L = \underline{\iint_R} f(x, y)\,dx\,dy. \qquad \text{...(5)}$$

Therefore

$$\underline{\iint_R} f(x, y)\, dx\, dy \le \int_{\underline{c}}^{d} \left\{ \int_a^b f(x, y)\, dx \right\} dy$$

$$\le \int_c^{\bar{d}} \left\{ \int_a^b f(x \quad)\, dx \right\} dy \le \overline{\iint_R} f(x, y)\, dx\, dy \qquad ...(6)$$

But by hypothesis

$$\overline{\iint_R} f(x, y)\, dx\, dy = \underline{\iint_R} f(x, y)\, dx\, dy. \qquad ...(7)$$

Now, from (6), (7), we have

$$\iint_R f(x, y)\, dx\, dy = \int_c^{\bar{d}} \left\{ \int_a^b f(x, y)\, dx \right\} dx$$

$$= \int_{\underline{c}}^{d} \left\{ \int_a^b f(x, y)\, dx \right\} dy = \int_c^d \left\{ \int_a^b f(x, y)\, dx \right\} dy.$$

Cor. *If a double integral exists, then the two repeated integrals cannot exist without being equal.*

Cor. *A function f is defined in [0, 1; 0, 1] as follows :*

$f(x, y) = \frac{1}{2}$, when y is rational, $f(x, y) = x$, when y is irrational; show that

$$\int_0^1 \left\{ \int_0^1 f(x, y)\, dx \right\} dy$$

exists and is equal to $\frac{1}{2}$, but the double integral and the second repeated integral do not exist.

EXAMPLES

1. *Prove that*

$$\int_0^1 \left\{ \int_0^1 \frac{x - y}{(x + y)^3}\, dy \right\} dx = \frac{1}{2} \ne -\frac{1}{2} = \int_0^1 \left\{ \int_0^1 \frac{x - y}{(x + y)^3}\, dx \right\} dy,$$

and show that this result is compatible with the theorem proved.

Sol. We suppose (x, y) to be any point of the rectangle [0, 1; 0, 1]. For any fixed value of $x \ne 0$, the function $f(x, y) = (x - y)/(x - y)^3$ is a bounded function of y and if $x = 0$, then $y = 0$ is a point of infinite discontinuity.

If $x \ne 0$,

$$\varphi(x) = \int_0^1 \frac{x - y}{(x + y)^3}\, dy = \int_0^1 \left\{ \frac{2x}{(x + y)^3} - \frac{1}{(x + y)^2} \right\} dy = \frac{1}{(1 + x)^2}$$

and φ (0) does not exist. Again,

$$\int_0^1 \varphi(x)\,dx = \lim_{\varepsilon \to 0} \int_\varepsilon^1 \frac{1}{(1+x)^2}\,dx - \lim_{\varepsilon \to 0}\left(-\frac{1}{2} + \frac{1}{1+\varepsilon}\right) = \frac{1}{2}.$$

Thus

$$\int_0^1 \left\{\int_0^1 \frac{x+y}{(x+y)}\,dy\right\} dx = \frac{1}{2}.$$

If $y \neq 0, \psi(y) = \int_0^1 \frac{x-y}{(x+y)^3}\,dx$

$$\int_0^1 \left\{\frac{1}{(x+y)^2} - \frac{2y}{(x+y)^3}\right\} dx = -\frac{1}{(1+y)^2}$$

and ψ (0) does not exist.

As before

$$\int_0^1 \psi(y)\,dy = -\frac{1}{2}.$$

Now, from (6), (7), we have

$$\iint_R f(x, y)\,dx\,dy = \int_c^{\bar{d}} \left\{\int_a^b f(x, y)\,dx\right\} dx$$

2. *Evaluate* $\iint xy(x+y)\,dx\,dy$ *over the area between* $y = x^2$ *and* $y = x$.

Sol. The area is bounded by the curves

$$y = f_1(x) = x^2,\ y = f_2(x) = x.$$

When $f_1(x) = f_2(x)$, $x^2 = x$, *i.e.*, $x = 0$, $x = 1$, *i.e.*, the area of integration is bounded by

$$y = x^2,\ y = x,\ x = 0,\ x = 1$$

$$\therefore \qquad \iint xy(x+y)\,dx\,dy = \int_0^1 \left[\int_{x^2}^x [xy(x+y)\,dy]\right] dx$$

$$= \int_0^1 \left[\frac{x^2y^2}{2} + \frac{xy^3}{3}\right]_{x^2}^x dx$$

$$= \int_0^1 \left(\frac{5x^4}{6} - \frac{x^6}{2} - \frac{x^7}{3}\right) dx$$

$$= \frac{3}{56}.$$

EXERCISES

1. Evaluate the following integrals :

(*i*) $\iint xy\left(x^2+y^2\right) dx\, dy$ over $[(0, a; 0, b)]$

(*ii*) $\iint ye^{xy}\, dx\, dy$ over $[0, a; 0, b]$.

(*iii*) $\iint \frac{x-y}{x+y}\, dx\, dy$ over $[0, 1; 0, 1]$.

[The integral is bounded and (0, 0) is its only point of discontinuity in the square [0, 1; 0, 1].

(*iv*) $\iint \frac{dx\, dy}{\sqrt{\left[c^2+(x-y)^2\right]}}$ over $[0, a; 0, a]$.

2. Show that

$$\int_0^1 \left\{\int_0^1 \frac{x^2-y^2}{x^2+y^2}\, dy\right\} dx = \int_0^1 \left\{\int_0^1 \frac{x^2-y^2}{x^2+y^2}\, dx\right\} dy.$$

3. Show that if $0 < h < 1$.

$$\int_h^1 \left\{\int_h^1 f(x, y)\, dx\right\} dy = \int_h^1 \left\{\int_h^1 f(x, y)\, dy\right\} dx = 0,$$

but

$$\int_0^1 \left\{\int_0^1 f(x, y)\, dx\right\} dy \neq \int_0^1 \left\{\int_0^1 f(x, y)\, dx\right\} dx,$$

where

$$f(x, y) = \left(y^2 - x^2\right) / \left(y^2 + x^2\right)^2.$$

4. Show that

$$\varphi(h) = \int_h^1 \left\{\int_h^1 \frac{x-y}{(x+y)^3}\, dy\right\} dx$$

is not continuous for $h = 0$.

5. Is it true that

$$\int_1^\infty dx \int_1^\infty \frac{x-y}{(x+y)^3}\, dy = \int_1^\infty dy \int_1^\infty \frac{x-y}{(x+y)^3}\, dx\ ?$$

If no, why so ?

6. Show that

$$\iint_R \varphi(x)\,\phi(y)\,dx\,dy = \left\{\int_a^b \varphi(x)\,dx\right\}\left\{\int_c^d \varphi(y)\,dy\right\}$$

where R is the rectangle $[a, b;\ c, d]$

16.7. SIMPLE CLOSED CURVES

Jordan Curves. We have seen that a curve in $\mathbf{R}^2$ is the range of **a** vector valued function whose domain is an interval $[a, b] \subset \mathbf{R}$ and whose range is contained in $\mathbf{R}^2$. Thus if $\mathbf{f} = (\phi, \varphi)$ where ϕ, ψ are real valued functions with domain $[\alpha, \beta]$, then the range of the function **f** *i.e.*, the set of points

$$\{[\phi(t), \psi(t)] : t \in [\alpha, \beta]\}$$

is **a** curve in $\mathbf{R}^2$. Again the curve is said to be closed if

$$\mathbf{f}(\alpha) = \mathbf{f}(\beta)$$

which is equivalent to saying that

$$[\phi(\alpha), \psi(\alpha)] = [\phi(\beta), \psi(\beta)] \Leftrightarrow \phi(\alpha) = \phi(\beta), \psi(\alpha) = \psi(\beta).$$

The curve is said to be *simple,* if the correspondence brought about by **f** between the semi-closed interval $]\alpha, \beta]$ and the curve is one-one. This means that the curve is simple if

$$t_1 \in [\alpha, \beta], t_2 \in\,]\alpha, \beta] \Rightarrow [\phi(t_1), \psi(t_1)] \neq [\phi(t_2), \psi(t_2)].$$

(Intuitively a curve is simple if it does not intersect itself.)

Finally the curve is said to be continuous if the functions ϕ and ψ are continuous.

Jordan Curves. Def. *A simple closed continuous curve is called a Jordan curve.*

Illustrations.

(*i*) The curve (circle) given by

$$\phi(t) = \cos t, \psi(t) = \sin t, t \in [0, 2\pi]$$

is a Jordan curve.

(*ii*) The curve (ellipse) given by

$$\phi(t) = a\cos t, \psi(t) = b\sin t;\ t \in [0, 2\pi]$$

is a Jordan curve.

Ex. Give some examples of curves which are Jordan curves and some examples of curves which are not Jordan curves.

Regions bounded by Jordan curves. Jordan's theorem. One may intuitively see that a closed curve which does not intersect itself divides the points of the plane not belonging to the curve into two disjoint components having the given curve as their common boundary. Of course the intuitive evidence of this result is based on our experience with some well known curves such as circle, ellipse, polygon.

The rigorous analytical proof is, however, not simple. The necessity for the proof was first percieved by C. Jordan in 1892. He also gave a proof but this was rather incomplete. Incomplete proofs were also given by other mathematicians. The first correct proof was, however, given by Veblen in 1905.

We shall only content ourselves by giving a precise statement the theorem without giving the proof. Before giving the statement, we state what is meant by the Boundary set of a given set.

Let S be a given set. Then a point is called a *Boundary point* of S if every neighbourhood of the point contains points belonging to S and not belonging to S.

Thus a given point is a boundary point of S, if it is a limit point of S as also of the complement of S.

For example a, b are boundary points of each of the intervals.

$$[a, b]\ ]a, b],\ ]a, b[,\ [a, b[.$$

Ex. What are the boundary points of the rectangle

$$[a, b; c, d].$$

The set of all boundary points of a set is called the Boundary of the set. Often the boundary of a set S is denoted by the symbol δS.

Jordan Theorem. *Let C be a Jordan curve in $\mathbf{R}^2$ and let S be the complement of C in $\mathbf{R}^2$, i.e., consist of points of $\mathbf{R}^2$ not belonging to C.*

Then there exist two open non-void and disjoint open sets A and B such that

$$S = A \cup B$$

and that C is the boundary of A as well as B.

Also exactly one of the two sets A and B is bounded.

Of the two sets A, B, the one which is bounded is called the *Interior* of the curve C and the other its *Exterior.*

If A be the interior of a simple closed curve C, we say that A is bounded by the curve C.

Note. In the following, whenever, we talk of a domain bounded by a curve or curves, it will be understood that the given curve or the union of given curves is a Jordan curve and that the domain in question is the union of the curve and of its interior.

16.8. INTEGRABILITY AND INTEGRAL OF A BOUNDED FUNCTION f OVER ANY BOUNDED DOMAIN E

Since the given region E is bounded there must exist a rectangle R which completely encloses E. We define a function F over R as follows :

$$F(x, y) = \begin{cases} f(x, y) & \text{at all points of } E. \\ 0, & \text{elsewhere.} \end{cases}$$

Def. *A function f is said to be integrable over E, if F is integrable over the rectangle R. Also then*

$$\iint_E f(x, y)\, dx\, dy = \iint_R F(x, y)\, dx\, dy.$$

16.8.1. An Important Case of Integrability

If f is continuous in a domain E which is bounded by a finite number of continuous curves of the form $y = \varphi(x), x = \psi(y)$, etc., then

$$\iint_E f(x, y)\, dx\, dy$$

exists.

The result follows from the fact that the only possible points of discontinuity of F, defined as above, are the points of the curves $y = \varphi(x)$, etc.

16.8.2. Calculation of Double Integrals

If f is continuous in a domain E which is bounded by the curves

$$y = \varphi(x),\ y = \psi(x);\ x = a;\ x = b;$$

where ψ, ϕ are continuous and $\psi(x) \le \phi(x)\ \forall\ x \in [a, b]$ then

$$\iint_E f(x, y)\, dx\, dy = \int_a^b \left\{ \int_{\psi(x)}^{\phi(x)} f(x, y)\, dy \right\} dx.$$

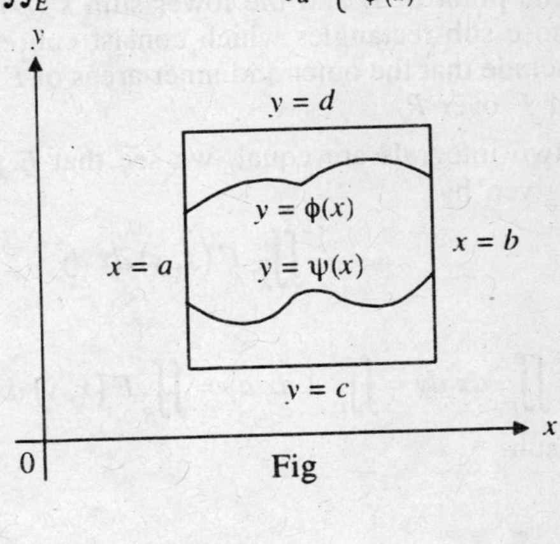

Fig

Let $[a, b; c, d]$ enclose E and let F be defined over R as in § 16.8. We have

$$\iint_E f(x, y)\,dx\,dy = \iint_R F(x, y)\,dx\,dy$$

$$= \int_a^b \left\{ \int_c^d F(x, y)\,dy \right\} dx$$

$$= \int_a^b \left\{ \int_c^{\psi(x)} F\,dy + \int_{\psi(x)}^{\phi(x)} F\,dy + \int_{\phi(x)}^{d} F\,dy \right\} dx$$

$$= \int_a^b \left\{ \int_{\psi(x)}^{\phi(x)} F\,dy \right\} dx;$$

each of the remaining two integrals being equal to zero.

Note. If a function f is continuous in a domain E which is bounded by the curves $x = \phi(y), x = \psi(y); y = c, y = d$, where ϕ, ψ are continuous and $\psi(y) \le \phi(y), \forall\, y \in [c, d]$, then

$$\iint_E f(x, y)\,dx\,dy = \int_c^d \left\{ \int_{\psi(y)}^{\phi(y)} f(x, y)\,dx \right\} dy.$$

16.8.3. Area of a Set

A sufficient condition for a set E to possess an area is that it is bounded by a finite number of continuous curves of the form $y = \phi(x), x = \psi(y)$, *and the area is given by the double integral*

$$\iint_E dx\,dy.$$

We enclose E in a rectangle R and define a function F in R as follows :

$F(x, y) = 1$ *at points of E and* $F(x, y) = 0$ *elsewhere.*

Corresponding to any division D of R into sub-rectangles, the upper sum $S(D)$ for F is the sum of the areas of those sub-rectangles which contain at least one point of E and the lower sum $s(D)$ for F is the sum of the areas of those sub-rectangles which consist entirely of points of E. From this we conclude that the outer and inner areas of E are the upper and lower integrals of F over R.

Since these two integrals are equal, we see that E possesses an area and the same is given by

$$\iint_R F(x, y)\,dx\,dy.$$

But, by def.

$$\iint_E dx\,dy = \iint_E 1.dx\,dy = \iint_R F(x, y)\,dx\,dy.$$

Hence the result.

EXAMPLES

1. *The double integral*

$$\iint f(x, y)\, dx\, dy$$

where the domain of integration is $\{(x, y); x^2 + y^2 \le 1\}$ is equivalent to the repeated integral

$$\int_{-a}^{+a} dx \int_{-\sqrt{(a^2 - x^2)}}^{+\sqrt{(a^2 - x^2)}} f(x, y)\, dy.$$

2. *Change the order of integration in the double integral*

$$\int_0^{2a} \int_{\sqrt{(2ax - x^2)}}^{\sqrt{(2ax)}} f(x, y)\, dx\, dy.$$ (*Lucknow 97*)

The given domain of integration is described by a line which starts from $x = 0$ and moving parallel to itself, goes over to $x = 2a$; the extremities of the moving line on the parts of the parabola $y^2 = 2ax$ and the circle $x^2 + y^2 = 2ax$ in the first quadrant.

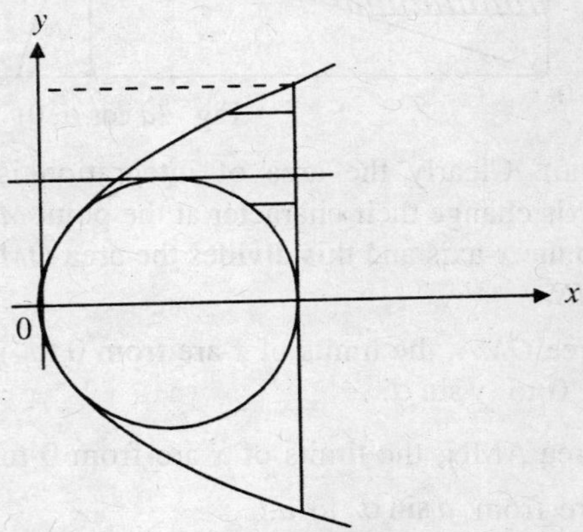

Fig

We have now to regard the same region as described by a line moving parallel to x-axis instead of y-axis.

In this way the domain of integration is sub-divided into three sub-regions to each of which corresponds a double integral. Thus we have

$$\int_0^{2a} \int_{\sqrt{(2ax - x^2)}}^{\sqrt{(2ax)}} f\, dx\, dy = \int_0^{a} \int_{y^2/2a}^{a - \sqrt{(a^2 - y^2)}} f\, dy\, dx$$

$$+ \int_0^{a} \int_{a + \sqrt{(a^2 - y^2)}}^{2a} f\, dy\, dx + \int_a^{2a} \int_{y^2/2a}^{2a} f\, dy\, dx.$$

3. *Change the order of integration in*

$$\int_0^{a\cos a}\int_{\tan\alpha}^{\sqrt{(a^2-x^2)}} f(x, y)\,dx\,dy$$

and verify result when $f(x, y) = 1$. (*Avadh 92, 97*)

The limits of integration are given by $y = x\tan\alpha$ (a straight line), $y=\sqrt{(a^2-x^2)}$, *i.e.*, $x^2+y^2=a^2$ (a circle); $x = 0$ (y-axis) and $x = a\cos\alpha$

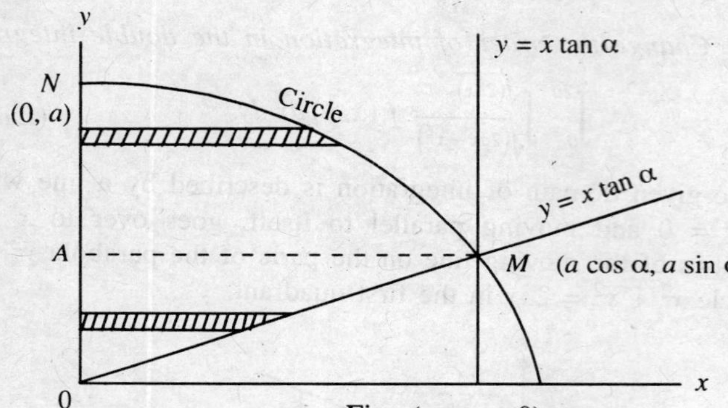

Fig (*a* cos α, 0)

(a straight line). Clearly the area of integration is *OMNO*. The strip parallel to x-axis change their character at the point M. Draw a straight line *AM* parallel to the x-axis and this divides the area *OMNO* into two portions *OMA* and *AMN*.

For the area *OMA*, the limits of x are from 0 to $y\cot\alpha$ and the limit of y are from 0 to $y\sin\alpha$.

For the area AMN, the limits of x are from 0 to $\sqrt{(a^2-y^2)}$ and the limits for y are from $a\sin\alpha$ to a.

Hence changing the order of integration, we have

$$\int_0^{a\cos\alpha}\int_{x\tan\alpha}^{\sqrt{(a^2-x^2)}} f(x, y)\,dx\,dy$$

$$=\int_0^{a\sin\alpha}\int_0^{y\cot\alpha} f(x, y)\,dy\,dx$$

$$+\int_{a\sin\alpha}^{a}\int_0^{\sqrt{(a^2-y^2)}} f(x, y)\,dy\,dx$$

To verify the result when $f(x, y) = 1$,

$$\text{L.H.S.} = \int_0^{a\cos\alpha}\int_{x\tan\alpha}^{\sqrt{(a^2-x^2)}} dx\,dy$$

$$= \int_0^{a\cos\alpha} \left\{ \sqrt{(a^2 - x^2)} - x \tan\alpha \right\} dx$$

$$= \frac{a}{2} \cos\alpha . a \sin\alpha + \frac{a^2}{2} \sin^{-1}(\cos\alpha) - \frac{a^2 \cos^2 dx}{2} \tan\alpha$$

$$= \frac{1}{2} a^2 \sin^{-1}\left\{ \sin\left(\frac{\pi}{2} - \alpha\right) \right\} = \frac{1}{2} a^2 \left(\frac{\pi}{2} - \alpha\right).$$

$$\text{R.H.S.} = \int_0^{a\sin\alpha} \int_0^{y\cot\alpha} dy\, dx + \int_{a\sin\alpha}^{a} \int_0^{\sqrt{(a^2-y^2)}} dy\, dx$$

$$= \int_0^{a\sin\alpha} y \cot\alpha\, dy + \int_{a\sin\alpha}^{0} \sqrt{(a^2 - y^2)}\, dy$$

$$= \frac{1}{2} a^2 \sin\alpha \cot\alpha + \frac{a^2}{2} \cdot \frac{\pi}{2} - \left\{ \frac{a}{2} \sin\alpha . a \cos\alpha + \frac{a^2}{2} \cdot \alpha \right\}$$

$$= \frac{a^2}{2} \left(\frac{\pi}{2} - \alpha \right).$$

4. *Change the order of integration in the double integral* $\int_0^{\infty} \int_x^{\infty} \frac{e^{-y}}{y}\, dx\, dy$ *and hence find the value. (Lucknow 94, 96; Avadh 96)*

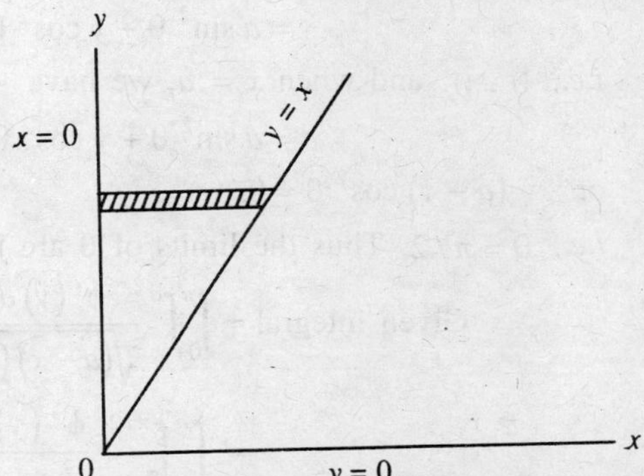

The limits of integration are given by the straight line $y = x$, $y = \infty$, $x = 0$ and $x = \infty$, *i.e.*, the region of integration is bounded by $x = 0$, $y = x$ and an infinite boundary. Hence taking the strips parallel to x-axis, the limits for y are from 0 to ∞. Hence changing the order of integration, we have

$$\int_0^{\infty} \int_x^{\infty} \frac{e^{-y}}{y}\, dx\, dy = \int_0^{\infty} \int_0^{y} \frac{e^{-y}}{y}\, dy\, dx$$

$$= \int_0^{\infty} \frac{e^{-y}}{y} [x]_0^y\, dy = \int_0^{\infty} e^{-y}\, dy$$

Fig.

$$= 1.$$

5. *Change the order of integration in double integral*

$$\int_0^a \int_0^x \frac{\phi'(y)\, dx\, dy}{\sqrt{(a-x)(x-y)}}$$

and hence find its value.

The region of integration is ONM.

The limit for x are from y to a and limits for y are from 0 to a.

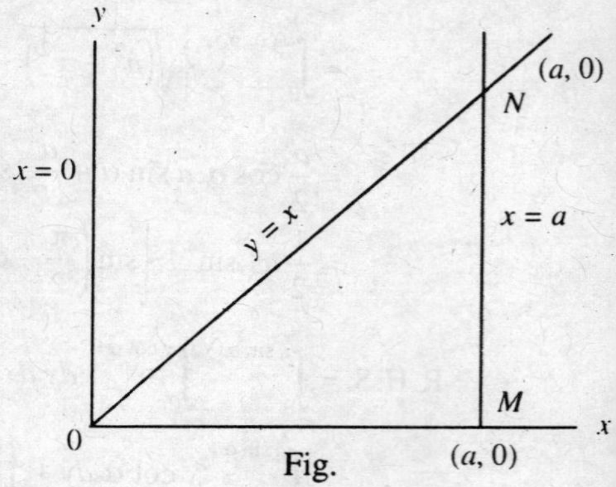

Fig.

Hence

$$\int_0^a \int_0^x \frac{\phi'(y)\,dx\,dy}{\sqrt{(a-x)(x-y)}}$$

$$= \int_0^a \int_y^a \frac{\phi'(y)\,dy\,dx}{\sqrt{(a-x)(x-y)}}$$

To find the value, let $x = a\sin^2\theta + y\cos^2\theta$

$$\therefore \qquad dx = 2(a-y)\sin\theta\cos\theta\,d\theta$$

also $\qquad (a-x) = (a-y)\cos^2\theta,\ x-y = (a-y)\sin^2\theta.$

For the limits of θ, when $x = y$, we have

$$y = a\sin^2\theta + y\cos^2\theta \text{ or } (y-a)\sin^2\theta = 0,$$

i.e., $\theta = 0$, and when $x = a$, we have

$$a = a\sin^2\theta + y\cos^2\theta$$

or $\quad (a-y)\cos^2\theta = 0,$

i.e., $\theta = \pi/2$. Thus the limits of θ are from 0 to $\pi/2$.

$$\therefore \quad \text{Given integral} = \int_0^a \int_y^a \frac{\phi'(y)\,dy\,dx}{\sqrt{(a-x)(x-y)}}$$

$$= \int_0^a \int_0^{\pi/2} \frac{\phi'(y).2(a-y)(\sin\theta\cos\theta)}{(a-y)(\sin\theta\cos\theta)}\,dy\,d\theta$$

$$= 2\int_0^a \int_0^{\pi/2} \phi'(y)\,dy\,d\theta$$

$$= 2\int_0^a \phi'(y)\cdot\frac{\pi}{2}\,dy$$

$$= \pi\left[\phi(y)\right]_0^a = \pi\left[\phi(a) - \phi(0)\right]$$

6. *Change the order of integration in*

$$\int_0^{2a}\int_0^{\sqrt{(2ax-x^2)}} \frac{\phi'(y)\left(x^2+y^2\right)x\,dx\,dy}{\sqrt{\left\{4a^2x^2-\left(x^2+y^2\right)\right\}}}$$

and hence evaluate it.

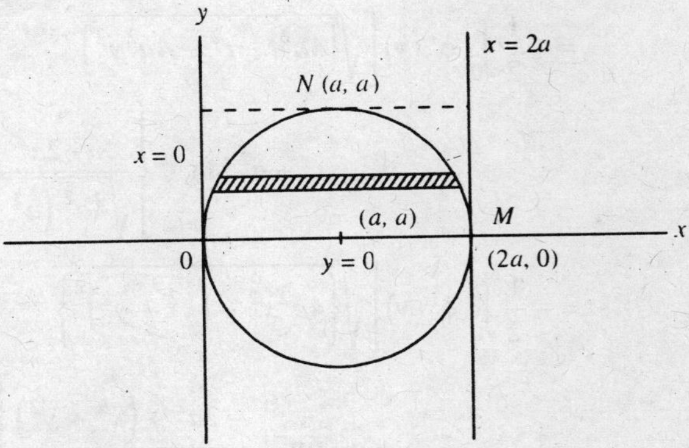

Fig

Clearly, the area of integrating is OMNO.

Solving $x^2 - 2ax + y^2 = 0$ for x, we get

$$x = a \pm \sqrt{(a^2 - y^2)}$$

Hence limits of x are from $a - \sqrt{(a^2 - y^2)}$ to $a + \sqrt{(a^2 - y^2)}$.

The limits of y are from 0 to a.

Hence, given integral becomes

$$= \int_0^a \int_{a-\sqrt{(a^2-y^2)}}^{a+\sqrt{(a^2-y^2)}} \frac{\phi'(y)\left(x^2+y^2\right) x\, dy\, dx}{\sqrt{4a^2x^2 - \left(x^2+y^2\right)^2}}.$$

To evaluate the integral, put $x^2 + y^2 = t$.

$\therefore\ 2x\, dx = dt$. Then

$$\int_0^a \int_{a-\sqrt{(a^2-y^2)}}^{a+\sqrt{(a^2-y^2)}} \frac{\phi'(y)\left(x^2+y^2\right) x\, dy\, dx}{\sqrt{\left\{4a^2x^2 - \left(x^2+y^2\right)^2\right\}}}$$

$$= \frac{1}{2}\int_0^a \int_{t_1}^{t_2} \frac{\phi'(y)\, t\, dy\, dt}{\sqrt{\left\{4a^2\left(t-y^2\right) - t^2\right\}}}$$

$$= -\frac{1}{4}\int_0^a \int_{t_1}^{t_2} \frac{\phi'(y)\left(4a^2 - 2t\right) dy\, dt}{\sqrt{\left(4a^2t - t^2 - 4a^2y^2\right)}}$$

$$+ \int_0^a \int_{t_1}^{t_2} \frac{\phi'(y).a^2\, dy\, dt}{\sqrt{\left\{4a^2\left(a^2-y^2\right) - \left(2a^2+t^2\right)\right\}}}$$

$$= -\frac{1}{2}\int_0^a \phi'(y)\left[\sqrt{\left\{4a^2t - t^2 - 4a^2y^2\right\}}\right.$$

$$\left. + a^2 \sin^{-1}\left\{\frac{-2a^2 + t}{\sqrt{4a^2\left(a^2 - y^2\right)}}\right\}\right]_{t_1}^{t_2} dy$$

$$= -\frac{1}{2}\int_0^a \phi'(y)\left[\sqrt{\left\{4a^2x^2 - \left(x^2 + y^2\right)^2\right\}}\right.$$

$$\left. + a^2 \sin^{-1}\frac{-2a^2 + \left(x^2 + y^2\right)}{\sqrt{\left\{4a^2\left(a^2 - y^2\right)\right\}}}\right]_{a - \sqrt{a^2 - y^2}}^{a + \sqrt{a^2 - y^2}} dy$$

$$= \pi a^2 \int_0^a \phi'(y)\, dy = \pi a^2 \left\{\phi(a) - \phi(0)\right\}.$$

7. *Show how the change in order of integration leads to the evaluation of* $\int_0^\infty \frac{\sin rx}{x}\, dx$ *from* $\int_0^\infty \int_0^\infty e^{-xy} \sin rx\, dx\, dy.$

Let $\qquad I = \int_0^\infty \int_0^\infty e^{-xy} \sin rx\, dx\, dy$

$$= \int_0^\infty \sin rx \left[\frac{e^{-xy}}{-x}\right]_0^\infty dx$$

$$= \int_0^\infty \frac{\sin rx}{x}\, dx \qquad \text{...(i)}$$

Again, $\qquad I = \int_0^\infty \int_0^\infty e^{-xy} \sin rx\, dx\, dy$

$$= \int_0^\infty \left[\int_0^\infty e^{-xy} \sin rx\, dx\right] dy$$

[Limits being constants]

$$= \int_0^\infty \frac{r\, dy}{\left(r^2 + y^2\right)} = \left[\tan^{-1}\left(\frac{y}{r}\right)\right]_0^\infty$$

$$= \pi/2 \qquad \text{...(ii)}$$

Hence from (*i*) and (*ii*), we have

$$\int_0^\infty \frac{\sin rx}{x}\, dx = \pi/2.$$

EXERCISES

1. Express, as a repeated integral, the double integral

$$\iint f(x, y)\, dx\, dy,$$

taken over the interior of the quadrilateral bounded by the lines

$$x + y = 0,\ x - y = 0,\ 2x - y = 1,\ 2x - 3y + 5 = 0$$

taken in order.

2. Evaluate

$$\iint x^2y^2\, dx\, dy \text{ over the domain}$$

$$\{(x, y) : x \geq 0, y \geq 0, (x^2 + y^2) \leq 1\}.$$

3. Show that the value of

$$\iint \sqrt{(4y - x^2)}\, dx\, dy$$

taken over the interior of the circle $x^2 + y^2 = 2y$ is $\pi + \dfrac{8}{3}$.

4. Change the order of integration in

$$\int_0^{2a} \int_{x^2/4a}^{3a - x} \varphi(x, y)\, dx\, dy.$$

5. Show that

$$\int_a^b \int_{a^2/x}^{x} F\, dx\, dy = \int_{a^2/b}^{a} \int_{a^2/y}^{b} F\, dx\, dy + \int_a^b \int_y^b F\, dx\, dy$$

6. In the integral

$$\int_2^4 \int_{4/x}^{(20 - 4x)/(8 - x)} (4 - y)\, dy\, dx,$$

change the order of integration, and evaluate the integral.

7. Change the order of integration in the integral

$$\int_0^1 dx \int_0^{\sqrt{(1 - x^2)}} \frac{dy}{(1 + e^y)\sqrt{(1 - x^2 - y^2)}}.$$

and hence evaluate.

8. Prove that

$$\int_0^1 dx \int_x^{1/x} \frac{y\, dy}{(1 + xy)^2 (1 + y^2)} = \frac{\pi - 1}{4}.$$

9. Evaluate

$$\iint (x^2 + y^2)\, dx\, dy \text{ over the domain bounded by}$$

$xy = 1$, $y = 0$, $y = x$, $x = 2$.

10. Evaluate $\iint x^2 \, dx \, dy$ over the domain bounded by $x^2 - y^2 = 1$, $x^2 + y^2 = 4$ which contains the origin.

11. Evaluate

$$\iint (x + y + a) \, dx \, dy.$$

taken over $\{(x, y) : x^2 + y^2 \le a^2\}$.

12. Prove that

$$\int_0^a dx \int_0^x f(x, y) \, dy = \int_0^a dy \int_y^a f(x, y) \, dx.$$

Deduce Dirichlet's formula

$$\int_0^t dx \int_0^x f(y) \, dy = \int_0^t (t - y) f(y) \, dy.$$

Def. A domain E will be said to be *Quadratic* with respect to y-axis, if it is bounded by curves of the form

$$y = \varphi(x), \; y = \psi(x); \; x = a, \; x = b,$$

where φ, ψ are continuous and $\varphi(x) \ge \psi(x) \; \forall \; x \in [a, b]$.

Thus a domain which is quadratic with respect to y-axis is such that a line parallel to y-axis and lying between $x = a$, $x = b$ meets the boundary of E in just two points.

Similarly we may have domains which are quadratic with respect to x-axis.

16.9. GREEN'S THEOREM

Connection between double and line integrals. Consider a region E bounded by curves

$$x = a, \; x = b, \; y = \phi(x), \; y = \psi(x)$$

such that $\psi(x) \le \phi(x) \; \forall \; x \in [a, b]$.

Let, f be a real valued continuous function defined in E and admitting of continuous partial derivative f_y in E.

We have

$$\iint_E f_y(x, y) \, dx \, dy = \int_a^b \left\{ \int_{\psi(x)}^{\phi(x)} f_y(x, y) \, dy \right\} dx$$

$$= \int_a^b f[x, \phi(x)] \, dx - \int_a^b f[x, \psi(x)] \, dx$$

Let C_1, C_2, C_3, C_4 denote the four parts of C taken in the positive sense, *i.e.*, in such a way that the interior of the region lies to the left a the contour is described in the counter-clockwise sense.

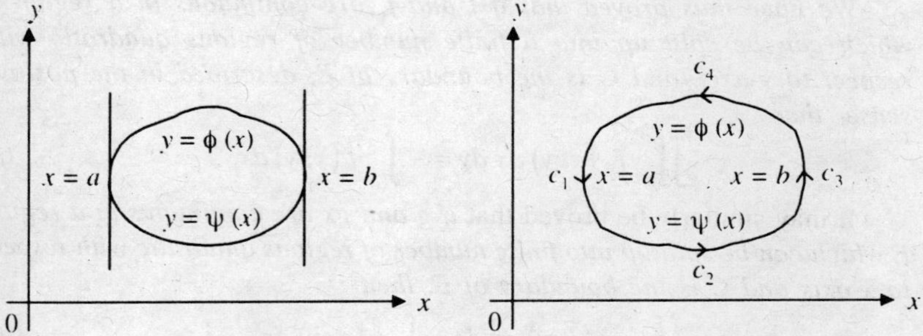

Fig.

We have

$$\int_C f(x, y)\,dx = \int_{C_1} f(x, y)\,dx + \int_{C_2} f(x, y)\,dx + \int_{C_3} f(x, y)\,dx + \int_{C_4} f(x, y)\,dx.$$

But $$\int_{C_1} f(x, y)\,dx = 0 = \int_{C_3} f(x, y)\,dx,$$

$$\int_{C_2} f(x, y)\,dx = 0 = \int_a^b f[x, \psi(x)]\,dx,$$

and $$\int_{C_4} f(x, y)\,dx = \int_b^a f[x, \phi(x)]\,dx = -\int_a^b f[x, \phi(x)]\,dx.$$

$$\iint_E f_y(x, y)\,dx\,dy = -\int_C f(x, y)\,dx.$$

∴ We, now, *generalise* the above result.

Let a region E be such as can be divided into a finite number of sub-regions $E_1, E_2, ..., E_n$ each of which is quadratic with respect to y-axis.

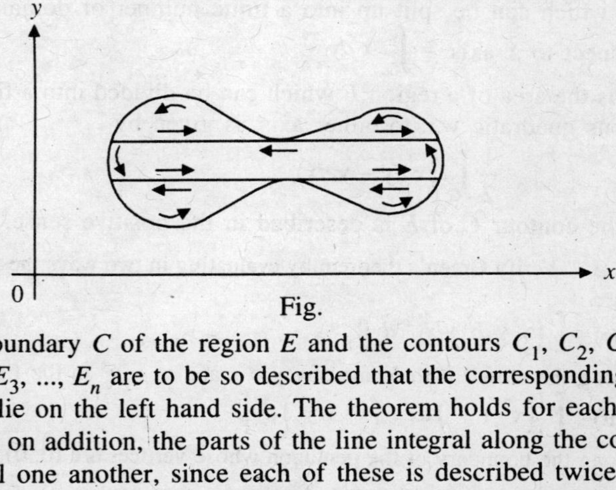

Fig.

The boundary C of the region E and the contours $C_1, C_2, C_3, ..., C_n$ of $E_1, E_2, E_3, ..., E_n$ are to be so described that the corresponding regions constantly lie on the left hand side. The theorem holds for each separate region, and on addition, the parts of the line integral along the connecting lines cancel one another, since each of these is described twice, once in each direction, and we arrive at the theorem for the whole region.

We have thus proved that if f and f_y are continuous in a region E which can be split up into a finite number of regions quadratic with respect to y-axis, and C is the boundary of E, described in the positive sense, then

$$\iint_E f_y(x, y)\, dx\, dy = -\int_C f(x, y)\, dx. \qquad ...(i)$$

It may similarly be proved that *if g and gx are continuous in a region E which can be split up into finite number of regions quadratic with respect to x-axis and C is the boundary of E, then*

$$\iint_E g_x(x, y)\, dx\, dy = \int_C g(x, y)\, dy. \qquad ...(ii)$$

Finally, on subtracting (*i*) from (*ii*), we see that

If we suppose that f, g, f_y and g_x are continuous in a domain E which can be split up in finite number of regions quadratic with respect to either axis, then

$$\int_C [f(x, y)\, dx + g(x, y)\, dy] = \iint_E [g_x(x, y) - f_y(x, y)]\, dx\, dy,$$

where the integral on the left is a line integral round the boundary C of the region taken in such a way that the interior of the region remains on the left as the boundary is described.

This may be written as

$$\int_C f\, dx + g\, dy = \iint_E (g_x - f_y)\, dx\, dy.$$

A particular case. Taking $f(x, y) = y$, we see from (*i*) that the area of a domain which can be split up into a finite number of domains quadratic with respect to Y-axis $= -\int_C y\, dx$.

Similarly, putting $g(x, y) = x$, we see from (*ii*) that the area of domain which can be split up into a finite number of domains quadratic with respect to X-axis $= \int_C x\, dy$.

Thus the area of a region E which can be divided into a finite number of regions quadratic w.r. to either axis, is given by

$$\frac{1}{2}\int_C (x\, dy - y\, dx),$$

where the contour C of E is described in the positive sense.

Ex. 1. Verify Green's theorem by evaluating in two ways the following line integrals :

(*i*) $\int (x^2 y\, dx + xy^2\, dy)$,

taken along the closed path formed by $y = x$ and $x^2 = y^3$ in the first quadrant.

(*ii*) $\int [(x^3 + y^2)\, dx + (x^2 + y^3)\, dy]$,

taken along the boundary of the pentagon whose vertices are (0, 0), (1, 0), (2, (0, 1).

2. Evaluate

$$\int \left(xy\,dx + xy^2\,dy\right)$$

taken round the positively oriented square with vertices (± 1, 0), (0, ± 1). Verify the result by using Green's Theorem.

3. Evaluate

$$\int_C \left(x^4\,dy - y^4\,dx\right),$$

where C is the semi-circumference of the ellipse

$$\frac{(x-a)^2}{a^2} + \frac{y^2}{b^2} = 1,$$

which runs from the origin to $(2a, 0)$ in the first quadrant.

16.10. DOUBLE INTEGRAL AS A LIMIT

Let f be continuous in a region E. Let the region E be divided into sub-regions $E_1, E_2, \ldots, E_n$ with areas $A_1, A_2, \ldots, A_n$. Let (ξ_r, η_r) be any point of the region E_r. Form the sum

$$\Sigma f(\xi_r, \eta_r) A_r.$$

By the diameter of a sub-region E_r will be meant the supremum of the set of the distances between pairs of points on its boundary and the greatest of the diameters of all the sub-regions will be called the *norm* of the division.

It can now be proved that *as the norm $\delta \to 0$, the limit of the sum* (1) *is the integral of $f(x, y)$ over E, i.e.,*

$$\lim_{\delta \to 0} \Sigma f(\xi_r, \eta_r) A_r = \iint_E f(x, y)\,dx\,dy.$$

The details of the proof will not be given here.

Cor. *If f be continuous in a region E with area A, then there exists point $(\xi, \eta) \in E$ such that*

$$\iint_E f(x, y)\,dx\,dy = Af(\xi, \eta).$$

16.11. CHANGE OF VARIABLE IN A DOUBLE INTEGRAL

First Proof

Lemma. Let

$$x = \varphi(u, v),\ y = \psi(u, v) \qquad \ldots(1)$$

where φ and ψ are two functions defined in a certain region E_1 of the u, plane bounded by a curve C_1. We suppose that

(*i*) the two functions possess continuous first order partial derivatives all points of E_1 and C_1.

(*ii*) the equations (1) transform the region E_1 bounded by C_1, into a region E of the xy plane bounded by a curve C in such a way that a one-to-one correspondence exists between the two regions and their boundaries.

(*iii*) the Jacobian, $F = \partial(x, y)/\partial(u, v)$, does not change sign in E_1, though it may vanish at certain points of C_1.

As the point (u, v) describes the boundary C_1 in the positive sense, then the point (x, y) may describe C in the positive or else in the negative sense without ever changing the direction of motion. The transformation will be said to be direct or inverse respectively in the two cases.

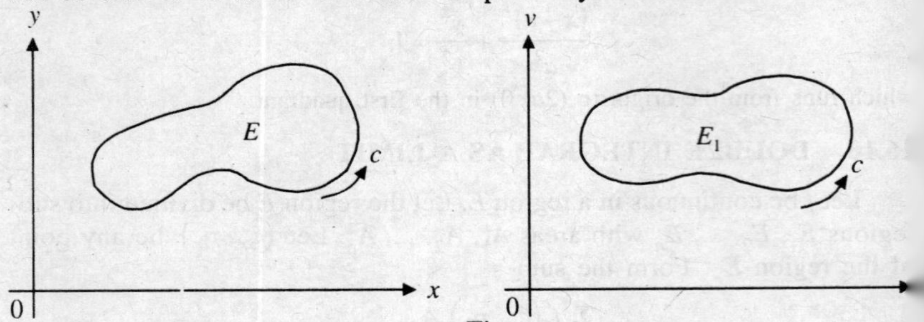

Fig

We shall now obtain a formula connecting the areas A and A_1 of th regions E, E_1.

Let C_1 be given by

$$u = u(t),\ v = v(t),$$

where t varies from α to β, say.

Then C is given by

$$x = \varphi\left[u(t), v(t)\right] = x(t), \text{ say}$$

$$y = \psi\left[u(t), v(t)\right] = y(t), \text{ say.}$$

We have

$$A = \int_C x\, dy,$$

taken along C in the positive sense.

Expressing the line integral as an ordinary integral we have

$$A = \int_\alpha^\beta x(t) \frac{dy}{dt}\, dt$$

$$= \int_\alpha^\beta x(t)\left[y_u u_t + y_v v_t\right] dt.$$

Expressing this again as a line integral, we see that

$$A = \pm \int_{C_1} (xy_u\, du + xy_v\, dv)$$

$$= \pm \int_C \varphi(u, v)(\psi_u \, du + \psi_v \, dv)$$

where the new integral is to be taken along the positive sense of C_1 and the sign is + or – according as the transformation is direct or inverse.

Applying Green's theorem, we have

$$\int_{C_1} (\varphi\psi_u \, du + \varphi\psi_v \, dv) = \int_{E_1}\int \left[(\varphi\psi_v)_u - (\varphi\psi_u)_v\right] du \, dv$$

$$= {}^* \int_{E_1}\int \frac{\partial(\varphi, \psi)}{\partial(u, v)} du \, dv = A_1 \left[\frac{\partial(\varphi, \psi)}{\partial(u, v)}\right]_{\xi, \eta}$$

(*Cor. of § 16.10, p. 482*)

where (ξ, η) is a point of E_1.

$$\therefore \qquad A = \pm A_1 \left[\frac{\partial(\varphi, \phi)}{\partial(u, v)}\right] = \pm A_1 [J]_{\xi, \eta} = A_1 \, | J |_{\xi, \eta}.$$

Since A, A_1 are essentially positive, we see that the sign, +, or, –, should be taken according as J is positive or negative. This shows that *the transformation is direct or inverse according as the Jacobian is positive or negative.*

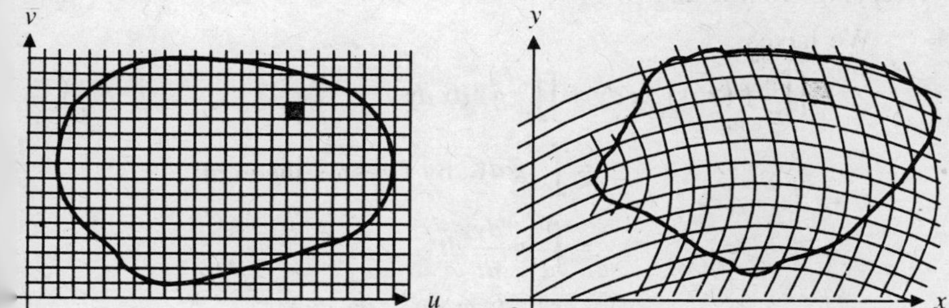

Fig

Main Theorem. Let f be continuous in the region E. We divide the region E_1 by lines parallel to the u-axis and v-axis. This division of E_1 gives rise to a curvilinear division of the region E. Let E'_{rs} be any sub-region of E_1 and E_{rs} the corresponding sub-region of E and let w'_{rs}, w_{rs} be their areas.

By the lemma, we have

$$w_{rs} = w'_{rs} \, | J |_{\xi_{rs}, \eta_{rs}}$$

where (ξ_{rs}, η_{rs}) is some point of E'_{rs}. Let (x_{rs}, y_{rs}) be the corresponding point of E_{rs}. We have

$$f(x_{rs}, y_{rs}) \, w_{rs} = f\left[\varphi(\xi_{rs}, \eta_{rs}), \, | \, \psi(\xi_{rs}, \eta_{rs}) \, |\right] | \, J \, | (\xi_{rs}, \eta_{rs}) \, w'_{rs}.$$

---

*It is assumed that $\psi_{uv} = \psi_{vu}$.

A similar equality will be obtained for each pair of corresponding sub-regions. Adding them and letting the norm of the divisions tend to zero, we see that

$$\iint_E f(x, y)\, dx\, dy = \iint_{E_1} f\left[\varphi(u, v), \psi(u, v)\right] |J|\, du\, dv.$$

Another proof. We shall now give another proof which is independent of the lemma.

We have supposed that f is continuous in E.

If we define

$$g(x, y) = \int_a^x f(x, y)\, dy,$$

we have,

$$g_x = f(x, y).$$

We suppose that C_1 is given by

$$u = u(t), v = v(t);$$

and C by

$$x = x(t), = y(t);$$

t varying from α and β.

We have

$$\iint_E f(x, y)\, dx\, dy = \iint_E g_x\, dx\, dy$$

$$= \int_C g\, dy, \text{ by Green's theorem}$$

$$= \int_\alpha^\beta g \frac{dy}{dt}\, dt$$

$$= \int_\alpha^\beta g\left(\frac{\partial y}{\partial u}\frac{du}{dt} + \frac{\partial y}{\partial v}\frac{dv}{dt}\right) dt$$

$$= \pm \int_{C_1} \left(g \frac{\partial y}{\partial u}\, du + g \frac{\partial y}{\partial v}\, dv\right),$$

where we take + or – sign according as the transformation is direct or inverse.

Again, applying Green's theorem, we see that

$$\int_C \left(g \frac{\partial y}{\partial u}\, du + g \frac{\partial y}{\partial v}\, dv\right) + \iint_E \left[\frac{\partial}{\partial u}\left(g \frac{\partial y}{\partial v}\right) - \frac{\partial}{\partial v}\left(g \frac{\partial y}{\partial u}\right)\right] du\, dv$$

Now

$$\frac{\partial}{\partial u}\left(g \frac{\partial y}{\partial v}\right) - \frac{\partial}{\partial v}\left(\frac{\partial y}{\partial u}\right)$$

$$= \frac{\partial g}{\partial u}\frac{\partial y}{\partial v} + g\frac{\partial^2 y}{\partial u \partial v} - \frac{\partial g}{\partial v}\frac{\partial y}{\partial u} - g\frac{\partial^2 y}{\partial v \partial u}$$

$$= \left(\frac{\partial g}{\partial x}\frac{\partial x}{\partial u} + \frac{\partial g}{\partial y}\frac{\partial y}{\partial u}\right)\frac{\partial y}{\partial v} - \left(\frac{\partial g}{\partial x}\frac{\partial x}{\partial v} + \frac{\partial g}{\partial y}\frac{\partial y}{\partial v}\right)\frac{\partial y}{\partial u}$$

$$= \frac{\partial g}{\partial x}\left(\frac{\partial x}{\partial u}\frac{\partial y}{\partial v} - \frac{\partial x}{\partial v}\frac{\partial y}{\partial u}\right)$$

$$= f(x, y)\frac{\partial (x, y)}{\partial (u, v)}.$$

Thus we obtain

$$\iint_E f\,dx\,dy = \pm \iint_{E_1} f\frac{\partial (x, y)}{\partial (u, v)}\,du\,dv = \iint_{E_1} f\left|\frac{\partial (x, y)}{\partial (u, v)}\right|\,du\,dv.$$

Note. The principle of the second proof is contained in the following steps :

I. By Green's theorem, express the given double integral over, E, as a line integral around C.

II. Express the integral around C, as a line integral around C_1.

III. By Green's theorem, express the line integral C_1 as a double integral over E_1.

Ex. 1. $x = \dfrac{u}{1+v^2}$ and $y = \dfrac{uv}{1+v^2}$; find the region in the (u, v) plane which corresponds to the region

$$\{(x, y) : x^2 - x + y^2 \le 0 \le y\}.$$

Ex. 2. If $x = \frac{1}{2}u(1+v)$, $y = \frac{1}{2}u(1-v)$, find the region in the (u, v) plane which corresponds to the first quadrant in the (x, y) plane.

EXAMPLES

1. *Evaluate*

$$\iint \frac{\sqrt{(a^2b^2 - b^2x^2 - a^2y^2)}}{\sqrt{(a^2b^2 + b^2x^2 + a^2y^2)}}\,dx\,dy.$$

the field of integration being the positive quadrant of the ellipse

$$x^2/a^2 + y^2/b^2 = 1.$$

Sol. Changing the variables x, y to X, Y where

$$x = aX,\ y = bY$$

we see that, since $\partial(x,y)/\partial(X,Y)=ab$, the integral

$$= ab\iint \sqrt{\left(\frac{1-X^2-Y^2}{1+X^2+Y^2}\right)}\,dX\,dY,$$

the new field of integration being the positive quadrant of the circle

$$X^2+Y^2=1.$$

Changing X, Y to r, θ where

$$X=r\cos\theta,\ Y=r\sin\theta,$$

so that $\partial(X,Y)/\partial(r,\theta)=r$, we see that the integral

$$= ab\iint \frac{\sqrt{(1-r^2)}}{\sqrt{(1+r^2)}}\,r\,dr\,d\theta.$$

It is easily seen that the positive quadrant of the circle $X^2+Y^2=1$, will be described if θ varies from 0 to $\pi/2$ and corresponding to each value of θ between 0 and $\pi/2$, r varies from 0 to 1. This new domain of inegration, therefore, is the rectangle $\left[0,1;0,\frac{1}{2}\pi\right]$. Thus

$$\text{the integral} = ab\int_0^{\frac{1}{2}\pi} d\theta \int_0^1 \frac{\sqrt{(1-r^2)}}{\sqrt{(1+r^2)}}\,r\,dr$$

$$=\frac{\pi}{2}ab\int_0^1 \frac{\sqrt{(1-r^2)}}{\sqrt{(1+r^2)}}\,r\,dr=\frac{1}{2}\pi(\pi-2)\,ab$$

where the integral has been evaluated by putting $r^2=\cos t$.

2. *Integrate the function $1/xy$ over the area bounded by the four circles $x^2+y^2=ax, a'x, by\ b'y$ where a, a', b, b' are all positive.*

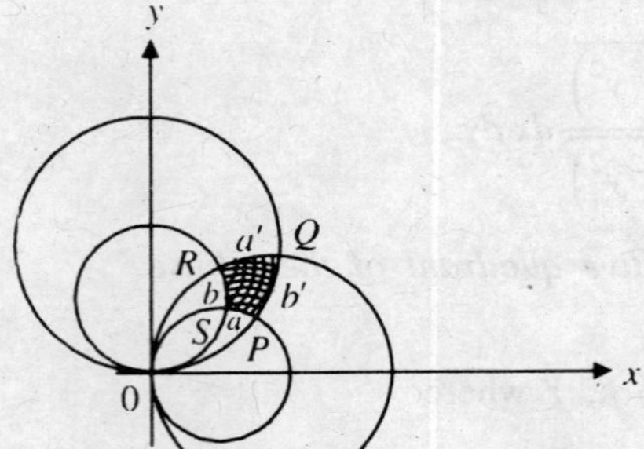

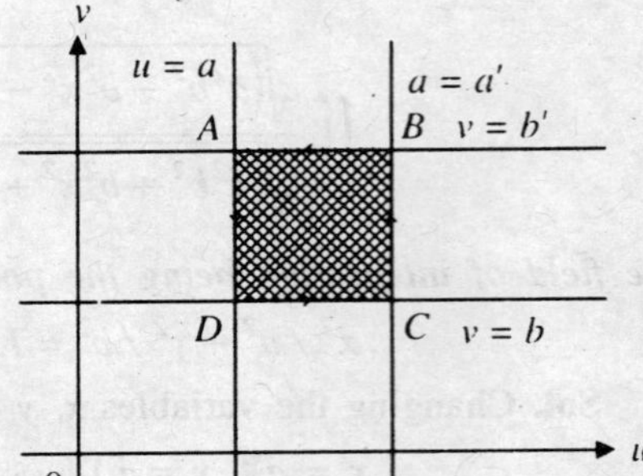

Fig

Sol. The integration is to be carried over the shaded area shown in the figure. We have supposed $a' > a$ and $b' > b$.

The region of integration is defined by

$$ax \le x^2 + y^2 \le a' x; \; by \le x^2 + y^2 \le b'y.$$

We change the variables to u, v, where

$$u = \left(x^2 + y^2\right) / x, \; v = \left(x^2 + y^2\right) / y.$$

$$\Rightarrow \qquad x = \frac{uv^2}{u^2 + v^2}; \; y = \frac{u^2 v}{u^2 + v^2}.$$

It is easy to see that

$$\partial(\partial, v) / \partial(x, y) = -\left(x^2 + y^2\right)^2 / x^2 y^2$$

$$\therefore \qquad {}^*\partial(x, y) / \partial(u, v) = -x^2 y^2 / \left(x^2 + y^2\right)^2.$$

Since the Jacobian is negative, the transformation is inverse. This fact may also be directly verified. The new field of integration is determined by the boundaries $u = a$, $u = a'$, $v = b$, $v = b'$, and is, therefore, the rectangle $[a, a'; b, b']$. Thus we see that

$$\text{the integral} = \iint \frac{1}{xy} | J | \, du \, dv$$

$$= \iint \frac{xy}{\left(x^2 + y^2\right)^2} du \, dv$$

$$= \iint \frac{1}{uv} du \, dv = \int_a^{a'} \frac{1}{u} du \int_b^{b'} \frac{1}{v} dv = \log \frac{a'}{a} \cdot \log \frac{b'}{b}.$$

3. *Prove that*

$$B(m, n) = \frac{\Gamma(m) \Gamma(n)}{\Gamma(m+n)}. \qquad (m > 0, \; n > 0)$$

Sol. We have putting $x = r \cos^2 \theta, t = r^2$

---

* If $x = \varphi(\mu, v)$ and $y = \psi(u, v)$ then looking upon u, v as functions of x we have $y = \psi(u, v)$

$$1 = \frac{\partial x}{\partial u} \cdot \frac{\partial u}{\partial x} + \frac{\partial x}{\partial v} \cdot \frac{\partial v}{\partial x} \qquad 0 = \frac{\partial x}{\partial u} \cdot \frac{\partial u}{\partial x} + \frac{\partial x}{\partial v} \cdot \frac{\partial v}{\partial x}$$

$$0 = \frac{\partial y}{\partial u} \cdot \frac{\partial u}{\partial x} + \frac{\partial y}{\partial v} \cdot \frac{\partial v}{\partial x} \qquad 1 = \frac{\partial y}{\partial u} \cdot \frac{\partial u}{\partial y} + \frac{\partial y}{\partial v} \cdot \frac{\partial v}{\partial y}$$

$$\frac{\partial(x, y)}{\partial(u, v)} \cdot \frac{\partial(u, v)}{\partial(x, y)} = \begin{vmatrix} \frac{\partial x}{\partial u} & \frac{\partial x}{\partial v} \\ \frac{\partial y}{\partial u} & \frac{\partial y}{\partial v} \end{vmatrix} \begin{vmatrix} \frac{\partial u}{\partial x} & \frac{\partial v}{\partial x} \\ \frac{\partial u}{\partial y} & \frac{\partial v}{\partial y} \end{vmatrix} = \begin{vmatrix} 1 & 0 \\ 0 & 1 \end{vmatrix} = 1$$

$$B(m,n)=\int_0^1 x^{m-1}(1-x)^{n-1}\,dx=2\int_0^{\frac{1}{2}\pi}\cos^{2m-1}q\,\sin^{2n-1}\theta\,d\theta \quad ...(1)$$

$$\Gamma(m)=\int_0^{\infty} t^{m-1}e^{-t}\,dt=2\int_0^{\infty} r^{2m-1}e^{-r^2}\,dr \quad ...(2)$$

From (2), we see that the integral

$$\iint_E x^{2m-1}y^{2n-1}e^{-x^2-y^2}\,dx\,dy$$

where E is the square $[0, R; 0, R]$ tends as its limit, to $\Gamma(m)\Gamma(n)$ as $R\to\infty$.

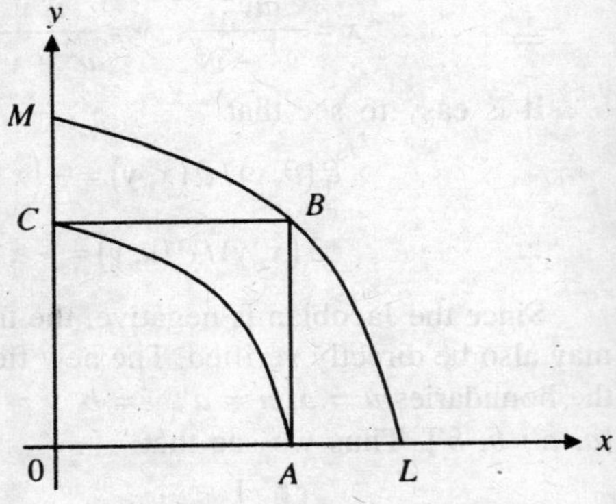

Fig.

The positive quadrant of the circle $x^2+y^2=R^2$ is a part of the square E which again, is a part of the positive quadrant of the circle $x^2+y^2=2R^2$. We denote these positive quadrants by E_1, E_2 respectively. The integrand being positive, we have

$$4\iint_{E_1} x^{2m-1}y^{2n-1}e^{-x^2-y^2}\,dx\,dy\le 4\iint_E x^{2m-1}y^{2n-1}e^{-x^2-y^2}\,dx\,dy$$

$$\le 4\iint_{E_1} x^{2m-1}y^{2n-1}e^{-x^2-y^2}\,dx\,dy.$$

But changing the variables to r, θ, where $x=r\cos\theta$, $y=r\sin\theta$, we have

$$4\iint_{E_1} x^{2m-1}y^{2n-1}e^{-x^2-y^2}\,dx\,dy$$

$$=4\int_0^{\frac{1}{2}\pi}\cos^{2m-1}\theta\,\sin^{2n-1}\theta\,d\theta\int_0^R e^{-r^2}r^{2m+2n-1}\,dr$$

$$=2B(m,n)\int_0^R e^{-r^2}r^{2m+2n-1}\,dr.$$

Similarly

$$4\iint_{E_2} x^{2m-1}y^{2n-1}e^{-x^2-y^2}\,dx\,dy=2B(m,n)\int_0^{\sqrt{2R}} e^{-r^2}r^{2m+2n-1}\,dr.$$

$$\therefore\ 2B(m,n)\int_0^R e^{-r^2}r^{2m+2n-1}\,dr\le 4\iint_E x^{2m-1}y^{2n-1}e^{-x^2-y^2}\,dx\,dy$$

$$\le 2B(m,n)\int_0^{\sqrt{2R}} e^{-r^2}r^{2m+2n-1}\,dr.$$

Letting $R \to \infty$, we obtain the required result.

Ex. 1. Show that

$$(i)\ \Gamma\left(\frac{1}{2}\right) = \sqrt{\pi}, \qquad (ii)\ \int_0^{\frac{1}{2}\pi} \frac{d\theta}{\sqrt{(\sin\theta)}} \cdot \int_0^{\frac{1}{2}\pi} \sqrt{(\sin\theta)}\, d\theta = \pi.$$

Ex. 2. If $I = \int_0^a e^{-x^2}\, dx$, show that

$$\frac{\pi}{4}\left(1 - e^{-a^2}\right) < I^2 < \frac{\pi}{4}\left(1 - e^{-2a^2}\right)$$

and deduce that

$$\int_0^\infty e^{-x^2\, dx} = \frac{\sqrt{\pi}}{2}.$$

4. *By substituting $x + y = u$, $x = uv$, prove that the value of*

$$\iint \sqrt{[xy\,(1 - x - y)]}\, dx\, dy$$

taken over the interior of the triangle bounded by the lines $x = 0$, $y = 0$, $x + y - 1 = 0$ is $2\pi/105$.

Sol. Now

$$x = uv,\ y = u(1 - v) \Rightarrow \partial(x, y)/\partial(u, v) = -u,$$

so that the Jacobian ≤ 0.

The Jacobian vanishes when $u = 0$, *i.e.*, when $x = y = 0$, but not

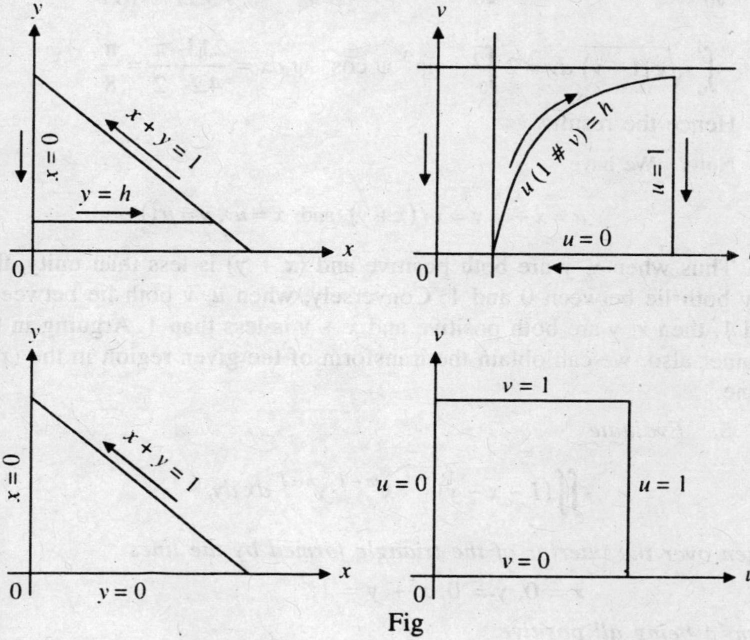

Fig

otherwise. It is easy to see that to the origin of the xy-plane corresponds the whole line $u = 0$ of the uv-plane so that the correspondence ceases to be one to one. In order to exclude $x = 0$, $y = 0$, we look upon the given integral, which certainly exists, as the limit, when $h \to 0$, of the integral over the region bounded by

$$x + y = 1, x = 0, y = h. \qquad (h > 0)$$

The transformed region is, then, bounded by the lines

$$u = 1, v = 0, u(1 - v) = h,$$

which correspond to the three boundaries of the region in the xy-plane.

When $h \to 0$, this new region of the uv-plane tends, as its limit, to the square bounded by the lines

$$u = 1, v = 1, u = 0, v = 0.$$

Thus the integral

$$= \iint \sqrt{[uv.u(1-v)(1-u)\,du\,dv]}$$

$$= \int_0^1 u^2 \sqrt{1-u}\,du \int_0^1 \sqrt{[v(1-v)]}\,dv.$$

Putting $u = \sin^2 \theta$ and $v = \sin^2 \psi$, we see that

$$\int_0^1 u^2 \sqrt{(1-u)}\,du = 2\int_0^{\frac{1}{2}\pi} \sin^5 \theta \cos^2 \theta\,d\theta = \frac{2.4.2.1}{7.5.3.1} = \frac{16}{105},$$

$$\int_0^1 \sqrt{v(1-v)}\,dv = 2\int_0^{\frac{1}{2}\pi} \sin^3 \psi \cos^2 \psi\,dx = \frac{2.1.1}{4.2}\cdot\frac{\pi}{2} = \frac{\pi}{8}.$$

Hence the result.

Note. We have

$$u = x + y, v = x/(x+y) \text{ and } x = uv, y = u(1-v).$$

Thus when x, y are both positive and $(x + y)$ is less than unity, then u, v both lie between 0 and 1. Conversely, when u, v both lie between 0 and 1, then x, y are both positive and $x + y$ is less than 1. Arguing in this manner also, we can obtain the transform of the given region in the (x, y) plane.

5. *Evaluate*

$$\iint (l - x - y)^{l-1} x^{m-1} y^{n-1}\,dx\,dy,$$

taken over the interior of the triangle formed by the lines

$$x = 0, y = 0, x + y = 1;$$

l, m, n being all positive.

Employing the transformation

$$x + y = u,\ x = uv,$$

as in Ex. 4, we see that the given integral

$$= \int_0^1 \int_0^1 (1-u)^{l-1} u^{m+n-1} v^{m-1} (1-v)^{n-1} \, du \, dv$$

$$= \left\{ \int_0^1 (1-u)^{l-1} u^{m+n-1} du \right\} \left\{ \int_0^1 v^{m-1} (1-v)^{n-1} dv \right\}$$

$$= B(l, m+n)\, B(m, n)$$

$$= \frac{\Gamma(l)\,\Gamma(m+n)}{\Gamma(l+m+n)} \cdot \frac{\Gamma(m)\,\Gamma(n)}{\Gamma(m+n)} = \frac{\Gamma(l)\,\Gamma(m)\,\Gamma(n)}{\Gamma(l+m+n)}.$$

EXERCISES I

1. Evaluate

$$\iint \sqrt{\left[x(2a-x) + y(2b-y)\right]} \, dx\, dy,$$

over the interior of the circle $x^2 + y^2 - 2ax - 2by = 0$.

2. Evaluate

$$\iint \sqrt{\left(a^2 - x^2 - y^2\right)} \, dx\, dy,$$

over the interior of the circle $x^2 + y^2 = ax$.

3. Show that the value of the integral

$$\iint (1-x-y)^{3/2}\, x^{\frac{1}{2}}\, y^{\frac{1}{2}} \, dx\, dy,$$

taken over the interior of the triangle whose vertices are the origin and the point (0, 1) and (1, 0) is $\pi/480$.

4. Show that

$$\iint x^{\frac{1}{2}}\, y^{\frac{1}{2}}\, (1-x-y)\, dx\, dy = B\left(\frac{17}{6}, \frac{5}{4}\right) B\left(\frac{3}{2}, \frac{4}{3}\right),$$

over the interior of the triangle bounded by the lines $x = 0$, $y = 0$, $x + y = 1$.

5. Show that

$$\iint \sqrt{\left[x^3 y^4 \left(1 - x^2 - y^2\right)\right]} \, dx\, dy = \frac{1}{4} B\left(\frac{11}{4}, \frac{3}{2}\right) B\left(\frac{5}{4}, \frac{3}{2}\right)$$

over the positive quadrant of the circle $x^2 + y^2 = 1$.

6. Show that

$$\int_0^2 \int_0^x \left\{(x+y+1)^2 - 4xy\right\}^{-\frac{1}{2}} dx\, dy = \frac{1}{2} \log(16/e),$$

by means of the transformation

$$x = u(1+v),\ y = v(1+u),$$

$\partial(x, y)/\partial(u, v) = 1 + u + v$, which is positive, for non-negative values of u, v. The field of integration in the xy-plane is bounded by $y = 0$, $x = 2$, $y = x$. Taking into consideration only non-negative values of u and v, we see that the corresponding region of the uv plane lies in the positive quadrant and is bounded by

$$v = 0,\ u\ (1 + v) = 2,\ u = v$$

Therefore, the integral

$$= \int_0^1 \int_0^u du\, dv + \int_1^2 \int_0^{(2-u)/u} du\, dv.$$

7. Transform the integral

$$= \int_0^{\frac{1}{2}\pi} \int_0^{\frac{1}{2}\pi} \sqrt{\frac{\sin\varphi}{\sin\theta}}\, d\varphi\, d\theta,$$

by the substitution $x = \sin\varphi \cos\theta,\ y = \sin\varphi \sin\theta$, and find its value.

8. Change the variables in the integral

$$\int_0^{2a} dx \int_{\sqrt{(2ax-x^2)}}^{\sqrt{(4ax-x^2)}} \left(1 + \frac{y^2}{x^2}\right) dy,$$

to r and θ, where $x = r\cos^2\theta,\ y = r\sin\theta\cos\theta$ and show that the value of the integral is $\left(\pi + \frac{8}{3}\right)a^2$.

9. Transform the integral

$$\int_0^c \int_0^{c-x} V(x, y)\, dx\, dy,$$

by the substituting $x + y = u$, $y = uv$,

10. In the double integral $\iint f(x, y)\, dx\, dy$, change the variables from x, y to u, v where $x^2 + y^2 = u,\ xy = v$ and find the limits in the new integral when the integration is performed over the positive quadrant of the circle $x^2 + y^2 = a^2$.

11. By using the transformation, $x^2 = u^2 - v^2,\ y = 2uv$ or otherwise, evaluate

$$\iint \frac{dx\, dy}{\sqrt{(x^2 + y^2)}},$$

taken over the region enclosed by arcs of the confocal parabolas

$$y^2 = 4a_r(x + a_r),\ (r = 1, 2, 3)$$

where $a_1 > a_2 > 0,\ a_3 < 0$.

12. Find the area of the curvilinear quadrilateral bounded by the four confocal conics of the system

$$x^2/\lambda + y^2/(\lambda - c^2) = 1,$$

which are determined by giving λ the values $\frac{1}{3}c^2, \frac{2}{3}c^2, \frac{4}{3}c^2, \frac{5}{3}c^2$ respectively.

[Transform into confocal co-ordinates, *i.e.*, express x and y in terms of λ, μ the parameters of the two confocals which pass through (x, y).]

13. Express the integral

$$\int_0^3 dx \int_0^{\sqrt{(5-5x^2/9)}} dy$$

in terms of the variables λ and μ defined by the equations

$$\lambda + \mu = \sqrt{\left[(x+2)^2 + y^2\right]}, \lambda - \mu = \sqrt{\left[(x-2)^2 + y^2\right]}$$

and thus verify that the value of the integral $\frac{3}{4}\pi\sqrt{5}$.

14. Evaluate $\iint xy\,dx\,dy$; the field of integration being the region common to the interiors of the circles

$$x^2 + y^2 = x,\ x^2 + y^2 = y.$$

[Change x, y to u, v where $u = \left(x^2 + y^2\right)/x, v = \left(x^2 + y^2\right)/y$. The new field in the uv plane is the square [0, 1; 0, 1].

15. Evaluate

$$\iint \left(xy^3 - x^3y\right) dx\,dy,$$

where the field of integration lies in the first quadrant and is bounded by the central conics,

$$ax^2 + by^2 = l, ax^2 + by^2 = m, ax^2 - by^2 = n, ax^2 - by^2 = p,$$

$$(l > m > 0,\ n > p > 0).$$

16. Evaluate

$$\iint \left(x^4 - y^4\right) dx\,dy$$

over the part of the positive quadrant in which

$$1 \le \left(x^2 - y^2\right) \le 2, 1 \le xy \le 2.$$

EXERCISES II

1. Show in a diagram the field of integration of the integral

$$\int_0^1 dx \int_x^{1/x} \frac{y^2 dy}{(x+y)^2 \sqrt{(1+y^2)}}.$$

and by changing the order of integration, show that the value of the integral is $\sqrt{2} - \frac{1}{2}$.

2. Change the order of integration in the integral

$$\int_0^1 dy \int_{1-\sqrt{y-1}}^{1+\sqrt{1-y}} \frac{dx}{\left(x^2 - 2x + y - 3\right)^2}$$

and hence evaluate it.

3. Evaluate

$$\iint \left(e^x + e^{2y}\right) dx\, dy$$

over the square [0, 1; 0, 1].

(*i*) directly;

(*ii*) as the limit of a sum, first dividing the square into n^2 equal squares and then making $n \to \infty$.

4. Evaluate

$$\iint_D x^2 y\, dx\, dy \text{ and } \iint_D xy^2\, dx\, dy$$

where D is the region

$$\left\{(x, y) : x^2 + y^2 \le x\right\}.$$

5. Evaluate

$$\iint_D \frac{y^2}{x} dx\, dy$$

where D, is given by

$$\left\{(x, y) ; x^2 + y^2 \le x\right\}.$$

6. Evaluate

$$\iint (x+y)^n\, xy\, dx\, dy,$$

over $\{(x, y); x \ge 0, y \ge 0, x + y \le 1\}$; n being positive.

7. Evaluate

$$\iint_D (2x + 3y)^2\, dx\, dy$$

(*i*) where D is the triangular region bounded by the lines

$$x = 0,\ y = 0,\ 2x + 3y = 1.$$

(*ii*) where D is $\left\{(x, y) : y \geq 0, x^2 + y^2 \leq 1\right\}$.

8. Evaluate

$$\iint_D \sin \pi x \sin \pi y \, dx \, dy,$$

where the region D is

$$\{(x, y) : 0 \leq x,\ 0 \leq y,\ x + y \leq 1\}$$

9. Evaluate

$$\iint_A y^{-1} \, dx \, dy$$

where A is the part of the (x, y) plane that lies between the lines $x = 1$, $x = 2$ and also between the line $y = 4x$ and the curve $y = x^2$.

10. Evaluate

$$\iint \frac{xy\,(x + y)^2}{x^2 + y^2} \, dx \, dy$$

over the sector in the first quadrant bounded by the straight lines

$$y = 0,\ y = x \text{ and the circle } x^2 + y^2 = 1.$$

11. Evaluate

$$\iint \sin \pi \left(x^2 + y^2\right) dx \, dy$$

over the interior of the circle

$$x^2 + y^2 = 1.$$

12. Evaluate

$$\iint_D \sqrt{(xy)} \, dx \, dy$$

where D is

$$\{(x, y) : 0 \leq x, 0 \leq y, x + y \leq 1\}.$$

13. Show that

$$\iint \sqrt{\left(4a^2 - x^2 - y^2\right)} \, dx \, dy = \frac{4}{9}(3\pi - 4)\, a^3$$

taken over the upper half of the circle

$$x^2 + y^2 = 2ax.$$

14. Evaluate

$$\iint_S \left(x^2 + y^2\right) dx \, dy$$

where S is the part of the xy-plane defined by the relations

$$2a \geq x^2 + y^2 \geq a, x \geq 0, a \geq x^2 - y^2 \geq -a, y \geq 0,$$

by means of the transformation

$$x^2 + y^2 = u, \ x^2 - y^2 = v.$$

15. By means of the substitution

$$x = u(1 - v), y = uv$$

prove that

$$\iint \varphi'(x + y) \frac{(xy)^{n-1}}{(x + y)^{2n-1}} dx\, dy, (n \geq 1)$$

over the region defined by

$$x \geq 0, y \geq 0, \ 1 \leq x + y \leq 2$$

within which $\varphi'(x + y)$ is given to be continuous, is equal to

$$\{\varphi(2) - \varphi(1)\} I$$

where

$$I = 2 \int_0^{\pi/2} \sin^{2m-1} \theta \cos^{2n-1} \theta \, d\theta.$$

16. Calculate the integral

$$\iint \frac{(x - y)^2}{x^2 + y^2} dx\, dy.$$

both over the triangular region given by

$$0 \leq x \leq 1, 0 \leq y \leq x$$

and over the circular region given by

$$x^2 + y^2 \leq 1.$$

17. Evaluate

$$\iint_D (x^2 + y^2)\, dx\, dy$$

where D is

$$\left\{(x, y) : \frac{x^2}{a^2} + \frac{y^2}{b^2} \leq 1\right\},$$

18. Evaluate

$$\iint_D \sin x \sin y \sin(x + y)\, dx\, dy$$

where D is

$$\{(x, y) : x \geq 0, y \geq 0, x + y \leq \pi/2\}.$$

19. Find the area in the first quadrant between the curves

$$x^2 + y^2 = a^2, x^2 + y^2 = 2a^2,$$
$$x^2 - y^2 = a^2, x^2 - y^2 = -a^2.$$

20. By means of a suitable substitution, or otherwise, evaluate the integral

$$\int_0^1 dx \int_0^{1-x} e^{(y-x)/(y+x)}\, dy.$$

21. Evaluate

$$\iint \frac{dx\, dy}{a^2 + x^2 + y^2},$$

taken over the circular region $x^2 + y^2 \le 1$ and hence, by equating the double integral to an equivalent line integral, show that

$$\int_0^{2\pi} \frac{\sin\theta}{\sqrt{(1 - k^2 \sin^2\theta)}} \sin^{-1}\left[\sin(k \sin\theta)\right] d\theta = -\frac{\pi}{4} \log\left(1 - k^2\right), 0 < k < 1.$$

22. Prove that

$$\iint_D x^{m-1} y^{n-1} f(x + y)\, dx\, dy = \frac{\Gamma(m)\Gamma(n)}{\Gamma(m+n)} \int_0^1 t^{m+n-1} f(t)\, dt,$$

where D is

$$\{(x, y) : x, y \ge 0, x + y \le 1\}.$$

23. Prove that if $a, b, h, ab - h^2$ are all positive.

$$\int_0^{\infty} \int_0^{\infty} e^{-\left(ax^2 + 2hxy + by^2\right)}\, dx\, dy = \frac{\pi}{2\sqrt{(ab - h^2)}}.$$

24. Prove that

$$\iint_D x^n\, dx\, dy = \frac{2}{(n+1)(n+2)} H_n \,\|D\|$$

where D is the interior of the triangle with vertices

$$(x_1, y_1), (x_2, y_2), (x_3, y_3),$$

$\|D\|$ is its area and H_n is the sum of the homogeneous products of degree n in x_1, x_2, x_3.

25. Prove that

$$\int_0^{\frac{1}{2}\pi} \int_0^{\frac{1}{2}\pi} \frac{m^2 \cos^2\varphi + n^2 \cos^2\theta}{\left[\left(1 - m^2 \cos^2\varphi\right)\left(1 - n^2 \cos^2\theta\right)\right]}\, d\theta\phi = \frac{\pi}{2}.$$

OBJECTIVE QUESTIONS

Note. *For each of the following questions, four alternatives are given for the answer. Only one of them is correct. Choose the correct alternative.*

1. $\int_{AC} f(x, y)\, dx =$

(*a*) $-\int_{CA} f(x, y)\, dx$ (*b*) $\int_{CA} f(x, y)\, dx$

(*c*) both (*a*) and (*b*) are true. (*d*) none of (*a*) or (*b*) is true

2. The supremum of the set of sums $S(D)$ is called

(*a*) inner area (*b*) outer area

(*c*) area (*d*) none of these

3. A function *f* is said to be integrable if

(*a*) upper integral > lower integral (*b*) lower integral > upper integral

(*c*) lower integral = upper integral (*d*) none of these

4. If a double integral exists, then the two repeated integrals :

(*a*) do not exist (*b*) exist but may not be equal

(*c*) exist and are equal (*d*) may or may not exist.

5. A simple closed continuous curve is called

(*a*) simple curve (*b*) continuous curve

(*c*) Jordan curve (*d*) none of these

6. If *f* be continuous is a region *E* with area *A*, then there exists a point $(\xi, \eta) \in E$ such that

(*a*) $\iint_E A f(x, y)\, dx\, dy = f(\xi, \eta)$ (*b*) $\iint_E A f(x, y)\, dx\, dy \neq f(\xi, \eta)$

(*c*) $\iint_E f(x, y)\, dx\, dy = A f(\xi, \eta)$ (*d*) $\iint_E f(x, y)\, dx\, dy \neq A f(\xi, \eta)$

7. The double integral $\int_0^1 \int_0^1 (x^2 + y^2)\, dx\, dy$ is equal to :

(*a*) 0 (*b*) 1 (*c*) $\frac{1}{3}$ (*d*) $\frac{2}{3}$

8. The area bounded by the curve $y = \phi(x)$, the *x*-axis and the lives $x = a$ and $x = b$ $(a < b)$ is given by :

(*a*) $\int_a^b \int_0^{\phi(x)} y\, dx\, dy$ (*b*) $\int_a^b \int_0^{\phi(x)} dx\, dy$

(*c*) $\int_a^b \int_0^{\phi(x)} x\, dx\, dy$ (*d*) none of these

9. Which of the following relations is true?

(*a*) $\iint_A f(x, y)\, dx\, dy = \iint_A f(r\cos\theta, r\sin\theta)\, d\theta\, dr$

(*b*) $\iint_A f(x, y)\, dx\, dy = \iint_A f(r\sin\theta, r\cos\theta)\, d\theta\, dr$

(c) $\iint_A f(x, y)\, dx\, dy = \iint_A f(r\cos\theta, r\sin\theta)\, r\, d\theta\, dr$

(d) $\iint_A f(x, y)\, dx\, dy = \iint_A f(r\cos\theta, r\sin\theta)\, r^2\, d\theta\, dr$

10. The volume of a region represented as :

$a \le x \le b,\ y_1(x) \le y \le y_2(x),\ z_1(x, y) \le z \le z_2(x, y)$ is given by :

(a) $\int_a^b \int_{y_1(x)}^{y_2(x)} \int_{z_1(x,y)}^{z_2(x,y)} dx\, dz\, dy$ (b) $\int_a^b \int_{y_1(x)}^{y_2(x)} \int_{z_1(x,y)}^{z_2(x,y)} dx\, dy\, dz$

(c) $\int_a^b \int_{y_1(x)}^{y_2(x)} \int_{z_1(x,y)}^{z_2(x,y)} dz\, dx\, dy$ (d) none of these

11. The integral $\iint \left(\frac{1 - x^2 - y^2}{1 + x^2 + y^2} \right) dx\, dy$, extended over all positive values of x and y such that $x^2 + y^2 \le 1$ can be evaluated by changing the variables x and y by using the substitution :

(a) $x + y = u,\ xy = v$ (b) $x = r\sin\theta,\ y = r\cos\theta$.

(c) $x = r\cos\theta,\ y = r\sin\theta$ (d) none of these

12. If $x = f_1(u, v, w)$, $y = f_2(u, v, w)$ and $z = f_3(u, v, w)$, then $dx\, dy\, dz$ is equal to :

(a) $du\, dv\, dw$ (b) $u\, v\, w\, du\, dv\, dw$

(c) $\left[\frac{\partial(u, v, w)}{\partial(x, y, z)} \right] du\, dv\, dw$ (d) $\left[\frac{\partial(x, y, z)}{\partial(u, v, w)} \right] du\, dv\, dw$

13. To change a given double integral from Cartesian to polar coordinates; we use the formula

(a) $dx\, dy = d\theta\, dr$ (b) $dx\, dy = r\, d\theta\, dr$

(c) $dx\, dy = r^2\, d\theta\, dr$ (d) $r\, dx\, dy = d\theta\, dr$

14. For transformation $x + y = u,\ y = uv$, the value of $dx\, dy$ is :

(a) $du\, dv$ (b) $u^2\, du\, dv$

(c) $\frac{1}{u} du\, dv$ (d) $u\, du\, dv$

15. Value of $\int_{-c}^{c} \int_{-b}^{b} \int_{-a}^{a} (x^2 + y^2 + z^2)\, dx\, dy\, dz$ is :

(a) 0 (b) $\frac{2}{3} abc(a^2 + b^2 + c^2)$

(c) $\frac{8}{3} abc(a^2 + b^2 + c^2)$ (d) none of these

ANSWERS

1. (b)	**2.** (a)	**3.** (c)	**4.** (c)	**5.** (c)
6. (c)	**7.** (d)	**8.** (b)	**9.** (c)	**10.** (b)
11. (c)	**12.** (d)	**13.** (b)	**14.** (d)	**15.** (c)

17

Curve Lengths, Surface Areas

17.1. RECTIFIABILITY OF A CURVE

Consider

$$x = f(t),\ y = g(t),\ z = h(t)$$

where *f, g, h* are three functions of *t* defined and continuous in an interval $[a, b]$. Then the set

$$\{(x, y, z) : x = f(t),\ y = g(t),\ z = h(t);\ t \in [a, b]\}$$

is called a continuous curve in the three-dimensional space $\mathbf{R}^3$. We shall be concerned with continuous curves only.

Let

$$D\{a = t_0, t_1, \ldots, t_{r-1},\ t_r, \ldots, t_n = b\}, \qquad \ldots(1)$$

be any division of $[a, b]$.

We write

$$x_r = f(t_r),\ y_r = g(t_r),\ z_r = h(t_r)$$

so that corresponding to a division (1) D of $[a, b]$, we get an ordered set of points

$$(x_0, y_0, z_0), \ldots, (x_r, y_r, z_r), \ldots, (x_n, y_n, z_n)$$

on the curve. This set of points may be referred to as a polygon inscribed in the curve. We write

$$\pi(D) = \sum_{r=1}^{n} \sqrt{\left[(x_r - x_{r-1})^2 + (y_r - y_{r-1})^2 + (z_r - z_{r-1})^2\right]},$$

so that $\pi(D)$ denotes the length of inscribed polygon in question.

Def. *The curve is said to be rectifiable, if the set of numbers* $\pi(D)$ *obtained on considering all possible division D of* $[a, b]$ *is bounded and, in the event of boundedness, the supremum is known as the length of the curve.*

Characterisation of Rectifiable curves. Theorem. *Necessary and sufficient conditions for the curve*

$$x = f(t),\ y = g(t),\ z = h(t), \qquad a \le t \le b$$

to be rectifiable are that the functions f, g, h are all of bounded variation over [a, b].

The conditions are necessary. Let the curve be rectifiable and let l denote the length of the curve.

We have, for any division D,

$$\left|x_r - x_{r-1}\right| \le \sqrt{\left[\left(x_r - x_{r-1}\right)^2 + \left(y_r - y_{r-1}\right)^2 + \left(z_r - z_{r-1}\right)^2\right]},$$

so that summing over r, we obtain

$$\sum_{r=1}^{n} \left|x_r - x_{r-1}\right| \le \sum_{r=1}^{n} \sqrt{\left[\left(x_r - x_{r-1}\right)^2 + \left(y_r - y_{r-1}\right)^2 + \left(z_r - z_{r-1}\right)^2\right]} \le l$$

$$\Rightarrow \qquad \sum_{r=1}^{n} \left|f(t_r) - f\left(t_{r-1}\right)\right| \le l, \qquad \text{...(i)}$$

for each division D of $[a, b]$. Thus the function f is of bounded variation over $[a, b]$.

Similarly the functions g and h are of bounded variation over $[a, b]$.

The conditions are sufficient. Let the three functions be of bounded variation.

We have, as may be easily shown,

$$\sqrt{\left[\left(x_r - x_{r-1}\right)^2 + \left(y_r - y_{r-1}\right)^2 + \left(z_r - z_{r-1}\right)^2\right]} \le \left|x_r - x_{r-1}\right| + \left|y_r - y_{r-1}\right| + \left|z_r - z_{r-1}\right|$$

so that on summing up, we easily obtain

$$\pi(D) \le \sum_{r=1}^{n} \sqrt{\left[\left(x_r - x_{r-1}\right)^2 + \left(y_r - y_{r-1}\right)^2 + \left(z_r - z_{r-1}\right)\right]^2}$$

$$\le \sum_{r=1}^{n} \left|x_r - x_{r-1}\right| + \sum_{r=1}^{n} \left|y_r - y_{r-1}\right| + \sum_{r=1}^{n} \left|z_r - z_{r-1}\right|,$$

$$\le V_f\left[a, b\right] + V_g\left[a, b\right] + V_h\left[a, b\right], \qquad \text{...(ii)}$$

where $V_f\left[a, b\right] + V_g\left[a, b\right], V_h\left[a, b\right]$, denote the variations of the functions f, g, h over $[a, b]$.

Thus $\pi(D)$ is bounded above and the curve is rectifiable.

Cor. 1. *If a curve*

$$x = f(t),\ y = g(t),\ z = h(t), \qquad a \le t \le b$$

is rectifiable and, l, denotes the length thereof then from (i) and (ii), i follows that

$$V_f[a, b] \le l,\ V_g[a, b] \le l,\ V_h[a, b] \le l \qquad ...(1$$

$$l \le V_f[a, b] + V_g[a, b] + V_h[a, b] \qquad ...(2$$

Cor. 2. *The curve*

$$x = f(t),\ y = g(t),\ z = h(t),$$

is rectifiable if the functions f, g, h are derivable with bounded derivative

It may be shown that under the given conditions, the functions f, g, h are of bounded variation.

17.2. PROPERTIES OF RECTIFIABLE CURVES

Arc of a curve. If

$$x = f(t),\ y = g(t),\ z = h(t) \qquad a \le t \le$$

is a curve C and $[c, d]$ is any sub-interval of $[a, b]$, then the curve give by

$$x = f(t),\ y = g(t),\ z = h(t)$$

when, t, $\in [c, d]$ is said to be an *arc* of the curve C.

Theorem. *Each arc of a rectifiable curve is rectifiable.*

This follows from the fact that f is of bounded variation over $[a, b$ then it is also of bounded variation over each sub-interval thereof.

Theorem. *If a curve C*

$$x = f(t),\ y = g(t),\ z = h(t), \qquad a \le t \le b$$

is rectifiable and c, is any point of $[a, b]$, *then*

$$l[a, b] = l[a, c] + l[c, b],$$

where l [a, c] denotes the length of the arc of the curve arising for t lyi in [a, c] and similar meanings are given to l [c, b] and l [a, b].

Let D be *any* division of $[a, b]$ and let D' denote the division of $[a,$ consecutive to D with c as the only additional point, if it were not alrea a point of D. The division D' of $[a, b]$ gives rise to divisions of $[a, c]$ a $[c, b]$ which we may denote by D_1 and D_2 respectively. We have then

$$\pi(D) \le \pi(D') = \pi(D_1) + \pi(D_2).$$

Also

$$\pi(D_1) \le l[a, c],\ \pi(D_2) \le l[c, b],$$

$\therefore \quad \pi(D) \leq l[a, c] + l[c, b].$

This gives

$$l[a, b] \leq l[a, c] + l[c, b]. \quad \ldots(1)$$

Now suppose that $\varepsilon > 0$ is any given number.

There exists divisions D_3 and D_4 of $[a, c]$ and $[c, b]$ respectively such that

$$\pi(D_3) > l[a, c] - \frac{1}{2}\varepsilon, \pi(D_4) > l[c, b] - \frac{1}{2}\varepsilon.$$

Let D'' denote the division of $[a, b]$ arising from the divisions D_3 and D_4 of $[a, c]$ and $[c, b]$ respectively. We have then

$$\pi(D'') = \pi(D_3) + \pi(D_4) > l[a, c] + l[c, b] - \varepsilon.$$

$$\therefore \quad l[a, b] \geq l[a, c] + l[c, b] - \varepsilon.$$

As $\varepsilon > 0$ is arbitrary, we obtain

$$l[a, b] \geq l[a, c] + l[c, b]. \quad \ldots(2)$$

From (1) and (2), we obtain

$$l[a, b] = l[a, c] + l[c, b],$$

as was to be proved.

Theorem. *The function with function value*

$$l[a, t] \cdot t \in [a, b],$$

is a monotonically increasing continuous function in $[a, b]$.

Let t_1, t_2 be two numbers such that $a \leq t_1 < t_2 \leq b$.

We have

$$l[a, t_2] = l[a, t_1] + l[t_1, t_2],$$

$$\Rightarrow \quad l[a, t_2] \geq l[a, t_1], \text{ for } l[t_1, t_2] \geq 0.$$

Thus the function with function value $l[a, t]$ at t is a monotonically increasing function of t.

Suppose, now, that c, $c + h$ are any two points of $[a, b]$.

We have, by Cor. 1 of page 501,

(*i*) if $h > 0$,

$$\begin{aligned} 0 &\leq l[c, c+h] \\ &\leq V_f[c, c+h] + V_g[c, c+h] + V_h[c, c+h] \\ &= \left[V_f[a, c+h] - V_f[a, c]\right] + \left[V_g[a, c+h] - V_g[a, c]\right] \\ &\qquad + \left[V_h[a, c+h] - V_h[a, c]\right] \end{aligned}$$

$$=\left[V_f(c+h)-V_f(c)\right]+\left[V_g(c+h)-V_g(c)\right]+\left[V_h'(c+h)-V_h(c)\right]$$

(*ii*) if $h<0$,

$$0\le l[c+h,c]\le\left[V_f(c)-V_f(c+h)\right]+\left[V_g(c)-V_g(c+h)\right]$$
$$+\left[V_h(c)-V_h(c+h)\right]$$

From these, we see that

$$l[a,c+h]-l[a,c]\le\left|V_f(c+h)-V_f(c)\right|+\left|V_g(c+h)-V_g(c)\right|$$
$$+\left|V_h(c+h)-V_h(c)\right|$$

Now, as shown in § 4.8.6, V_f, V_g and V_h are continuous at, c, and accordingly for given $\varepsilon>0$, there exists $\delta>0$ such that for $|h|\le\delta$,

$$\left|V_f(c+h)-V_f(c)\right|<\frac{1}{3}\varepsilon,\left|V_g(c+h)-V_g(c)\right|<\frac{1}{3}\varepsilon$$

$$\left|V_h(c+h)-V_h(c)\right|<\frac{1}{3}\varepsilon.$$

Thus for $|h|\le\delta$, we have

$$\left|l[a,c+h]-l[a,c]\right|<\varepsilon,$$

so that $l\ [a,\ t]$ is continuous for $t=c$.

Hence the theorem.

Note. 1. It can be seen that *the function with value $l\ [a,\ t]$ is a strictl[y] monotonically increasing function of t, unless f, g, h are all constant over som[e] sub-interval of* $[a, b]$. Now

$$l[a,t_1]=l[a,t_2]. \qquad t_2>t_1$$

$$\Rightarrow \qquad l[t_1,t_2]=0.$$

$$\Rightarrow \qquad V_f[t_1,t_2]=0,V_g[t_1,t_2]=0,V_h[t_1,t_2]=0.$$

[Cor. 1, § 17.1, page 501]

$\Rightarrow$ f, g, h are constant over $[t_1, t_2]$.

Note. 2. It is usual to write $s\ (t)=l\ [a,\ t]$, so that $s\ (t)$ denotes the lengt[h] of the arc corresponding to the sub-interval $[a,\ t]$ of $[a,\ b]$.

Supposing that f, g, h are such that there is no sub-interval of $[a,\ b]$ for which they are all constant, s is a *strictly* monotonically increasin[g] continuous function, and increases from 0 to l as t increases from a to [b]. By § 4.7.1., page 88, we see that, t, can itself be regarded as a continuou[s] function of s. If we write

$$t=\varphi(s), \qquad 0\le s\le 1.$$

we see that

$$x=f(t)=f[\varphi(x)]=F(s),\text{ say}$$

$$y = g(t) = g[\varphi(s)] = G(s),$$
$$z = h(t) = h[\varphi(s)] = H(s),$$

so that the given rectifiable curve is as well given by

$$x = F(s), y = g(s), z = H(s),$$

where the *parameter is now the arc length s.*

17.3. INTEGRAL EXPRESSION FOR THE LENGTH OF A CURVE

Theorem. *The length of the curve*

$$x = f(t), y = g(t), z = h(t), a \le t \le b$$

is

$$\int_a^b \sqrt{[f'^2(t) + g'^2(t) + h'^2(t)\,dt]}$$

if f, g, h are derivable with continuous derivatives and for no value of t,

$$f'^2(t) + g'^2(t) + h'^2(t) = 0.$$

Clearly f, g, h are under the given conditions, functions with bounded lerivatives, and accordingly they are functions with bounded variation. Thus the curve is rectifiable.

We shall now proceed to find an integral expression for the length of he curve.

Let

$$D = \{a = t_0, ..., t_{r-1}, t_r, ..., t_n = b\},$$

y any division of $[a, b]$.

We have, by Lagrange's mean value theorem,

$$x_r - x_{r-1} = f(t_r) - f(t_{r-1})$$
$$= (t_r - t_{r-1}) f'(\alpha_r),$$
$$y_r - y_{r-1} = (t_r - t_{r-1}) g'(\beta_r),$$
$$z_r - z_{r-1} = (t_r - t_{r-1}) h'(\gamma_r)$$

here $\alpha_r, \beta_r, \gamma_r$ belong to the sub-interval $[t_{r-1}, t_r]$.

$$\therefore \quad \pi(D) = \sum_{r=1}^{n} \sqrt{\left[(x_r - x_{r-1})^2 + (y_r - y_{r-1})^2 + (z_r - z_{r-1})^2\right]}$$
$$= \sum_{r=1}^{n} (t_r - t_{r-1}) \sqrt{[f'^2(\alpha_r) + g'^2(\beta_r) + h'^2(\gamma_r)]}.$$

Let ξ_r be any point of $[t_{r-1}, t_r]$.

We write $\quad a_r = f'(\alpha_r) \quad b_r = g'(\beta_r), \quad c_r = h'(\gamma_r);$

$$a_r = f'(\xi_r); \qquad b_r' = g'(\xi_r), \qquad c_r' = h(\xi_r).$$

We have

$$\sqrt{\left(a_r^{\,2} + b_r^{\,2} + c_r^{\,2}\right)} - \sqrt{a_r'^2 + b_r'^2 + c_r'^2}$$

$$= \frac{\left(a_r^{\,2} + b_r^{\,2} + c_r^{\,2}\right) - \left(a_r'^2 + b_r'^2 + c_r'^2\right)}{\sqrt{\left(a_r^{\,2} + b_r^{\,2} + c_r^{\,2}\right)} + \sqrt{\left(a_r'^2 + b_r'^2 + c_r'^2\right)}}.$$

$$= \frac{(a_r - a_r')(a_r + a_r') + (b_r - b_r')(b_r + b_r') + (c_r - c_r')(c_r + c_r')}{\sqrt{\left(a_r^{\,2} + b_r^{\,2} + c_r^{\,2}\right)} + \sqrt{\left(a_r'^2 + b_r'^2 + c_r'^2\right)}}.$$

Suppose that k is a number such that $\forall\, t \in [a, b]$

$$|f'(t)| \le k, |g'(t)| \le k, |h'(t)| \le k$$

Also since

$$[f'(t)]^2 + [g'(t)]^2 + [h'(t)]^2 \neq 0,$$

for any value of t, and is continuous, there exists a positive number K^2 viz the lower bound of the same, such that $\forall\, t \in [a, b]$.

$$[f'(t)]^2 + [g'(t)]^2 + [h'(t)]^2 \ge K^2.$$

Thus we obtain

$$\left|\left[\sqrt{\left(a_r^{\,2} + b_r^{\,2} + c_r^{\,2}\right)} - \sqrt{\left(a_r'^2 + b_r'^2 + c_r'^2\right)}\right]\right|$$

$$\le \frac{2k}{K}\,|(a_r - a_r')| + |(b_r - b_r')| + |(c_r - c_r')|.$$

We write

$$S(D) = \sum_{r=1}^{n} (t_r - t_{r-1}) \sqrt{\left[f'^2(\xi_r) + g'^2(\xi_r) + h'^2(\xi_r)\right]}$$

$$\therefore \qquad |\pi(D) - S(D)| \le \sum_{r=1}^{n} |t_r - t_{r-1}| \frac{2k}{K} \{|f'(\alpha_r) - f'(\xi_r)|$$

$$+ |g'(\beta_r) - g'(\xi_r)| + |h'(\gamma_r) - h'(\xi_r)|\}$$

Let $\varepsilon > 0$ be given. Because of the uniform continuity of $f', g',$ in $[a, b]$, there exists $\delta > 0$ such that

$$|f'(t') - f'(t'')| < \varepsilon, |g'(t') - g'(t'')| < \varepsilon, |h'(t') - h'(t'')| \le \varepsilon$$

when $|t'' - t'| < \delta$.

Now choosing D as any division of norm $\leq \delta$, we see that

$$\left|\pi(D) - S(D)\right| \leq \sum_{r=1}^{n} \left|t_r - t_{r-1}\right| \frac{2k}{K} \cdot 3\varepsilon = \frac{6k\varepsilon}{K}(b-a).$$

Thus

$$\lim\left[\pi(D) - S(D)\right] = 0,$$

as the norm of D tends to 0.

Also we know that

$$\lim S(D) = \int_a^b \sqrt{\left[f'^2(t) + g'^2(t) + h'^2(t)\right]}\, dt,$$

at the norm of D tends to zero. Thus we have

$$\lim \pi(D) = \int_a^b \sqrt{\left[f'^2(t) + g'^2(t) + h'^2(t)\right]}\, dt \qquad ...(1)$$

as the norm of D tends to zero.

We denote the integral in (1) by I, and show that I is the supremum of the numbers $\pi(D)$.

Firstly the supremum cannot be less than I, for we can choose $\pi(D)$ as near I as we like.

Secondly the supremum cannot exceed I, for, while on the one hand $\pi(D)$ increases on making D finer, it can also be made as near I as we like by choosing sufficiently fine divisions.

[The reader may also work out a straight proof.]

Smooth curves. Def. *A curve*

$$x = f(t), y = g(t), z = h(t), a \leq t \leq b$$

is said to be smooth, if the three functions f, g, h possess continuous derivatives in [a, b] and at no point thereof

$$f'^2(t) + g'^2(t) + h'^2(t),$$

is zero.

Note. The condition of the non-vanishing of

$$f'^2(t) + g'^2(t) + h'^2(t)$$

is equivalent to saying that for no value t, the three derivatives

$$f'(t), g'(t), h'(t)$$

simultaneously vanish.

The reader having some knowledge of differential geometry of curves may recall that geometrically interpreted, this condition means that the curve has a definite direction at each point thereof and is free from multiple points.

Ex. Find the length of the following curves.

1. $x = a \cos \theta, y = a \sin \theta, z = a\theta;$ $\quad 0 \le \theta \le 2\pi.$
2. $x = at^2, y = 2at, z = at;$ $\quad 0 \le t \le 1,$
3. $x = a(\theta - \sin \theta), y = a(1 - \cos \theta), z = a\theta,$ $\quad -\pi \le \theta \le \pi.$

17.4. SURFACES

We have seen that a curve in $\mathbf{R}^3$ is the range of a vector valued function whose domain is a sub-set of $\mathbf{R}$ and range a sub-set of $\mathbf{R}^3$.

We shall now extend this idea and define a *surface* in $\mathbf{R}^3$.

The range of a vector valued function whose domain is a sub-set of $\mathbf{R}^2$ *and range of sub-set of* $\mathbf{R}^3$ *is called a surface* $\mathbf{R}^3$.

We consider three real valued functions f, g, h with some domain $D \subset \mathbf{R}^2$ and write

$$x = f(u, v), y = g(u, v), z = h(u, v); (u, v) \in D.$$

Then the set

$$\{[f(u, v), g(u, v), h(u, v)] : (u, v) \in D\}$$

is a surface in $\mathbf{R}^3$.

It is usual to say that while a curve in $\mathbf{R}^3$ is a set with one degree of freedom, a surface in $\mathbf{R}^3$ is a set with two degrees of freedom.

Associated with the concept of surface is that of surface area just as curve length is associated with a curve. The subject of surface area is, however, beset with difficulties and it is not possible to adopt a straight generalisation of the method employed for defining the lengths of curves by means of inscribed polygons. Instead of adopting an elementary but rather difficult approach to the concept of surface area, we shall take to the more sophisticated one as given below.

17.4.1. SURFACE AREA

Def. *The area of the surface*

$$x = f(u, v), y = g(u, v), z = h(u, v); (u, v) \in D$$

is the double integral

$$\iint_D \sqrt{\left\{\left[\frac{\partial(y, z)}{\partial(u, v)}\right]^2 + \left[\frac{\partial(z, x)}{\partial(u, v)}\right]^2 + \left[\frac{\partial(x, y)}{\partial(u, v)}\right]^2\right\}}\, du\, dv,$$

assuming that the functions f, g, h possess continuous first order partial derivatives in D and at no point of D,

$$\left[\frac{\partial(y,z)}{\partial(u,v)}\right]^2+\left[\frac{\partial(z,x)}{\partial(u,v)}\right]^2+\left[\frac{\partial(x,y)}{\partial(u,v)}\right]^2=0.$$

(Compare with the expression for curve length in § 17.3, page 504).

Note. Here, we do not propose to *derive* the expression as given above by basing the derivation on some *elementary* notion of surface area. Lest the definition as given appear very arbitrary, we outline some considerations to call forth reader's *faith* in the same.

I. Let the surface be plane. We take

$$x = u,\ y = v,\ z = 0$$

where (u, v) ranges over a domain D in the XY-plane. In fact, the surface coincides with D in the present case.

We have

$$\frac{\partial(x,y)}{\partial(u,v)}=\begin{vmatrix}1 & 0\\ 0 & 1\end{vmatrix}=1,\ \frac{\partial(y,z)}{\partial(u,v)}=0,\ \frac{\partial(z,x)}{\partial(u,v)}=0,$$

$$\Rightarrow \quad \iint_D \sqrt{\left\{\left[\frac{\partial(y,z)}{\partial(u,v)}\right]^2+\left[\frac{\partial(z,x)}{\partial(u,v)}\right]^2+\left[\frac{\partial(x,y)}{\partial(u,v)}\right]^2\right\}}\,du\,dv$$

$$=\iint_D du\,dv = \text{area of } D.$$

[Refer to an article by T.Rado, '*What is an Area of the surface*' in American mathematical monthly, vol. 50, pp. 139—141 and to an article by J.W.T. Young, '*Curves and Surfaces*' in American mathematical monthly, vol. 51, pp. 1 — 11, 1944].

Thus the definition of surface area as now given agrees with that of areas of plane regions as given in (§ 16.3 p. 459).

II. Consider the surface given by

$$z = h\,(x, y).$$

We take it as given by

$$z = u,\ y = v,\ z = h\,(u, v)$$

where $(u, v) = (x, y) \in D \subset \mathbf{R}^3$.

We have

$$\frac{\partial(y,z)}{\partial(u,v)}=-\frac{\partial z}{\partial u}=-\frac{\partial z}{\partial z},$$

$$\frac{\partial(z,x)}{\partial(u,v)}=-\frac{\partial z}{\partial v}=-\frac{\partial z}{\partial y}.$$

$$\frac{\partial(x,y)}{\partial(u,v)}=1.$$

$$\therefore \quad \text{surface area} = \iint_D \left[1 + \left(\frac{\partial z}{\partial x}\right)^2 + \left(\frac{\partial x}{\partial y}\right)^2\right]^{\frac{1}{2}} dx\, dy. \qquad ...(1)$$

The reader aquainted with the elements of Differential Geometry knows that

$$-\frac{\partial z}{\partial x}, -\frac{\partial z}{\partial y}, 1$$

are the direction ratios of the normal to the surface at any point (x, y, z) thereof, so that if we suppose $\delta\sigma$ to be an element of surface lying on the tangent plane at the point, the projection area δA of $\delta\sigma$ on the XY-plane is given by

$$\delta A = \frac{\delta\sigma}{\sqrt{\left[1 + \left(\frac{\partial z}{\partial x}\right)^2 + \left(\frac{\partial z}{\partial y}\right)^2\right]}}$$

$$\Rightarrow \quad \delta\sigma = \sqrt{\left[1 + \left(\frac{\partial z}{\partial x}\right)^2 + \left(\frac{\partial z}{\partial y}\right)^2\right]}\, \delta A$$

$$= \sqrt{\left[1 + \left(\frac{\partial z}{\partial x}\right)^2 + \left(\frac{\partial z}{\partial y}\right)^2\right]}\, \delta x\, \delta y.$$

This suggests, on summation, the truth of (1).

Note. The reader may easily show that

$$\left[\frac{\partial(y, z)}{\partial(u, v)}\right]^2 + \left[\frac{\partial(z, x)}{\partial(u, v)}\right]^2 + \left[\frac{\partial(x, y)}{\partial(u, v)}\right]^2 = EG - F^2,$$

where

$$E = \left(\frac{\partial x}{\partial u}\right)^2 + \left(\frac{\partial y}{\partial u}\right)^2 + \left(\frac{\partial z}{\partial u}\right)^2,$$

$$F = \frac{\partial x}{\partial u} \cdot \frac{\partial x}{\partial v} + \frac{\partial y}{\partial u} \cdot \frac{\partial y}{\partial v} + \frac{\partial z}{\partial u} \cdot \frac{\partial z}{\partial v},$$

$$G = \left(\frac{\partial x}{\partial v}\right)^2 + \left(\frac{\partial y}{\partial v}\right)^2 + \left(\frac{\partial z}{\partial v}\right)^2.$$

Smooth surface. Def. *A surface given by*

$$x = f(u, v),\ y = g(u, v),\ z = h(u, v);\ (u, v) \in D$$

is said to be smooth, if f, g, h possess continuous first order partia derivatives at each point of D and at no point thereof

$$\frac{\partial(y,z)}{\partial(u,v)}, \frac{\partial(z,x)}{\partial(u,v)}, \frac{\partial(x,y)}{\partial(u,v)}$$

vanish simultaneously.

EXAMPLES

1. *Surface area of the sphere given by*

$$x = a\sin\theta\cos\varphi,\ y = a\sin\theta\sin\varphi,\ z = a\cos\theta$$

where

$$0 \le \theta \le \pi,\ 0 \le \varphi \le 2\pi.$$

It may be easily seen that

$$E = \left(\frac{\partial x}{\partial\theta}\right)^2 + \left(\frac{\partial y}{\partial\theta}\right)^2 + \left(\frac{\partial z}{\partial\theta}\right)^2 = a^2,$$

$$F = \frac{\partial x}{\partial\theta}\frac{\partial x}{\partial\varphi} + \frac{\partial y}{\partial\theta}\frac{\partial y}{\partial\varphi} + \frac{\partial z}{\partial\theta}\frac{\partial z}{\partial\varphi} = 0,$$

$$G = \left(\frac{\partial x}{\partial\varphi}\right)^2 + \left(\frac{\partial y}{\partial\varphi}\right)^2 + \left(\frac{\partial z}{\partial\varphi}\right)^2 = a^2\sin^2\theta.$$

$$\therefore \quad \text{Surface area} = \int_0^{2\pi}\int_0^{\pi}\sqrt{(EG - F^2)}\,d\theta\,d\varphi$$

$$= a^2\int_0^{2\pi}\int_0^{\pi}\sin\theta\,d\theta\,d\varphi = 4a^2\pi.$$

2. *Find the area of the surface of the sphere $x^2 + y^2 + z^2 = a^2$ which lies inside the cylinder $x^2 + y^2 = ay$.*

$$x^2 + y^2 + z^2 = a^2$$

$$\therefore \quad \frac{\partial z}{\partial x} = -\frac{x}{z},\ \frac{\partial z}{\partial y} = -\frac{y}{z}.$$

$$\therefore \quad \sqrt{1 + \left(\frac{\partial z}{\partial x}\right)^2 + \left(\frac{\partial z}{\partial y}\right)^2} = \sqrt{1 + \frac{x^2}{z^2} + \frac{y^2}{z^2}}$$

$$= \frac{1}{z}\sqrt{x^2 + y^2 + z^2} = \frac{a}{\sqrt{(a^2 - x^2 - y^2)}}$$

$$\therefore \quad S = 4\iiint \frac{a\,dx\,dy}{\sqrt{(a^2 - x^2 - y^2)}}$$

over half the cylinder $x^2 + y^2 = ay$.

Now, let $x = r\cos\theta,\ y = r\sin\theta,\ x^2 + y^2 = ay$ becomes $r = a\sin\theta$ and $dx\,dy = r\,d\theta\,dr$.

$$\therefore \quad S = 4a \int_0^{\pi/2} \int_0^{a \sin\theta} \frac{r\, d\theta\, dr}{\sqrt{(a^2 - r^2)}}$$

$$= 4a \int_0^{\pi/2} \left[-\sqrt{(a^2 - r^2)} \right]_0^{a \sin\theta} d\theta$$

$$= 4a^2 \int_0^{\pi/2} (1 - \cos\theta)\, du = 4a^2 \left[\theta - \sin\theta\right]_0^{\pi/2}$$

$$= 2a^2 (\pi - 2).$$

3. *Prove that the area of the surface $z^2 = 2xy$ included between the planes $x = 0$, $x = a$, $y = 0$, $y = b$ is $4\sqrt{(ab)}\,\dfrac{a+b}{3\sqrt{2}}$.*

$$z^2 = 2xy.$$

$$\therefore \quad \frac{\partial z}{\partial x} = \frac{y}{z}; \frac{\partial z}{\partial y} = \frac{x}{z}$$

$$\therefore \quad S = \int_0^a \int_0^b \sqrt{\left(1 + \frac{y^2}{z^2} + \frac{x^2}{z^2}\right)}\, dx\, dy$$

$$= \int_0^a \int_0^b \sqrt{\left(\frac{x^2 + y^2 + z^2}{z^2}\right)}\, dx\, dy$$

$$= \int_0^a \int_0^b \sqrt{\left(\frac{2xy + x^2 + y^2}{2xy}\right)}\, dx\, dy$$

$$= \int_0^a \int_0^b \frac{x + y}{\sqrt{2(xy)}}\, dx\, dy$$

$$= \frac{1}{\sqrt{2}} \int_0^a \left[\sqrt{x} \left(\frac{y^{1/2}}{1/2}\right) + \frac{1}{\sqrt{x}} \left(\frac{y^{3/2}}{3/2}\right) \right]_0^b dx$$

$$= \frac{1}{\sqrt{2}} \left[2b^{1/2} \frac{x^{3/2}}{3/2} + \frac{2}{3} b^{3/2} \frac{x^{1/2}}{1/2} \right]_0^a$$

$$= \frac{1}{\sqrt{2}} \left[\frac{4}{3} b^{1/2} a^{3/2} + \frac{4}{3} b^{3/2} a^{1/2} \right]$$

$$= \frac{4\sqrt{(ab)}}{3\sqrt{2}} (a + b).$$

4. *Prove that the area of the surface of the paraboloid $az = x^2 + y^2$ which lies between the planes $z = 0$, $z = a$ is $\frac{\pi}{6}\left[5\sqrt{5} - 1\right]a^2$.*

Here,
$$\frac{\partial z}{\partial x} = \frac{2x}{a}, \frac{\partial z}{\partial y} = \frac{2y}{a}$$

$$\therefore \quad S = \iint \sqrt{\left(1 + \frac{4x^2}{a^2} + \frac{4y^2}{b^2}\right)}\, dx\, dy$$

The projection on the plane $z = 0$ is $x^2 + y^2 = a$ or $r = a$ which is a circle between $\theta = 0$ and $\theta = 2\pi$ (changing to polars by putting $x = r\cos\theta$, $y = r\sin\theta$). Hence

$$S = \int_0^{2\pi} \int_0^a \sqrt{\left(1 + \frac{4}{a^2}r^2\right)}\, r\, d\theta\, dr$$

$$= \frac{1}{a}\int_0^{2\pi} \left[\frac{\left(a^2 + 4r^2\right)^{3/2}}{2.4.3/2}\right]_0^a d\theta$$

$$= \frac{1.a^3}{12\,a}\left[5\sqrt{5} - 1\right]\left[\theta\right]_0^{2\pi}$$

$$= \frac{\pi}{6}\left[5\sqrt{5} - 2\right]a^2.$$

5. *Find the area of the surface $x^2 + y^2 + z^2 = a^2$ inside the surface $(x^2 + y^2)^2 = a^2(x^2 - y^2)$.*

$$\frac{\partial z}{\partial x} = -\frac{x}{z}, \frac{\partial z}{\partial y} = -\frac{y}{z}$$

$$\therefore \quad \sqrt{1 + \left(\frac{\partial z}{\partial x}\right)^2 + \left(\frac{\partial z}{\partial y}\right)^2} = \sqrt{\left(\frac{x^2 + y^2 + z^2}{z^2}\right)}$$

$$= \frac{a}{z} = \frac{a}{\sqrt{\left(a^2 - x^2 - y^2\right)}} = \frac{a}{\sqrt{\left(a^2 - r^2\right)}}$$

by changing into polars.

Also, $\left(x^2 + y^2\right)^2 = a^2\left(x^2 - y^2\right)$ gives $r^2 = a^2\cos 2\theta$.

$$\therefore \quad S = 8\int_0^{\pi/4} \int_0^{a\sqrt{(\cos 2\theta)}} \frac{a}{\sqrt{\left(a^2 - r^2\right)}}\, r\, d\theta\, dr$$

[as surface is symmetrical in all the eight octants]

$$= 8a \int_0^{\pi/4} \left[-\sqrt{(a^2 - r^2)} \right]_0^{a\sqrt{(\cos 2\theta)}} d\theta$$

$$= 8a^2 \int_0^{\pi/4} \left[1 - \sqrt{2} \sin \theta \right] d\theta$$

$$= 8a^2 \left[\theta + \sqrt{2} \cos \theta \right]_0^{\pi/4} = 8a^2 \left[\frac{\pi}{4} + 1 - \sqrt{2} \right].$$

6. *Find the area of the surface $az = xy$ that lies inside the cylinder $(x^2 + y^2)^2 = 2a^2xy$.*

We have $az = xy$.

$$\therefore \quad \frac{\partial z}{\partial x} = \frac{y}{a}, \quad \frac{\partial z}{\partial y} = \frac{x}{a}$$

$$\sqrt{1 + \left(\frac{\partial z}{\partial x}\right)^2 + \left(\frac{\partial z}{\partial y}\right)^2} = \sqrt{\frac{x^2 + y^2 + a^2}{a^2}}$$

$$= \frac{1}{a} \sqrt{(a^2 + r^2)}, \text{ changing into polars.}$$

Also, $(x^2 + y^2)^2 = 2a^2xy$ becomes $r^2 = a^2 \sin 2\theta$

$$\therefore \quad S = 2 \int_0^{\pi/2} \int_0^{a\sqrt{(\sin 2\theta)}} \frac{\sqrt{(a^2 + r^2)}}{a} \, r \, d\theta \, dr$$

$$= \frac{2}{a} \int_0^{\pi/2} \left[\frac{(a^2 + r^2)^{3/2}}{2 \times 3/2} \right]_0^{a\sqrt{(\sin 2\theta)}} d\theta$$

$$= \frac{2}{3} a^2 \int_0^{\pi/2} \left\{ (1 + \sin 2\theta)^{3/2} - 1 \right\} d\theta$$

$$= \frac{2}{3} a^2 \int_0^{\pi/2} \left[(\sin \theta + \cos \theta)^3 - 1 \right] d\theta$$

$$= \frac{1}{9} (20 - 3\pi) \, a^2.$$

EXERCISES

1. Find the area of the part of the surface of the paraboloid

$$4ax = y^2 + z^2$$

enclosed by the cylinder

$$y^2 = ax,$$

and the plane

$$x = 3a.$$

2. Prove that the surface area of the cylinder

$$x^2 + z^2 = a^2,$$

by the cylinder

$$x^2 + y^2 = a^2.$$

3. Find the area of the surface cut from the cylinder

$$x^2 + z^2 = a^2,$$

by the cylinder

$$x^2 + y^2 = a^2.$$

4. Find the area of that part of the cone

$$z^2 = x^2 + y^2,$$

which lies inside the cylinder

$$z^2 + y^2 = 2x.$$

5. Show that the surface area of the torus

$$x = (r - \cos v)\cos u,\ \ y = (r - \cos v)\sin u,\ z = \sin v,$$

where $-\pi \le u \le \pi, -\pi \le v \le \pi, r > 1$

$$4\pi^2 r.$$

6. Show that the surface area of

$$z = 2 - x^2 - y,$$

where (x, y) ranges over the triangle formed by the lines

$$x = 0,\ y = 1,\ y = x$$

$$\frac{1}{2}\log\left(\sqrt{2} + \sqrt{3}\right) + \sqrt{2/6}$$

7. Show that the surface of the sphere

$$x^2 + y^2 + z^2 = 1$$

that lies inside the cylinder

$$2x^2\left(x^2 + y^2\right) = 3\left(x^2 - y^2\right)$$

$$2\pi - 4\sqrt{2}\left[\sqrt{3}\log\left(\sqrt{3} + \sqrt{2}\right) - 2\log\left(1 + \sqrt{2}\right)\right].$$

8. If a surface is given by the equations

$$x = c\sin u,\ y = c\cos v,\ z = c(\cos u + \cos v),$$

prove that its area bounded by

$$u = 0, u = \pi/2;\ v = 0, v = \pi/2$$

is

$$\frac{\pi c^2}{4}\left[1 - \sum_{n=1}^{\infty} \frac{p_{2n}^3}{2n-1}\right]$$

where

$$p_{2n} = \frac{(2n-1)(2n-3)\dots 1}{2n(2n-2)\dots 2}.$$

9. Show that if x and y co-ordinates of any point on the paraboloid

$$2z = x^2/a + y^2/b$$

are expressed in the form

$$x = a\tan\theta\cos\phi,\ y = b\tan\theta\sin\phi$$

the angle θ is the inclination of the normal at any point to the axis of z, prove that the area of the cap of the surface cut off by the curve $0 = \lambda$ is

$$\frac{2\pi ab}{3}\left(\sec^2\lambda - 1\right).$$

OBJECTIVE QUESTIONS

Note. *For each of the following questions, four alternatives are given for the answer. Only one of them is correct. Choose the correct alternative.*

1. Curve $x = f(t), y = g(t), z = h(t), a \le t \le b$ is rectifiable if
   (*a*) f, g, h are all continuous
   (*b*) f, g, h are all bounded over $[a, b]$
   (*c*) f, g, h are all of bounded variation over $[a, b]$
   (*d*) f, g, h are neither bounded nor continuous.

2. If *a* curve *c*
   $x = f(t), y = g(t), z = h(t)\ a \le t \le b$ is rectifiable and c, is any point of $[a, b]$ then
   (*a*) $l[a, b] + l[c, b] = l[a, c]$ (*b*) $l[a, b] = l[a, c] + l[c, b]$
   (*c*) $l[a, b] > l[a, c] + l[c, b]$ (*d*) none of these

3. The function whose function value is $l[a, t]$. $t \in [a, b]$ is
   (*a*) monotonically increasing continuous function in $[a, b]$
   (*b*) monotonically increasing discontinuous function in $[a, b]$
   (*c*) monotonically decreasing continuous function in $[a, b]$

(*d*) none of these.

4. A curve $x = f(t)$, $y = g(t)$, $z = h(t)$, $a \leq t \leq b$ in which f, g, h possess continuous derivatives in $[a, b]$ and at no point thereof $f'^2(t) + g'^2(t) + h'^2(t)$ is zero, is known as

(*a*) rectifiable curve (*b*) non-rectifiable curve

(*c*) smooth curve (*d*) rough curve

5. In smooth curves, we have

(a) infinite multiple points (*b*) finite multiple points

(*c*) no multiple points (d) none of these

ANSWERS

1. (*c*) 2. (*b*) 3. (*a*) 4. (*c*) 5. (*c*)

18

Integration in $\mathbf{R}^2$ Gauss' and Stokes' Theorem

Introduction. This chapter will be devoted to developing the notions of line, surface and volume integrals in the three dimensional Euclidean space $\mathbf{R}^3$.

Integral transformation theorems due to Gauss and Stokes will also be taken up.

18.1. LINE INTEGRAL

Consider a curve C

$$x = f(t),\ y = g(t),\ z = h(t);\ t \in [a, b].$$

Also let

$$P, Q, R$$

be three functions defined in a domain $D \subset \mathbf{R}^3$, containing the curve C.

We shall now define the symbol

$$\int_C (P\,dx + Q\,dy + R\,dz).$$

Let $D = \{a = t_0., t_1, \ldots, t_{r-1}, t_r, \ldots, t_n = b\}$

be any division of $[a, b]$.

Also let ξ_r be any value of t belonging to the sub-interval $[t_{r-1}, t_r]$.

We write

$$x_r = f(t_r), y_r = g(t_r), z_r = h(t_r), \qquad 1 \le r \le n$$
$$\alpha_r = f(\boldsymbol{x}_r), \beta_r = g(\boldsymbol{x}_r), \gamma_r = h(\boldsymbol{x}_r).$$

Consider the sum

$$\sum_{r=1}^{n} \Big[P(\alpha_r, \beta_r, \gamma_r)(x_r - x_{r-1}) + Q(\alpha_r, \beta_r, \gamma_r)(y_r - y_{r-1}) + R(\alpha_r, \beta_r, \gamma_r)(z_r - z_{r-1})\Big]$$

If, as the norm of D tends to zero, the sum tends to a finite limit which is independent of the choice of points ξ_r, then we denote the limit by symbol

$$\int_C (P\,dx + Q\,dy + R\,dz).$$

18.1.1. Sufficient Condition for the Existence of

$$\int_C (P\,dx + Q\,dy + R\,dx). \qquad ...(1)$$

As in § 16.3.1., p. 360, it can be shown that if

(*i*) *f, g, h possess continuous derivatives in* $[a, b]$; *and*

(*ii*) *P, Q, R are continuous in D*

then the line integral (*i*) *exists and is equal to the ordinary integral*

$$\int_b^a (Pf' + Qg' + Rh')\,dt.$$

Here $\qquad P = P\left[f(t), g(t), h(t)\right]$, etc.

18.1.2. Line Integral w.r. to Arc Length

Suppose that the curve C given by

$$x = f(t), y = g(t), z = h(t),$$

is smooth. Let, l, denote the length of the curve. As seen in Note 2, p. 378 we can express x, y, z as functions of s where s varies from 0 to l. Let

$$x = x(s), y = y(s), z = z(s).$$

Also let P be a continuous function defined in a domain containing C.

Let

$$D = \{0, s_1, s_2, ..., s_{r-1}, s_r, ..., s_n = l\},$$

be any division of $[0, l]$. Let ξ_r be any number between s_{r-1} and s_r.

We write

$$\alpha_r = x(\xi_r), \beta_r = y(\xi_r), \gamma_r = z(\xi_r).$$

Consider the sum

$$\sum_{r=1}^{n} (s_r - s_{r-1})\,P(\alpha_r, \beta_r, \gamma_r).$$

It can be shown that under the given conditions, this sum tends to a limit when the norm of D tends to zero. This limit is denoted by the symbol

$$\int_C P\,ds.$$

This is equal to the ordinary integral

$$\int_a^b P\left[x(s), y(s), z(s)\right]\frac{ds}{dt}\,dt.$$

Note. Vectorial formulation. Assuming knowledge of the Elements of Vector Calculus, we give vectorial formulation of the concept of line integral.

We write

$$\mathbf{r} = \mathbf{i}x + \mathbf{j}y + \mathbf{k}z$$

$$\mathbf{i}f(t) + \mathbf{j}g(t) + \mathbf{k}h(t)$$

so that **r** denotes the position vector of any point x, y, z on the curve and **i, j, k** are unit vectors along the three co-ordinate axes.

Also, we consider the vector function

$$\mathbf{F}(x, y, z) = \mathbf{i}P(x, y, z) + \mathbf{j}Q(x, y, z) + \mathbf{k}R(x, y, z).$$

Considering the scalar product, we have

$$\mathbf{F}\cdot\frac{dr}{dt} = P\frac{dx}{dt} + Q\frac{dy}{dt} + R\frac{dz}{dt}$$

$$\Rightarrow \quad \int_a^b \left(P\frac{dx}{dt} + Q\frac{dy}{dt} + R\frac{dz}{dt}\right)dt = \int_a^b \left(\mathbf{F}\cdot\frac{d\mathbf{r}}{dt}\right)dt$$

and we write

$$\int_a^b \left(\mathbf{F}\cdot\frac{d\mathbf{r}}{dt}\right)dt = \int_C \mathbf{F}\cdot d\mathbf{r}.$$

Note. It is known from Differential Geometry, that if θ, φ, ψ denote the angles which the tangent at any point of the curve drawn in the direction of s, increasing makes with the three co-ordinate axes, then

$$\cos q = \frac{dx}{ds}, \cos j = \frac{dy}{ds}, \cos y = \frac{dz}{ds}.$$

Thus we have

$$\int_C (P\,dx + Q\,dy + R\,dz) = \int_0^l \left(P\frac{dx}{ds} + Q\frac{dy}{ds} + R\frac{dz}{ds}\right)ds$$

$$= \int_0^l (P\cos\theta + Q\cos\varphi + R\cos\psi)\,ds.$$

EXERCISES

1. Prove that

$$\int \frac{x^2 + y^2}{p}\,ds$$

taken round the ellipse

$$\frac{x^2}{a^2} + \frac{y^2}{b^2} = 1.$$

is equal to

$$\frac{1}{4}\pi ab\left\{4+\left(a^2+b^2\right)\left(a^{-2}+b^{-2}\right)\right\},$$

where, p, is the perpendicular from the centre to the tangent

2. Evaluate

$$\int_C \frac{x^4}{p}\,dx,$$

where C is the ellipse,

$$\frac{x^2}{a^2}+\frac{y^2}{b^2}=1,$$

and p, is the perpendicular from the origin to the tangent.

3. Show that the value of

$$\int \frac{dx}{y^2},$$

taken along the whole curve

$$x^2+y^2=y^3,$$

is 2π,

[The curve is given by

$$x=\frac{1+t^2}{t^3},\ y=\frac{1+t^2}{t^2}$$

where $t \in]-\infty, \infty[$.

4. Evaluate

$$\int_C (xy\,dx+yz\,dy+zx\,dz)$$

where the curve C is given by

$$x=t,\ y=t^2,\ z=t^3; \qquad \textit{(Gorakhpur 95)}$$

t varying from -1 to $+1$.

5. Show that

$$\int_C \left(y^2+z^2\right)dx+\left(z^2+x^2\right)dy+\left(x^2+y^2\right)dz=-2\pi ab^2,$$

where the curve C is the part for which $x \geq 0$ of the intersection of the surfaces

$$x^2+y^2+z^2=2ax,\ x^2+y^2=2bx,\ a>b>0,$$

the curve begins at the origin and runs at first in the positive octant.

$$\left[x=b\,(1+\cos\theta),\ y=b\sin\theta,\ z=2\sqrt{[(a-b)\,b]}\cos\frac{1}{2}\theta;\ -\pi\leq\theta\leq\pi\right]$$

18.2. ORIENTED CURVES AND SURFACES

Associated with each *type* of integral, there is that '*Path*' or '*Region*' of integration. Thus the path of integration associated with an ordinary integral

$$\int_a^b f(x)\,dx,$$

is the interval $[a, b]$. It is, however, clear that the mere mention of the interval is not enough. We have to further mention whether the path extends from a to b or from b to a so that we get into integrals, *viz.*,

$$\int_a^b f(x)\,dx, \int_b^a f(x)\,dx$$

in the two cases. Thus there arises what we may say the concept of an *Oriented* path. We have two possible orientations and the reversal of the orientation is equivalent to multiplication with – 1.

In the case of line integral along any curve C, we may again distinguish between the two possible oriented paths, one of which corresponds to the direction in which the parameter increases and the other to the direction in which the parameter decreases.

Also if C denotes any one of the two oriented paths, then $-C$ denotes the same path with its orientation reversed and the integral along, $-C$, is equal to that along, C, multiplied with – 1. While considering a line integral along any curve, we must thus necessarily specify the orientation also.

We now consider the concept of an *oriented surface* prior to considering surface integrals wherefor the *region* of integration is a surface.

A surface is referred to as *bilateral*, if it is possible to distinguish one of its sides from the other. It is interesting to note that not all surfaces are bilateral and we have instances of *unilateral* surfaces. One example of a unilateral surface is provided by what is known as a *Mobius strip* which may be constructed by taking a rectangular sheet $ABCD$ of paper, and by pasting the side BC to the side AD in such a way that the point C coincides with the point A and the point B with D.

In the following, we shall be exclusively concerned with bilateral surfaces only.

Oriented plane regions. A plane region will be considered oriented in respect of the orientation of its bounding curve so that orientation of a region will be reversed on reversing the orientation of its boundary curve. It is usual to consider a plane region as positively (negatively) oriented if the orientation of its boundary curve is such that it lies to the left (right) while the curve is being described.

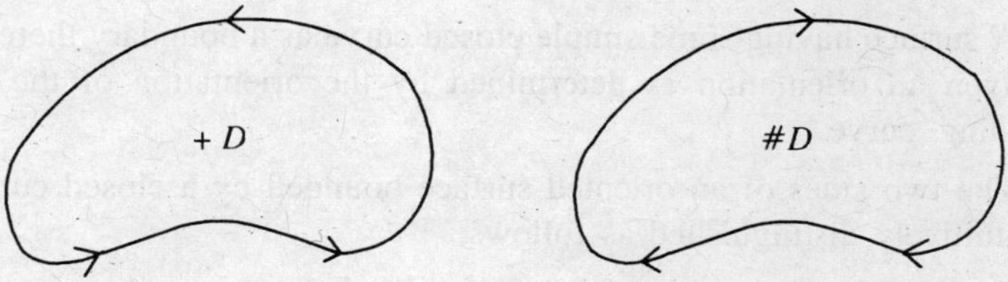

Fig

Also if D be any plane region and f any continuous function over the same so that

$$\iint_D f(x, y)\, dx\, dy$$

denotes the double integral of f over D, we shall find it *convenient* to write

$$\iint_{+D} f(x, y)\, dx\, dy = \iint_D f(x, y)\, dx\, dy,$$

and

$$\iint_{-D} f(x, y)\, dx\, dy = -\iint_D f(x, y)\, dx\, dy.$$

If D_1 be any oriented region in the uv-plane and we transform the same one-one by some relations

$$x = x(u, v),\; y = y(u, v),$$

then the new region D in the xy-plane which is the transform of D_1 acquires what we may term as **Induced** orientation.

The orientation of D will be the same as that of D_1 or different according as the Jacobian

$$\frac{\partial(x, y)}{\partial(u, v)}, \text{ is everywhere positive or negative.}$$

Adopting the above convention in respect of oriented plane regions and double integrals over the same, we have the relation

$$\iint_D f(x, y)\, dx\, dy = \iint_{D_1} f(x, y) \frac{\partial(x, y)}{\partial(u, v)}\, du\, dv,$$

irrespective of the sign of the Jacobian.

Oriented surfaces. We shall consider here only those surfaces which have as their boundary some simple closed curve or which are themselves closed.

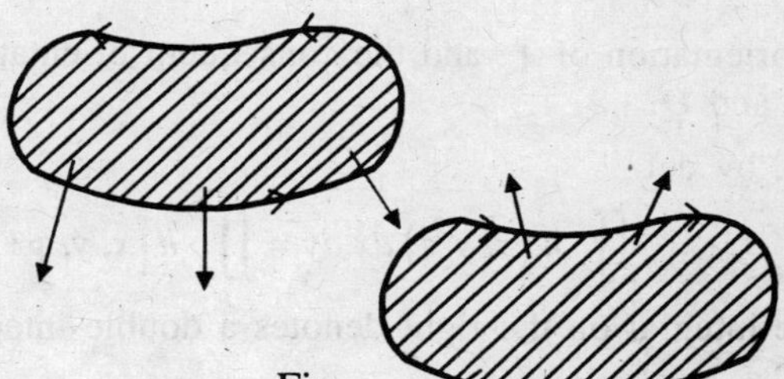

Fig

A surface having some simple closed curve as a boundary thereof will be given an orientation as determined by the orientation of the closed bounding curve.

The two sides of an oriented surface bounded by a closed curve can be intuitively distinguished as follows.

Consider any one side of the surface and draw at each point thereof a half line normal to the surface and lying on the side of the surface under consideration. Then the orientation of the bounded curve is either right-handed or left-handed with respect to each of semi-lines. (See figures above).

Note. The reader may see that here we have adopted only an intuitive manner of introducing the notion of oriented surfaces and have adopted the rigorous and analytical way for the purpose.

18.3. SURFACE INTEGRALS

Firstly we consider the case of a surface S of the form

$$z = \varphi(x, y),$$

where (x, y) ranges over some domain D of the xy-plane.

Let h be any function of (x, y, z) which is continuous in some domain containing the given surface.

Let C, denote the boundary of D and Γ the boundary curve of the surface.

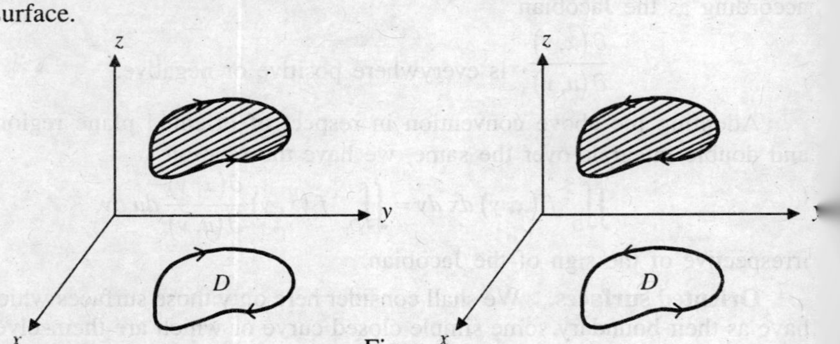

Fig

The orientation of Γ and the consequent orientation of S determines that of C and D.

Then, by def.,

$$\iint_S h(x, y, z)\,dx\,dy = \iint_D h[x, y, \varphi(x, y)]\,dx\,dy,$$

where the integral on the right denotes a double integral on the oriented domain D.

Now suppose that the surface S is given by

$$x = x(u, v), y = y(u, v), z = z(u, v)$$

where $(u, v) \in$ the domain D_1 of the uv-plane. Then clearly the domain D of the xy-plane is the image of the domain D_1 of the uv-plane. We suppose that D_1 is itself oriented in a manner that the orientation of D corresponding to that of D_1. We thus have

$$\iint_D h(x, y, \varphi)\, dx\, dy = \iint_{D_1} h(x, y, \varphi) \frac{\partial(x, y)}{\partial(u, v)} du\, dv$$

$$\Rightarrow \qquad \iint_S h(x, y, z)\, dx\, dy = \iint_{D_1} h(x, y, z) \frac{\partial(x, y)}{\partial(u, v)} du\, dv.$$

Now we consider the case of any oriented surface S which can be split up into a finite number of orientable sub-surfaces each of which is given by an equation of the form $z = \varphi(x, y)$. Then, by def. the surface integral over S is the sum of the surface integrals over the different sub-surfaces.

Then if the *entire* surface S is given by

$$x = x(u, v),\ y = y(u, v),\ z = z(u, v)$$

where $(u, v) \in D_1$ whose orientation corresponds to that of S, the result

$$\iint_S h(x, y, z)\, dx\, dy = \iint_{D_1} h(x, y, z) \frac{\partial(x, y)}{\partial(u, v)} du\, dv,$$

holds in respect of the entire surface and we do not need to split S into sub-surfaces of the requisite type.

We may similarly define the surface integrals

$$\iint_S f(x, y, z)\, dy\, dz, \quad \iint_S g(x, y, z)\, dz\, dx,$$

and show that

$$\iint_S f(x, y, z)\, dy\, dz = \iint_{D_1} f(x, y, z) \frac{\partial(y, z)}{\partial(u, v)} du\, dv,$$

$$\iint_S g(x, y, z)\, dz\, dx = \iint_{D_1} g(x, y, z) \frac{\partial(z, x)}{\partial(u, v)} du\, dv,$$

Also, we have the general result as follows :

$$\iint_S \left[f(x, y, z)\, dy\, dz + g(x, y, z)\, dz\, dx + h(x, y, z)\, dx\, dy\right]$$

$$= \iint_{D_1} \left[f \frac{\partial(y, z)}{\partial(u, v)} + g \frac{\partial(z, x)}{\partial(u, v)} + h \frac{\partial(x, y)}{\partial(u, v)}\right] du\, dv,$$

where on the right hand side, we have ordinary double integral.

Note. 1. If θ, ϕ, ψ denote the angles which the semi-normal drawn at any point of the surface and lying on the side of the surface under consideration makes with the co-ordinate axes, we have

$$\iint_S h(x, y, z)\, dx\, dy = \iint_S h(x, y, z) \cos\psi\, dS,$$

$$\iint_S g(x, y, z)\, dz\, dx = \iint_S g(x, y, z) \cos\phi\, dS,$$

$$\iint_S f(x, y, z)\, dy\, dz = \iint_S f(x, y, z) \cos\theta\, dS$$

where dS denotes *an element* of the surface.

Note. 2. Vectorial formulation.

Let $\mathbf{r} = \mathbf{i}x + \mathbf{j}y + \mathbf{k}z,$

be the position vector of any point of the surface and

$$\mathbf{F}(x, y, z) = \mathbf{i}f(x, y, z) + \mathbf{j}g(x, y, z) + \mathbf{k}h(x, y, z),$$

be a vector function with f, g, h as its components.

Also let, n denote the unit normal vector at any point of the surface on the side of the surface under consideration so that

$$\mathbf{n} = \mathbf{i}\cos\theta + \mathbf{j}\cos\varphi + \mathbf{k}\cos\psi.$$

$$\iint_S \mathbf{F}.\mathbf{n}\, dS = \iint_S (f\cos\theta + g\cos\varphi + h\cos\psi)\, dS$$

$$= \iint_S (f\, dy\, dz + g\, dz\, dx + h\, dx\, dy).$$

We thus see that the surface integral

$$\iint_S (f\, dy\, dz + g\, dz\, dx + h\, dx\, dy)$$

can be compactly written as

$$\iint_S \mathbf{F}.\mathbf{n}\, dS.$$

EXERCISES

1. Show that

$$\iint_S (yz\, dy\, dz + zx\, dz\, dx + h\, dx\, dy) = \frac{3}{8},$$

where S is the surface of the sphere

$$x^2 + y^2 + z^2 = 1,$$ *(Kanpur 94)*

in the first quadrant.

2. Evaluate the surface integral

$$\iint_S \left(x^3\, dy\, dz + y^3\, dz\, dx + x^3\, dx\, dy\right)$$

where S is the outer surface of the sphere

$$x^2 + y^2 + z^2 = 1.$$

3. Evaluate

$$\iint_S (x\,dy\,dz + y\,dz\,dx + z\,dx\,dy)$$

where S is the outer surface of the part of the ellipsoid

$$\frac{x^2}{a^2} + \frac{y^2}{b^2} + \frac{z^2}{c^2} = 1.$$

lying above the xy-plane.

4. If S is the ellipsoid

$$\frac{x^2}{a^2} + \frac{y^2}{b^2} + \frac{z^2}{c^2} = 1,$$

prove that

$$\iint_S \frac{x^2}{p}\,dS = \frac{4\pi abc}{15}\left[6 + \left(\Sigma\, a^2\right)\left(\Sigma\, a^{-2}\right)\right].$$

$$\int p\,ds = 4\pi abc$$

$$\int \frac{ds}{p} = \frac{4}{3}\pi abc$$

$$I_n = \int \frac{dS}{p^n} = \frac{1}{3}\left(\Sigma \frac{1}{a^2}\right) I_{n-2},$$

where p is the length of the perpendicular from the origin to the tangent plane at the point (x, y, z).

5. Find the value of

$$\iint_S p\left(x^4 + y^4 + z^4\right) dS,$$

where S is the surface

$$\frac{x^2}{a^2} + \frac{y^2}{b^2} + \frac{z^2}{c^2} = 1,$$

and p, is the perpendicular from the origin to the tangent plane at the point (x, y, z).

6. Evaluate

$$\iint_S (x + y + z)(lx + my + nz)\,dS,$$

where S is the boundary of the region

$$x^2 + y^2 \le a^2, 0 \le z < h.$$

18.4. FIRST GENERALISATION OF GREEN'S THEOREM

Stoke's theorem. It is possible to generalise in two ways the Green's theorem which expresses a double integral over a plane region in terms of a line integral over the boundary curve of the same. One generalisation arises when we replace the plane region by any surface and express a surface integral taken over the same by a line integral taken over the boundary curve of the surface. This generalisation will now be stated and proved in the form of *Stoke's theorem.* A second generalisation arises when we replace double integral by a triple integral and express a triple integral taken over a three dimensional region in terms of a surface integral taken over the bounding surface of the region. This will be given as Gauss's theorem after we have introduced the concept of Triple integration.

Stoke's theorem. *If, S, denotes an original surface relative to the sense of description of its boundary curve C and f, g, h are three functions of x, y, z which are continuous and which possess continuous first order partial derivatives at each point of a domain containing S, then,*

$$\int_C (f\,dx + g\,dy + h\,dz)$$

$$= \iint_S \left[\left(\frac{\partial h}{\partial y} - \frac{\partial g}{\partial z}\right) dy\,dz + \left(\frac{\partial f}{\partial z} - \frac{\partial h}{\partial x}\right) dz\,dx + \left(\frac{\partial g}{\partial x} - \frac{\partial f}{\partial y}\right) dx\,dy\right].$$

Suppose that the oriented surface is given by

$$x = x(u, v),\ y = y(u, v),\ z = z(u, v),$$

where (u, v) ranges over a certain oriented domain D of the uv-plane. Also let Γ denote the oriented boundary curve of the domain D. We suppose that the curve Γ is given by

$$u = u(t),\ v = v(t).$$

t varying from a to b.

The principle of the proof is contained in the following steps :

I. Express the line integral along C in terms of an ordinary integral.

II. Express the ordinary integral in I as a line integral along the curve Γ.

III. Express, by Green's theorem, the line integral along the curve Γ, in terms of a double integral over D.

IV. Express the double integral over D in terms of a surface integral over the surface S.

Now we carry out the programme indicated above.

We have

$$\int_C (f\,dx + g\,dy + h\,dz)$$

$$= \int_b^a \left[f\left(\frac{\partial x}{\partial u}\frac{du}{dt} + \frac{\partial x}{\partial v}\frac{dv}{dt}\right) + g\left(\frac{\partial y}{\partial u}\frac{du}{dt} + \frac{\partial y}{\partial v}\frac{dv}{dt}\right) + h\left(\frac{\partial z}{\partial u}\frac{du}{dt} + \frac{\partial z}{\partial v}\frac{dv}{dt}\right)\right] dt$$

$$= \int_\Gamma \left[\left(f\frac{\partial x}{\partial u} + g\frac{\partial y}{\partial u} + h\frac{\partial z}{\partial u}\right) du + \left(f\frac{\partial x}{\partial v} + g\frac{\partial y}{\partial v} + h\frac{\partial z}{\partial v}\right) dv\right]$$

$$= \iint_D \left[\frac{\partial}{\partial u}\left(f\frac{\partial x}{\partial v} + g\frac{\partial y}{\partial v} + h\frac{\partial z}{\partial v}\right) - \frac{\partial}{\partial v}\left(f\frac{\partial x}{\partial u} + g\frac{\partial y}{\partial u} + h\frac{\partial z}{\partial u}\right)\right] du\,dv. \quad ...(1)$$

We now have

$$\frac{\partial}{\partial u}\left(f\frac{\partial x}{\partial v}\right) = \left(\frac{\partial f}{\partial x}\frac{\partial x}{\partial u} + \frac{\partial f}{\partial y}\frac{\partial y}{\partial u} + \frac{\partial f}{\partial z}\frac{\partial z}{\partial u}\right)\frac{\partial x}{\partial v} + f\frac{\partial^2 x}{\partial u \partial v},$$

and also we may compute the other expressions obtained in the integrand.

Carrying out these straightforward steps, we see that the double integral over D on the right hand side of (1) is

$$= \iint_D \left[\left(\frac{\partial h}{\partial y} - \frac{\partial g}{\partial z}\right)\frac{\partial(y, z)}{\partial(u, v)} + \left(\frac{\partial f}{\partial z} - \frac{\partial h}{\partial x}\right)\frac{\partial(z, x)}{\partial(u, v)} + \left(\frac{\partial g}{\partial x} - \frac{\partial f}{\partial y}\right)\frac{\partial(x, y)}{\partial(u, v)}\right] du\,dv$$

which again is equal to the following surface integral over S,

$$\iint_S \left[\left(\frac{\partial h}{\partial y} - \frac{\partial g}{\partial z}\right) dy\,dz + \left(\frac{\partial f}{\partial z} - \frac{\partial h}{\partial x}\right) dz\,dx + \left(\frac{\partial g}{\partial x} - \frac{\partial f}{\partial y}\right) dx\,dy\right].$$

Thus we have proved that

$$\int_C (f\,dx + g\,dy + h\,dz) = \iint_S \left[\left(\frac{\partial h}{\partial y} - \frac{\partial g}{\partial z}\right) dy\,dz + \left(\frac{\partial f}{\partial z} - \frac{\partial h}{\partial x}\right) dz\,dx + \left(\frac{\partial g}{\partial x} - \frac{\partial f}{\partial y}\right) dx\,dy\right].$$

Hence the theorem.

Vectorial Formulation. We write

$$\mathbf{F} = \mathbf{i}f + \mathbf{j}g + \mathbf{k}h$$

$$\Rightarrow \qquad \text{curl } \mathbf{F} = \mathbf{i}\left(\frac{\partial h}{\partial y} - \frac{\partial g}{\partial z}\right) + \mathbf{j}\left(\frac{\partial f}{\partial z} - \frac{\partial h}{\partial x}\right) + \mathbf{k}\left(\frac{\partial g}{\partial x} - \frac{\partial f}{\partial y}\right).$$

Thus Stoke's theorem can be re-written as

$$\int_C \mathbf{F}.d_\gamma = \iint_S \text{curl } \mathbf{F}.\mathbf{n}\, dS.$$

Ex. 1. *Verify Stoke's theorem for the line integral*

$$\int_C \left(x^2\, dx + y\, x\, dy\right),$$

where C denotes the square in the plane $z = 0$ *with sides along the lines*

$$x = 0,\ y = 0,\ x = a,\ y = a. \qquad \textit{(Poorv. 93, 95; Kumaon 92)}$$

Ex. 2. *Apply Stoke's theorem to prove that*

$$\int_C (y\, dx + z\, dy + x\, dz) = -2\sqrt{2\pi}\, a^2,$$

where C is the curve given by

$$x^2 + y^2 + z^2 - 2ax - 2ay = 0,\ \ x + y = 2a$$

and begins at a point (2*a*, 0, 0) *and goes at first below the z-plane.*

Ex. 3. *Show that*

$$\iint_S \left[(y - z)\, dy\, dz + (z - x)\, dz\, dx + (x - y)\, dx\, dy\right] = a^3\pi$$

where S is the portion of the surface

$$x^2 + y^2 - 2ax + az = 0$$

for which $z \geq 0$.

18.5. VOLUME INTEGRALS

The treatment of volume integrals, also known as triple integrals, in R^3 is a simple and straight extension of the ideas in respect of double integrals. Thus we shall only briefly indicate the various stages of development of the theory of triple integrals.

I. **Volume of a rectangle**

$$[a, b; c, d; e, f] \subset \mathbf{R}^3$$

is, by definition,

$$(b - a)(d - c)(f - e).$$

Note. A rectangle in $\mathbf{R}^3$ also called a three-dimensional rectangle is often referred to as a rectangular parallelopiped.

II. **Integrability and integral of a bounded function f over a rectangle**

$$R = [a, b; c, d; e, f].$$

Let

$$D_1 = \{a = x_0, x_1, ..., x_{r-1}, x_r, ..., x_m = b\}$$
$$D_2 = \{c = y_0, y_1, ..., y_{s-1}, y_s, ..., y_n = d\}$$
$$D_3 = \{e = z_0, z_1, ..., z_{t-1}, z_t, ..., z_p = f\}$$

be divisions of the intervals

$$[a, b], [c, d], [e, f]$$

respectively. These three divisions D_1, D_2, D_3 gives rise to a division say, D of the rectangle R into mnp, sub-rectangles

$$[x_{r-1}, x_r\ ; y_{s-1}, y_s\ ; z_{t-1}, z_t] \qquad ...(1)$$

r, s, t varying from 1 to m, 1 to n and 1 to p respectively.

We write

$$w_{rst} = (x_r - x_{r-1})(y_s - y_{s-1})(z_t - z_{t-1}),$$

which is the volume of the rectangle (1).

Also let

$$M_{rst}, w_{rst}$$

denote the supremum and the infimum of f in the rectangle (1).

We write

$$S(D) = \sum_{t=1}^{p} \sum_{s=1}^{n} \sum_{r=1}^{m} M_{rst}, w_{rst},$$
$$s(D) = \sum_{t=1}^{p} \sum_{s=1}^{n} \sum_{r=1}^{m} M_{rst}, w_{rst}.$$

Then the infimum of the set of sums $S(D)$ and the supremum of the set of sums $s(D)$ are known respectively as the upper and lower integrals of f over R and denoted by

$$\overline{\iiint_R} f(x, y, z)\, dx\, dy\, dz, \quad \underline{\iiint_R} f(x, y, z)\, dx\, dy\, dz.$$

In case these two are equal, we say that the function is integrable over R and the common value, called the integral of f over R is denoted by

$$\iiint_R f(x, y, z)\, dx\, dy\, dz.$$

Norm of a division. By the norm of a division D is meant the max. of the norms of the divisions D_1, D_2 and D_3.

III. **Darboux's theorem and necessary and sufficient conditions for integrability.**

The statements and proofs are exactly the same as for double integrals.

IV. Particular classes of bounded integrable functions. We have the following three results in this connection :

(*i*) *Every continuous function is integrable.*

(*ii*) *A bounded function whose points of discontinuity can be enclosed in a finite number of rectangles, the sum of whose volumes is less than a given positive number, is integrable.*

(*iii*) *A bounded function whose points of discontinuity lie on a finite number of surfaces of the form* $z = \varphi(x, y)$, *etc., where* φ *is continuous, is integrable.*

The proofs of the corresponding results for double integrals are easily adaptable except that we need properties of continuous functions of three variables.

V. Calculation of Triple integrals. Reduction to Ordinary integrals

If

(*i*) $\iiint_R f(x, y, z)\, dx\, dy\, dz,$

and (*ii*) $\iint_S f(x, y, z)\, dx\, dy,$

where $e \le z \le f$ *and* $S = [a, b; c, d]$,

both exist, then

$$\int_e^f \left[\iint_S f(x, y, z)\, dx\, dy\right] dz,$$

also exists and is equal to the triple integral.

Let U and L denote the upper and lower integrals of $f(x, y, z)$ over R. Let $\varepsilon > 0$ be given.

There exists a division of R into mnp, sub-rectangles

$$\left[x_{r-1}, x_r; y_{s-1}, y_s; z_{t-1}, z_t\right],$$

such that

$$\sum_{t=1}^{p} \sum_{s=1}^{n} \sum_{r=1}^{m} M_{rst}\, w_{rst} < U + \varepsilon; \qquad \text{...(1)}$$

the symbols having their usual meanings.

For each fixed value of $z \in \left[z_{t-1}, z_t\right]$, we have

$$\iint_S f(x, y, z)\, dx\, dy$$

$$\le \sum_{s=1}^{n} \sum_{r=1}^{m} M_{rst} \left(x_r - x_{r-1}\right)\left(y_s - y_{s-1}\right) \qquad \text{...(2)}$$

$$\Rightarrow \quad \int_e^f \left[\iint_S f(x, y, z)\, dx\, dy \right] dz$$

$$\leq \sum_{t=1}^{p} \sum_{s=1}^{n} \sum_{r=1}^{m} M_{rst}\, w_{rst} < U + \varepsilon, \text{ by } (1)$$

As ε is an arbitrary positive number, we have

$$\int_e^f \left[\iint_S f(x, y, z)\, dx\, dy \right] dz \leq U.$$

Similarly, we may see that

$$\int_e^f \left[\iint_S f(x, y, z)\, dx\, dy \right] dz \geq L.$$

As

$$U = L,$$

the result follows.

Cor. If f is continuous over R, we have

$$\iiint_E f(x, y, z)\, dx\, dy\, dz = \int_e^f \left[\int_c^d \left[\int_a^b f(x, y, z)\, dx \right] dy \right] dz,$$

where we can interchange the order of x, y, z on the right hand side in any manner we like.

VI. **Integrability and integral over any bounded domain.** If f be defined in a bounded domain E, we enclose E in a rectangle R and define a function over the same as follows :

$$F(x, y, z) = \begin{cases} f(x, y, z) \text{ at all points of } E. \\ 0, \quad \text{elsewhere.} \end{cases}$$

Then f is defined to be integrable over E, if F is integrable over R and we write

$$\iiint_E f(x, y, z)\, dx\, dy\, dz = \iint_R F(x, y, z)\, dx\, dy\, dz.$$

It can be shown that a continuous function f is integrable over E, if E is bounded by a finite number of surfaces of the form $z = \varphi(x, y), \psi(x, y)$ etc., where φ, ψ etc., are continuous.

VII. **Calculation of Triple Integrals.** If f is continuous in a region E bounded by surfaces

$$z = \varphi(x, y);\ z = \psi(x, y);\ y = g(x),\ y = h(x);\ x = a,\ x = b$$

then

$$\iiint_E f(x, y, z)\, dx\, dy\, dz = \int_a^b \left[\int_{g(x)}^{h(x)} \left\{ \int_{\psi(x, y)}^{\varphi(x, y)} f(x, y, z)\, dz \right\} dy \right] dx.$$

The proof may be constructed as in the case of double integrals.

18.6. GAUSS'S THEOREM

Second Generalisation of Green's theorem. *If S be any closed surface enclosing a three-dimensional region E and if f, g, h are three functions of three variables each of which is continuous and possesses continuous first order partial derivatives*

$$f_x, g_y, h_z,$$

at each point of E and S, then

$$\iiint_E \left(f_x + g_y + h_z\right) dx\, dy\, dz = \iint_S \left(f\, dy\, dz + g\, dz\, dx + h\, dx\, dy\right),$$

where the surface integral of the right is taken over the exterior of the surface S.

Firstly we consider the case of a surface S which is such that a line parallel to each axis and having points in common with the interior of S meets the same in at the most two points.

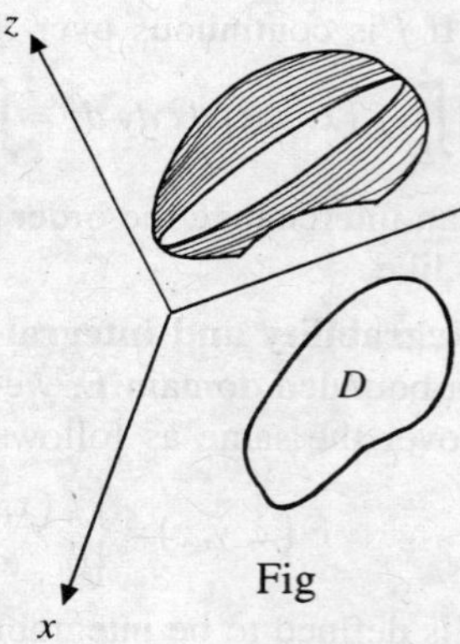

Fig

Then it is possible to split up the external surface S into two sub-surfaces S_1 and S_2 both of which project into the same region of the xy-plane. These two sub-surfaces have equations of the form

$$z = \varphi(x, y),\ z = \psi(x, y)$$

where (x, y) varies over the same region of the xy-plane.

These two sub-surfaces S_1 and S_2 have the same boundary curve. Relative to the external surface S under consideration and the consequently determined sides of S_1 and S_2, we see that the boundary curve of S_1 differs from that of S_2 in orientation only. Accordingly, therefore, if, D denotes the region of xy-plane on which S_1 projects, then $-D$ is the region on which S_2 projects.

Thus we have

$$\begin{aligned}\iiint_E h_z\, dx\, dy\, dz &= \int_D\int \left[\int_{\psi(x,y)}^{\varphi(x,y)} h_z\, dx\right] dx\, dy\\ &= \iint_D \left[h(x, y, \varphi) - h(x, y, \psi)\right] dx\, dy\\ &= \iint_D h(x, y, \varphi)\, dx\, dy - \iint_D h(x, y, \psi)\, dx\, dy\\ &= \iint_D h(x, y, \varphi)\, dx\, dy + \iint_{-D} h(x, y, \psi)\, dx\, dy\end{aligned}$$

$$= \iint_{S_1} h(x, y, z)\, dx\, dy + \iint_{S_2} h(x, y, z)\, dx\, dy$$

$$= \iint_{S} h(x, y, z)\, dx\, dy.$$

We may similarly show that

$$\iiint_E f_x\, dx\, dy\, dz = \iint_S f\, dy\, dz,$$

$$\iiint_E g_y\, dx\, dy\, dz = \iint_S g\, dz\, dx.$$

Adding we get

$$\iiint_E \left(f_x + g_y + h_z\right) dx\, dy\, dz = \iint_S (f\, dy\, dz + g\, dz\, dx + h\, dx\, dy).$$

To generalise, we suppose that the surface S is such that it can be divided by a finite number of smooth curves into a finite number of portions in such a way that each portion satisfies the assumptions made above. We then apply the theorem proved for each portion and add the results. On adding, we shall obtain on the left a triple integral over E and, on the right, some of the surface integrals combine to form the surface integral over S and others cancel one another in pairs. Hence the theorem.

Note. Vectorial formulation. Writing,

$$\mathbf{F} = \mathbf{i}f + \mathbf{j}g + \mathbf{k}h,$$

$$\Rightarrow \qquad \text{div } F = \frac{\partial f}{\partial x} + \frac{\partial g}{\partial y} + \frac{\partial h}{\partial z},$$

we see that Gauss's theorem can be re-written as

$$\iiint_E \text{div } \mathbf{F}\, d\,v = \iint_S \mathbf{F}.\mathbf{n}\, dS$$

where, d_v denotes element of volume.

EXERCISES

1. Show using Gauss's theorem that the surface integral

$$\iint_S \left[\left(x^3 - yz\right) dy\, dz - 2x^2 y\, dz\, dx + 2\, dx\, dy\right]$$

taken over the outer surface of the cube bounded by the planes

$$x = 0, x = a; y = 0, y = a; z = 0, z = a$$

is $\frac{1}{3} a^5$.

Also verify the result by direct evaluation of the surface integral.

2. Evaluate

$$\iint_S (x\, dy\, dz + y\, dz\, dx + z\, dx\, dy)$$

taken over the outer surface of the cube [0, a; 0, a; 0, a]

3. Evaluate

$$\iint_S \left(y^2z^2\,dy\,dz + z^2x^2\,dz\,dx + x^2y^2\,dx\,dy\right)$$

where S is the part of the surface of the sphere $x^2 + y^2 + z^2 = 1$, above the XY-plane.

4. Show that

$$\iint_S \left(x^2\,dy\,dz + y^2\,dz\,dx + z^2\,dx\,dy\right) = 0$$

where S denotes the surface of the ellipsoid $x^2/a^2 + y^2/b^2 + z^2/c^2 = 1$.

5. Show that

$$\iint_S (ax\,dy\,dz + by\,dz\,dx + cz\,dx\,dy) = \frac{4}{3}\pi\,(a+b+c),$$

where S is the surface of the sphere

$$x^2 + y^2 + z^2 = 1.$$

6. Evaluate

$$\iint_S \left(x\,dy\,dz + y\,dz\,dx + z^2\,dx\,dy\right)$$

where S denotes the closed surface bounded by the cone $x^2 + y^2 = z^2$ and the plane $z = 1$.

18.7. CHANGE OF VARIABLES IN A TRIPLE INTEGRAL

Let f be continuous in a domain E bounded by a surface S in xyz-space.

Also let

$$x = x(X, Y, Z),\ y = y(X, Y, Z),\ z = z(X, Y, Z)$$

define three functions of three variables X, Y, Z defined in a domain E bounded by a surface S_1 in XYZ-space.

We suppose that these three functions with values

$$x(X, Y, Z),\ y(X, Y, Z),\ z(X, Y, Z)$$

(i) possess continuous first order partial derivatives at each point of E_1 and S_1,

(ii) transform E_1 into E and S_1 into S,

(iii) the transformation is one-one,

(iv) the Jacobian

$$\frac{\partial(x, y, z)}{\partial(X, Y, Z)}$$

does not change sign at any point of E_1 even though it may vanish at some points of S_1.

It will then be proved that

$$\iiint_E f(x, y, z)\, dx\, dy\, dz$$

$$= \iiint_{E_1} f(x, y, z) \left| \frac{\partial(x, y, z)}{\partial(X, Y, Z)} \right| dX\, dY\, dZ,$$

where x, y, z have to be replaced by their values in terms of X, Y, Z in $f(x, y, z)$ in the integral on the right.

There exists a continuous function F such that at every point of E

$$F_z = f(x, y, z).$$

Also we suppose that the surface S is given by

$$x = x(u, v),\ \ y = y(u, v),\ z = z(u, v)$$

where (u, v) ranges over some domain D of the uv-plane.

We have

$$\iiint_E f(x, y, z)\, dx\, dy\, dz = \iiint_E F_z\, dx\, dy\, dz$$

$$= \iiint_S F(x, y, z)\, dx\, dy \qquad ...(1)$$

(By Gauss's theorem)

Expressing the surface integral on the right of (1) as a double integral on the domain D of the uv-plane, we have

$$\iint_S F(x, y, z)\, dx\, dy = \iint_D F(x, y, z) \frac{\partial(x, y)}{\partial(u, v)}\, du\, dv, \quad ...(2)$$

where x, y, z have to be replaced by their values in terms of u, v.

Now, as may be easily shown

$$\frac{\partial(x, y)}{\partial(u, v)} = \frac{\partial(x, y)}{\partial(X, Y)} \frac{\partial(X, Y)}{\partial(u, v)} + \frac{\partial(x, y)}{\partial(Y, Z)} \frac{\partial(Y, Z)}{\partial(u, v)} + \frac{\partial(x, y)}{\partial(Z, X)} \frac{\partial(Z, X)}{\partial(u, v)}.$$

$$\therefore \quad \iint_D F(x, y, z) \frac{\partial(x, y)}{\partial(u, v)}\, du\, dv$$

$$= \int_D F(x, y, z) \left[\frac{\partial(x, y)}{\partial(X, Y)} \frac{\partial(X, Y)}{\partial(u, v)} + \frac{\partial(x, y)}{\partial(Y, Z)} \frac{\partial(Y, Z)}{\partial(u, v)} + \frac{\partial(x, y)}{\partial(Z, X)} \frac{\partial(Z, X)}{\partial(u, v)} \right] du\, dv$$

$$\pm \iint_{S_1} F\frac{\partial(x, y)}{\partial(X, Y)} dX\, dY + F\frac{\partial(x, y)}{\partial(Y, Z)} dY\, dZ + F\frac{\partial(x, y)}{\partial(Z, X)} dZ\, dX, \quad ...(3)$$

where on the right we have a surface integral on S_1 and the sign to be taken is + or – according as the transformation is direct or inverse.

Applying Gauss's theorem to the integral on the right of (3), we see that the same is equal to

$$\pm \iiint_{E_1} \left\{\frac{\partial}{\partial Z}\left[F\frac{\partial(x, y)}{\partial(X, Y)}\right] + \frac{\partial}{\partial X}\left[F\frac{\partial(x, y)}{\partial(Y, Z)}\right] + \frac{\partial}{\partial Y}\left[F\frac{\partial(x, y)}{\partial(Z, X)}\right]\right\} dX\, dY\, dZ \quad ...(4)$$

Now $$\frac{\partial}{\partial Z}\left[F\frac{\partial(x, y)}{\partial(X, Y)}\right] = \left(\frac{\partial F}{\partial x}\frac{\partial x}{\partial Z} + \frac{\partial F}{\partial y}\frac{\partial y}{\partial Z} + \frac{\partial F}{\partial z}\frac{\partial z}{\partial Z}\right)\frac{\partial(x, y)}{\partial(X, Y)}$$
$$+ F\left(\frac{\partial^2 x\, \partial y}{\partial Z\, \partial X\, \partial Y} + \frac{\partial x\, \partial^2 y}{\partial X\, \partial Z\, \partial Y} - \frac{\partial^2 x\, \partial y}{\partial Z\, \partial Y\, \partial Z} - \frac{\partial x\, \partial^2 y}{\partial Y\, \partial Z\, \partial X}\right)$$

Similarly carrying out the computation of

$$\frac{\partial}{\partial X}\left[F\frac{\partial(x, y)}{\partial(Y, Z)}\right], \frac{\partial}{\partial Y}\left[F\frac{\partial(x, y)}{\partial(Z, X)}\right]$$

and some straightforward simplifications, we may see that the integral on the right of (4) becomes

$$\pm \iiint_{E_1} \frac{\partial F}{\partial z}\frac{\partial(x, y, z)}{\partial(X, Y, Z)} dX\, dY\, dZ$$
$$= \pm \iiint_{E_1} f\frac{\partial(x, y, z)}{\partial(X, Y, Z)} dX\, dY\, dZ.$$

Now assuming that the transformation is direct or inverse according as the Jacobian $\partial(x, y, z)/\partial(X, Y, Z)$ is positive or negative we see from above that

$$\iiint_{E_1} f\, dx\, dy\, dz = \iiint_{E_1} f\frac{\partial(x, y, z)}{\partial(X, Y, Z)} dX\, dY\, dZ,$$

as was to be proved.

EXAMPLES

1. *The triple integral*

$$\iiint f(x, y, z)\, dx\, dy\, dz$$

over the domain

$$\{(x, y, z) : x^2 + y^2 + z^2 \le a^2\}$$

is equivalent to the repeated integral

$$\int_{-a}^{+a} dx \int_{-\sqrt{(a^2 - x^2)}}^{+\sqrt{(a^2 - x^2)}} dy \int_{-\sqrt{(a^2 - x^2 - y^2)}}^{+\sqrt{(a^2 - x^2 - y^2)}} f(x, y, z)\, dz$$

2. *Evaluate*

$$\iiint (x + y + z + 1)^2\, dx\, dy\, dz,$$

over the region defined by

$$x \ge 0, y \ge 0, z \ge 0, x + y + z \le 1.$$

Thus region of integration is the interior of the tetrahedron bounded by the three co-ordinate planes and the plane

$$x + y + z = 1.$$

The given integral is equal to the repeated integral

$$\int_0^1 dx \int_0^{1-x} dy \int_0^{1-x-y} (x + y + z + 1)^2\, dz = \frac{31}{60}.$$

3. *Evaluate*

$$\iiint (1 - x - y - z)^{l-1}\, x^{m-1}\, y^{n-1}\, z^{p-1}\, dx\, dy\, dz,$$

over the interior of the tetrahedron bounded by the planes

$$x = 0, y = 0, z = 0, x + y + z = 1.$$

We employ the transformation defined by

$$x + y + z = u,\ \ x + y = uv,\ x = uvw$$

$$\Leftrightarrow \quad x = uvw,\ y = uv(1 - w),\ z = u(1 - v),$$

$$\Leftrightarrow \quad u = x + y + z,\ v = (x + y)/(x + y + z),\ w = x/(x + y).$$

When x, y, z are all positive and $x + y + z$ is less than 1, then u, v, w all lie between 0 and 1.

Conversely, when u, v, w all lie between 0 and 1, the x, y, z are all positive and $x + y + z$ is less than 1. Thus the given region transforms to the region bounded by the planes

$$u = 0, u = 1;\ v = 0, v = 1;\ w = 0, w = 1.$$

Also

$$\frac{\partial(x, y, z)}{\partial(u, v, w)} = -u^2 v$$

$\therefore$ the given integral

$$= \left[\int_0^1 (1-u)^{l-1} u^{m+n+p-1}\, du\right]\left[\int_0^1 v^{m+n-1} (1-v)^{p-1}\, dv\right] \times \left[\int_0^1 w^{m-1} (1-w)^{n-1}\, dw\right]$$

$$= B(l, m+n+p)\, B(m+n, p)\, B(m, n)$$

$$= \frac{\Gamma(l)\,\Gamma(m)\,\Gamma(n)\,\Gamma(p)}{\Gamma(l+m+n+p)}.$$

4. *Evaluate*

$$I = \iiint \sqrt{\left(a^2b^2c^2 - b^2c^2x^2 - c^2a^2y^2 - a^2b^2z^2\right)}\, dx\, dy\, dz$$

taken throughout the domain

$$\left\{(x, y, z) : \frac{x^2}{a^2} + \frac{y^2}{b^2} + \frac{z^2}{c^2} \le 1\right\}.$$

Change the variables x, y, z to X, Y, Z where

$$x = aX,\ y = bY,\ z = cZ,$$

so that

$$\partial(x, y, z)/\partial(X, Y, Z) = abc.$$

Thus

$$I = a^2b^2c^2 \iint \sqrt{\left(1 - X^2 - Y^2 - Z^2\right)}\, dX\, dY\, dZ$$

taken throughout $X^2 + Y^2 + Z^2 \le 1$.

Changing X, Y, Z to polar co-ordinates r, θ, φ so that

$$X = r \sin\theta \cos\varphi,\ Y = r \sin\theta \sin\varphi,\ Z = r\cos\theta,$$

We have, since

$$\partial(X, Y, Z)/\partial(r, \theta, \varphi) = r^2 \sin\theta,$$

$$I = a^2b^2c^2 \iiint \sqrt{\left(1 - r^2\right)}\, r^2 \sin\theta\, dr\, d\theta\, d\varphi.$$

It is easily seen that to describe the whole region

$$X^2 + Y^2 + Z^2 \le 1,$$

r varies from 0 to 1; θ varies from 0 to π ; φ varies from 0 to 2π.

Thus (r, θ, φ) varies in the rectangular parallelopiped.

$$[0, 1;\ 0, \pi;\ 0, 2\pi]$$

$$\therefore \quad I = a^2b^2c^2 \int_0^{2\pi} d\varphi \int_0^{\pi} \sin\theta\, d\theta \int_0^1 \sqrt{\left(1 - r^2\right)}\, r^2\, dr = \frac{1}{4} a^2b^2c^2\pi^2$$

EXERCISES

1. Show that

$$\iiint \frac{dx\,dy\,dz}{(x+y+z+1)^3} = \frac{1}{16}\log\frac{226}{e^5}$$

taken throughout the interior of the tetrahedron bounded by the planes

$$x=0,\ y=0,\ z=0,\ x+y+z=1.$$

2. Show that

$$\iiint (....+nz)^2\,dx\,dy\,dz = \frac{4}{15}\pi\left(l^2+m^2+n^2\right)$$

taken throughout the positive region

$$\left\{(x,y,z): x^2+y^2+z^2 \le 1\right\}.$$

3. Prove that the value of

$$\iiint \frac{xyz}{\sqrt{\left(x^2+y^2+z^2\right)}}\,dx\,dy\,dz,$$

taken throughout the positive octant of the ellipsoid $x^2/a^2+y^2/b^2+z^2/c^2=1$ is

$$a^2b^2c^2\,(bc+ca+ab)/15\,(b+c)(c+a)(a+b)$$

4. Show that

$$\iiint z\,dx\,dy\,dz = \frac{\pi}{4}h^4\cot\theta\cot\varphi,$$

taken throughout the volume bounded by the cone

$$z^2 = x^2\tan^2\theta + y^2\tan^2\varphi,$$

and the planes $z=0$, $z=h$.

5. Show that

$$\iiint \left(ax^2+by^2+cz^2\right)dx\,dy\,dz = \frac{4}{15}\pi\,(a+b+c)\,R$$

taken throughout

$$\left\{(x,y,z): x^2+y^2+z^2 \le R^2\right\}.$$

6. Show that

$$\iiint z^2\,dx\,dy\,dz = \frac{30\pi-32}{225}a^5,$$

the field of integration being the region bounded by the surfaces

$$x^2+y^2+z^2=a^2,\ x^2+y^2=ax.$$

7. Show that the integral of the function

$$\sqrt{\left(x^2/a^2+y^2/b^2+z^2/c^2\right)}$$

taken throughout the region

$$\left\{(x, y, z): x^2 / a^2 + y^2 / b^2 + z^2 / c^2 \le 1\right\}$$

is $4\pi abc(e-2)$.

8. Evaluate

$$\iiint \sqrt{\left(\frac{1-x^2-y^2-z^2}{1+x^2+y^2+z^2}\right)}\, dx\, dy\, dz$$

integral being taken over all positive value of x, y, z such that $x^2 + y^2 + z^2 \le 1$.

9. Prove that

$$\iiint \left(\frac{dx\, dy\, dz}{\sqrt{(1-x^2-y^2-z^2)}} = \frac{\pi^2}{8}\right)$$

integral being extended to all positive values of the variables for which the expression is real.

10. Integrate $1/xyz$ throughout the volume bounded by the six spheres,

$$x^2 + y^2 + z^2 = ax, a'x, by, b'y, cz, c'z,$$

where a, a', b, b', c, c' are positive.

$$\left[\text{Take } u = (x^2+y^2+z^2)/x,\ v = (x^2+y^2+z^2)/y,\ w = (x^2+y^2+z^2)/z.\right]$$

11. Prove that $\iiint xyz\, dx\, dy\, dz$ taken through the volume common to three spheres

$$x^2 + y^2 + z^2 = 2ax,\ x^2 + y^2 + z^2 = 2by,\ x^2 + y^2 + z^2 = 2cz$$

is $\frac{2}{15}\left(a^{-2} + b^{-2} + c^{-2}\right)^3$.

12. Evaluate

$$\iiint x^{l-1}\, y^{m-1}\, z^{n-1}\, dx\, dy\, dz$$

when the variables (x, y, z) are all positive and

(*i*) $x + y + z \le 1$. (*ii*) $(x/a)^p + (y/b)^q + (z/c)^r \le 1$

13. Show that

$$\iiint (x+y+z)\, x^2y^2z^2\, dx\, dy\, dz = \frac{1}{50400}$$

throughout the region

$$x + y + z \le 1,\ x \ge 0, y \ge 0, z \ge 0.$$

14. Evaluate

$$\iiint xyz^2\, dx\, dy\, dz \text{ for } 0 \le 4(x-2)^2 + (y-1)^2 + 9z^2 \le 36.$$

15. Evaluate

$$\iiint \left(y^2z^2 + z^2x^2 + x^2y^2\right) dx\, dy\, dz$$

taken through the volume of the cylinder $x^2 + y^2 - 2ax = 0$ between the sheets of the cone $z^2 = k^2\left(x^2 + y^2\right)$.

16. Prove that

$$\int_0^\infty \int_0^\infty \int_0^\infty \frac{dx\, dy\, dz}{\left(a^2 + x^2 + y^2 + z^2\right)^{\frac{9}{2}}} = \frac{4\pi}{105\, a^6}.$$

17. Find $\iiint z\, dx\, dy\, dz,$
throughout the region

$$x^2 + y^2 + z^2 \le 1,\; x^2 + y^2 \le z^2,\; z \ge 0.$$

18. Calculate

$$\iiint x^\alpha y^\beta z^\gamma (1 - x - y - z)^\delta\, dx\, dy\, dz,$$

over the interior of the tetrahedron formed by the co-ordinate planes and the plane

$$px + qy + rz = 1.$$

19. Find the value of

$$\iiint F\left[\left(\frac{x}{a}\right)^p + \left(\frac{y}{b}\right)^q + \left(\frac{z}{c}\right)^r\right] x^{l-1} y^{m-1} z^{n-1}\, dx\, dy\, dz$$

for all positive values of x, y, z subject to the condition that

$$\left[\frac{x}{a}\right]^p + \left[\frac{y}{b}\right]^q + \left[\frac{z}{c}\right]^r \le h.$$

20. Prove that

$$\iiint \frac{1}{x^2}\sqrt{\left(\frac{x^2}{a^2} + \frac{y^2}{b^2} + \frac{z^2}{c^2}\right)}\, dx\, dy\, dz = \frac{\pi}{lm}$$

taken over the smaller region bounded by the ellipsoid $x^2/a^2 + y^2/b^2 + z^2/c^2 = 1$ and the half-cone

$$z^2 = l^2x^2 + m^2y^2,\; z > 0.$$

21. Evaluate

$$\iiint \sqrt{(1-z)}\, dx\, dy\, dz$$

over the interior of the tetrahedron with faces $x = 0$, $y = 0$, $z = 0$, $x + y + z = 1$ and

$$\iiint \frac{dx\,dy\,dz}{a^2 + x^2 + y^2 + z^2}$$

over the interior of the sphere $x^2 + y^2 + z^2 = a^2$.

22. Show that

$$\iiint_W e^{a(2x+3y+4z)}\,dx\,dy\,dz = \frac{\left(e^s - 1\right)^3\left(3e^s + 1\right)}{24a^3}.$$

where Ω is the region bounded by

$$x = 0,\ y = 0,\ z = 0,\ x + y + z = 1.$$

Deduce that

$$\iiint_W (2x + 3y + 4z)^{2x+3y+4z}\,dx\,dy\,dz = \frac{\left(e^2 - 1\right)^2}{8}.$$

23. If A is the smaller of the two regions into which the interior of the ellipsoid $x^2/16 + y^2/9 + z^2/4 = 1$ is divided by the plane $x + y + z = 1$, determine the volume of A and evaluate

$$\iiint_A z^3\,dx\,dy\,dz.$$

24. Evaluate

$$\iiint_V x^2\,dx\,dy\,dz,$$

where V is the sphere $x^2 + y^2 + z^2 \le 1$.

25. If

$$I = \int\int\int\int_W e^{-x^2 - y^2 - z^2 - u^2}\,dx\,dy\,dz\,du,$$

where Ω is the finite region in 4 dimensions bounded by

$$x^2 + y^2 + z^2 + u^2 = a^2,$$

prove that

$$I = p\iint_C \left(e^{-x^2 - y^2}c^{-a^2}\right)dx\,dy$$

where C is the interior of the circle $x^2 + y^2 = a^2$.

Deduce that

$$I = p^2\left(1 - e^{-a^2} - a^2e^{-a^2}\right).$$

26. Prove that

$$\iiint dx\,dy\,dz = \frac{\frac{4}{3}\pi}{(1-m)/(1+2m)}$$

where the integration over all values of x, y, z such that

$$x^2 + y^2 + z^2 + 2m(xy + yz + zx)$$

does not exceed unity and

$$-\frac{1}{2} < m < 1.$$

18.8. EVALUATION OF VOLUMES

18.8.1. By Triple Integration

The triple integral

$$\iiint_E dx\, dy\, dz,$$

carried throughout a region E in space of three dimensions gives the volume of E.

18.8.2. By Double Integration

Let C be the boundary of a region E of the xy-plane and let a cylinder be constructed by lines through the points of C parallel to z-axis. Then the volume of the cylinder enclosed between the surfaces

$$z = \varphi(x, y), z = \psi(x, y), [\varphi(x, y) \geq \psi(x, y)]$$

is, as can be easily seen, given by the double integral

$$\iint_E (\varphi - \psi)\, dx\, dy,$$

EXAMPLES

1. *Find the volume enclosed by the surfaces* $x^2 + y^2 = cz$, $x^2 + y^2 = 2ax$, $z = 0$.

Sol. The limits of z are 0 to $(x^2 + y^2)/c$. The limits of y are from $-\sqrt{(2ax - x^2)}$ to $\sqrt{(2ax - x^2)}$ and the limits of x are from 0 to $2a$.

$\therefore$ The value

$$(V) = 2\int_0^{2a}\int_0^{\sqrt{(2a - x^2)}}\int_0^{(x^2 - y^2)/c} dx\, dy\, dz$$

$$= 2\int_0^{2a}\int_0^{\sqrt{(2ax - x^2)}} \frac{x^2 + y^2}{c}\, dx\, dy$$

$$= \frac{2}{c}\int_0^{2a}\left[x^2\sqrt{(2ax - x^2)} + \frac{1}{3}(2ax - x^2)^{3/2}\right] dx$$

$$= \frac{2}{c}\int_0^{2a}\left[x^2\sqrt{\{a^2 - (a - x)^2\}} + \frac{1}{3}\{a^2 - (a - x)^2\}^{3/2}\right] dx$$

Put $a - x = a \sin\theta$

$$= -\frac{2}{c}\left[\int_{\pi/2}^{-\pi/2} (a - a\sin\theta)^2 a^2 \cos^2\theta \, d\theta + \frac{1}{3}\int_{\pi/2}^{-\pi/2} a^4 \cos^4\theta \, d\theta\right]$$

$$= \frac{2a^4}{c}\int_{-\pi/2}^{\pi/2} \left(\cos^2\theta + \cos^2\theta \sin^2\theta - 2\cos^2\theta \sin\theta - \frac{1}{3}\cos^4\theta\right)$$

$$= \frac{3\pi a^4}{2c}.$$

2. *Find the volume bounded by*

$$y^2 + z^2 = 4ax, \; y^2 = ax, \; x = 3a.$$

Sol. Limits of $z = \pm\sqrt{(4ax - y^2)}$ from $y^2 + z^2 = 4ax$

Limits of $y = \pm\sqrt{(ax)}$ from $y^2 = ax$

Limits of $x = 0$ to $3a$

$$\therefore \quad V = \int_0^{3a} 2\int_0^{\sqrt{(ax)}} 2\int_0^{\sqrt{(4ax - y^2)}} dx\, dy\, dz$$

$$= 4\int_0^{3a}\int_0^{\sqrt{(ax)}} \sqrt{(4ax - y^2)}\, dx\, dy$$

$$= 4\int_0^{3a}\left[\frac{1}{2}y\sqrt{(4ax - y^2)} + \frac{1}{2}\cdot 4ax \sin^{-1}\frac{y}{2\sqrt{ax}}\right]_0^{\sqrt{(ax)}} dx$$

$$= \frac{4}{2}\int_0^{3a}\left[\sqrt{(ax)}\sqrt{(3ax)} + 4ax\cdot\frac{\pi}{6}\right] dx$$

$$= a^3\left(6\pi + 9\sqrt{3}\right).$$

3. *The space enclosed by the planes $x = 0$, $y = 0$, $x + y = 1$ and surface $zx = e^{x+y}$ is filled with matter whose density at any point (x, y, z) is given by $\rho = (x/y)^{2/3}$. Show that the whole mass is $2\pi(e-1)/\sqrt{3}$.*

Sol. Whole mass $= \iiint \rho \, dx\, dy\, dz$

$$= \int_0^1\int_0^{1-x}\int_0^{\frac{e^{x+y}}{x}} \left(\frac{x}{y}\right)^{2/3} dx\, dy\, dz$$

$$= \int_0^1\int_0^{1-x} \frac{e^{x+y}}{x}\cdot\frac{x^{2/3}}{y^{2/3}}\, dx\, dy$$

$$= \iint e^{x+y} \cdot x^{(2/3-1)} \; y^{(1/3-1)} \, dx\, dy$$

where $0 < x + y \leq 1$

$$= \frac{\Gamma\left(\frac{2}{3}\right)\Gamma\left(\frac{1}{3}\right)}{\Gamma\left(\frac{2}{3}+\frac{1}{3}\right)} \int_0^1 e^u . u^{\frac{2}{3}+\frac{1}{3}-1} \, du$$

by Liouville's Theorem

$$= \frac{\Gamma\left(\frac{1}{3}\right)\Gamma\left(1-\frac{1}{3}\right)}{1} \int_0^1 e^u \, du$$

$$= \frac{\pi}{\sin(\pi/3)} [e]_0^1 = \frac{2\pi}{\sqrt{3}} (e-1).$$

4. *Use the theorem* $V = \iiint dx\, dy\, dz = \iiint J\, du\, dv\, dw$ *to find the volume of the parallelopiped enclosed by the planes*

$$ax + by + cz = 0, \qquad ax + by + cz = d;$$
$$a_1x + b_1y + c_1z = 0, \qquad a_1x + b_1y + c_1z = d_1;$$
$$a_2x + b_2y + c_2z = 0, \qquad a_2x + b_2y + c_2z = d_2.$$

Sol. Put $u = ax + by + cz = 0$

$v = a_1x + b_1y + c_1z = 0$

and $w = a_2x + b_2y + c_2z = 0$

$\therefore$ limits of u are from 0 to d; of v from 0 to d_1 and of w from 0 to d_2, and

$$J = \frac{\partial(x, y, z)}{\partial(u, v, w)} = \frac{1}{\dfrac{\partial(u, v, w)}{\partial(x, y, z)}}$$

$$= \frac{1}{\begin{vmatrix} a & b & c \\ a_1 & b_1 & c_1 \\ a_2 & b_2 & c_2 \end{vmatrix}} = \frac{1}{D}, \text{ say}$$

$$\therefore \quad \text{Volume} = \iiint dx\, dy\, dz$$

$$= \iiint J\, du\, dv\, dw \qquad \text{(given)}$$

$$= \frac{1}{D} \int_0^d \int_0^{d_1} \int_0^{d_2} du\, dv\, dw$$

$$= \frac{d d_1 d_2}{D}.$$

5. *The sphere $x^2 + y^2 + z^2 = a^2$ is pierced by the cylinder $(x^2 + y^2)^2 = a^2 (x^2 - y^2)$; prove that the volume of the sphere that lies inside the cylinder is* $\frac{8}{3}\left(\frac{\pi}{4}+\frac{5}{3}-\frac{4\sqrt{2}}{3}\right)a^3$.

Here limits of z are from $-\sqrt{(a^2 - x^2 - y^2)}$ to $\sqrt{(a^2 - x^2 - y^2)}$ and therefore,

$$\text{the volume} = \iiint dx\, dy\, dz$$

$$= 2\iint \sqrt{(a^2 - x^2 - y^2)}\, dx\, dy.$$

Now the equation of the cylinder is

$$(x^2 + y^2)^2 = a^2 (x^2 - y^2)$$

Putting $x = r\cos\theta$, $y = r\sin\theta$, we get

$$r^2 = a^2 \cos 2\theta.$$

$\therefore$ The limits of r are from $-a\sqrt{(\cos 2\theta)}$ to $a\sqrt{(\cos 2\theta)}$ and limits of θ are from $-\pi/4$ to $\pi/4$.

$\therefore$ The required volume

$$= 2\int_{-\pi/4}^{\pi/4} \int_{-a\sqrt{(\cos 2\theta)}}^{a\sqrt{(\cos 2\theta)}} \sqrt{(a^2 - r^2)}\, r\, d\theta\, dr$$

$$= 8\int_0^{\pi/4} \frac{1}{2}\left[-\frac{(a^2 - r^2)^{3/2}}{3/2}\right]_0^{a\sqrt{(\cos 2\theta)}} d\theta$$

$$= \frac{8}{3}a^3 \int_0^{\pi/4} \left(1 - 2^{3/2}\sin^3\theta\right) d\theta$$

$$= \frac{8}{3}a^3 \int_0^{\pi/4} \left\{1 - 2^{3/2}\left(1 - \cos^2\theta\right)\sin\theta\right\} d\theta$$

$$= \frac{8}{3}a^3 \left[\frac{\pi}{4} - 2^{3/2}\left(-\frac{1}{\sqrt{2}} + \frac{1}{3\times 2\sqrt{2}} + 1 + \frac{1}{3}\right)\right]$$

$$= \frac{8}{3}a^3 \left[\frac{\pi}{4} + \frac{5}{3} - \frac{4\sqrt{2}}{3}\right].$$

EXERCISES

1. Find the volume included between the co-ordinate planes, the part of the right cylinder standing on the quadrant $y = \sqrt{(9 - x^2)}$ for which x and y are both positive and the plane $z = 3 - \frac{1}{3}x - \frac{3}{4}y$.

2. Show that the volume included between the elliptical paraboloid $2z = x^2/p + y^2/q$, the cylinder $x^2 + y^2 = a^2$ and the xy-plane is $\pi a^2 (p + q)/8pq$.

3. Calculate the volume bounded by the surface $z = kxy$, the plane OXY and the first quarter of the cylinder $x^2 + y^2 = a^2$.

4. Show that the volume of the solid bounded by the cylinder $x^2 + y^2 = 2ax$ and the paraboloid $y^2 + z^2 = 4ax$ is $\frac{2}{3}a^3(3\pi + 8)$.

5. Find the volume of the region bounded by the plane $z = x + y$ and the paraboloid $cz = x^2 + y^2$.

6. Show that the volume of the region bounded by the hyperboloid of one sheet $x^2/a^2 + y^2/b^2 - z^2/c^2 = 1$, its asymptotic cone $x^2/a^2 + y^2/b^2 - z^2/c^2 = 0$ and the planes $z = z_1$, $z = z_2$, $(z_2 > z_1)$ is $\pi ab(z_2 - z_1)$.

7. Prove that the volume in the positive octant bounded by the planes $x = 0$, $y = 0$, $z = h$ and the surface $(z/c)^m = (x/a)^m + (y/b)^m$ is equal to

$$\frac{\frac{1}{2}abh(h/c)^{2/m}\,\Gamma(1/m)\,\Gamma(1/m)}{(m+2)\,\Gamma(2/m)}$$

8. Show that the volume exposed by the surfaces defined by the equations $x^2 + y^2 = c^2$, $x^2 + y^2 = ax$, $z = 0$ is $3\pi a^4/32c$.

9. Show that the active volume bounded by the positive side of the three coordinate planes and the surface

$$(x/a)^{\frac{1}{2}} + (y/b)^{\frac{1}{2}} + (z/c)^{\frac{1}{2}} = 1, \text{ is } abc/90.$$

[Change the variables to r, θ, φ where

$$x/a = r^4\sin^4\theta\cos^4\varphi,\; y/b = r^4\sin^4\theta\sin^4\varphi,\; z/c = \cos^4\theta;$$

varies from 0 to 1, θ from 0 to $\frac{1}{2}\pi$ and φ from 0 to $\frac{1}{2}\pi$].

10. Show that entire volume of the solid

$$(x/a)^{\frac{2}{3}} + (y/b)^{\frac{2}{3}} + (z/c)^{\frac{2}{3}} = 1 \text{ is } 4\pi abc/35.$$

11. Show that the volume of the solid bounded by the cylinders

$bz^2 = c^2y$, $bz^2 = 2c^2y$, $cx^2 = a^2z$, $cx^2 = 2a^2z$; $ay^2 = b^2x$, $ay^2 = 2b^2x$ is $\frac{1}{7}abc$.

12. The area lying in the first quadrant which is enclosed by the curve

$y = ax^3, y = bx^3; x = cy^3, x = dy^3, (a > b, c > d)$

revolves about X-axis; obtain the volume of the solid generated.

[The required volume is

$$\iint 2\pi y \, dx \, dy;$$

change the variables to u, v where

$u = y/x^3$, $v = x/y^3$]

13. A plane lamina of thickness, k, is bounded by the lines

$$\frac{x}{a}+\frac{y}{b}=1, \frac{x}{a}+\frac{y}{b}=2,$$

and the rectangular axes $x = 0$, $y = 0$. The density at each point (x, y) is proportional to the nth power of the distance from the line $x/a + y/b = 1$.

Find the mass of the lamina.

14. Find the ratio in which the volume of the ellipsoid

$$\frac{x^2}{a^2}+\frac{y^2}{b^2}+\frac{z^2}{c^2}=k, \; k>1$$

is divided by the hyperboloid

$$\frac{x^2}{a^2}+\frac{y^2}{b^2}-\frac{z^2}{c^2}=1.$$

15. Prove that the volume of that part of the sphere $x^2 + y^2 + z^2 = 1$ lying inside the cone formed by joining the origin to the points of intersection of the sphere with the surface $36x^2 + 25y^2 - 16z = 16$ is $16\pi/45$.

16. Prove that the volume common to the two cylinders $x^2 + y^2 = a^2$ and $x^2 + z^2 = a^2$ is $16a^3/3$.

17. Find the volume of that part of the cylinder $x^2 + y^2 - 2ax = 0$ which is cut off by the solid $z^2 = 2ax$.

18. Find the volume of the part of the solid sphere $x^2 + y^2 + z^2 \le 4a^2$ enclosed by the cylinder

$$\left(x^2+y^2\right)^2 = 2a^2\left(2x^2+y^2\right).$$

19. Prove that the 'volume' and 'area' of n-dimensional sphere of radius are respectively

$$\frac{1}{2^n} a^n \left[\Gamma\left(\frac{1}{2}\right)\right]^n / \Gamma\left(\frac{1}{2}n+1\right) \text{ and } 2a^{n-1}\left[\Gamma\left(\frac{1}{2}\right)^n / \Gamma\left(\frac{1}{2}n\right)\right].$$

20. Prove that the volume of a cone which extends from the origin to surface $x = f(u, v)$, $y = g(u, v)$, $z = h(u, v)$ is given by

$$\frac{1}{3}\iint \Delta \, du \, dv$$

where

$$\Delta = \begin{vmatrix} x & y & z \\ x_u & y_u & z_u \\ x_v & y_v & z_v \end{vmatrix}$$

and suffixes denote partial derivatives

$$x_u = \frac{\partial x}{\partial u} \dots$$

Hence prove that the volume of one octant of the cone

$$x^4 + y^4 = x^4 \tan^2 \alpha$$

between its vertex and the surface

$$x^4 + y^4 + z^4 = 1$$

is

$$\frac{\left[\Gamma\left(\frac{1}{4}\right)\right]^2}{24\,\Gamma\left(\frac{1}{2}\right)} \int_0^x \frac{du}{\sqrt{(\cos u)}}.$$

MISCELLANEOUS EXERCISES

1. If d is a positive integral but not the square of an integer and

$$y = x\,(x^2 + 3d)/(3x^2 + d),$$

where x is a positive rational number, show that

$$y - x = \frac{2x\left(d - x^2\right)}{3x^2 + d} \text{ and } y^2 - d = \frac{\left(x^2 - d\right)^3}{\left(3x^2 + d\right)^2}.$$

Hence show that the section of the positive rational numbers determined by assigning to the upper class all rational numbers whose square is greater than d, and to the lower class all other rational numbers, is not generated by a rational number.

2. Show that an algebraic equation

$$x^n + a_1 x^{n-1} + a_2 x^{n-2} + \dots + a_n = 0,$$

with integral co-efficients, cannot have a rational but non-integral root.

Deduce that $\sqrt{3}$ and $\sqrt[3]{2}$ are not rational numbers.

3. If m/n is a good approximation to $\sqrt{2}$, prove that

$$(m + 2n)/(m + n)$$

is a better one, and that the errors in the two cases are in opposite direction. Apply this result to show that the limit of the sequence

$$\frac{1}{1}, \frac{3}{2}, \frac{7}{5}, \frac{17}{12}, \frac{41}{29}, \frac{99}{70}, \frac{239}{169}, \frac{577}{408} \dots$$

is $\sqrt{2}$.

4. Prove that, as $n \to \infty$,

$$\left[(2n)!/(n!)^2\right]^{\frac{1}{n}} \to \frac{1}{4}.$$

5. Prove that, as $n \to \infty$,

$$(n!)(a/n)^n \to 0 \text{ or } +\infty,$$

according as $a <$ or $> e$.

6. Show that, as $n \to \infty$.

(*i*) $\left(n^5/2^n\right) \to 0$. (*ii*) $\left(n^2\right)^{\frac{1}{n}} \to 1$,

7. Denoting

$$\lim\left(1 + \frac{1}{2} + \frac{1}{3} + \ldots + \frac{1}{n} - \log n\right),$$

by γ, known as Euler's constant, show that

$$\lim \frac{1}{n}\left[\frac{n}{1} + \frac{n-1}{2} + \frac{n-2}{3} + \ldots + \frac{1}{n} - \log(n!)\right] = \gamma.$$

8. Prove that if a and b are positive, then

$$\frac{1}{2}\left\{\left(a^{1/n} + b^{1/n}\right)\right\}^n \to \sqrt{(ab)}, \text{ as } n \to \infty.$$

9. If $S_n \to S$ as n tends to infinity, prove that the following sequences also tend to S.

(*i*) $\left[\frac{S_n}{1} + \frac{S_{n-1}}{2} + \frac{S_{n-2}}{3} + \ldots + \frac{S_1}{n}\right] / \log n.$

(*ii*) $\left(S_1 + 2S_2 + 2^2 S_3 + \ldots + 2^{n-1} S_n\right)^{2-n}$

10. Prove that if $x > 0$, then

$$n \log\left\{\frac{1}{2}\left(1 + x^{1/n}\right)\right\} \to -\frac{1}{2}\log x, \text{ when } n \to \infty.$$

11. If

$$f(x, y) = \left(xy^5 - x^2y^3\right)/\left(x^2 + xy^2 + y^4\right), \text{ when } (x, y) \neq (0, 0),$$

$$f(0, 0) = 0,$$

show that at the origin $f_{yx} \neq f_{xy}$ but that f_x is continuous and f_y differentiable the origin. Discuss the applicability of the condition of § 13.6.1.

12. Discuss the existence and the equality at the origin of f_{yx} and f_{xy} for

$$f(x, y) = \left(ax^2 + 2bxy + cy^2\right)^3 / \left(x^2 + y^2\right)^2 \quad \text{when} \quad (x, y) \neq (0, 0),$$

and $\quad f(0, 0) = 0.$

13. If at all points of the plane, f is continuous with respect to x and has partial derivative $f_y(x, y)$ and, if $f_y(x, y)$ is bounded in the whole plane, prove that f is continuous with respect to two variables at all points of the plane.

14. The number θ is defined by the Taylor's formula

$$f(a+h, b+k) = f(a, b) + hf_x(a+\theta h, b+\theta k) + kf_y(a+\theta h, b+\theta k).$$

Prove that if f_x and f_y are differentiable at (a, b) and if $(h, k) \to (0, 0)$ so that

$$\left(h^2 + k^2\right) / \left\{h^2 f_{xx}(a, b) + 2hkf_{xy}(a, b) + k^2 f_{yy}(a, b)\right\}$$

is bounded, then $\theta \to \dfrac{1}{2}$.

15. Discuss at (0, 0) the existence of the first order partial derivatives and the continuity and differentiability of f itself when $f(x, y)$ has the following forms :

(*i*) $x^2 y^2 \log\left(x^2 + y^2\right)$. (*ii*) $xy \log\left(x^2 + y^2\right)$,

$f(0, 0)$ being equal to zero in either case.

16. Show that, if f_x, f_y are differentiable at a point, then f is also differentiable at the point.

17. Find the values of x, y, z for which the function

$$e^{-u}(x - y + 2z)$$

is a maximum, where

$$u = x^2 + y^2 + 2z^2 - 2yz + 2xz - xy.$$

18. From the point $B(0, -b)$, of the ellipse $x^2/a^2 + y^2/b^2 = 1$, a chord BP of the ellipse is drawn. Show that if $a^2 \geq 2b^2$, the greatest length the chord can have is

$$a^2 / \sqrt{\left(a^2 - b^2\right)}.$$

What is the corresponding result when $a^2 < 2b^2$.

19. Find all the stationary values of the function

$$\frac{x}{y} + \frac{y}{x} - \frac{(x-y)^2}{a^2}$$

of the variables (x, y); discuss how their nature depends on the values of the parameter a. Consider in particular the point at which $x = y = a$.

20. P and Q are any two points on the parabolas

$$y^2 = 4x, y^2 = 2x - 6.$$

respectively. Show that the minimum distance between P and Q is $\sqrt{5}$.

21. If $m \geq 0$, prove that

$$\iint \left(1 - x^2 - y^2\right)^m f(Ax + By)\, dx\, dy$$

$$= B\left(\frac{1}{2}, m+1\right) \int_{-1}^{1} \left(1 - x^2\right)^{m+\frac{1}{2}} f(kx)\, dx, k^2 = A^2 + B^2,$$

when the integral on the left is taken over the circle $x^2 + y^2 = 1$.

22. By using the transformation

$$x = uv, y = u(1 - v),$$

prove that

$$\iint \frac{f(x+y)}{\sqrt{(xy)}}\, dx\, dy = \pi \int_0^{\alpha} f(u)\, du,$$

where the double integral extends over all positive values of x and y subject to

$$x + y = \alpha.$$

23. Show that volume of the wedge intercepted between the cylinder $x^2 + y^2 = 2ax$ and the planes $z = x \tan \alpha$, $z = x \tan \beta$ is $\pi (\tan \beta - \tan \alpha) a^3$.

24. Find the volume contained between the ellipsoid

$$x^2 / a^2 + y^2 / b^2 + z^2 / c^2 = 1$$

and the cylinder $x^2/a^2 + y^2/b^2 = x/a$.

25. Show that the volume obtained by one revolution of the loop of the Folium $x^3 + y^3 = 3axy$ about OX is $4\pi^3 a^3 / 3\sqrt{3}$.

[Change the variables to u, v where $u = x^2/y$, $v = y^2/x$]

26. If $x = r \cos \theta$, $y = r \sin \theta$, find the volume included by the surfaces whose equations are $r = a$, $z = 0$, $\theta = 0$, $z = mr \cos \theta$.

27. Using the transformation

$$x = c \cosh u \cos v, y = c \sinh u \sin v,$$

or otherwise, where $c^2 = a^2 - b^2$, show that the mean value of the product of the focal distances of a point inside the ellipse $x^2/a^2 + y^2/b^2 = 1$ is

$$\frac{1}{4}\left[a^2 + b^2 + \frac{\left(a^2 - b^2\right)^2}{a - b} \log \frac{a + b}{a - b}\right]$$

[The mean value $= \left[\iint (SPS'\, dx\, dy)\right] / \iint dx\, dy$].

28. Evaluate

$$\iint\left[\left(2x+y^2\right)/xy\right]dx\,dy$$

taken over the area in the positive quadrant of the xy-plane bounded by the curves

$$x^2+y^2=h^2,\ x^2+y^2=k^2,\ y^2=4ax,\ y^2=4bx.$$

29. The axes of two equal cylinders intersect at right angles. If, a, be their radius, show that the volume common to the cylinders is $(163)a^3$.

[If $x^2+y^2=a^2$ and $y^2+z^2=a^2$ be the two cylinders then the required volume

$$=8\int_0^a dx\int_0^{\sqrt{(a^2-x^2)}} dy\int_0^{\sqrt{(a^2-y^2)}} dz.$$

30. Prove that the graph of

$$y=f(x)=\frac{2}{\pi}\int_0^\infty \frac{\sin^2 t\cos xt}{t}\,dx,$$

consists of parts of the lines $y=-\frac{1}{4}$, $y=0$, $y=\frac{1}{2}$ together with four isolated points.

31. Show that, when $n\to\infty$,

$$\int_a^b f(x)\,|\sin nx|\,dx\to\frac{2}{\pi}\int_a^b f(x)\,dx.$$

32. Evaluate the limit of the sequence $\{u_n\}$, where

$$u_n=\frac{1}{\sqrt{n}}\left(\frac{1}{\sqrt{1}}+\frac{1}{\sqrt{2}}+\frac{1}{\sqrt{3}}+\ldots+\right),\ \text{when}\ n\to\infty.$$

33. Show that, as $n\to\infty$,

$$\frac{2.2.4.4.6.6\ldots(2n)(2n)}{1.3.3.5.5.7\ldots(2n-1)(2n+1)}\to\frac{p}{2}. \qquad \text{(Walli's formula)}$$

Deduce that

$$\left\{\left[(n!)^2 2^{2n}\right]/(2n)!\sqrt{n}\right\}\to\sqrt{\pi}\ \text{as}\ n\to\infty.$$

34. f is a non-negative function admitting an elementary infinite integral and another function $f_n(x)$ is defined thus :

$$f_n(x)=\begin{cases} f(x), & \text{if } f(x)<n,\\ n, & \text{if } f(x)\geq n,\end{cases}$$

show that

(*i*) if the interval of integration is bounded, then as $n\to\infty$

$$\int_a^b f_n(x)\,dy\to\int_a^b f(x)\,dx.$$

(*ii*) if the interval of integration is infinite, then as $n \to \infty$

$$\int_{-n}^{+n} f_n(x)\,dx \to \int_{-\infty}^{+\infty} f(x)\,dx.$$

35. If f, f' are continuous in $[a, b]$ and

$$S_n = \sum_{r=1}^{n} hf\,[a + rh],\ I = \int_a^b f(x)\,dx,$$

where $h = (b - a)/n$, then as $\to \infty$,

$$n(S_n - 1) \to \frac{1}{2}(b - a)\left[f(b) - f(a)\right].$$

36. If I, h have the same meaning as in the Ex. above, if but now $f''(x)$ is also continuous in $[a, b]$ and

$$S_n = \sum_{r=1}^{n} hf\left[a + \frac{1}{2}(2r - 1)\,h\right]$$

show that, as $n \to \infty$,

$$h^2(1 - S_n) \to \frac{1}{24}(b - a)^2\left[f'(b) - f'(a)\right].$$

37. If $f(t, x) = \pi t \sin \pi tk$ when $0 \le x \le 1/i$ and $f(t, x) = 0$ when $1/t \le x \le 1$, show that

$$\int_0^1 \left\{\lim_{t \to \infty} f(t, x)\right\} dx = 0 \ne 2 = \lim_{t \to \infty} \left\{\int_0^1 f(t, x)\,dx\right\}.$$

38. If $f(t, x) = tx/\left(1 + t^2x^4\right)$, show that

$$\int_0^1 \left\{\lim_{t \to \infty} f(t, x)\right\} dx = 0 \ne \frac{\pi}{4} = \lim_{t \to \infty} \left\{\int_0^1 f(t, x)\,dx\right\}.$$

39. Prove that if

(*i*) a is positive;

(*ii*) f is continuous except perhaps at the origin;

(*iii*) $\int_0^a f(x)\,dx = \lim_{e \to 0} \int_x^a f(x)\,dx$ where the limit exists

(*iv*) $g(x) = \int_x^a \frac{f(t)}{t}\,dx,$

then $\int_0^a g(x)\,dx = \int_0^a f(x)\,dt.$

[**Hint.** Change the order of integration.]

40. If $0 < a < b$ and f is continuous in $[0, b]$, prove that

$$\lim_{h \to (0+0)} \int_a^1 \frac{f(at) - f(bt)}{t}\,dt$$

exists and is equal to

$$f(0)\log\frac{b}{a}-\int_a^b\frac{f(x)}{x}\,dx$$

41. Show that

$$\int_{-1}^{+1}\frac{\sqrt{(1-x^2)}\,dx}{(1+\alpha^2+2\alpha x)(1+\beta^2+2\beta x)}=\begin{cases}\pi/2\,(1-\alpha\beta), \text{ if } \alpha^2<1, \beta^2<1,\\ \pi/2\alpha\,(\alpha-\beta), \text{ if } \alpha^2>1>\beta^2.\end{cases}$$

42. A is a fixed point on the axis OX and an ellipse is drawn with OA as its minor axis. Calculate the value of $f\,(x^2\,dy-y^2\,dx)$ taken along the elliptic are from O to A, y negative and show that if it has its max. values the eccentricity is $\sqrt{(1-64/9\pi^2)}$.

43. Show that

$$\tan^{-1}x=x-\frac{x^3}{3}+\frac{x^5}{5}-\frac{x^7}{7}+\ldots+\frac{(-1)^{n-1}\,x^{2n-1}}{2n-1}+(-1)^n x^n\int_0^x\frac{t^{2n}}{1+t^2}\,dt.$$

44. Show that, when $n\to\infty$

$$\int_0^1\frac{x^{n+1}}{1-x}\log x\,dx\to 0.$$

Sol. We write,

$$\int_0^1\frac{x^{n+1}}{1-x}\log x\,dx=\int_0^a\frac{x^{n+1}}{1-x}\log x\,dx+\int_a^1\frac{x^{n+1}}{1-x}\log x\,dx=I_1+I_2.$$

$$(0<a<1)$$

Now, $|I_1|\le a^{n+1}\int_0^a\left|\frac{\log x}{1-x}\right|dx$, which $\to 0$ as $n\to\infty$; a being <1.

Again, if we assign to the relation $\log x/(1-x)$, the value 1 for $x=1$, it becomes continuous in $[a, 1]$. Thus $\log x/(1-x)$ is bounded in $[a, 1]$. Therefore

$$|I_2|\le M\int_a^1 x^{n+1}\,dx \text{ which}\to 0, \text{ as } n\to\infty.$$

Hence the result.

45. From the identity, $\dfrac{1}{1-x}=1+x+x^9+\ldots+x^n+\dfrac{x^{n+1}}{1-x}$, show by integration in the interval [0, 1], that

$$\int_0^1\frac{\log x}{1-x}\,dx=-\sum_{n=1}^{\infty}\frac{1}{n^2}.$$

46. If p, q are positive, prove that

$$\int_0^{\infty}\left\{e^{-px}\left[1+(p+r)x\right]-e^{-qx}\left[1+(q+r)x\right]\right\}\frac{dx}{x^2}=q-p+r\log\frac{q}{p}.$$

47. Show that $\int_0^\infty \frac{x^{p-1}}{1+x}\,dx = \frac{\pi}{\sin p\pi}\cdot \quad (0<p<1).$

Sol. It is easy to see that the given improper integral converges, when $0 < p < 1$. We have

$$\int_0^\infty \frac{x^{p-1}}{1+x}\,dx = \int_0^1 \frac{x^{p-1}}{1+x}\,dx + \int_1^\infty \frac{x^{p-1}}{1+x}\,dx.$$

By means of the substitution $y = 1/x$, we see that

$$\int_1^\infty \frac{x^{p-1}}{1+x}\,dx = \int_0^1 \frac{y^{-p}}{1+y}\,dx = \int_0^1 \frac{x^{-p}}{1+x}\,dx.$$

$$\therefore \quad \int_0^\infty \frac{x^{p-1}}{1+x}\,dx = \int_0^1 \left\{\frac{x^{p-1}+x^{-p}}{1+x}\right\} dx$$

$$= \int_0^1 \left(x^{p-1}+x^{-p}\right)\left[1-x+x^2-\ldots+(-1)^n x^n+(-1)^{n+1}\frac{x^{n+1}}{1+x}\right] dx$$

$$= \sum_{k=0}^{n} (-1)^k \left(\frac{1}{k+p}+\frac{1}{k+1-p}\right) + R_n,$$

where $|R_n| = \int_0^1 \left(x^{p-1}+x^{-p}\right)\frac{x^{n+1}}{1+x}\,dx < \int_0^1 \left(x^{p-1}+x^{-p}\right)x^{n-1}\,dx$

$$= \left\{\frac{1}{n+p+1}+\frac{1}{n+2-p}\right\} \to 0 \text{ as } n\to\infty.$$

$$\therefore \quad \int_0^\infty \frac{x^{p-1}}{1+x}\,dx = \sum_{k=0}^{\infty} (-1)\,k\left(\frac{1}{k+p}+\frac{1}{k+p-1}\right) = \frac{\pi}{\sin p\pi}$$

48. Prove that $\Gamma(x+1) - 1\,\Gamma(x).$

49. Show that $\Gamma(a)\,\Gamma(1-a) = \pi \sin a\pi.$

We have

$$b(a, 1-a) = \frac{\Gamma(a)\,\Gamma(1-a)}{\Gamma(1)} = \Gamma(a)\,\Gamma(1-a).$$

Also

$$b(a, 1-a) = \int_0^1 x^{a-1}(1-x)^{-a}\,dx.$$

By means of the substitution $x = y\,(1+y)$, we may now show that

$$\int_0^1 x^{a-1}(1-x)^{-a}\,dx = \int_0^\infty \frac{y^{a-1}}{1+y}\,dy = \frac{\pi}{\sin a\pi}.$$

Hence the result.

50. Show that

$$\Gamma\left(\frac{1}{a}\right)\Gamma\left(\frac{2}{a}\right)\Gamma\left(\frac{3}{a}\right)\ldots\Gamma\left(\frac{a-1}{a}\right) = \left(\frac{(2\pi)^{a-1}}{a}\right)^{1/2}$$

51. Prove that

$$\sum_{k=1}^{n} \log \Gamma\left(\frac{k}{n}\right) = \frac{1}{2}(n-1)\log(2\pi)\frac{1}{2}\log n.$$

52. Show that $\int_0^1 \log \Gamma(x)\, dx = \frac{1}{2}\log(2\pi).$

Sol. The point '0' is a point of infinite discontinuity of the integral. We have

$$\Gamma(x) = \Gamma(x+1)/x,$$

so that $$\log \Gamma(x) = \log \Gamma(x+1) - \log x.$$

We know that the integral of $\log x$ converges at 0 and the integral of $\log \Gamma(x+1)$ is proper, and hence the integral of $\log \Gamma(x)$ is convergent. By means of the substitution $x = 1 - y$, we see that

$$\int_0^1 \log \Gamma(x)\, dx = \int_0^1 \log \Gamma(1-y)\, dy$$

$$= \int_0^1 \log \Gamma(1-x)\, dx$$

$$= \frac{1}{2}\int_0^1 \log\left[\Gamma(x)\,\Gamma(1-x)\right] dx$$

$$= \frac{1}{2}\int_0^1 \frac{\pi}{\sin \pi x} = \frac{1}{2}\log(2\pi).$$

53. Prove that $\left(\sqrt{x}\right)\Gamma(2x) = 2^{2x-1}\,\Gamma(x)\,\Gamma\left(x+\frac{1}{2}\right).$

[Duplication formula]

Sol. We have $$\frac{\Gamma(x)\,\Gamma\left(\frac{1}{2}\right)}{\Gamma\left(x+\frac{1}{2}\right)} = B\left(x, \frac{1}{2}\right) = \int_0^1 \frac{t^{x-1}\, dt}{\sqrt{(1+t)}}$$

and $$\frac{\Gamma(x)\,\Gamma(x)}{\Gamma(2x)} = B(x, x) = \int_0^1 t^{x-1}(1-t)^{x-1}\, dt.$$

In the latter integral, put $v = 2t - 1$ so that we have

$$B(x, x) = \frac{1}{2^{2x-1}}\int_{-1}^{+1}\left(1-v^2\right)^{x-1} dy$$

$$= \frac{1}{2^{2x-1}}\int_0^1 (1-w)^{x-1}\, w^{-\frac{1}{2}}\, dw, \quad \left(v^2 = w\right)$$

$$= \frac{1}{2^{2x-1}}\, B\left(\frac{1}{2}, x\right) = \frac{1}{2^{2x-1}}\, B\left(x, \frac{1}{2}\right).$$

Since $\Gamma\left(\frac{1}{2}\right) = \sqrt{\pi}$, we can now obtain the given result.

OBJECTIVE QUESTIONS

Note. *For each of the following questions, four alternatives are given for the answer. Only one of them is correct. Choose the correct alternative.*

1. Surface integral over a surface is denoted by a live integral taken over the boundary curve of the surface in

(*a*) Gauss's theorem (*b*) Stokes theorem

(*c*) Green's theorem (*d*) none of these

2. Volume of a rectangle $[a, b, c, d, e, f] \subset \mathbf{R}^3$ is

(*a*) $(b-a)(d-c)(f-e)$ (*b*) $(a-f)(b-c)(c-d)$

(*c*) $(a+b)(d+c)(f+e)$ (*d*) none of these

3. We express a triple integral taken over a three dimensional region in terms of a surface integral taken over the bounding surface of the region in

(*a*) Gauss's theorem (*b*) Stroke's theorem

(*c*) Green's theorem (*d*) none of these.

4. A surface in which it is possible to distinguish one of its sides from the other is :

(*a*) unilateral (*b*) bilateral

(*c*) multilateral (*d*) Mobius strip

5. Volume of the cylinder enclosed between the surface $z = \phi(x, y)$, $z = \psi(x, y)$ is

(*a*) $\int\int\int dx\, dy\, dz$ (*b*) $\int\int \phi\, dx\, dy + \int\int \psi\, dx\, dy$

(*c*) $\int\int (\phi - \psi)\, dx\, dy$ (*d*) none of these

ANSWERS

1. (*b*) **2.** (*a*) **3.** (*a*) **4.** (*b*) **5.** (*c*)

ANSWERS

Page 143

1. (*i*) $f(0) = 0$, (*ii*) $f(0)$ is not determinable (*iii*) $f(0) = a$.

Page 184

2. 0, – 1, For $f(x) = |x|$ and $c = 0$ the first limit does not exist but the second exists and is equal to 1. For $c \neq 0$, both the limits exist.

4. By Taylor's theorem, $f(0) - f(x) = -xf'(x) + 1/2\,(!)x^2 f''(\xi)$ so that $f'(x) - f(x)/x > 0$ and $\therefore d[f(x)/x]/dx$ is positive.

1. (*i*) continuous; (*ii*) $x = 0$ is point of discontinuity.

2. discontinuity at $x = 0, \frac{1}{2}, 1$.

3. (*a*) discontinuous for integral value of x; (*b*) discontinuous for integral values of x; (*c*) continuous for $x = \alpha$, α is an integer.

5. Continuous for integral values and at $x = n + \beta$, $\frac{1}{2} < \beta < 2$ and discontinuous at $x = x + \frac{1}{2}, n$ is an integer.

8. Discontinuous; 9. Discontinuous at $x = 0$,

10. Discontinuous; 11. Discontinuous at 0.

Page 264

12. Does not exist.

Page 289

(*i*) log 2. (*ii*) π. (*vi*) – 1. (*viii*) 2. The integrals (*iii*), (*iv*), (*v*) and (*vii*) do not exist.

Page 299

1. C stands for Convergent and N.C. for Not convergent.

(*i*) C for $n > 1$ (*ii*) C for $n < 2$

(*iii*) C for $a > 0$. (*iv*) $N.C.$

(*v*) C. (*vi*) C.

(*vii*) C for $m > 0$ and $n > 0$, (*viii*) C for $a > 0$. (*ix*) C

(*x*) C for $p > -1$. (*xi*) $N.C.$ (*xii*) $N.C.$

2. C for $n > -1$ and $m > 0$.

Page 305

1. (*i*) C for $p > (1 + m + n)$; m, n being > 0. Consider also the case when either m or n or both are negative.

(*ii*) C for $(n-m) > \frac{1}{2}$, (*iii*) C for $0 < m, n, < 1$,

(*iv*) C for $n > 0$ (*v*) C for $m < -1$

(*vi*) C for $0 < (a+1) < b$.

5. Convergent.

Page 308

(*i*) C (*ii*) C for $a > 0$.

Page 312

1. (*i*) C (*ii*) C. (*iii*) C. (*iv*) N.C.

(*v*) C for $(1-m) < n < (1+m)$. (*vi*) C for $0 < a < 4$.

(*vii*) C for $-1 < n < 2$.

(*viii*) C for $n > 0$, $-1 < m < n$ and for $n < 0$, $0 > m > (n-1)$.

2. (*i*) C. (*ii*) C. (*iii*) C for $p > 1$. (*iv*) C for $-3 < p < 2$.

Page 318

3. $\frac{\pi}{2} \log \frac{a}{b}$.

Page 319

11. The integral is $2/a$, or $-2/a$ according as $a > 1$, $a > -1$ or $-1 \le a \le 1$.

12. $\pi/2$ if $2m\pi < a < (2m+1)\pi$;

$-\pi/2$ if $(2m+1)\pi < a < (2m+2)\pi$

and 0 if $a = 2m\pi$ or $(2m+1)\pi$; m being any integer.

Page 320

15. $-\pi/4n$.

Page 347

4. $\frac{4}{\pi} \Sigma \frac{\sin \frac{1}{2} n\pi \sin nx}{n^2}$.

6. $\frac{2}{\pi} \Sigma \frac{4 \cos 2nx}{(2n^2-1)\pi}$.

7. (*a*) $\cosh \pi$.

(*b*) $-e^{-x}$ for $-\pi < x < 0$ and 0 for $x = 0, \pm\pi$ and e^{-x} for $-\pi \le x \le 0$.

Page 348

9. $\Sigma \left[\frac{-2(-1)^n \pi}{n} - \frac{4}{\pi n^3} \left[1 - (-1)^n \right] \right] \sin nx$. **12.** 0.

13. (*i*) $\dfrac{4lk}{\pi^2} \Sigma \dfrac{1}{n^2} \sin \dfrac{n\pi}{2} \sin \dfrac{n\pi x}{l}$.

(*ii*) $\dfrac{kl}{4} + \dfrac{4kl}{n^2} \Sigma \dfrac{1}{n^2}\left(\cos \dfrac{n\pi}{2} - \cos^2 \dfrac{n\pi}{2}\right) \cos \dfrac{n\pi x}{l}$.

14. $\Sigma \left[\dfrac{2(-1)^{n-1}}{n\pi} - \dfrac{4}{n^2\pi^2} \sin \dfrac{n\pi}{2} \sin \dfrac{n\pi x}{2} \right]$

16. $\dfrac{4-2\sqrt{2}}{\pi} + \dfrac{4\sqrt{2}}{\pi} \Sigma \left[\dfrac{(-1)^n - \sqrt{2}}{16n^2 - 1} \right] \cos 4nx$.

Page 373

3. (*i*) Does not exist. (*ii*) Does not exist. (*iii*) 0. (*iv*) 0.

Page 375

2. (*i*) Discontinuous. (*ii*) Continuous.

5. No.

Page 397

3. (*i*) Not differentiable. (*ii*) Not differentiable.

5. Differentiable at all points except (0, 0).

6. $k > 0$.

Page 402

3. (*i*) Equal. (*ii*) Unequal.

4. Equal.

Page 415

(*i*) Min. at $\left(\dfrac{9}{3}, -\dfrac{4}{3}\right)$.

(*ii*) Min. at (1, 2) and (– 1, 2); max. at (0, 0).

(*iii*) Min. at (0, 0) if $a > \dfrac{1}{4}$.

(*iv*) Min. at $\left[\dfrac{1}{3}(b-2a), \dfrac{1}{3}(a-2b)\right]$, (*v*) max. at (2, 1),

(*vi*) Max. or min. at $\left(\dfrac{1}{3}a, \dfrac{1}{3}a\right)$ according as $a > 0$ or $a < 0$.

(*vii*) Min. at (1, 0) and max. at (– 3, 2).

(*viii*) Min. at $\left(\pm a/\sqrt{2}, 0\right)$.

(*ix*) Min. at $\left(\pm 3/\sqrt{2}, 0\right)$.

(*x*) Min. at every point of the line $y = x$.

(*xi*) Max. at (0, 0) and min. at $(\pm 1, 2)$.

(*xii*) Extremes at (0, 0), (– 1, 0), $\left(-\frac{1}{2}, 0\right)$.

6. Max. at (0, 0); min. at $\left(\pm\sqrt{5}, \pm\sqrt{5}\right)$. Neither max. nor min. at $(\pm 1, \mp 1)$.

7. The positions of the corresponding points *P, Q* are

$\left(3a, \pm\sqrt{12a}\right), \left(3a, \pm\sqrt{12a}\right); (0, 0), \left(\pm\sqrt{12a+3a}, 0\right);$

$(a, 3a), \left(3a \mp \sqrt{6a}, \pm\sqrt{6a}\right); (a, -2a), \left(3a\sqrt{\pm\sqrt{6a}}, \pm\sqrt{6a}\right),$

8. (*i*) Min. at (0, 0, 0). (*ii*) Min. at (0, 0, – 1).

Page 434

3. $\varphi(x)$ is not unique.

4. (*i*) 0. (*ii*) 0. (*iii*) $2\sqrt{3/9}$. (*iv*) – 1.

5. (*i*) $\varphi/(x)$ unique near (1, – 1) but not near (0, 0).

(*ii*) $\varphi(x)$ unique near (1, 1). (*ii*) $\varphi(x)$ not unique but the equation possesses a unique continuous solution $x = \psi(y)$ near (1, 1).

(*i*) Does not exist.

Page 458

1. $\frac{7}{3}a^3, \frac{46}{15}a^3$. **3.** 2 log 2.

5. $\frac{1}{84}$. **6.** (*i*) $38a^4/\sqrt{3}$. **7.** -2π.

Page 468

1. (*i*) $\frac{1}{5}a^2b^2\left(a^2+b^2\right)$. (*ii*) $\left(e^{ab}-1\right)/(a-b)$.

(*iii*) 0. (*iv*) $2\left[a\sinh^{-1}(a/c) + c - \sqrt{\left(a^2+c^2\right)}\right]$.

Page 478

1. $\int_{-1}^{0} dx \int_{-x}^{\frac{1}{3}(2x+5)} f\,dy + \int_{0}^{1} dx \int_{x}^{\frac{1}{3}(2x+5)} f\,dy + \int_{1}^{2} dx \int_{(2x-1)}^{\frac{1}{3}(2x+5)} f\,dy.$

2. $\pi/96$. **4.** $\int_0^a dy \int_0^{\sqrt{(4ay)}} \varphi\,dx + \int_a^{3a} dy \int_0^{3a-y} \varphi\,dx.$

6. 12 – 16 log 2. **7.** $\frac{1}{2}\pi\log\left[2e/(1-e)\right]$. **9.** 47/24.

10. $8\sin^{-1}\left(\sqrt{10/4}\right) + \frac{1}{2}\log\left[\left(\sqrt{3}+\sqrt{5}\right)/\sqrt{2}\right] - 3\sqrt{15/4}$.

11. $a^3\pi$.

Page 492

1. $\frac{2}{3}\pi\left(a^2+b^2\right)^{\frac{3}{2}}$ **2.** $\frac{1}{9}a^3(3\pi-4)$. **7.** $\frac{1}{4}\beta\left(\frac{3}{4},\frac{1}{2}\right)\beta\left(\frac{1}{2},\frac{1}{4}\right)$

8. In the $r\theta$ plane the field of integration is bounded by the lines $r=2a$, $r=4a$; $a=\frac{1}{2}\pi$, $r=2a\sec^2\theta$.

9. $\int_0^1\int_0^c V\left[u(1-v), uv\right]u\,du\,dv$.

10. To the two parts of the given region in the (x, y) plane determined by the line $y=x$, corresponds the same region of the (u, v) plane bounded by the lines $u=0$, $u=a^2$, $u=2v$. The transformations for the two parts are given by

$$\left.\begin{aligned} x+y&=(u+2v)^{\frac{1}{2}} \\ x-y&=(u-2v)^{\frac{1}{2}} \end{aligned}\right\} \qquad \begin{aligned} x+y&=(u+2v)^{\frac{1}{2}} \\ x-y&=-(u-2v)^{\frac{1}{2}} \end{aligned}$$

11. $8\left(\sqrt{a_1}-\sqrt{a_2}\right)\sqrt{-a_3}$.

12. If λ, μ denote the values of the parameter for the confocal ellipse and hyperbola respectively through (x, y), then

$$\lambda+\mu=\left(x^2+y^2+c^2\right),\ \lambda\mu=c^2x^2$$

so that $x=\sqrt{(\lambda\mu)/c}$, $y=\sqrt{\left(\lambda-c^2\right)\left(c^2-\mu\right)/c}$.

$\therefore$ the required area is

$$\iint dx\,dy=\iint\frac{\partial(x, y)}{\partial(\lambda, \mu)}d\lambda\,d\mu=\frac{1}{4}\iint\frac{\mu-\lambda}{\sqrt{\lambda\mu\left(\lambda-c^2\right)\left(c^2-\mu\right)}}d\lambda\,d\mu,$$

$$=\frac{1}{6}c^2\left(\sqrt{10}-2\right)\sin^{-1}\frac{1}{3},$$

where for the double integral

$$\frac{1}{3}c^2\le\mu\le\frac{2}{3}c^2, \frac{4}{3}c^2\le\lambda\le\frac{5}{3}c^2.$$

13. The field of integration in the xy-plane in the positive quadrant of the ellipse $x^2/9+y^2/5=1$. As (x, y) moves along x-axis from (0, 0) to (2, 0), the point (λ, μ) moves in the $\lambda\mu$ plane along the line $\lambda=2$ from (2, 0) to (2, 2); as (x, y) moves along x-axis from (2, 0) to (3, 0), (λ, μ) moves along $\mu=2$ from (2, 2) to (3, 2); as (x, y) moves along the arc of the ellipse from (3, 0) to $\left(0, \sqrt{5}\right)$, (λ, μ) moves along $\lambda=3$ from (3, 2) to (3, 0); finally as (x, y) moves along $x=0$ from $\left(0, \sqrt{5}\right)$ to (0, 0), (λ, μ) (λ, μ) moves along $\mu=0$ from (3, 0) to (2, 0). Thus the region in $\lambda\mu$ plane is the rectangle [2, 3; 0, 1].

Since, as may easily be seen,

$$x = \frac{1}{2}\lambda\mu, \quad y = \frac{1}{2}\sqrt{(\lambda^2 - 4)(4 - \mu^2)}$$

$\partial(x, y)/\partial(\lambda, \mu)$ can be calculated.

14. $\frac{1}{96}$. The equation may be solved by the substitution $x = r\cos\theta$, $y = r\sin\theta$ also.

15. $(n-p)(l-m)\left[(l+m)(a+b)+(n+p)(a-b)\right]/32a^2b^2$.

16. $\frac{3}{4}$.

Page 495

2. $\frac{2}{3} + \frac{1}{4}\log\frac{1}{9}$. **3.** $\frac{1}{3}(e^2 + 2e - 3)$.

4. $0, \pi/128$. **5.** $\pi/24$.

6. $1/6(n+4)$. **7.** $1/30, 12$.

8. $2\pi^{-2}$. **9.** 1.

10. $(\pi + 4)/64$. **11.** 2.

12. $\pi/24$. **14.** $\pi a^2/24 + \sqrt{3a^2/4}$.

16. $\frac{1}{2}\log\frac{e}{2}, \pi$. **17.** $\frac{\pi ab(a^2+b^2)}{4}$.

18. $\pi/16$. **19.** $\frac{a^2}{2}\left[-\frac{\pi}{6} + \log(2+\sqrt{3})\right]$.

20. $(e^2 - 1)/4e$. **21.** $\pi \log\frac{1+a^2}{a^2}$.

Page 514

1. $2\sqrt{2a\pi}$. **2.** $a(12 + 5\log 5)/8$.

3. $2a\int_0^{\pi}\sqrt{(3 - 2\cos\theta)}\,d\theta$.

Page 518

1. $56\pi a^2/9$. **3.** $4a^2\pi$.

4. $2\sqrt{2\pi}$.

Page 529

2. $12\pi/5$. **3.** $2\pi abc$.

5. $\frac{4}{5}\pi abc(a^4 + b^4 + c^4)$. **6.** $\frac{1}{3}\pi ah\left[3(l+m)a^2 + 3nah + 2nh^2\right]$.

Page 534

2. $3a^2$. **3.** $\pi/24$. **6.** $5\pi/6$.

Page 540

8. $\frac{1}{8}\pi\left[\beta\left(\frac{3}{4},\frac{1}{2}\right)-\beta\left(\frac{5}{4},\frac{1}{2}\right)\right]$. **10.** $\log(a'/a)\log(b'/b)\log(c'/c)$.

12. $\Gamma(l)\,\Gamma(m)\,\Gamma(n)/\Gamma(l+m+n+1)$.

14. $384\pi/5$. **15.** $\frac{8192}{735}a^7 k\left(k^2+\frac{8}{33}\right)$.

17. $\pi/8$.

19. $$\frac{a^l b^m c^n}{pqr}(h)\frac{l}{p}+m+\frac{n}{r}\frac{\Gamma\left(\frac{l}{p}\right)\Gamma\left(\frac{m}{q}\right)\Gamma\left(\frac{n}{r}\right)}{\Gamma\left(\frac{l}{p}+\frac{m}{q}+\frac{n}{r}\right)}\int_0^1 F(hu)\,n^{\Sigma\frac{l}{p}-1}\,du$$

21. $\frac{1}{7}.a\pi(4-\pi)$. **24.** $4\pi/3$.

Page 543

1. $9(3\pi-4)/4$. **3.** $\frac{1}{8}ka^4$. **5.** $\frac{1}{3}\pi c^2$.

12. $\frac{1}{3}\frac{6}{5}\pi\left(b^{-\frac{5}{8}}-a^{-\frac{5}{8}}\right)\left(d^{-\frac{6}{7}}-c^{-\frac{7}{8}}\right)$.

13. $\frac{(2n+3)k}{(n+1)(n+2)}\cdot\frac{a^{n+1}\,b^{n+1}}{\left(a^2-b^2\right)^{n/2}}$.

14. $\frac{4k\sqrt{(2k)}-(k+5)-\sqrt{(k-1)}}{(k+5)\sqrt{(k-1)}}$.

17. $128a^3/15$. **18.** $32a^3\left(3\pi-\sqrt{2}\right)/9$.

Page 549

11. $f_{yx}(0,0)=1\neq 0=f_{xy}(0,0)$.

12. $f_{xy}(0,0)=6a^2b\neq 6c^2b=f_{yx}(0,0)$.

15. $f(x, y)$ is differentiable.

17. $\left(0,0,\frac{1}{2}\right)$. **18.** If $a^2<2b^2$, the greatest length is $2b$.

24. $\frac{2}{9}abc(3\pi-4)$. **26.** $\frac{2}{3}ma^3$.

28. $\frac{1}{2}\left(k^2-h^2\right)\log(b/a)$.

32. If $S_n = \left(\frac{1}{\sqrt{1}} + \frac{1}{\sqrt{2}} + \frac{1}{\sqrt{2}}\right)$, then since

$$1/\sqrt{r} < \int_{r-1}^{r} \left(1/\sqrt{x}\right) dx < 1/\sqrt{(r-1)}$$

we have, on putting $r = 2, ..., n$,

$$S_n - 1 < 2\left(\sqrt{n} - 1\right) < S_n - 1/\sqrt{n},$$

i.e., $$2 + \frac{1}{n} - \frac{2}{\sqrt{n}} < \frac{S_n}{\sqrt{n}} < 2 - \frac{1}{\sqrt{n}} \text{ so that } S_n / \sqrt{n} \to 2.$$

33. The formula follows from the following 3 results :

(*i*) $$\frac{\pi}{2} = \frac{2.2.4.4.6.6\ldots(2n)(2n)}{1.3.3.5.5.7\ldots(2n-1)(2n-1)} \frac{\int_0^{\frac{1}{2}\pi} \sin^{2n} x \, dx}{\int_0^{\frac{1}{2}\pi} \sin^{2n+1} x \, dx}.$$

(*ii*) $0 \le \sin^{2n+1} x \le \sin^{2n} x \le \sin^{2n-1} x.$

(*iii*) $$\frac{\int_0^{\frac{1}{2}\pi} \sin^{2n-1} x \, dx}{\int_0^{\frac{1}{2}\pi} \sin^{2n+1} x \, dx} = \frac{2n+1}{2n}.$$

APPENDIX

Everywhere Continuous Non-derivable Function

Weierstrass had drawn attention to the fact that there exist functions which are continuous for every value of x but do not possess a derivative for any value. We now consider the celebrated function given by Weierstrass to show this fact. It will be shown that if

$$f(x) = \sum_{n=0}^{\infty} b^n \cos\left(a^n \pi n\right) \qquad \text{...(1)}$$

$$= \cos \pi x + b \cos a\pi x + b^2 \cos a^2 \pi x + \ldots$$

where a is an odd positive integer, $0 < b < 1$ and $ab > 1 + \frac{3}{2}\pi$, then the function f is continuous $\forall\, x$ but not finitely derivable for any value of x.

We have

$$\left|b^n \cos\left(a^n \pi x\right)\right| \le b^n,$$

and $\Sigma\, b^n$ is convergent. Thus, by M = test, the series (1) is uniformly convergent in every interval. Hence f is continuous $\forall\, x$.

Again, we have

$$\frac{f(x+h) - f(x)}{h} = \sum_{n=1}^{\infty} b^n \frac{\cos\left[a^n \pi (x+h)\right] - \cos a^n \pi x}{h} \qquad \text{...(2)}$$

Let, now, m be any positive integer. Also let S_m denote the sum of the m terms and R_m, the remainder after m terms, of the series (2) so that

$$\sum_{n=0}^{\infty} b^n \frac{\cos\left[a^n \pi (x+h)\right] - \cos a^n \pi x}{h} = S_m + R_m.$$

By Lagrange's mean value theorem, we have

$$\left|\cos\left[a^n \pi (x+h)\right] - \cos a^n \pi x\right| = \left|a^n \pi h \sin a^n \pi (x + \theta h)\right| \le a^n \pi |h|,$$

$$|S_m| \le \sum_{n=0}^{m-1} b^n a^n \pi = \pi \frac{a^m b^m - 1}{ab - 1} < \pi \frac{a^m b^m}{ab - 1}.$$

We shall now consider R_m.

So far we have taken h arbitrary but we shall now choose it as follows : We write

$$a^m x = \alpha_m + \xi_m,$$

where α_m is the integer nearest to $a^m x$ and $-\frac{1}{2} \leq \xi_m < \frac{1}{2}$.

$$\therefore \quad a^m (x+h) = \alpha_m + \xi_m + ha^m.$$

We choose, h, so that

$$\xi_m + ha^m = 1,$$

i.e., $$h = \frac{1-\xi_m}{a^m} \text{ which} \to 0 \text{ as } m \to \infty,$$

for

$$0 < h \leq \frac{3}{2a^m}. \qquad \text{...(3)}$$

Now $$a^n \pi (x+h) = a^{n-m} a^m \pi (x+h)$$
$$= a^{n-m} \pi (\alpha_m + \xi_m + 1 - \xi_m)$$
$$= a^{n-m} \pi (\alpha_m + 1).$$

$$\therefore \quad \cos\left[a^n \pi (x+h)\right] = \cos\left[a^{n-m} (\alpha_m - 1)\pi\right] = (-1)^{\alpha_m + 1}$$

$$\cos\left(a^n \pi x\right) = \cos\left[a^{n-m}\left(a^m \pi x\right)\right]$$
$$= \cos\left[a^{n-m} (\alpha_m + \xi_m)\pi\right]$$
$$= \cos a^{n-m} \alpha_m \pi \cos a^{n-m} \xi_m \pi$$
$$- \sin a^{n-m} \alpha_m \pi \sin a^{n-m} \xi_m \pi$$
$$= (-1)^{\alpha_m} \cos a^{n-m} \xi_m \pi,$$

for a, is an odd integer and α_m is an integer.

$$\therefore \quad R_m = \frac{(-1)^{\alpha_m} + 1}{h} \sum_{n=m}^{\infty} b^n \left[2 + \cos\left(a^{n-m} \xi_m \pi\right)\right] \qquad \text{...(4)}$$

Now each term of series in (4) is ≥ 0 and, in particular, the first term is positive,

$$|R_m| > \frac{b^m}{|h|} > \frac{2a^m b^m}{3}. \qquad \text{by (3)}$$

Thus

$$\left|\frac{f(x+h) - f(x)}{h}\right| = |R_m + S_m|$$

$$\geq |R_m| - |S_m| > \left(\frac{2}{3} - \frac{\pi}{ab-1}\right)a^m b^m.$$

As
$$ab > 1 + \frac{3}{2}\pi,$$

therefore

$$\left(\frac{3}{2} - \frac{\pi}{ab-1}\right) \text{is positive.}$$

Thus we see that when $m \to \infty$ so that $h \to 0$, the expression $\{f(x+h) - f(x)\}/h$ takes arbitrary large values. Hence $f'(x)$ does not exist or is at least not finite.

MATHEMATICS AND STATISTICS BOOKS FOR DEGREE CLASSES

A COURSE OF MATHEMATICAL ANALYSIS

Shanti Narayan

Designed for students who desire to proceed to a course preparatory to a serious study of the purely arithmetical theory of functions of real variables in its general aspects.

CONTENTS: Real Numbers • Bounded Sets, Open and Closed Sets • Real Sequences • Real Valued Functions of a Single Real Variable, Limit and Continuity • Real Valued Functions of a Single Real Variable • Derivability • Riemann Integrability • Sequences of Functions Point-wise and Uniform Convergence • Elementary Functions • Improper Integrals • Fourier Series • Euclidean Spaces • Open and Closed Sets • Compact Sets • Real Valued Functions of Several Real Variables. • Limit • Continuity • Partial Derivatives • Invertible Functions Implicit Functions • Integrals as Functions of a Parameter • Integration in R^2 Line Integrals. Double Integrals • Curve Lengths. Surface Areas • Integration in R^3 Gauss's and Stoke's Theorems Mix-Exercises • Answers • Appendix.

14 045 ISBN: 81-219-0472-2

MATHEMATICAL STATISTICS

J.N. Kapur & H.C. Saxena

This book is on the same model as books on calculus, differential equation, statics, dynamics, etc. A special feature is the large number of exercises especially constructed to illustrate the theory.

CONTENTS: Introduction • Frequency Distribution and Measures of Location • Measures of Dispersions, Skewness & Kurtosis, Moments of Frequency Distributions • Theory of Probability • Discrete Probability Distributions • Special Discrete Probability Distributions • Univariate Continuous Probability Distributions • Special Continuous Probability Distributions • Principle of Least Squares, Fittings of Curves & Orthagonal Polynomials • Correlation and Regression • Multiple and Partial Correlation • Theory of Sampling • Exact Sampling Distributions • Tests of Significance Based on the *T*, *F* and *Z* Distributions • Tests of Significance Based on the Chi-Square Distribution • Statistical Theory of Point Estimation • Testing of EHypotheses, Sequential Analysis Distribution-Free Methods, Statistical Decision Theory • Elements of Stochastic Processes • Appendices • Index • Log Tables.

14 046 ISBN: 81-219-1246-6

ANALYTICAL SOLID GEOMETRY

Shanti Narayan

This book is intended as an introduction to Analytical Solid Geometry and is suitable for students of B.A., B.Sc., Pass and Honours courses. The book contains numerous exercises of varied types in a graded form.

CONTENTS: Co-ordinates • The Plane • Right Line • Interpretation of Equations-Loci • Transformation of Co-ordinates • The Sphere • Cones, and Cylinders • Homogeneous Cartesian Co-ordinates-Elements at infinity • The Conicoid • Plane Sections of Conicoids • Generating Lines of Conicoids • General Equation of the Second Degree: Reduction to Canonical forms • Appendix • Index.

14 047 **ISBN:81-219-0473-0** **pp. 320**

A TEXTBOOK ON DYNAMICS

M. Ray & G.C. Sharma

A thoroughly revised edition where the chapters on Central Forces, Moment of Inertia and D'Alembert's Principle have been substantially retouched for the B.A. and B.Sc. students of Dynamics.

CONTENTS: Kinematics and Kinetics • Rectilinear Motion • Uniplanar Motion • Work Energy and Impulse • Impact • Circular and Harmonic Motions • Hodograph • Central Forces • Resisting Medium • Constrained Motion • Moment of Inertia • D'Almbert's Principle and the Equations of Motion.

14 048 **ISBN: 81-219-0342-4** **pp. 328**

DIFFERENTIAL CALCULUS

Shanti Narayan

The text is marked with unified and graphical presentation. The complicated issues are well illustrated.

CONTENTS: Real Numbers • Functions and Graphs-Elementary Functions • Continuity and Limit • Differentiation • Successive Differentiation • Tangents and Normals • Mean Value Theorem • Maxima and Minima • Indeterminate Forms • Partial Differentiation • Equality of Repeated Derivatives • Concavity and Points of Inflexion • Curvature & Evolutes Asymptotes • Singular Points • Curve Tracing • Envelopes • Partial Differentiation • Equality of Repeated Derivatives.

14 049 **ISBN: 81-219-0471-4** **pp. 400**

INTEGRAL CALCULUS

Shanti Narayan

Has a new chapter on `Centre of Gravity and Moment of Inertia'. Many illustrations are given so as to enlighten the subject-matter.

CONTENTS: Table of Standard Results • Methods of Integration • Integration of Algebraic Rational Functions • Integration of Trigonometric Functions • Integration of Irrational Functions • Definite Integral as the Limit of a Sum • Areas of Plane Regions • Rectification • Lengths of Plane Curves • Volumes and Surfaces of Revolution • Centre of Gravity: Moment of Inertia • Differential Equations of First Order and First Degree • Equations of First Order but not of the First Degree • Trajectories of a Family of Curves • Linear Equations • Appendices Mise. Exercises.

14 050 ISBN: 81-219-0681-4 pp. 360

A TEXTBOOK OF VECTOR CALCULUS

Shanti Narayan & J.N. Kapur

Gives a systematic and self- contained treatment to the subject- matter. Good many illustrations are there to make understanding comprehensive.

CONTENTS: Vector Valued Functions of Scalar Variables • Differential Geometry I—Curves • Differential Geometry II—Surfaces • Mechanics I—Particles • Mechanics II—Rigid Body • Differential Operators—Point Functions • Integral Transfor-mations—Line, Surface, Volume Integrals • Hydrodynamics • Electromagnetic Theory.

14 051 ISBN:81-219-0161-8 pp. 328

ORDINARY AND PARTIAL DIFFERENTIAL EQUATIONS

M.D. Raisinghania

Caters to the needs of the honours and postgraduate students. Part I deals with Elementary Differential Equations. Part II deals with ordinary differential equations, and Part III deals with partial differential equations.

CONTENTS: Part I: Elementary Differential Equations— Introduction • Equations of first order and first degree • Trajectories • Linear Equations with constant coefficients • Homogeneous linear equations or Cauchy-Euler equations • Equations of the first order but not of the first degree and singular solutions- Extraneous loci • Ordinary Simultaneous differential equations • **Part II: Ordinary Differential Equations**—Numerical Integration •Existence and uniqueness theorems • Independence of solutions of linear differential equations-Wronskian • Exact differential equations and equations of special forms • Linear Equations of second order • Simultaneous equations of the form dx/P = dy/Q = dz/R • Total differential equations • Riccati's equations • Series solutions of linear differential equations • Legendre Polynomials and Functions • Legendre functions of second kind • Bessel functions • Hermite Polynomials • Laguerre Polynomials • Hyper-geometric Function • **Part III:** Partial Differential Equations— Linear partial differential equations of order one • Non-linear partial-differential equations of Order one • Partial differential equation of the second order • Linear partial differential equations with constant coefficients • Monge's methods • Appendices.

14 282 **ISBN:81-219-0892-2**
pp. 576

LAPLACE AND FOURIER TRANSFORMS

M.D. Raisinghania

This book has been written for the use of students of B.A., B.Sc., M.A. and M.Sc. (Mathematics and Physics). It caters to the needs of engineering students also. Students preparing for various competitive examinations will find this book particularly useful.

CONTENTS: Laplace Transform • Inverse Laplace Transform • Application of Laplace Transforms • Fourier Transforms • Finite Fourier Transform • Index.

14 357 **ISBN:81-219-0884-1**
pp. 240

EXAMPLES IN FINITE DIFFERENCES AND NUMERICAL ANALYSIS

H.C. Saxena

This book comprises of twelve chapters with an introduction to important concepts and the list of important formulae needed in the solution of the examples.

CONTENTS: Differences, Operators, Interpolation with Equal Intervals • Interpolation for Unequal Intervals of the Argument • Central Difference Interpolation Formulae • Numerical Differentiation • Inverse Interpolation • Numerical Quadrature • Summation of Series • Difference Equations • Roots of Algebraic and Transcendental Equations • Simultaneous Linear Algebraic Equations • Numerical solution of Ordinary Differential Equations • Linear Programming • Log Tables.

14 175 **ISBN:81-219-0753-5**
pp. 352

ADVANCED DIFFERENTIAL EQUATIONS
(Including Boundary Value Problems and Difference Equations)

M. D. Raisinghania

This book has been written for Honours, M.A., M.Sc. (Mathematics and Physics) and B.E. students of various universities. The students preparing for P.C.S., I.A.S., A.M.I.E. will also find it useful.

CONTENTS: Part I: Ordinary Differential Equations • Numerical Integration • Existence and Uniqueness Theorem • Independence of solutions of Linear Differential Equations – Wronskian • Exact Differential Equations and Equations of Special Forms • Linear Equations of Second Order • Simultaneous Equations of the form dx/P = dy/Q = dz/R • Total Differential Equations • Riccati's Equation • Series Solutions of Linear Differential Equations • Legendre Polynomials and Functions • Legendre Functions of Second Kind • Bessel Functions • Hermite Polynomials • Languerre Polynomials • Hypergeometric Function • • **Part II: Partial Differential Equations** • Linear Partial Differential Equations of Order One • Non-linear Partial Differential Equations of the Second Order • Linear Partial Differential Equations with Constant Coefficients • Monge's Methods • More an Partial Differential Equations • **Part III: Boundary Value Problems** • **Part IV: Difference Equations.**

14 271 6th Edn. 1997 ISBN:81-219-0893-0 pp. 648

A TEXTBOOK OF VECTOR ALGEBRA

Shanti Narayan

An attractively well-written book which deals rigorously and systematically with three dimensional Vector algebra and its application to Euclidean Geometry and Statics.

CONTENTS: Multiplication of Vectors by Scalars and Addition of Vectors • Geometry with Vectors Affine Geometry • Scalar Product • Applications to Metric Geometry • Vector Product and Scalar Triple Product • Application of Scalar Triple Products to Geometry • Geometry with Cartesian Co-ordinates • Some Miscellaneous Topics • Some Properties of Tetrahedra • Statics with Vectors • Answers.

14 064 8th Edn. Rep. 1998 ISBN: 81-219-0952-X pp. 240

FLUID DYNAMICS

M.D. Raisinghania

The present text has been well-organised and made up-to-date in the light of the latest syllabi. A detailed index has been provided at the end of the book and more solved examples have been added so that the readers may gain confidence in the technique of problem solving. Almost all the chapters have been rewritten so that in the present form, readers will easily understand the subject matter.

CONTENTS: Introduction • Kinematics of Fluids in Motion • Equations of Motion of Inviscid Fluid • One-Dimensional Inviscid Incompressible Flow (Bernoulli's equation and its applications) • Motion in Two-Dimensions and Sources and Sinks • General Theory of Irrotational Motion • Motion of Cylinders • Irrotational Motion in three Dimensions

(Motion of a sphere) Stoke's Stream Function • Vortex Motion (Rectilinear vortices) • Waves • General Theory of Stress and Rate of Strain • The Navier-Stokes Equations and the Energy Equation • Dimensional Analysis • Dimensionless Constants (numbers) • Similarity of Flows • Laminar Flow of Viscious Incompressible Fluids • Boundary Layer Theory • Discontinuous Motion • Index.

14 349 2nd Edn. Rep.1998

ISBN: 81-219-0869-8 pp. 872

INTEGRAL TRANSFORMS

M.D. Raisinghania

This book is useful for Honours, M.A., M.Sc. (Mathematics) and Physics and B.E. students. It is also useful for students preparing for A.M.I.E., I.A.S. and other competitive examinations. Various types of problems have been fully illustrated by solved examples.

CONTENTS: Laplace Transform • Inverse Laplace Transform • Applications of Laplace Transforms • Fourier Transforms • Finite Fourier Transform • Hankel Transforms • The Finite Hankel Transforms • The Mellin Transform • Index.

14 283 2nd Edn. 1995

ISBN:81-219-0886-8 pp. 296

RAMANUJAN NUMBERS (Mathematical Thoughts and Ideas)

M. Meyyappan

This book is intended to provide a modest knowledge of mathematics with an excursion in the number world. The book has a lot of illustrations and through activities the students are helped to acquire basic concepts and facts.

The standard of students in the subject mathematics is gradually deteriorating in our country. Popular Science books may interest and help divert students' attention on to developing skills but, this book in its humble effort tries to rejuvenate the mathematical inclination of Indian students.

To remove the fear of mathematics at an early stage and to create strong interest in this vital subject, the mother of all Sciences, mathematics, may be introduced as fun and entertainment. In this book such an attempt has been made.

The book also presents algebra and problem solving in an exciting and amusing manner. On the whole the book endeavours to arouse the curiosity in the students and thereby develop interest.

CONTENTS: Ramanujan—Mathematician Extraordinary • Second Order Ramanujan Numbers • Third Order Ramanujan Numbers • Fourth Order Ramanujan Numbers • Appendix.

14 368 **1st Edn. Rep. 1998** **ISBN: 81-219-1272-5**
pp. 128

ELEMENTS OF REAL ANALYSIS

Shanti Narayan

This book is an attempt to make presentation of Elements of Real Analysis more lucid. The book contains examples and exercises meant to help a proper understanding of the text.

CONTENTS : Sets and Statements • The Real Numbers • Limit Points of a Set • Sequences • Infinite Series with positive terms • Infinite Series with positive and negative terms • Real Functions—Limit and Continuity • Real Functions— The Derivative • Riemann Integrability.

14 052 **6th Edn. Rep. 1998** **ISBN: 81-219-0306-8**
pp. 216

VECTOR ALGEBRA

M.D. Raisinghania

The present edition of the book has been thoroughly revised, and updated in the light of latest syllabi of various universities and institutions of India. More solved examples have been added and a detailed index has been provided at the end of the book inorder to make it more useful to students and teachers.

CONTENTS: Vector Addition • Centroid • Vector equations of straight line and plane (Parametric form) • Scalar products of two vectors • Vector product of two vectors • Product of three and four vectors (Multiple products) • Vector equations of straight line and plane (non-parametric forms) • Volume of tetrahedron • Solution of vector equation • Vector equation of sphere • Applications of vectors to mechanics (Dynamics and Statics) • Dynamics with vectors • statics with vectors **Section I: Concurrent Forces • Section II : Vector moment (or torque) and work • Section III : Parallel Forces • Section IV Couple • Section V : Coplaner non-concurrent forces acting on a rigid body • Section VI : Forces in three dimensions** • Index.

14 363 **3rd Edn. 1997** **ISBN : 81-219-4447-7** **pp. 312**

FINITE DIFFERENCES AND NUMERICAL ANALYSIS

H.C. Saxena

This thoroughly revised edition of the book completely covers the syllabi in the calculus of Finite Differences of various Indian universities. Examples given at the end of each chapter have been specially constructed, taken from university papers, and standard books.

CONTENTS : Part I—Calculus of Finite Differences — Finite Differences, Interpolation with Equal Intervals • Interpolation of Unequal Intervals of the Argument • Central Difference Interpolation Formulae • Inverse

Interpolation • Numerical Differentiation • Numerical Quadrature • Summation of Series • Difference Equations • Generating Functions and Theorems • Bernoulli and Euler Polynomials • Remainder Terms in (or Errors associated with) • Interpolation Formulas **Part II —Numerical Analysis—** Eigen Value and Eigen Vector • Roots of Polynomial and Transcendental equation in one variable • Simultaneous Linear Algebraic Equations • Numerical Solution of Ordinary Differential Equations • Errors • Partial Differential Equations • Appendices • Answers • Index • Log Tables.

14 056 **14th Rev. Edn. 1998** **ISBN: 81-219-0339-4**
pp. 420